T0180212

# Lecture Notes in Electrical Engineering

## Volume 785

The book series *Lecture Notes in Electrical Engineering* (LNEE) publishes the latest developments in Electrical Engineering - quickly, informally and in high quality. While original research reported in proceedings and monographs has traditionally formed the core of LNEE, we also encourage authors to submit books devoted to supporting student education and professional training in the various fields and applications areas of electrical engineering. The series cover classical and emerging topics concerning:

- Communication Engineering, Information Theory and Networks
- Electronics Engineering and Microelectronics
- Signal, Image and Speech Processing
- Wireless and Mobile Communication
- Circuits and Systems
- Energy Systems, Power Electronics and Electrical Machines
- Electro-optical Engineering
- Instrumentation Engineering
- Avionics Engineering
- Control Systems
- Internet-of-Things and Cybersecurity
- Biomedical Devices, MEMS and NEMS

For general information about this book series, comments or suggestions, please contact leontina.dicecco@springer.com.

To submit a proposal or request further information, please contact the Publishing Editor in your country:

**China**

Jasmine Dou, Editor (jasmine.dou@springer.com)

**India, Japan, Rest of Asia**

Swati Meherishi, Editorial Director (Swati.Meherishi@springer.com)

**Southeast Asia, Australia, New Zealand**

Ramesh Nath Premnath, Editor (ramesh.premnath@springernature.com)

**USA, Canada:**

Michael Luby, Senior Editor (michael.luby@springer.com)

**All other Countries:**

Leontina Di Cecco, Senior Editor (leontina.dicecco@springer.com)

**\*\* This series is indexed by EI Compendex and Scopus databases. \*\***

More information about this series at http://www.springer.com/series/7818

Sunil Bhooshan

# Fundamentals of Analogue and Digital Communication Systems

 Springer

Sunil Bhooshan
Department of Electrical and Electronics
Engineering
Mahindra University
Hyderabad, Telangana, India

ISSN 1876-1100              ISSN 1876-1119   (electronic)
Lecture Notes in Electrical Engineering
ISBN 978-981-16-4279-1       ISBN 978-981-16-4277-7   (eBook)
https://doi.org/10.1007/978-981-16-4277-7

This Springer imprint is published by the registered company Springer Nature Singapore Pte Ltd.
The registered company address is: 152 Beach Road, #21-01/04 Gateway East, Singapore 189721,
Singapore

*To*

*Roma, Devaki and Agastya*

# Preface

Engineering Communication is a rapidly expanding, highly mathematical subject. I felt the need to introduce the subject to the average student in a simple and understandable manner. Other texts treat the subject with little or no regard for the *basis* of the subject, which was investigated starting from the early nineteenth century.

In the late twentieth century and in the twenty-first century, Digital Communication has gained tremendous importance. Because of this, the treatment of Analogue Communication, in most books, is sketchy and superficial. However, on giving some thought to the matter, it was apparent that Digital Communication is just an extension of Analogue Communication, and if the latter is not understood properly, there is little probability of understanding Digital Communication, and the subject, to a normal student, would be incomprehensible. The book may be used in a two- or three-semester course: one course on Analogue Communication, another course on Digital Communication (I) and a third on Digital Communication (II).

A note to the student. First, the chapter and material should be read *at least four times*. Next, the solved problems should be worked out without skipping steps, and finally the problems at the back of each chapter may then be worked out. By this method, some grasp of the subject may be obtained.

Hyderabad, Telangana, India
April 2021

Sunil Bhooshan

# About This Book

This book is divided into three parts, namely Part I which introduces all the mathematical and some circuit concepts required for the rest of the book, Part II which is Analogue Communication and Part III which consists of Digital Communication.

Part I, Chaps. 2–4, may be read as and when required. Much of it may be read as subsidiary material when attending a course of Signals and Systems. This material is useful for communication systems, but the z-transform or Discrete Time Fourier Transform has not been included. Part II, Chaps. 5 and 6, consists of Amplitude Modulation, and is followed by Frequency and Phase Modulation schemes. These modulations must be read thoroughly and understood well to proceed to Digital Communication.

Part III is mainly Digital Communication (baseband and passband) with additional material on Information Theory and Error Control Coding.

# Contents

# About the Author

**Dr. Sunil Bhooshan** is a Professor in the Electrical Engineering Department at Mahindra University École Centrale School of Engineering. Dr. Bhooshan did his joint MS, Ph.D in Electrical Engineering Department from University of Illinois at Urbana-Champaign, USA, and B.Tech from IIT Delhi, New Delhi.

Prior to joining Mahindra University, he worked as a Professor at Jaypee University of Information Technology, Waknaghat and briefly at IIT Kanpur, Kanpur, UP. He has published more than seventy research papers published in prestigious international journals and conference proceedings.

# Nomenclature and Abbreviations

| | |
|---|---|
| $\tilde{a}(t)$ | A sine or cosine function |
| AM | Amplitude modutation |
| ASK | Amplitude shift keying |
| AWGN | Additive white Gaussian noise |
| **b** | Parity bits; see Eq. (13.1) |
| $B$ | Sometimes used for bandwidth |
| $b(t)$ | Bit sequence |
| BCH | Bose–Choudhuri–Hocquenghem |
| BFSK | Binary frequency shift keying |
| BPF | Bandpass filter |
| BPSK | Binary phase shift keying |
| **c** | Code word; see Eq. (13.1) |
| $C$ | Channel capacity |
| $c(t)$ | PN sequence |
| $c(t)$ | Carrier signal |
| $C_N(t_1, t_2)$ | Covariance function; see Eq. (3.95) |
| CDF, cdf | Cumulative distribution function |
| $\delta(t)$ | The impulse function |
| $d(\mathbf{c}_i, \mathbf{c}_j)$ | Hamming distance |
| DM | Delta modulation |
| DPCM | Differential pulse code modulation |
| DS | Direct sequence |
| DSB | Amplitude modutation, double sideband |
| DSB/SC | Amplitude modutation, double sideband, supressed carrier |
| DSSS | Direct sequence (DS) spread spectrum |
| $\eta$ | Coding efficiency; see Eq. (12.9) |
| $\mathbb{E}$ | Energy of a signal; see Eq. (2.5) |
| $E$ | An event |
| $E[X]$ | Average or expected value of the random variable $X$ ; see Eq. (3.44) |
| $E^c$ | Complement of the set $E$ |
| $E_b$ | Energy per bit |

| | |
|---|---|
| $\mathcal{F}\{\cdots\}$ | Fourier transform |
| $\mathcal{F}^{-1}\{\cdots\}$ | Inverse Fourier transform |
| $\Phi_X(v)$ | Charcteristic function |
| $f(t)$ | An analogue signal or function |
| $f[n]$ | A digital function |
| $F_X(x)$ | Cumulative distribution function of the random variable $x$ |
| $f_X(x)$ | Probability density function of the random variable $x$ |
| FDM | Frequency division multiplexing |
| FM | Frequency modulation |
| FSK | Frequency shift keying |
| FT | Fourier transform |
| **G** | Generator matrix |
| $g(x)$ | Generator polynomial |
| $\mathbf{h}(X)$ | Differential entropy; see Eq. (12.51) |
| **H** | Parity check matrix |
| $\mathcal{H}\{x(t)\} = x_h(t)$ | Hilbert Transform of $x(t)$ |
| $H(\mathcal{S})$ | Entropy of a source; see Eq. (12.5) |
| HPF | Highpass filter |
| $\mathbf{I}(X; Y)$ | Mutual information, continuous RVs; see Eq. (12.57) |
| $\mathfrak{J}[\cdots]$ | Imaginary part |
| $I(\mathcal{X}; \mathcal{Y})$ | Mutual information, page 606 |
| $I(s)$ | Information in bits about a symbol $s$; see Eq. (12.1) |
| $I_{th}(t)$ | Thermal noise current |
| IFT | Inverse Fourier transform |
| ISI | Intersymbol interference |
| $j(t)$ | Jamming signal, page 573 |
| $k_a$ | Amplitude sensitivity; see Eq. (5.39) |
| $k_f$ | Frequency sensitivity; see Eq. (6.3) |
| $k_p$ | Phase sensitivity; see Eq. (6.7) |
| $l_k$ | The lengths of the binary code words |
| LMS | Least mean squared |
| LTI | Linear time invariant |
| LZ | Lempel–Ziv |
| **m** | Message bits; see Eq. (13.1) |
| $\mu$ | Modulation index; see Eq. (5.43) |
| $m(t)$ | Message signal |
| $m_X$ | Average (mean) or expected value of the random variable $X$ |
| $M_{ij}$ | Mutual inductor; see Eq. (4.14) |
| ML | Maximum Likelihood |
| ML | Maximum lemgth |
| MMSE | Minimum mean squared equaliser |
| MSK | Minimum shift keying |
| $\mathcal{N}$ | Noise power |
| $N_0/2$ | Magnitude of the PSD of white noise |
| $n_{\text{NB}}(t)$ | Narrowband noise |

| | |
|---|---|
| NRZ | Non-return to zero |
| $\omega_0$ | Fundamental frequency of a Fourier series |
| $\mathbb{P}$ | Power of a signal; see Eq. (2.7) |
| $\phi$ | The null set |
| $P(\cdot)$ | Probability of an event |
| $p(t)$ | Any kind of pulse |
| $P_e$ | Probability of error |
| $p_s(t)$ | Single pulse of width $2\tau$ |
| $x_+(t)$ | Pre-envelope of a function $x(t)$ |
| $\phi_j(t)$ | Orthonormal functions |
| PAM | Pulse amplitude modulation |
| PCM | Pulse code modulation |
| PDF | Probability density function |
| PLL | Phase locked loop |
| PM | Phase modulation |
| PPM | Pulse position modulation |
| PSD | Power spectral density |
| PSK | Phase shift keying |
| PWM | Pulse width modulation |
| QAM | Quadrature amplitude modulation |
| QPSK | Quadrature phase shift keying |
| $\text{rect}(t)$ | The rect function; see Eq. (2.32) |
| $\Re[\cdots]$ | Real part |
| $r$ | Code rate; see Eq. (12.46) |
| $R_N(\tau)$ | Autocorrelation function of the random process $N(t)$ |
| $R_b$ | Bit rate |
| rms | Root mean square |
| RV | Random variable |
| $\|s_i\|_v$ | Vector norm; see Eq. (10.22) |
| $\|s_i(t)\|_f^2$ | Function norm; see Eq. (10.21) |
| $\mathbf{s}_i$ | Signal vector |
| $\mathcal{S}$ | Sample space |
| $S$ | Signal power |
| $\mathcal{S}$ | Symbol set |
| $\mathcal{S}$ | A set |
| $\text{sgn}(t)$ | The signum function; see Eq. (2.24) |
| $\text{sinc}(\omega)$ | The sinc function; see Eq. (2.26) |
| $\sigma_X$ | Standard deviation of a random variable |
| $\sigma_X^2$ | Variance of the random variable $X$; see Eq. (3.48) |
| $s_k$ | Source symbols |
| $S_X(\omega)$ | Power spectral density of the random process $X(t)$ |
| SNR | Signal-to-noise ratio |
| SSB | Amplitude modutation, single sideband |
| $T$ | Fundamental period of a Fourier series |
| $T_b$ | Bit width in sec |

| | |
|---|---|
| TDM | Time division multiplexing |
| $\mathbf{u}_j$ | Unit vector |
| $u(t)$ | The unit step function; see Eq. (2.19) |
| $V_{th}(t)$ | Thermal noise voltage |
| VSB | Amplitude modutation, vestigeal sideband |
| $\omega_s$ | Sampling frequency; see Eq. (7.14) |
| $w(\mathbf{c}_i)$ | Hamming weight |
| $w(t)$ | White noise |
| WSS | Wide-sense stationary process |
| $x(t)$ | Generally an input signal |
| $x_T(t)$ | A periodic function with period $T$ |
| $x_+(t)$ | Pre-envelope of a function $x(t)$ |
| $x_h(t)$ | Hilbert transform of a function $x(t)$ |
| $\langle g_1(t) \cdot g_2(t) \rangle$ | Inner product of two functions |
| $\oplus$ | Modulo 2 addition; see Eq. (11.1) |
| $f(t) * g(t)$ | Convolution; see Eq. (2.32) |

# Part I
# Introduction to Communication

# Chapter 1
# General Introduction

## 1.1 Historical Developments

Communication is as fundamental to humans as sleeping, eating and drinking. This is so because we humans are social creatures and the need for communication is essential to our nature.

Early in the development of man, a need was felt for the written word for the purpose of communication. From 3500 to 2500 B.C., the alphabet and writing were developed by various cultures: the Phoenicians developed an alphabet, the Sumerians, cuneiform writing on clay tablets, and the Egyptians developed hieroglyphic writing. Around 1770–1780 B.C., Greeks used a phonetic alphabet written from left to right and by 1400 B.C. the oldest record of writing—on bones—has been found in China. By 900 B.C., the first postal service, for government use, was in place in China. From 500 to 170 B.C., papyrus rolls and early parchments made of dried reeds were developed in Egypt, which were the first portable and light writing surfaces. From 200 B.C. to 100 B.C., human messengers on foot or horseback were common in Egypt and China using messenger relay stations; sometimes, fire messages were used to relay messages from relay station to station instead of humans carrying the messages. Some of the important landmarks in the history of communication, from the engineering point of view, are given below.

**200 B.C. to 1000 A.D.**

$c$ 105 BC Paper invented in China.

37 Heliography—use of mirrors and use of the Sun's rays to send messages by the Romans.

$c$ 305 Wooden printing press invented in China.

**1000 A.D. to 1800 A.D.**

1455 Gutenberg invents a printing press with metal movable type.[1]

1560 Camera obscure invented.

---

[1] There are other claims to this invention.

S. Bhooshan, *Fundamentals of Analogue and Digital Communication Systems*,
Lecture Notes in Electrical Engineering 785,
https://doi.org/10.1007/978-981-16-4277-7_1

1714 Typewriter invented.

1793 First telegraph line.

**1800 to 1850**

1814 First photographic image.

1821 The first microphone.

1835 Samuel Morse invents Morse code.

**1850 to 1900**

1876 Invention of the telephone.[2]

1885 James Clerk Maxwell predicts the existence of radio waves.

1887 Heinrich Hertz verifies the existence of radio waves.

1889 Invention of the automatic telephone exchange.

1895 Marconi invents the first radio transmitter/receiver.

1899 Loudspeakers invented.

**1900 to 1950**

1902 Marconi transmits radio signals across the Atlantic Ocean.

1905 The first audio broadcast.

1906 Lee Deforest invents the triode.

1910 First motion picture.

1914 First cross-continental telephone call made.

1916 First radios with tuners—different stations.

1923 First television camera.

1927 First television broadcast.

1935 Invention of the FM.

**1950–the present**

1957 The first artificial satellite.

1958 Integrated circuits invented.

1969 The first Internet started.

1971 The microprocessor invented.

1979 First cellular phone communication network started.

1985 CD-ROMs in computers.

1994 WWW is born.

## 1.2   Analogue Communication: Signals, Systems and Modulation

Communication starts with a signal.[3] The simplest method of communication is visual—the policeman standing at an intersection directing traffic with signals. Another level of communication—using technology—is the concept of signals sent via Morse code over telegraph lines and received at a distant post office. This method

---

[2] There are other claims to this invention.

[3] Some terms and concepts in this chapter may be unfamiliar. The student is requested to re-read it after completing the first few chapters of the text.

was very popular during the 1800s and sending a telegram was the only way to quickly send information over long distances. In the modern context, the signal in electronic communication has come to mean in particular an electrical, information- bearing signal, and the information content is voice, images or text.

The simplest signal is mathematically represented by a function of time which, for example, may be a time-varying message (like a sound signal) converted to a time-varying *electrical* voltage. Such an electrical signal may be represented by $x(t)$ (or $m(t)$, when we talk about a message signal). If the concept of a signal is not clear here, it will become clearer as you proceed in reading this text.

In an engineering communication system, it is important to note the following steps to convert raw information to the final signal which is made ready for communication, and after the communication process is over, to reconvert the signal back to the raw information:

1. The raw data or information which we want to transmit—which is generally analogue in nature. Most naturally occurring signals are analogue, like the diurnal temperature or ECG, EEG or voice.
2. Conversion of the raw information to electrical signals:

   (a) Conversion of the data into electrical signals, perhaps by use of a transducer. The resulting signal we call $m(t)$ for analogue communication.
   (b) Conversion from $m(t)$ to $m[n]$ for digital communication.

3. Processing $m(t)$ (or $m[n]$) by a communication system so that it may be conveniently transmitted over a *channel* such as

   (a) space or
   (b) electrical wires or
   (c) a fibre optic cable or
   (d) any other medium.

4. Extraction

   (a) Receiving the signal and extracting the electrical signal $m(t)$ (or $m[n]$);
   (b) Converting the digital signal, $m[n]$, back to the analogue signal, $\tilde{m}(t) \approx m(t)$ (see Fig. 1.1).

The communication system processes the signal and makes it 'ready' for transmission. This process may be done repeatedly so that the final system may be a conglomerate of many subsystems, each of which would perform a specific function. The connection between a signal and a system is diagrammatically represented in Fig. 1.2. The input signal here is $x(t)$ which is processed by the system $H$ to give an output $y(t)$. The relation between the input and output may be written in the most general fashion to be

$$y(t) = H\,[x(t)] \tag{1.1}$$

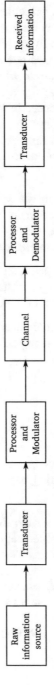

**Fig. 1.1** Figure showing a basic communication system block diagram

**Fig. 1.2** A system, $H$,
showing an input $x(t)$ and
output $y(t)$

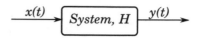

the input signal may be quite general in nature with some restriction on its *bandwidth*,[4]
in many cases the relation may take the form

$$y(t) = \int_0^\infty h(t - \tau)x(\tau)d\tau \tag{1.2}$$

where $h(t)$ is called the *impulse* response of the system $H$. The system in this case
is a *linear time-invariant* system (LTI).

In general, before the raw signal (which may be, for example, an audio signal) can
be transmitted, it must be 'modulated' which means that the signal must be suitably
electrically modified so that it may be conveniently transmitted. Thus, for example, if
we were to broadcast the raw information via an antenna, then two possibilities exist
from the point of view of the communication system: firstly, we may find that the
transmitted signal may be too heavily attenuated at the receiver end; and secondly if
many people sent such raw information simultaneously then, only garbled nonsense
would reach each receiver.

The modulation process in analogue communication in essence is changing the
parameters of a high-frequency signal in accordance with a low-frequency message
signal and then transmitting that signal. This process is such that the same communication channel may be used by many transmitters and many receivers. Also,
modulation may be used so that the channel may be used more effectively in some
sense.

For example, in analogue communication, generally the signal to be transmitted
is

$$f(t) = A \cos(\omega t + \phi) \tag{1.3}$$

where $A$ is the amplitude of the signal, $\omega$ is the frequency of the signal and $\phi$ the
phase of the signal.

1. By allowing the amplitude, $A$, to be time-varying in accordance with the
   information-bearing signal, $m(t)$ (keeping $\omega = \omega_c$ and $\phi = \phi_0$ fixed), we get
   *amplitude modulation*:

$$f(t) = A(t) \cos(\omega_c t + \phi_0)$$
$$A(t) \propto m(t)$$

2. by allowing $\omega$, the frequency of the sinusoid, to be time-varying in accordance
   with the information-bearing signal (keeping $A = A_0$ and $\phi = \phi_0$ fixed), we get
   *frequency modulation*:

---

[4] The concept of bandwidth will be explained a little later on in the book.

$$f(t) = A_0 \cos [\omega(t) + \phi_0]$$
$$\omega(t) \propto m(t)$$

and
3. by allowing $\phi$ to be time-varying in accordance with the information- bearing signal (keeping $A = A_0$ and $\omega = \omega_c$ fixed), we get *phase modulation*:

$$f(t) = A_0 \cos [\omega_c t + \phi(t)]$$
$$\phi(t) \propto m(t)$$

On reception, of course, we have to extract the original information. For purposes of illustration, these three types of modulation are shown in Fig. 1.3. If for some reason we want to send the information-bearing or message signal, $m(t) = 0.1t^2$, where $t$ is time, then the three cases of modulation are shown in this figure. We have used this message signal to visually clarify the three types of modulations.

In all three cases the modulating signal is $0.1t^2$, which is the message signal. Therefore in (a), the signal is an amplitude-modulated signal and the waveform is

$$f(t) = \underbrace{(1 + 0.1t^2)}_{\text{Amplitude}} \cos(3t) \tag{1.4}$$

while in (b) and (c), the signals are frequency-modulated and phase- modulated, respectively, and the two signals are given (in that order) to be

$$f(t) = \cos \left[ \underbrace{3t + \frac{0.1t^3}{3}}_{\text{Frequency}} \right] \tag{1.5}$$

$$f(t) = \cos \left[ 3t + \underbrace{0.1t^2}_{\text{Phase}} \right] \tag{1.6}$$

In the first of these three equations, the amplitude is varying proportional to the message signal. In the second equation, the phase of the signal is

$$\theta(t) = 3t + \frac{0.1t^3}{3}$$

and the frequency is

$$\omega(t) = \frac{d}{dt}\theta(t) = \frac{d}{dt}\left[ 3t + \frac{0.1t^3}{3} \right]$$
$$= 3 + 0.1t^2$$

which is proportional to $m(t)$. In the third equation, the phase varies as $m(t)$.

**Fig. 1.3** Introduction to three types of modulations. **a** Amplitude modulation, **b** frequency modulation and **c** phase modulation

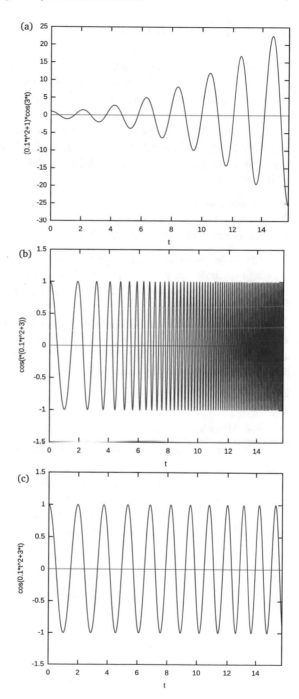

## 1.3   A Typical Communication System

A communication system, in the more modern sense of the term, consists of four different blocks: a transmitter, a receiver, a channel and a source of noise. This is shown in Fig. 1.4. In this book, communication *essentially* will mean to be *electrical* communication.

If we take a look at the individual parts of a communication system, then we find that the transmitter consists of a source of information and a converter which converts the raw information into suitable electrical signals. Examples of the source of information are

(a) an audio signal,
(b) pixel values in an image and
(c) characters such as 'a', '%' and 'L'.

These sources of information are important in that in the first case, a speech message or music may comprise the audio signal. In the second case, one may want to send a digital photograph to a friend so the individual pixels constitute the information, and so on.

The channel is the medium used for communication. The importance of the channel is that the channel properties must be such that the channel must be a 'good' medium to convey electrical signals and must not impede the progress of the signal which is transmitted in any manner. Examples of the channel are

(a) air,
(b) an optical cable and
(c) a coaxial line.

Each of these channels of communication has its own advantages and disadvantages.

The third block of a communication channel is the source of noise. In any communication system, noise 'smears' the signal in transit. Sources of noise are all those electrical sources which tend to confuse the signal being transmitted through the channel. An example of electrical noise is shown in Fig. 1.5.

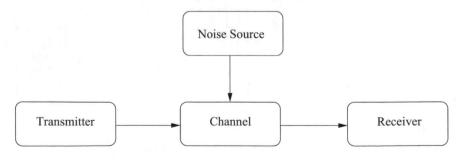

**Fig. 1.4**  A typical communication system showing a transmitter, receiver, channel and noise source

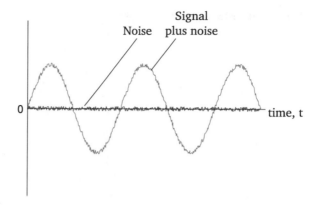

**Fig. 1.5** Noise $n(t)$ as a function of time and signal plus noise, $s(t) + n(t)$

The figure shows noise as it appears at any instant on the screen of an oscilloscope. Two aspects are immediately apparent: the amplitude of the noise varies randomly while noise seems to undergo rapid shifts from positive to negative and therefore the average value is zero. The same diagram also shows the signal plus noise. Here, the signal is distorted slightly by the noise, but if the signal amplitude is small, and the noise power is large, then the distortion may prove to be fatal, and the signal may never be recovered from the noise.

## 1.4 Analogue Communication, Digital Communication

Analog or analogue[5] communication is the transmission and reception of continuous-time signals. A typical continuous-time signal, or analogue signal, is one which is defined for all values of time and a representative sample is shown in Fig. 1.6.

Examples of analogue communication are

1. The transmission of an audio signal for AM or FM radio,
2. The transmission of audio and video signals for television and
3. Transmission of voice through telephone wires prior to the present digital revolution (*c* 1960).

On the other hand, examples of digital transmission are

1. Transmission of signals over the Internet,
2. Information sent over a LAN and
3. Mobile communication.

In the years from 1910 up to 1960, most communication was of the analogue variety. Three types of communication systems were in use: radio, television and telephone. Radio and television used the atmosphere above the ground as the communication channel while telephone used wires and the mode was analogue communication in

---

[5] Analog is the American spelling while analogue is the British spelling.

**Fig. 1.6** An analogue signal

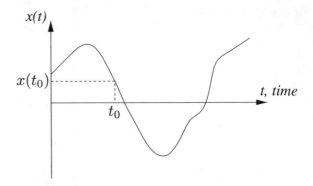

all three. Today, telephone communication is mainly digital but radio and TV are still analogue. We expect developments in the future whereby all communication is heading towards a digital revolution. For example, better communication performance occurs when we use digital systems in television as in the case of HDTV.

Therefore, the natural question which arises is that why study analogue communication at all? The real answer to this question is that all digital systems in the most essential sense are *analogue* in nature, and the design of these systems is buried in analogue communication theory. Consequently, if we do not study analogue communication, then our understanding of digital communication would be incomplete.

What are the advantages and disadvantages of analogue communication? The fact of the matter is that analogue communication is very simple in its essentials and very inexpensive to implement, though some parts of the design may be complicated. But although analogue systems are less expensive in many cases than digital ones for the same application, digital systems offer better performance and overall greater flexibility. But the greatest advantage of digital communication is *the immunity from noise*. Digital communication, through complex hardware design, can virtually overcome the problem of noise cluttering the signal.

With the increased speed of digital hardware, the development of increasingly efficient theory of digital communication and the ability to form a good communication infrastructure, digital communication is *indeed* now the best choice for many situations.

## 1.5   Self-Assessment

### 1.5.1   Review Questions

1. Why is the invention of photography important from the point of view of communication systems?
2. What is the importance of a microphone from the point of view of communication systems?

3. With the Morse code, what type of communication is possible?
4. How do radio waves impact communication?
5. What is the importance of Marconi's experiment?
6. How do artificial satellites help communication?
7. How do you convert voice into electrical signals?

# Chapter 2
# Analysis of Signals

## 2.1 Introduction

An electronic signal, as we understand in electrical communication engineering, is generally a travelling voltage waveform which may be an analogue or digital signal which carries information. For example, a person may speak into a microphone, and the output is an extremely weak voltage signal (of the order of mV or less). Such a signal is an audio signal converted into an electrical one whose representation is shown in Fig. 2.1. The $x$-axis shows one division of the order of a few hundred milliseconds, while the $y$-axis is the normalised voltage.

By looking at the figure, it is clear that human speech is highly oscillatory in nature—something like a sinusoidal signal—which oscillates quickly from positive to negative and back about a few thousand times a second. This is an example of a type of waveform which we will be using for communication purposes.

Other examples of signals are digital signals. An example of a digital signal created from an analogue signal is shown in Fig. 2.2. Here, we see that the function $f[n]$, the digital function, is defined through the relation

$$f[n] = \sin{(\omega_0 t)}|_{t=nT} \qquad n = -N_1, \ldots, 0, \ldots N_2 \qquad (2.1)$$

where $T$ is the sampling time and $N_1$, $N_2$ and $n$ are integers.

## 2.2 Signal Transformations

When we study engineering communication, a great deal of manipulation of signals have to be performed. With this in mind, we need to understand the transformation of signals, both of the dependent variable and independent variable. We take a look at these two types of transformations.

© The Author(s), under exclusive license to Springer Nature Singapore Pte Ltd. 2022    15
S. Bhooshan, *Fundamentals of Analogue and Digital Communication Systems*,
Lecture Notes in Electrical Engineering 785,
https://doi.org/10.1007/978-981-16-4277-7_2

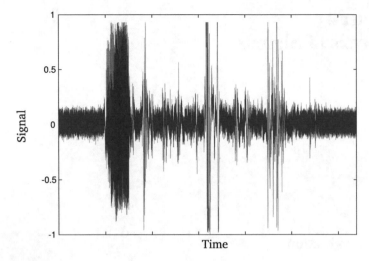

**Fig. 2.1**  An audio signal with noise (generated from a .wmv file)

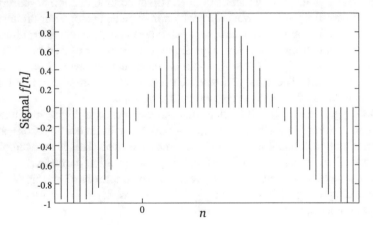

**Fig. 2.2**  Digitised sine function

If we consider a real signal $y(t)$, which is a function of time, then $y$ is the dependant variable and $t$ is the independent variable. Generally, the independent variable is on the x-axis while the dependent variable is on the y-axis. The independent variable values may be chosen in an *independent* manner, while *the ones for the dependent variable are obtained from the independent variable according to a rule*. (i.e. the function $y(t)$).

We will be considering two types of transformations:

1. Transformations of the dependent variable.
2. Transformations of the independent variable.

## 2.2.1 Transformation of the Dependent Variable

The simple transformations of the dependent variables are

1. Multiplication of the signal $f(t)$ by a constant: $af(t)$, where $a$ may be a complex number.
2. Addition or subtraction of a constant $f(t) \pm a$ where $a$ in general may be complex.

These types of transformations are shown in Fig. 2.3. In the (a) part of the figure, the original signal, $f(t)$, is shown, in the (b) part, a real positive constant $a$ is added to $f(t)$ and in the (c) part, the signal $af(t)$ is shown where $a$ is a real number greater than one.

Similarly, two complex signals, $f_1(t)$ and $f_2(t)$, may added, subtracted or multiplied:

1. $f(t) = f_1(t) \pm f_2(t)$ and
2. $f(t) = f_1(t) f_2(t)$

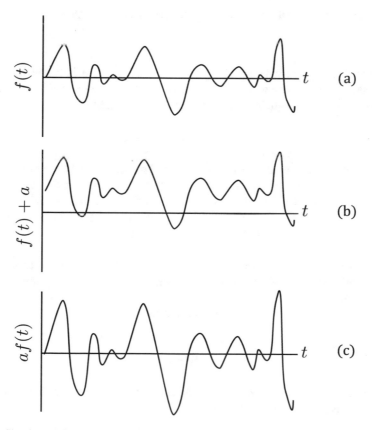

**Fig. 2.3** Simple transformations of the dependent variable, $y(t)$

A simple case where two signals are added is shown in Fig. 1.5(b), where noise is added to a signal when it passes through a channel. Similarly, the multiplication of two signals is shown in Fig. 1.3, where amplitude modulation is shown, and the two signals $f_1(t) = (1 + 0.1t^2)$ and $f_2(t) = \cos(3t)$ are multiplied.

We may also consider more general cases where complicated operations may be done on the dependant variable, $y = f(t)$, such as squaring of a given real signal

$$g(t) = [f(t)]^2$$

or taking the logarithm of a given complex signal

$$g(t) = \ln [|f(t)|]$$

## 2.2.2  Transformation of the Independent Variable

We may also transform the real independent variable, $t$. Three cases are of paramount importance:

1. The transformation where $t$ is replaced by '$at$' where $a$ is a real number. As an example, we may concentrate on two functions, $\sin(t)$ and $\sin(2t)$. When we consider these two functions and look at a point $t_0$, then

$$\sin(t)|_{t=t_0} = \sin(2t)|_{t=t_0/2} \tag{2.2}$$

since then the arguments of two functions will be equal. In this equation, (2.2), the point $t_0$ is *any point*, therefore, whatever the value of the function $\sin(t)$ for any particular value of $t$, ($= t_0$) will be the value of the function $\sin(2t)$ at $t = t_0/2$. This implies that if we graph the two functions, the second function will appear compressed by a factor of two. This is so since the value of $\sin(t)$ will be the same as $\sin[2 \times (t/2)]$.

This is shown in Fig. 2.4. In general, if we consider the general real function $f(t)$ and another function $f(at)$ where $a > 0$, then the two values are identical for the two points

$$t_0 \text{ and } t_0/a$$

If $a > 1$ then

$$t_0/a < t_0$$

therefore the graph of $f(at)$ is compressed when compared to that of $f(t)$. On the other hand, if $0 < a < 1$ then

$$t_0/a > t_0$$

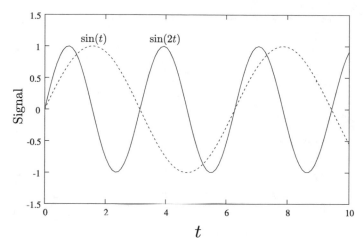

**Fig. 2.4**  Functions $\sin(t)$ and $\sin(2t)$

and the graph of $f(at)$ is expanded when compared to that of $f(t)$. In the figure, compare $0 < t < \pi$ in the $\sin(2t)$ graph and $0 < t < 2\pi$ in the $\sin(t)$ graph and one will notice the obvious compression.

2. The second case which is of importance is the case when $t$ is replaced by $-t$. Let us observe the function defined by

$$f(t) = \begin{cases} e^{-t} & t > 0 \\ 0 & t < 0 \end{cases} \tag{2.3}$$

which is shown plotted by the solid line in Fig. 2.5. And then compare this to the curve

$$f(-t) = \begin{cases} e^{t} & -t > 0 \\ 0 & -t < 0 \end{cases}$$

where we have substituted $-t$ wherever $t$ occurs. We can see that each function is a mirror image of the other. This is so, because if we were to consider any point $t_0$, then

$$f(t)|_{t=t_0} = f(-t)|_{t=-t_0} \tag{2.4}$$

3. The third case of interest is when the variable $t$ is shifted by $\tau$. In this case, the two functions are $f(t)$ and $f(t - \tau)$. Now consider the two points $t_0$ and $t_1$ such that

$$f(t_0) = f(t_1 - \tau)$$
$$\text{therefore } t_0 = t_1 - \tau$$
$$\text{or } t_1 = t_0 + \tau \tag{2.5}$$

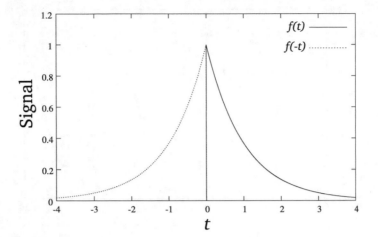

**Fig. 2.5** $f(t)$ and $f(-t)$

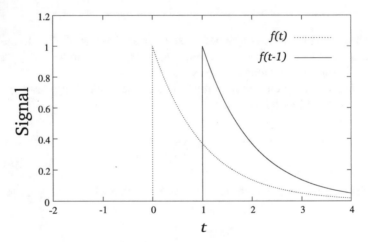

**Fig. 2.6** $f(t)$ and $f(t-1)$

or $f(t)$ at $t_0$ is shifted to $t_0 + \tau$ in the function $f(t - \tau)$ (assuming $\tau > 0$).
The graph of $f(t)$ is shifted to the right by $\tau$. Considering Fig. 2.6, we can see
the obvious shifting. The function $f(t)$ is the function given in Eq. (2.3), a few
equations back, which is plotted alongside $f(t - 1)$.

**Example 2.1** Write the periodic function given in Example 2.14.

The function is

$$\frac{1}{|t|}$$

in the interval

$$-T/2 \leq t \leq T/2$$

After a little thought, it is clear that we have to use the property of time shifting given above. So

$$f(t) = \sum_{n=-\infty}^{n=\infty} f_n(t)$$

$$\text{where } f_n(t) = \frac{1}{|t - nT|} \quad nT - \frac{T}{2} < t < nT + \frac{T}{2}$$

**Example 2.2** Given the pulse function

$$p(t) = \begin{cases} 0.25 & \text{for } 1 \leq t \leq 2 \\ 0 & \text{everywhere else} \end{cases}$$

plot $4p\,[3(t-3)]$.

Let us look at the argument of $p(3(t-3))$ which is derived from $p(t)$. $3(t_0 - 3) = 3t_0 - 9 = t_1$. If $t_1$ satisfies $1 \leq t_1 \leq 2$, then $p(t_1)$ is 0.25. That is, if the inequality

$$1 \leq \underbrace{3t_0 - 9}_{=t_1} \leq 2$$

is satisfied, then $p(t_1)$ is 0.25. On simplifying, we get

$$10 < 3t_0 \leq 11$$

or on further simplification, we get

$$\frac{10}{3} \leq t_0 \leq \frac{11}{3}$$

therefore

$$p\,[3(t-3)] = \begin{cases} 0.25 & \text{for } \frac{10}{3} \leq t \leq \frac{11}{3} \\ 0 & \text{elsewhere} \end{cases}$$

or multiplying by four,

$$4p\,[3(t-3)] = \begin{cases} 1 & \text{for } \frac{10}{3} \leq t \leq \frac{11}{3} \\ 0 & \text{elsewhere} \end{cases}$$

Next, we check our solution: substitute 10/3 and 11/3 in $3(t-3)$. At the first point it is 1, and at the second point it is 2.

The plots of the two pulses are shown in Fig. 2.7.

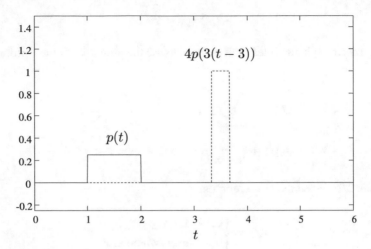

**Fig. 2.7**  Figure for Example 2.2

## 2.3  Energy and Power

We progress next to considering the energy and power in a signal. If a voltage $f(t)$ is applied to an $R$ $(\Omega)$ resistor, then the power consumed at any instant of time would be

$$\frac{[f(t)]^2}{R} \text{ (W)}$$

and the total energy consumed by the $R$ $(\Omega)$ resistor would be

$$\mathbb{E}_R = \frac{1}{R} \int_{-\infty}^{\infty} [f(t)]^2 \, dt \text{ (J)}$$

Similarly, if $f(t)$ is a current signal, then the power consumed at any instant of time would be

$$R\,[f(t)]^2 \text{ (W)}$$

and the total energy consumed would be

$$R \int_{-\infty}^{\infty} [f(t)]^2 \, dt \text{ (J)}$$

To consider energy for all signals, and taking both definitions into account, we normalise the energy, and consider energy consumed by a $R = 1$ $(\Omega)$ resistor only.

Keeping these definitions in mind, the energy of a signal is formally defined as

$$\boxed{\mathbb{E}\{f(t)\} \stackrel{\Delta}{=} \int_{-\infty}^{\infty} [f(t)]^2 \, dt} \tag{2.6}$$

where the symbol $\stackrel{\Delta}{=}$ has the meaning, "by definition is equal to". As we know that $f(t)$ being real may be positive or negative, but $[f(t)]^2$ will always be positive, therefore, the energy of a signal, Definition (2.6), will always be positive.

Similarly, we can extend this definition to complex signals. if the signal $f(t)$ is complex, then $f(t)f(t)^*$ is real and positive (where the star ($*$) denotes complex conjugation). Therefore, the energy of a complex signal is

$$\boxed{\mathbb{E}\{f(t)\} \stackrel{\Delta}{=} \int_{-\infty}^{\infty} f(t)f(t)^* dt = \int_{-\infty}^{\infty} |f(t)|^2 \, dt} \tag{2.7}$$

**Example 2.3** Let us compute the energy of the signal

$$f(t) = \begin{cases} Ae^{-at} & \text{for } t > 0; \ \text{where } a, A > 0 \\ 0 & \text{for } t < 0 \end{cases}$$

The energy in the signal by our earlier definition is

$$\mathbb{E} = \int_{-\infty}^{\infty} [f(t)]^2 \, dt = A^2 \int_{0}^{\infty} e^{-2at} dt = -\frac{A^2}{2a} e^{-2at} \Big|_{0}^{\infty} = \frac{A^2}{2a}$$

**Example 2.4** Find the energy in the signal

$$f(t) = \begin{cases} A \sin(\omega_0 t) & 0 \le t \le NT \\ 0 & \text{everywhere else} \end{cases}$$

where

$$T = 2\pi/\omega_0$$

Using the definition of energy,

$$\mathbb{E} = \int_{-\infty}^{\infty} [f(t)]^2 \, dt = A^2 \int_{0}^{NT} \sin^2(\omega_0 t) \, dt$$

$$= \frac{t}{2} - \frac{\sin(2t\omega_0)}{4\omega_0} \Big|_{0}^{NT} \times A^2 \quad (\text{using } \sin^2 \theta = \frac{1 - \cos 2\theta}{2})$$

$$= A^2 \frac{NT}{2} \quad (\text{since } \sin(2NT\omega_0) = 0)$$

**Example 2.5**  Find the energy in the complex signal

$$f(t) = \frac{A}{a + jt} \qquad -\infty < t < \infty$$

and where $A$ and $a$ are real.

The energy in the signal is given by

$$\mathbb{E} = \int_{-\infty}^{\infty} |f(t)|^2 \, dt$$

$$= \int_{-\infty}^{\infty} \frac{A^2}{a^2 + t^2} \, dt \qquad (\text{using } |1/(a + jt)|)$$

$$= \frac{A^2}{a} \tan^{-1}\left(\frac{t}{a}\right)\Big|_{-\infty}^{\infty}$$

$$= \frac{A^2 \pi}{a}$$

Notice that the energy of this example and that of Example 2.3 differ only by a factor of $2\pi$. This result will be of significance at a later point in this chapter.

**Example 2.6**  Find the energy in the time-limited signal

$$f(t) = \begin{cases} A \sin(\omega_0 t) & 0 < t < NT \\ 0 & \text{otherwise} \end{cases}$$

where $N$ is a positive integer and $T = 2\pi/\omega_0$.

For certain signals, the energy of the signal turns out to be infinite. In Example 2.6, if $N$ is infinity, then the energy is infinite. For these types of signals, we define the *average power* of a signal:

$$\boxed{\mathbb{P}\{f(t)\} \triangleq \lim_{T \to \infty} \left\{ \frac{1}{T} \int_{-T/2}^{T/2} [f(t)]^2 \, dt \right\}} \qquad (2.8)$$

**Example 2.7**  Find the power in the signal which is not time-limited,

$$f(t) = \begin{cases} A \sin(\omega_0 t) & 0 < t < \infty \\ 0 & \text{otherwise} \end{cases}$$

The power in the signal is given by definition to be

$$\mathbb{P} = \lim_{T \to \infty} \left\{ \frac{1}{T} \int_{-T/2}^{T/2} [f(t)]^2 \, dt \right\}$$

$$= \lim_{T \to \infty} \left\{ \frac{1}{T} \int_{-T/2}^{T/2} [A \sin(\omega_0 t)]^2 \, dt \right\}$$

$$= \lim_{T \to \infty} \left\{ \frac{A^2}{T} \left[ \frac{t}{2} - \frac{\sin(2t\omega_0)}{4\omega_0} \Bigg|_0^{T/2} \right] \right\}$$

$$= \frac{A^2}{4}$$

In a similar manner, we may define the power of a complex power signal, $f(t)$

$$\mathbb{P}\{f(t)\} \triangleq \lim_{T \to \infty} \left\{ \frac{1}{T} \int_{-T/2}^{T/2} |f(t)|^2 \, dt \right\} \qquad (2.9)$$

where $|f(t)|^2 = f(t) f(t)^*$ is the squared magnitude of the power signal.

**Example 2.8** Find the power in the complex signal $Ae^{j\omega_0 t}$ where $A$ and $\omega_0$ are real constants.

The power by definition is

$$\mathbb{P}\{f(t)\} \triangleq \lim_{T \to \infty} \left\{ \frac{1}{T} \int_{-T/2}^{T/2} |f(t)|^2 \, dt \right\}$$

$$- \lim_{T \to \infty} \left\{ \frac{1}{T} \int_{-T/2}^{T/2} \left( Ae^{j\omega_0 t} \right) \left( Ae^{-j\omega_0 t} \right) dt \right\}$$

$$= A^2$$

**Example 2.9** From Example 2.8, find the power in the two signals $A \sin \omega_0 t$ and $A \cos \omega_0 t$.

Since

$$\left( Ae^{j\omega_0 t} \right) \left( Ae^{-j\omega_0 t} \right) = A^2 (\sin^2 \omega_0 t + \cos^2 \omega_0 t)$$

therefore the power in each of these signals is $A^2/2$.

## 2.4 Some Special Functions

Before we proceed, we will take a look at some new functions (and some which have already been studied earlier) which are of importance in analogue communication.

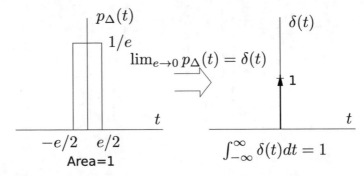

**Fig. 2.8** Figure used in the discussion of the delta or impulse function, $\delta(t)$

### 2.4.1   Impulse Function

Let us take a look at a new kind of function shown in Fig. 2.8. The function $p_\Delta(t)$ is defined as

$$p_\Delta(t) = \begin{cases} 1/e & -e/2 < t < e/2 \\ 0 & |t| > e/2 \end{cases} \tag{2.10}$$

There are a few points which we can notice straight away about this function.

1. The value of this pulse is linked to the width of the pulse. That is, as $e$ becomes smaller the y value increases in the interval $-e/2 \le t \le e/2$.
2. The area under the pulse is always equal to 1. The width is $e$ and the height is $1/e$, so $e \times 1/e = 1$.
3. In the limit as $e \to 0$, the value $1/e \to \infty$. Since we are talking about the limit, this means that the width $e$ *approaches zero but never becomes zero*. Also, *the area under the function remains equal to 1*. These ideas can be put mathematically as

$$\lim_{e \to 0} p_\Delta(t) = \delta(t)$$

$$\int_{-\infty}^{\infty} p_\Delta(t)dt = 1$$

$$\int_{-\infty}^{\infty} \delta(t)dt = 1 \tag{2.11}$$

the first of these equations says that as $e \to 0$ the function $p_\Delta(t)$ becomes another function, namely the impulse or delta function, $\delta(t)$. Note that the delta function[1] is shown as an arrow with value 1. The '*1*' means that the area under the arrow is one. The second and third of these equations reiterate that the area under $p_\Delta(t)$ or $\delta(t)$ is one.

---

[1] Also called the Dirac delta function or the unit impulse.

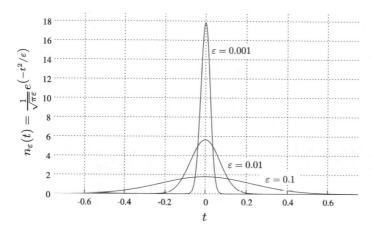

**Fig. 2.9** The impulse function as a limiting case of the $\exp[-x^2]$ function

More advanced readers may object to this function since it is not continuously differentiable. Let us look at the function

$$n_\varepsilon(t) = \frac{1}{\sqrt{\pi\varepsilon}}e^{(-t^2/\varepsilon)} \tag{2.12}$$

if we integrate this function, then

$$\int_{-\infty}^{\infty} \frac{1}{\sqrt{\pi\varepsilon}}e^{(-t^2/\varepsilon)}dt = 1 \tag{2.13}$$

and as $\varepsilon \to 0$ it approaches the impulse function. The function is shown for three values of $\varepsilon$ in Fig. 2.9, where we observe that in the limiting case,

$$\delta(t) = \lim_{\varepsilon \to o} n_\varepsilon(t) \tag{2.14}$$

### 2.4.1.1 Properties of the Impulse Function

There are some special properties about the delta function which we can notice straight away.

1. If we multiply the delta function by a real number $a$, then it has a value equal to $a\delta(t)$. If we take the pulse definition (Eq. (2.10)) of the impulse function, then

$$a \times \lim_{e \to 0}[p_\Delta(t)] = a\delta(t) \tag{2.15}$$

2. The function $\delta(t)$ is an even function. (Remember that $p_\triangle(t)$ was an even function.)
3. If we multiply the function $\delta(t)$ by another function $x(t)$ (which is smooth at $t = 0$), and then integrate, the integral

$$\int_{-\infty}^{\infty} x(t)\delta(t)dt = x(0) \int_{-\infty}^{\infty} \delta(t)dt = x(0) \qquad (2.16)$$

Why? This is so because $\delta(t)$ exists only at $t = 0$. At all other points, $x(t)$ is multiplied by zero. Or we may write more conveniently

$$x(t)\delta(t) = x(0)\delta(t) \qquad (2.17)$$

4. Similarly,
$$x(t)\delta(t - t_0) = x(t_0)\delta(t - t_0) \qquad (2.18)$$

5. Let us take a look at $\delta(at)$. It can be shown that

$$\delta(at) = \frac{1}{|a|}\delta(t) \qquad (2.19)$$

which is given as Problem 8 in Sect. 2.19.2.

### 2.4.2 The Unit Step Function

The unit step function is defined as

$$u(t) = \begin{cases} 1 & t > 0 \\ 0 & t < 0 \\ 0.5 & t = 0 \end{cases} \qquad (2.20)$$

The depiction of the unit step function is shown in Fig. 2.10.

**Fig. 2.10** The unit step function

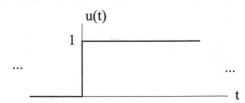

**Fig. 2.11** RC circuit with switch

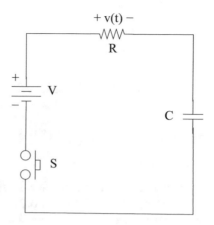

The unit step function is very important in engineering. It is used perpetually in cases when a signal is applied for $t \geq 0$. That is, if $x(t)$ is the input, then

$$x(t) = \begin{cases} f(t) & t \geq 0 \\ 0 & t < 0 \end{cases}$$

or

$$x(t) = f(t)u(t)$$

Consider Fig. 2.11, which shows a DC source with voltage $V$, connected in series with a resistor $R$ and capacitor $C$. The switch is depressed at $t = 0$. Therefore

$$V \equiv Vu(t)$$

Before the switch is depressed, the voltage $v(t)$ across the resistor is zero. But after it is depressed, the voltage is given by $Ve^{-t/RC}$. In other words,

$$v(t) = \begin{cases} 0 & t < 0 \\ Ve^{-t/RC} & t > 0 \end{cases}$$

These ideas can be put succinctly into the simple form

$$v(t) = Ve^{-t/RC}u(t)$$

The multiplication of the function $Ve^{-t/RC}$ by $u(t)$ erases all outputs for $t < 0$.

The delta function and the unit step functions are linked in the following manner. Consider the integral

$$\int_{-\infty}^{t} \delta(\tau)d\tau \tag{2.21}$$

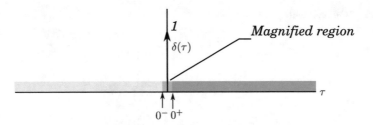

**Fig. 2.12** Integration of $\delta(\tau)$

This is an indefinite integral. Notice that as long as $t < 0$, the integral is zero.
This so since $\delta(t) = 0$ for $t < 0$. Now let us take the value of $t = 0^+$ that is slightly
greater than zero. We notice then that the integral is then equal to 1, since $\delta(t)$ is then
included. We can clarify these ideas by observing Fig. 2.12. As long as $t$ lies in the
lightly shaded region, the integral is zero and as soon as $t$ is in the region of heaviest
shading, the integral is equal to 1. The main contribution comes from integration
from $0^-$ to $0^+$. Hence,

$$\int_{-\infty}^{t} \delta(\tau)d\tau = \begin{cases} 0 & t < 0 \\ 1 & t > 0 \end{cases} \tag{2.22}$$

or

$$\int_{-\infty}^{t} \delta(\tau)d\tau = u(t) \tag{2.23}$$

and conversely

$$\frac{d}{dt}u(t) = \delta(t) \tag{2.24}$$

### 2.4.3  The Signum Function

This function is defined as

$$\text{sgn}(t) = \begin{cases} 1 & t > 0 \\ 0 & t = 0 \\ -1 & t < 0 \end{cases} \tag{2.25}$$

and is shown in Fig. 2.13. A little reflection tells us that

$$\text{sgn}(t) = 2u(t) - 1 \tag{2.26}$$

**Fig. 2.13**  The signum
function

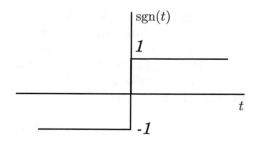

## 2.4.4  The Sinc Function

Another function which makes an appearance in communication is the sinc($\omega$) function, which is defined as

$$
\text{sinc}(\omega) = \begin{cases} 1 & \omega = 0 \\ \sin(\omega)/\omega & \omega \neq 0 \end{cases}
\tag{2.27}
$$

The function is shown in Fig. 2.14. At $\omega = 0$, the function can be written as a limit:

$$
\lim_{\omega \to 0} \frac{\sin(\omega)}{\omega} = 1
\tag{2.28}
$$

Observing the plot of the sinc function, we can see that the zero crossings occur at

$$
\omega = \pm\pi, \pm2\pi, \pm3\pi, \ldots
$$

since the zeros of $\sin(\omega)$ are the zeros of sinc($\omega$) except the zero at $\omega = 0$.

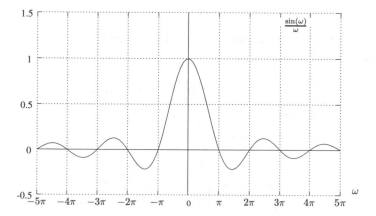

**Fig. 2.14**  The sinc function

The envelope of the sinc function is proportional to $\pm 1/\omega$ as $\omega$ increases in the positive direction and decreases in the negative direction. The minus sign occurs since the sine function oscillates between $\pm 1$. The sinc function is an even function since it consists of an odd function divided by another odd function.

Similarly, if we plot the function $\mathrm{sinc}(\omega \tau)$, then the zero crossings exist when the numerator is zero,

$$\sin(\omega \tau) = 0$$

or when

$$\omega_{z,m} \tau = m\pi \qquad \text{for } m = \pm 1, \pm 2, \ldots \tag{2.29}$$

which excludes $\omega = 0$ for the reason that both the numerator and denominator both tend to zero. Or

$$\omega_{z,m} = \frac{m\pi}{\tau} \qquad m = \pm 1, \pm 2, \ldots \tag{2.30}$$

and the sinc(.) graph expands and contracts as per the value of $\tau$: if $\tau > 1$ it contracts; if $\tau < 1$ it expands, as given in the section on 'Transformation of the Independent Variable.' Generally, we will be interested in the case when $\tau < 1$, when we realise from Eq. (2.30) (given above) the zeros are widely spaced.

### 2.4.5  The Rect Function

The rect function is defined by

$$\mathrm{rect}(t) = \begin{cases} 1 & |t| < 0.5 \\ 0.5 & |t| = 0.5 \\ 0 & |t| > 0.5 \end{cases} \tag{2.31}$$

The rect(.) plays a very important role in communication since it is a pulse. The rect function may be written as

$$\mathrm{rect}(t) = u(t + 0.5) - u(t - 0.5) \tag{2.32}$$

and is shown in Fig. 2.15. To describe a pulse of width $\tau$, we consider the function $\mathrm{rect}(t/\tau)$ which can be written as

$$\mathrm{rect}(t/\tau) = \begin{cases} 0 & t < -\tau/2 \\ 1 & -\tau/2 < t < \tau/2 \\ 0 & t > \tau/2 \end{cases}$$

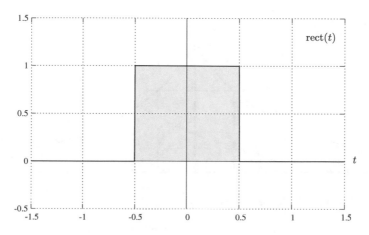

**Fig. 2.15**  The rect($t$) function

## 2.5  Convolution

Convolution is a special operation between two functions involving integration to obtain a third function.[2] Convolution has various applications in communication and therefore has to be understood thoroughly before we proceed.

To be specific, if the two time-dependant functions are $f(t)$ and $g(t)$, then the convolution 'product' is represented by the mathematical formula

$$f(t) * g(t) = h(t) = \int_{-\infty}^{\infty} f(t - \tau)g(\tau)d\tau \tag{2.33}$$

where the $*$ represents convolution. Notice that integration  is over the dummy variable $\tau$, while $t$ is treated as a constant. The value of $t$ lies in the interval, $-\infty < t < \infty$. From the above remarks, it is obvious that the result of the convolution is a function of time. From Eq. (2.33), we observe that if the integration is performed numerically, then an integration has to be performed for every value of $t$.

If we observe Eq. (2.33), then we see that the function $f(t - \tau)$ is involved in the integration of the variable $\tau$. Let us determine what the function $f(t - \tau)$ looks like.

Figure 2.16 shows three plots. The first plot is of $f(\tau)$ shown as a solid curve. For convenience, the function is considered to be zero when $\tau \leq 0$. The second plot with the dashed lines is the function $f(-\tau)$ which is the mirror image of $f(\tau)$ or the function $f(\tau)$ folded about the y-axis. The third trace, $f(t - \tau)$, is the dot-dash line which is the function $f(-\tau)$ shifted to the right by $t$. In Fig. 2.16 where the three functions are shown, the line AB cuts the three functions at different points, $\tau_0$, $\tau_1$ and $\tau_2$ such that each of the functions has the value of $f(\tau_0)$. Therefore, in general

---

[2] It is assumed that convolution has been studied by the student in an earlier course. What follows is a very abridged exposition.

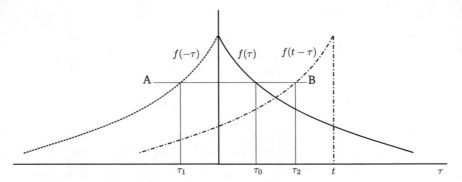

**Fig. 2.16** Illustration for the function $f(t - \tau)$

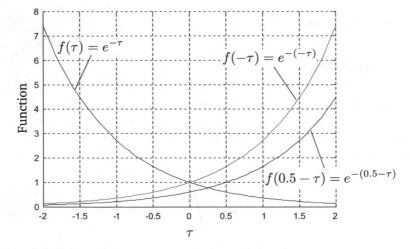

**Fig. 2.17** Plot for example, $e^{-\tau}$

$$f(\tau) = f(-\tau) = f(t - \tau) \tag{2.34}$$

In particular, the value $\tau = 0$ of $f(\tau)$ maps to 0 in $f(-\tau)$ and $t$ in $f(t - \tau)$.

**Example 2.10** If $f(\tau) = \exp\{-\tau\}$, then plot $f(\tau)$, $f(-\tau)$ and $f(0.5 - \tau)$ in the range $-2 \leq \tau \leq 2$.

The plot is shown in Fig. 2.17. If we examine the three functions, we may notice that $f(\tau)$ and $f(-\tau)$ are mirror images of each other and $f(0.5 - \tau)$ is the function $f(-\tau)$ *shifted to the right by* 0.5.

Now that we have considered the function $f(t - \tau)$, we can look at convolution in more detail. Equation (2.33) says that the convolution of two functions is the integral of the product $f(t - \tau)$ and $g(\tau)$. Let us look at a concrete example.

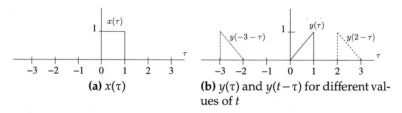

**Fig. 2.18** Particulars of $x(\tau)$ and $y(\tau)$

**Example 2.11** Let the two functions be

$$x(\tau) = \begin{cases} 1 & \text{for } 0 \le \tau \le 1 \\ 0 & \text{otherwise} \end{cases}$$

$$y(\tau) = \begin{cases} \tau & \text{for } 0 \le \tau \le 1 \\ 0 & \text{otherwise} \end{cases} \qquad (2.35)$$

These functions are depicted in Fig. 2.18. The (a) part of the figure shows $x(\tau)$, while the (b) part of the figure shows the functions $y(\tau)$ and $y(t - \tau)$ for $t = -2$ and $t = 3$, respectively. We can now start the convolution procedure which is shown diagrammatically in Fig. 2.19.

On observing the figure, we can see that in (a), the two functions $x(\tau)$ and $y(t - \tau)$ are shown for $t = 0$ and we can see that there is no overlap between the two functions, so that if we set $g(t) = x(t) * y(t)$, then $g(0) = 0$. In fact

$$g(t) = 0 \qquad \text{for } -\infty < t < 0 \qquad (2.36)$$

since for these values of $t$, there is no overlap between the two functions. In (b) and (c) parts of the figure, the value of $t$ is between 0 and 2 and there obviously is an overlap between the two functions, and for ease of computation we may divide the overlap part into two sub-parts:

$$\text{I} \qquad 0 \le t \le 1$$
$$\text{II} \qquad 1 \le t \le 2$$

For $0 \le t \le 1$ as shown in Fig. 2.19b,

$$g(t) = \int_{-\infty}^{\infty} \underbrace{x(\tau)}_{=1} \underbrace{y(t - \tau)}_{t-\tau} d\tau = \int_{0}^{t} \underbrace{x(\tau)}_{=1} \underbrace{y(t - \tau)}_{t-\tau} d\tau$$

$$= \int_{0}^{t} (t - \tau) d\tau = t\tau - \frac{\tau^2}{2} \Big|_{0}^{t} = \frac{t^2}{2} \qquad (2.37)$$

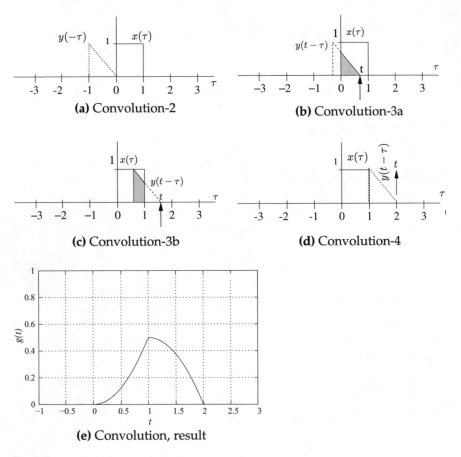

**Fig. 2.19** Details of the convolution for the example

and for $1 \le t \le 2$ as shown in Fig. 2.19c

$$g(t) = \int_{-\infty}^{\infty} \underbrace{x(\tau)}_{=1} \underbrace{y(t-\tau)}_{t-\tau} d\tau = \int_{t-1}^{1} \underbrace{x(\tau)}_{=1} \underbrace{y(t-\tau)}_{t-\tau} d\tau$$

$$= \int_{t-1}^{1} (t-\tau) d\tau = t\tau - \frac{\tau^2}{2} \Big|_{t-1}^{1} = t\left(1 - \frac{t}{2}\right) \qquad (2.38)$$

For $t > 2$, shown in Fig. 2.19d, $g(t) = 0$ once more since there is no overlap between the two functions. The result of the convolution is shown in Fig. 2.19e.

**Example 2.12** Find the convolution of $\delta(t)$ with $x(t)$.

By the definition of convolution,

$$\int_{-\infty}^{\infty} x(t-\tau)\delta(\tau)d\tau = x(t)\int_{-\infty}^{\infty}\delta(\tau)d\tau \qquad \text{(since } \delta(\tau) \text{ exists only at } \tau = 0\text{)}$$
$$= x(t)$$

From this example, it is clear that $\delta(t)$ behaves like the identity element:

$$\boxed{x(t) * \delta(t) = x(t)} \tag{2.39}$$

## 2.6 Periodic Signals and Trigonometric Fourier Series

The science of analysis of signals treats signals in a special manner. Let us consider a periodic signal $x_T(t)$ with period $T$ as shown in Fig. 2.20. Since the function is periodic

$$x_T(t) = x_T(t+T) \tag{2.40}$$

We stipulate that this function may be written as a sum of sine and cosine functions in the following manner:

$$x_T(t) = a_0 + \sum_{k=1}^{\infty} a_k \cos(k\omega_0 t) + \sum_{k-1}^{\infty} b_k \sin(k\omega_0 t) \tag{2.41}$$

where

$$\frac{2\pi}{T} \equiv \omega_0 \tag{2.42}$$

is called the fundamental frequency, and each $a_k$ and $b_k$ is a constant. The idea behind this expansion is to convert an arbitrary periodic signal in terms of a 'standard' set of other, 'known,' periodic signals, which in this case are sine and cosine functions, and the only characterisation feature comprises the constant coefficients $a_k$s and $b_k$s.

**Fig. 2.20** A periodic signal

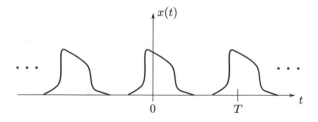

The idea is somewhat like depicting any n-space vector in terms of n-space unit vectors. Thus

$$(\mathbf{a} =) \begin{bmatrix} a_1 \\ a_2 \\ \vdots \\ a_n \end{bmatrix} = a_1 \begin{bmatrix} 1 \\ 0 \\ \vdots \\ 0 \end{bmatrix} + a_2 \begin{bmatrix} 0 \\ 1 \\ \vdots \\ 0 \end{bmatrix} + \cdots + a_n \begin{bmatrix} 0 \\ 0 \\ \vdots \\ 1 \end{bmatrix} \qquad (2.43)$$

where the unit vectors are

$$\mathbf{u}_1 = \begin{bmatrix} 1 \\ 0 \\ \vdots \\ 0 \end{bmatrix}, \ \mathbf{u}_2 = \begin{bmatrix} 0 \\ 1 \\ \vdots \\ 0 \end{bmatrix}, \ \cdots \mathbf{u}_n = \begin{bmatrix} 0 \\ 0 \\ \vdots \\ 1 \end{bmatrix} \qquad (2.44)$$

which are the standard vectors, and

$$\mathbf{a} = \begin{bmatrix} a_1 \\ a_2 \\ \vdots \\ a_n \end{bmatrix}$$

is any n-space vector. If we consider Eq. (2.43), then $a_1$ can be determined by taking the dot product of that equation with

$$\mathbf{u}_1 = \begin{bmatrix} 1 \\ 0 \\ \vdots \\ 0 \end{bmatrix}$$

so that

$$\begin{bmatrix} a_1 \\ a_2 \\ \vdots \\ a_n \end{bmatrix} \bullet \begin{bmatrix} 1 \\ 0 \\ \vdots \\ 0 \end{bmatrix} = a_1 \begin{bmatrix} 1 \\ 0 \\ \vdots \\ 0 \end{bmatrix} \bullet \begin{bmatrix} 1 \\ 0 \\ \vdots \\ 0 \end{bmatrix} + a_2 \begin{bmatrix} 0 \\ 1 \\ \vdots \\ 0 \end{bmatrix} \bullet \begin{bmatrix} 1 \\ 0 \\ \vdots \\ 0 \end{bmatrix} + \cdots + a_n \begin{bmatrix} 0 \\ 0 \\ \vdots \\ 1 \end{bmatrix} \bullet \begin{bmatrix} 1 \\ 0 \\ \vdots \\ 0 \end{bmatrix}$$
$$(2.45)$$

and we know that all the terms on the right side of the equation are zero, except the first one. That is,

$$\mathbf{u}_1 \bullet \mathbf{u}_j = \begin{cases} 0 & \forall \ j \neq 1 \\ 1 & \text{for } j = 1 \end{cases} \qquad (2.46)$$

so

$$a_1 = \begin{bmatrix} a_1 \\ a_2 \\ \vdots \\ a_n \end{bmatrix} \bullet \begin{bmatrix} 1 \\ 0 \\ \vdots \\ 0 \end{bmatrix} = \mathbf{a} \bullet \mathbf{u}_1 \tag{2.47}$$

or in general

$$a_k = \mathbf{a} \bullet \mathbf{u}_k \qquad k = 1, \cdots, n \tag{2.48}$$

Going back to (2.41), as a first step let us find out whether $x_T(t)$ given by this representation is in fact a periodic function with period $T$. So obviously,

$$x_T(t + T) = a_0 + \sum_{k=1}^{\infty} a_k \cos \{k\omega_0 (t + T)\} + \sum_{k=1}^{\infty} b_k \sin \{k\omega_0 (t + T)\}$$

$$= a_0 + \sum_{k=1}^{\infty} a_k \cos \left\{ k \left( \frac{2\pi}{T} \right) t \right\} + \sum_{k=1}^{\infty} b_k \sin \left\{ k \left( \frac{2\pi}{T} \right) t \right\}$$

$$= x_T(t) \tag{2.49}$$

where we have used the relationship between $\omega_0$ and $T$. In this case, the basis functions are

$$1, \cos \left\{ k \left( \frac{2\pi}{T} \right) t \right\}, \sin \left\{ k \left( \frac{2\pi}{T} \right) t \right\}, \ k = 0, 1, \ldots$$

The next question is: how to determine the $a_k$s and $b_k$s? We proceed in a similar manner as in the case of vectors in n-space: we take the 'function dot product':

$$\langle f(t) \bullet g(t) \rangle = \int_{\text{time period}} f(t) g(t) dt \tag{2.50}$$

Multiply (2.41) by $\cos(m\omega_0 t)$ and integrate both sides over a time interval equal to $T$,

$$\int_{t_0}^{t_0+T} x_T(t) \cos(m\omega_0 t) dt = a_0 \int_{t_0}^{t_0+T} \cos(m\omega_0 t) dt + \sum_{k=1}^{\infty} a_k \int_{t_0}^{t_0+T} \cos(k\omega_0 t) \cos(m\omega_0 t) dt$$

$$+ \sum_{k=1}^{\infty} b_k \int_{t_0}^{t_0+T} \sin(k\omega_0 t) \cos(m\omega_0 t) dt \tag{2.51}$$

and using

$$\int_{t_0}^{t_0+T} \cos(k\omega_0 t) \cos(m\omega_0 t)\, dt = \begin{cases} 0 & \text{for } k \neq m \\ T/2 & \text{for } k = m \neq 0 \end{cases}$$

$$\int_{t_0}^{t_0+T} \sin(k\omega_0 t) \cos(m\omega_0 t)\, dt = 0 \text{ for all values } k, m \qquad (2.52)$$

we get

$$a_0 = \frac{1}{T} \int_{t_0}^{t_0+T} x_T(t)\, dt$$

$$a_1 = \frac{2}{T} \int_{t_0}^{t_0+T} x_T(t) \cos(\omega_0 t)\, dt$$

$$\ldots = \ldots$$

$$a_m = \frac{2}{T} \int_{t_0}^{t_0+T} x_T(t) \cos(m\omega_0 t)\, dt \qquad (2.53)$$

Similarly, we can obtain the $b_k$s by multiplying (2.41), by $\sin(m\omega_0 t)$ and then integrating over a time interval equal to $T$.

$$b_m = \frac{2}{T} \int_{t_0}^{t_0+T} x_T(t) \sin(m\omega_0 t)\, dt \qquad (2.54)$$

and where we have used

$$\int_{t_0}^{t_0+T} \sin(k\omega_0 t) \sin(m\omega_0 t)\, dt = \begin{cases} 0 & \text{for } k \neq m \\ T/2 & \text{for } k = m \neq 0 \end{cases} \qquad (2.55)$$

We can recreate $x_T(t)$ by (2.41):

$$x_T(t) = a_0 + a_1 \cos(\omega_0 t) + a_2 \cos(2\omega_0 t) \cdots + b_1 \sin(\omega_0 t) + b_2 \sin(2\omega_0 t) \cdots$$
$$(2.56)$$

What does this equation, (2.56), say? The general periodic signal with the fundamental period $T$ consists actually of signals with various frequencies. Namely, a DC component with magnitude $a_0$ and higher order harmonics ($k\omega_0$) with periods $T/2$, $T/3$, ... and with magnitudes $a_2, b_2$; $a_3, b_3$; ... which are coefficients of the cosines and sines, respectively.

**Exercise 2.1** Do the integrations to show that

$$\int_{t_0}^{t_0+T} \cos(k\omega_0 t) \cos(m\omega_0 t)\, dt = \begin{cases} 0 & \text{for } k \neq m \\ T/2 & \text{for } k = m \neq 0 \end{cases}$$

$$\int_{t_0}^{t_0+T} \sin(k\omega_0 t) \cos(m\omega_0 t)\, dt = 0 \text{ for all values } k, m$$

$$\text{and } \int_{t_0}^{t_0+T} \sin(k\omega_0 t) \sin(m\omega_0 t)\, dt = \begin{cases} 0 & \text{for } k \neq m \\ T/2 & \text{for } k = m \neq 0 \end{cases}$$

(use Appendix A.3 for trigonometric formulae).

### 2.6.1 Fourier Series of a Pulse Train

**Example 2.13** Find the Fourier series of the pulse train depicted in Fig. 2.21.

A pulse train shown in Fig. 2.21 is written in terms of sines and cosines in the interval $-T/2 \leq t \leq T/2$ as

$$p_T(t) = a_0 + \sum_{k=1}^{\infty} a_k \cos\left\{ k\left(\frac{2\pi}{T}\right) t \right\} + \sum_{k=1}^{\infty} b_k \sin\left\{ k\left(\frac{2\pi}{T}\right) t \right\}$$

by virtue of the fact these trigonometric functions form a complete orthogonal set of functions over the period $T$. By (2.53),

$$a_0 = \frac{1}{T} \int_{-\tau}^{\tau} dt = \frac{2\tau}{T} \tag{2.57}$$

$$a_m = \frac{2}{T} \int_{-\tau}^{\tau} \cos(m\omega_0 t) dt \qquad m = 1, 2, \ldots \tag{2.58}$$

$$= \frac{2}{mT\omega_0} [\sin(m\omega_0 t)]_{-\tau}^{\tau}$$

$$= \frac{2}{m\pi} \sin(m\omega_0 \tau) \tag{2.59}$$

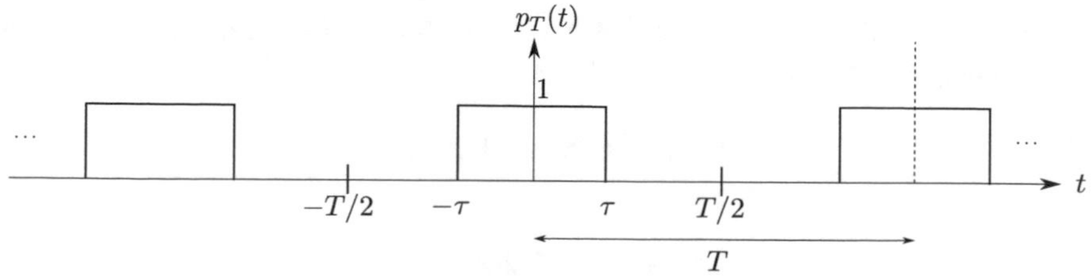

**Fig. 2.21** A pulse train

and by (2.54),

$$b_m = \frac{2}{T} \int_{-\tau}^{\tau} \sin(m\omega_0 t)dt \quad m = 1, 2, \ldots$$
$$= 0 \tag{2.60}$$

so

$$p_T(t) = \frac{2\tau}{T} + \sum_{m=1}^{\infty} \frac{2}{m\pi} \sin\left(\frac{2\tau m\pi}{T}\right) \cos(m\omega_0 t) \tag{2.61}$$

which may be also written as

$$p_T(t) = \frac{2\tau}{T} + \sum_{m=1}^{\infty} \left(\frac{4\tau}{T}\right) \left(\frac{T}{2\tau m\pi}\right) \sin\left(\frac{2\tau m\pi}{T}\right) \cos(m\omega_0 t)$$

$$= \frac{2\tau}{T} + \sum_{m=1}^{\infty} \left(\frac{4\tau}{T}\right) \left\{ \frac{\sin\left(\frac{2\tau m\pi}{T}\right)}{\left(\frac{2\tau m\pi}{T}\right)} \right\} \cos(m\omega_0 t)$$

$$= \frac{2\tau}{T} + \sum_{m=1}^{\infty} \left(\frac{4\tau}{T}\right) \text{sinc}\left(\frac{2\tau m\pi}{T}\right) \cos(m\omega_0 t) \tag{2.62}$$

and where

$$\text{sinc}(x) = \frac{\sin(x)}{x} \tag{2.63}$$

For the special case,

$$2\tau = T/2 \tag{2.64}$$

That is, when the pulses are half the width of the time period,

$$p_T(t) = 0.5 + \sum_{k=1}^{\infty} \frac{2}{k\pi} \sin\left(\frac{k\pi}{2}\right) \cos(k\omega_0 t)$$

$$= 0.5 + \sum_{k=1,\ 3,\ \ldots}^{\infty} \frac{2}{k\pi} \sin\left(\frac{k\pi}{2}\right) \cos(k\omega_0 t) \tag{2.65}$$

To show how the Fourier series converges, we observe Fig. 2.22. In the figure, we are assuming that $\tau = 0.25$ and $T = 1$. As a result,

$$k\omega_0 \tau = k \times (2\pi/1) \times 0.25$$

or

$$a_k = (2/k\pi) \sin(k\pi/2)$$

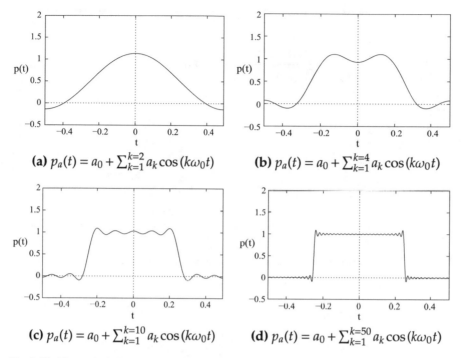

**Fig. 2.22** Figure depicting how the Fourier series for a pulse $p(t)$ converges. $T = 1$; $\tau = 0.25$

while

$$b_k = 0$$

with

$$a_0 = 0.5$$

In the (a) part of the figure, only three terms are summed: $k = 0$, 1 and 2. In the (b) part of the figure, five terms are considered: $k = 0, \ldots 4$. In part (c) eleven terms are summed, while part (d) shows the pulse when 51 terms are used. We can see how the summation of the series gives us the original pulse. Note that at the edges of the pulse where there is a sharp discontinuity, there are undershoots and overshoots, which is commonly known as the *Gibbs phenomenon*.

In part (a), the sum hardly looks like a pulse. But as we increase the number of terms, the pulse appears to take shape in better and better terms as shown in part (d).

If we plot the coefficients of the series as shown in Fig. 2.23, then we find that coefficients are

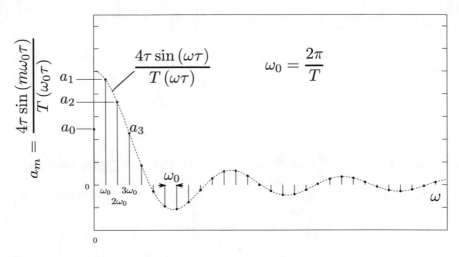

**Fig. 2.23** Plot of the coefficients $a_m$ of the Fourier series of a pulse train

$$a_0 = \frac{2\tau}{T}$$

$$a_k = \left(\frac{4\tau}{T}\right) \text{sinc}\left(\frac{2\tau k\pi}{T}\right) = \left(\frac{4\tau}{T}\right) \text{sinc}\left(\tau k\omega_0\right) \qquad (2.66)$$

With the exception of $a_0$, all lie in an envelope of

$$\frac{4\tau}{T} \text{sinc}(\tau\omega)$$

shown by the broken line in the figure. Whenever $\tau$ is reduced or $T$ is increased, the envelope reduces in value. Furthermore, a coefficient appears whenever

$$\omega = k\omega_0 \qquad k = \pm 1, \ \pm 2, \dots$$

The horizontal distance between coefficients is given by $\omega_0$ which is shown in the figure. This implies that if $T$ becomes large and $\tau$ is unchanged, then the distance between the coefficients $a_k$ is reduced (since $\omega_0 \propto 1/T$).

## 2.6.2  Dirichlet Conditions

The Dirichlet conditions are *sufficient* conditions for a *real-valued, periodic function* $f_T(x)$ to be equal the sum of its Fourier series at each point where $f$ is *continuous:*

1. $f_T(x)$ must have a finite number of extrema in any given interval.
2. $f_T(x)$ must have a finite number of discontinuities in any given interval.

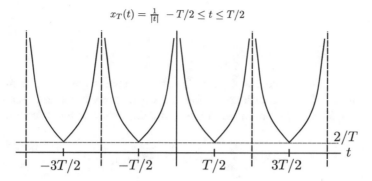

Fig. 2.24  The periodic function consisting of $1/|t|$

3. $f_T(x)$ must be absolutely integrable over a period.
4. $f_T(x)$ must be bounded.

**Example 2.14**  For example, the periodic function $f(t)$ in the interval $-T/2 < t < T/2$

$$f(t) = \frac{1}{|t|} \text{ for } -T/2 < t < T/2 \qquad (2.67)$$

as shown in Fig. 2.24 has no Fourier series, since the Dirichlet conditions are not satisfied.

## 2.7  Orthogonal Functions

Two real functions $g_1(t)$ and $g_2(t)$ are orthogonal on the interval $t_0 \le t \le t_0 + T$ if

$$\langle g_1(t) \bullet g_2(t) \rangle = \int_{t_0}^{t_0+T} g_1(t)g_2(t)dt = 0 \qquad (2.68)$$

Thus, we can see from Eqs. (2.52) and (2.55) that the set of trigonometric functions:

$$\left\{ \sin\left(k\frac{2\pi}{T}t\right), \cos\left(k\frac{2\pi}{T}t\right) \text{ and } 1 \quad k = 1, 2, \ldots \right\} \qquad (2.69)$$

are an infinite set of orthogonal functions over any interval spanning $T$. That is, more formally, the set of orthogonal functions

$$\left\{ g_k(t) \ k = 1, 2, \ldots \ : \forall i, j, \ g_i(t), g_j(t) \text{ satisfy } (2.68) \right\} \qquad (2.70)$$

The property of this set is that any two functions of this set satisfy Eq. (2.68).

Also

$$\int_{t_0}^{t_0+T} [g_i(t)]^2 \, dt = \alpha_i \tag{2.71}$$

where $\alpha_i$ is a positive constant, since $[g_i(t)]^2$ is always positive. An important property of an infinite set orthogonal functions is that any function in the interval $t_0 \le t \le t_0 + T$ can be represented in terms of a linear sum of these functions:

$$x(t) = \sum_{k=1}^{\infty} a_k g_k(t) \qquad t_0 \le t \le t_0 + T \tag{2.72}$$

To obtain the $a_k$s, multiply (2.72) by $g_m(t)$ and integrate over the interval $T$ :

$$\int_{t_0}^{t_0+T} x(t)g_m(t)dt = \sum_{k=1}^{\infty} a_k \int_{t_0}^{t_0+T} g_k(t)g_m(t)dt \qquad \begin{cases} t_0 \le t \le t_0 + T \\ m = 1, 2, \ldots \end{cases}$$

In the RHS of the equation, all integrals are zero except when $k = m$. Then by (2.71), we have

$$a_m \alpha_m = \int_{t_0}^{t_0+T} x(t)g_m(t)dt$$

which then gives

$$a_k = \frac{\int_{t_0}^{t_0+T} x(t)g_k(t)dt}{\int_{t_0}^{t_0+T} [g_k(t)]^2 \, dt} \tag{2.73}$$

## 2.8 Exponential Fourier Series

Another set of orthogonal functions comprises the complex exponential functions

$$g_k(t) = e^{jk\omega_0 t} \qquad \omega_0 = 2\pi/T, \ k = 0, \pm 1, \ldots \tag{2.74}$$

Thus, when we integrate $g_m(t)g_k^*(t)$ over the interval $t_0 \le t \le t_0 + T$, we find that

$$\int_{t_0}^{t_0+T} g_m(t)g_k^*(t)dt = \int_{t_0}^{t_0+T} e^{jm\omega_0 t} e^{-jk\omega_0 t} dt = \begin{cases} 0 & m \ne k \\ T & m = k \end{cases} \tag{2.75}$$

The definition of orthogonality is somewhat different from (2.68). In the case of complex orthogonal functions, the dot product is obtained by multiplying each function $g_k(t)$ by the complex conjugate of any other function of the same set of orthogonal functions, $g_m^*(t)$.

$$\int_{t_0}^{t_0+T} g_m(t)g_k^*(t)dt = \begin{cases} 0 & m \neq k \\ \alpha_m & m = k \end{cases} \tag{2.76}$$

To prove this result (Eq. (2.75)), when $k \neq m$

$$\int_0^T e^{j\frac{2\pi}{T}(k-m)t}dt = \left[\frac{e^{j\frac{2\pi}{T}(k-m)t}}{j\frac{2\pi}{T}(k-m)}\right]_0^T \quad (\text{for } k \neq m)$$

$$= \underbrace{\frac{1}{j\frac{2\pi}{T}(k-m)}}_{\neq 0}\left[\underbrace{e^{j\frac{2\pi}{T}(k-m)T}-1}_{=1}\right] = 0$$

For convenience, we set $t_0 = 0$. And when $k = m$,

$$\int_0^T e^{j\frac{2\pi}{T}(k-m)t}dt = \int_0^T 1dt \quad (for\ k = m)$$

$$= T$$

With the above discussion in view, we may write a Fourier series in the form

$$f_T(t) = \sum_{k=-\infty}^{k=\infty} F_k e^{jk\omega_0 t} \tag{2.77}$$

where $f(t)$ is a periodic function with a fundamental period $T$, and $\omega_0 = 2\pi/T$ is the fundamental frequency. The $F_k$s are obtained from

$$F_k = \frac{1}{T}\int_{t_0}^{t_0+T} f_T(t)e^{-jk\omega_0 t}dt \tag{2.78}$$

This Fourier series is the exponential series and is of *paramount* importance in the analysis of periodic signals.

## 2.8.1 Exponential Fourier Series of a Pulse Train

**Example 2.15** Find the exponential Fourier series for the pulse train of Fig. 2.21.

From the definition of a Fourier series,

$$p_T(t) = \sum_{k=-\infty}^{\infty} F_k e^{jk\omega_0 t} \quad k = 0, \pm 1, \pm 2, ... \tag{2.79}$$

Using the procedure outlined earlier, multiply this equation, (2.79), by $\exp(-jm\omega_0 t)$ and integrate in the region $-T/2 \le t \le T/2$:

$$\int_{-T/2}^{T/2} p_T(t)e^{-jm\omega_0 t}\,dt = \sum_{k=-\infty}^{\infty} \int_{-T/2}^{T/2} F_k e^{j(k-m)\omega_0 t}\,dt = \sum_{k=-\infty}^{\infty} F_k \int_{-T/2}^{T/2} e^{j(k-m)\omega_0 t}\,dt$$

Using the property (2.75), we get

$$F_m T = \int_{-T/2}^{T/2} p(t)e^{-jm\omega_0 t}\,dt$$

$$= \int_{-\tau}^{\tau} e^{-jm\omega_0 t}\,dt$$

$$= \frac{2\tau \sin(m\omega_0\tau)}{(m\omega_0\tau)} \tag{2.80}$$

or

$$\boxed{F_k = \frac{2\tau \sin(k\omega_0\tau)}{T\,(k\omega_0\tau)} \qquad k = \dots, -1,\ 0,\ 1,\ \dots} \tag{2.81}$$

So

$$\boxed{p_T(t) = \cdots + F_{-1}e^{-j\omega_0 t} + F_0 + F_1 e^{j\omega_0 t} + \cdots} \tag{2.82}$$

where

$$\cdots F_{-1} = \frac{2\tau}{T} \times \frac{\sin(\omega_0\tau)}{\omega_0\tau}; \quad F_0 = \frac{2\tau}{T}; \quad F_1 = \frac{2\tau}{T} \times \frac{\sin(\omega_0\tau)}{\omega_0\tau} \cdots$$

**Exercise 2.2**  Show that

$$\int_{t_0}^{t_0+T} e^{jm\omega_0 t} e^{-jk\omega_0 t}\,dt = \begin{cases} 0 & m \ne k \\ T & m = k \end{cases}$$

when $t_o \ne 0$.

## 2.9  Parseval's Theorem for Fourier Series

Let us find the energy in the signal for one time period of the trigonometric Fourier series

$$f_T(t) = a_0 + \sum_{k=1}^{\infty} a_k \cos(k\omega_0 t) + \sum_{k=1}^{\infty} b_k \sin(k\omega_0 t) \tag{2.83}$$

where $f_T(t)$ is real. The energy in one time period of the signal is

$$\int_T f_T^2(t)dt = \int_T \left[ a_0 + \sum_{k=1}^{\infty} a_k \cos{(k\omega_0 t)} + \sum_{k=1}^{\infty} b_k \sin{(k\omega_0 t)} \right] \times$$
$$\left[ a_0 + \sum_{k=1}^{\infty} a_k \cos{(k\omega_0 t)} + \sum_{k=1}^{\infty} b_k \sin{(k\omega_0 t)} \right] dt$$

where $\int_T$ represents integration over one time period. Observing the expression, we find that there are four types of terms on the right side which contain $t$

$$\int_T dt,$$
$$\int_T \cos{(k\omega_0 t)} \cos{(l\omega_0 t)}\, dt,$$
$$\int_T \cos{(k\omega_0 t)} \sin{(l\omega_0 t)}\, dt,$$
$$\text{and } \int_T \sin{(k\omega_0 t)} \sin{(l\omega_0 t)}\, dt$$

where $k$, and $l = 1, 2 \ldots$. Then using the results of Eqs. (2.52) and (2.55),

$$\int_T f_T^2(t)dt = \mathbb{E} = T \left\{ a_0^2 + \frac{1}{2}\left[a_1^2 + a_2^2 + \cdots \right] + \frac{1}{2}\left[b_1^2 + b_2^2 + \cdots \right] \right\} \quad (2.84)$$

where $\mathbb{E}$ is the energy of one time period $T$. The power in the interval $-nT/2 \le t \le nT/2$,

$$\mathbb{P} = \lim_{n \to \infty} \frac{1}{nT} \int_{-nT/2}^{nT/2} f_T^2(t)dt$$
$$= \lim_{n \to \infty} \left\{ \frac{n\mathbb{E}}{nT} \right\}$$
$$= a_0^2 + \frac{1}{2}\left[a_1^2 + a_2^2 + \cdots \right] + \frac{1}{2}\left[b_1^2 + b_2^2 + \cdots \right] \quad (2.85)$$

We may apply the same approach to the exponential Fourier series,

$$f_T(t) = \sum_{k=-\infty}^{\infty} F_k e^{jk\omega_0 t}$$

and if $f_T(t)$ is in general complex, then

$$\mathbb{P} = \frac{1}{T} \int_{-T/2}^{T/2} f_T(t) f_T^*(t) dt$$

$$= \frac{1}{T} \int_{-T/2}^{T/2} \left[ \sum_{k=-\infty}^{\infty} F_k e^{jk\omega_0 t} \right] \left[ \sum_{l=-\infty}^{\infty} F_l e^{jl\omega_0 t} \right]^* dt$$

$$= \frac{1}{T} \left\{ \sum_{k=-\infty}^{\infty} \sum_{l=-\infty}^{\infty} \left[ \int_{-T/2}^{T/2} F_l^* F_k e^{j(k-l)\omega_0 t} dt \right] \right\} \tag{2.86}$$

where the summation and integration signs may be interchanged, and

$$\int_{-T/2}^{T/2} e^{j(k-l)\omega_0 t} dt = \begin{cases} T & \text{when } k = l \\ 0 & \text{when } k \neq l \end{cases}$$

therefore

$$\mathbb{P} = \sum_{k=-\infty}^{\infty} |F_k|^2 \tag{2.87}$$

## 2.10   Fourier Transform

In our treatment of the Fourier series, it was clear that every periodic function with period $T$ and frequency $\omega_0 (= 2\pi/T)$ consists of a sum of periodic functions of the form

$$e^{j(m\omega_0)t} \qquad m = 0, \pm 1, \ldots$$

where $m\omega_0$ is the frequency of the complex periodic function. When $m = 0$ it means that it is the constant component, when $m = 1$, the exponential function is a complex number rotating in the counter- clockwise sense with a frequency $\omega_0$. When $m = -1$, the exponential function is a complex number rotating in the clockwise sense with a frequency $\omega_0$ and so on. The original function, $f_T(t)$ (or whichever periodic function we are considering), can be written as

$$f_T(t) = \underset{\text{first term}}{F_0} + \underset{\text{second term}}{F_1 e^{j\omega_0 t}} + \underset{\text{third term}}{F_{-1} e^{-j\omega_0 t}} + \underset{\text{fourth term}}{F_2 e^{j2\omega_0 t}} + \underset{\text{fifth term}}{F_{-2} e^{-j2\omega_0 t}} + \ldots$$

The first term is the constant term. The second and third terms represent the components at the fundamental frequencies; the fourth and fifth terms represent the components at $\pm 2\omega_0$ and so on. The question then is to ask if it is possible to represent a 'Fourier series' of an aperiodic function? If we think on this matter for a while, then we can conclude that an aperiodic function is one whose period is 'infinity' or very large. Using these ideas, we find that the Fourier series leads in a natural way to the Fourier transform, which is the 'Fourier series' of an aperiodic function.

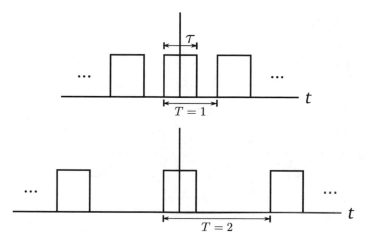

**Fig. 2.25** Pulses with $T = 1$ and $T = 2$

The natural question we can ask is that what happens when we *increase* the fundamental period $T$ of a Fourier series given by?

$$F_k = \frac{1}{T} \int_{\text{over } T} f(t) e^{-jk\omega_0 t} dt$$

$$f(t) = \sum_k F_k e^{jk\omega_0 t} \tag{2.88}$$

1. The coefficients, $F_k$, become smaller, since they are proportional to $1/T$.
2. $\omega_0$ become smaller, since $\omega_0 \propto 1/T$.

Taking a concrete example of the pulse, take two pulses both with $\tau = 1/3$ but with different time periods, one with $T = 1$ and the other with $T = 2$ as shown in Fig. 2.25.

The Fourier coefficients for $T = 1$ are

$$F_k = \left(\frac{2}{3}\right) \frac{\sin\left[k(2\pi/3)\right]}{(2\pi/3)} \quad at \ \omega = k(2\pi/3)$$

and for $T = 2$, they are

$$F_k = \left(\frac{1}{3}\right) \frac{\sin\left[k(\pi/3)\right]}{(\pi/3)} \quad at \ \omega = k(\pi/3)$$

The coefficients are depicted in Fig. 2.26. The figure shows two graphs with the coefficients clearly shown as given by the dots. The solid line is the sinc function. There are several points of interest:

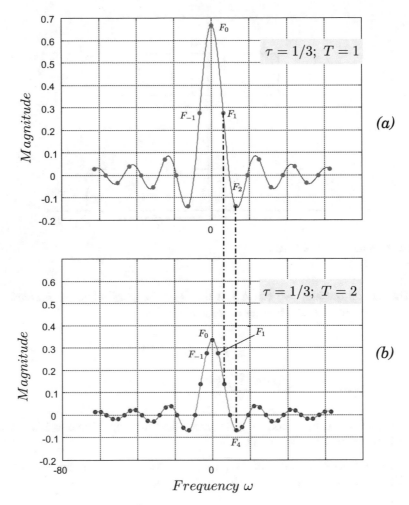

**Fig. 2.26** Fourier coefficients of two pulses, both with $\tau = 1/3$ but with different time periods: $T = 1$ and $T = 2$

1. The coefficients lie on the sinc($\cdot$) function.
2. The coefficients of the function with $T = 1$ are of double the value of that of $T = 2$.
3. Between every two coefficients $F_k$ and $F_{k+1}$ of the function with $T = 1$, there is an extra coefficient of the function with $T = 2$. The frequency domain is more densely populated in the second case.
4. From this, we can infer that as $T \to \infty$, the coefficients become 'almost continuous' along the frequency domain but their magnitude tends to go to zero.

We then proceed along the following lines to define the Fourier series with infinite time periods. We define a new coefficient function

$$F'(k\omega_0) = T F_k \tag{2.89}$$

so that

$$F'(k\omega_0) = \lim_{T \text{ very large}} \int_{-T/2}^{T/2} f(t)e^{-jk\omega_0 t} dt$$

$$f(t) = \frac{\omega_0}{2\pi} \sum_k F'(k\omega_0)e^{jk\omega_0 t}$$

In the case given above, we know that

$$f(t) = \frac{\omega_0}{2\pi} \sum_{k=-\infty}^{\infty} F'(k\omega_0)e^{jk\omega_0 t} \Rightarrow \frac{1}{2\pi} \int_{-\infty}^{\infty} F'(\omega)e^{j\omega t} d\omega \tag{2.90}$$

where $\omega = k\omega_0$ and $\omega_0 \simeq d\omega$. We are now in a position to define the Fourier transform.

The Fourier transform pair is Fourier transform (FT) and inverse Fourier transform (IFT):

$$F(\omega) = \int_{-\infty}^{\infty} f(t)e^{-j\omega t} dt$$

$$f(t) = \frac{1}{2\pi} \int_{-\infty}^{\infty} F(\omega)e^{j\omega t} d\omega \tag{2.91}$$

where we have dropped the prime on $F$. The pair is written in short notation as

$$f(t) \Leftrightarrow F(\omega) \tag{2.92}$$

The FT may also be seen from another angle, and that is

$$F(\omega) = \int_{-\infty}^{\infty} e^{-j\omega t} f(t)dt$$

$$\approx \sum_k \underbrace{e^{-j\omega t_k}}_{phasor} \underbrace{f(t_k)\Delta t}_{amplitude}$$

so the FT is the sum of clockwise rotating phasors (shown in Fig. 2.27) multiplied by the amplitudes $f(t_k)\Delta t$. The inverse FT may be given a similar interpretation in terms of anti-clockwise rotating phasors.

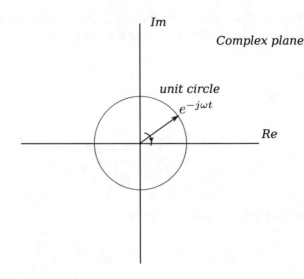

**Fig. 2.27** Phasor rotating in the clockwise sense

The other point that is of importance is that $f(t)$ and $F(\omega)$ are equivalent representations since one representation may be uniquely obtained from the other one.

A general non-periodic signal consists actually of signals with various frequencies in a continuous spectrum. The magnitude of each frequency, $\omega$, is $F(\omega)$ and multiplies all the frequencies in a small interval $d\omega$, since

$$f(t) = \frac{1}{2\pi} \int_{-\infty}^{\infty} F(\omega) e^{j\omega t} d\omega$$

One special modification may also be remembered. If we take the frequency $f = \omega/2\pi$ as the variable of integration, then

$$F(f) = \int_{-\infty}^{\infty} f(t) e^{-j2\pi f t} dt \tag{2.93}$$

and

$$f(t) = \int_{-\infty}^{\infty} F(f) e^{j2\pi f t} d\frac{\omega}{2\pi} = \int_{-\infty}^{\infty} F(f) e^{j2\pi f t} df \tag{2.94}$$

can be considered as an FT pair. In the Fourier transform, the function $f(t)$ must satisfy the Dirichlet conditions given in Sect. 2.6.2 where of course the independent variable $x$ is to be replaced by $t$.

## 2.11 Examples of Fourier Transforms

The Fourier transform is quite straightforward to calculate. Let us calculate the FT and inverse FT of some simple functions. See Table 2.1 for some most often used FT pairs.

### 2.11.1 FT of the $\delta(t)$

The Fourier transform of the impulse function is often used in calculations in communication theory.

**Table 2.1** Some Fourier transform pairs

| Time function | Fourier transform | Notes |
|---|---|---|
| $\delta(t)$ | $1$ | |
| $1$ | $2\pi\delta(\omega)$ | Duality |
| $u(t)$ | $\pi\delta(\omega) + \frac{1}{j\omega}$ | |
| $e^{-at}u(t)\ (a>0)$ | $\frac{1}{a+j\omega}$ | |
| $e^{at}u(-t)\ (a>0)$ | $\frac{1}{a-j\omega}$ | |
| $e^{-a|t|}\ (a>0)$ | $\frac{2a}{a^2+\omega^2}$ | |
| $te^{-at}u(t)\ a>0$ | $\frac{1}{(a+j\omega)^2}$ | |
| $t^n e^{-at}u(t)\ (a>0)$ | $\frac{n!}{(a+j\omega)^{n+1}}$ | |
| $e^{j\omega_0 t}$ | $2\pi\delta(\omega-\omega_0)$ | |
| $\cos\omega_0 t$ | $\pi[\delta(\omega-\omega_0)+\delta(\omega+\omega_0)]$ | |
| $\sin\omega_0 t$ | $j\pi[\delta(\omega+\omega_0)-\delta(\omega-\omega_0)]$ | |
| $\mathrm{sgn}(t)=2u(t)-1$ | $\frac{2}{j\omega}$ | |
| $\cos\omega_0 t\ u(t)$ | $\frac{\pi}{2}[\delta(\omega-\omega_0)+\delta(\omega+\omega_0)] + \frac{j\omega}{\omega_0^2-\omega^2}$ | |
| $\sin\omega_0 t\ u(t)$ | $\frac{\pi}{2j}[\delta(\omega-\omega_0)-\delta(\omega+\omega_0)] + \frac{\omega_0}{\omega_0^2-\omega^2}$ | |
| $e^{-at}\sin\omega_0 t\ u(t)\ (a>0)$ | $\frac{\omega_0}{(a+j\omega)^2+\omega_0^2}$ | $a>0$ |
| $e^{-at}\cos\omega_0 t\ u(t)\ (a>0)$ | $\frac{a+j\omega}{(a+j\omega)^2+\omega_0^2}$ | $a>0$ |
| $\mathrm{rect}\left(\frac{t}{\tau}\right)$ | $\tau\,\mathrm{sinc}\left(\frac{\omega\tau}{2}\right)$ | |
| $\frac{W}{\pi}\,\mathrm{sinc}(Wt)$ | $\mathrm{rect}\left(\frac{\omega}{2W}\right)$ | |
| $\Delta\left(\frac{t}{\tau}\right)$ | $\frac{\tau}{2}\,\mathrm{sinc}^2\left(\frac{\omega\tau}{4}\right)$ | |
| $\frac{W}{2\pi}\,\mathrm{sinc}^2\left(\frac{Wt}{2}\right)$ | $\Delta\left(\frac{w}{W}\right)$ | |
| $\sum_{n=-\infty}^{\infty}\delta(t-nT)$ | $\omega_0\sum_{n=-\infty}^{\infty}\delta(\omega-n\omega_0)$ | $\omega_0=\frac{2\pi}{T}$ |
| $e^{-t^2/2\sigma^2}$ | $\sigma\sqrt{2\pi}\,e^{-\sigma^2\omega^2/2}$ | |

**Example 2.16** Find the FT of the impulse function, $\delta(t)$.

Let us take the function $\delta(t)$, then the FT is

$$\mathcal{F}\{\delta(t)\} = \Delta(\omega)$$

$$= \int_{-\infty}^{\infty} \delta(t) \underbrace{e^{-j\omega t}}_{=1 \; at \; t=0} \; dt$$

$$= \int_{-\infty}^{\infty} \delta(t)dt$$

$$= 1 \tag{2.95}$$

where we have used the properties,

$$x(t)\delta(t) = x(0)\delta(t)$$

$$\int_{-\infty}^{\infty} \delta(t)dt = 1$$

The FT pair is shown in Fig. 2.28. In other words the FT of the impulse function is a constant equal to 1. The meaning of this result is that the delta function contains all frequencies equally since for any value of $\omega$, the FT is equal to 1.

$$\boxed{\mathcal{F}\{\delta(t)\} = 1} \tag{2.96}$$

Please note that the symbol $\mathcal{F}$, $\mathcal{F}^{-1}$ and $\Leftrightarrow$ are the symbols which can be read as "The FT of ...", "The inverse FT of ..." and "The FT pair is ...", respectively.

**Fig. 2.28** FT of $\delta(t)$

## 2.11.2  IFT of $\delta(\omega)$

**Example 2.17**  Find the inverse FT of $\delta(\omega)$.

From the definition of the inverse FT

$$\mathcal{F}^{-1}\{\delta(\omega)\} = \frac{1}{2\pi}\int_{-\infty}^{\infty}\delta(\omega)\ \underbrace{e^{j\omega t}}_{=1\ at\ \omega=0}\ d\omega$$

$$= \frac{1}{2\pi}\int_{-\infty}^{\infty}\delta(\omega)d\omega$$

$$= \frac{1}{2\pi} \qquad\qquad (2.97)$$

This is shown in Fig. 2.29. Since $\delta(\omega)$ exists only at $\omega = 0$, therefore it contains only one frequency, namely $\omega = 0$. Therefore, the constant term $1/2\pi$ consists of only one 'frequency,' namely $\omega = 0$!

$$\boxed{\mathcal{F}^{-1}\{\delta(\omega)\} = \frac{1}{2\pi}} \qquad\qquad (2.98)$$

or

$$1 \Leftrightarrow 2\pi\,\delta(\omega)$$

**Example 2.18**  Find the FT of two functions which are connected by the relation

$$f_1(t) = f_2(t) + c$$

where $c$ is a constant.

Taking the previous equation and taking FT on both sides, then

$$F_1(\omega) = F_2(\omega) + 2\pi c\delta(\omega)$$

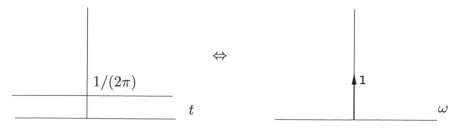

**Fig. 2.29**  Inverse FT of $\delta(\omega)$

This applies to relations like

$$u(t) = 0.5\big[\text{sgn(t)} + 1\big]$$

or

$$\mathcal{F}\,[u(t)] = 0.5\mathcal{F}\,\big[\text{sgn(t)}\big] + \pi\delta(\omega) \tag{2.99}$$

### 2.11.3  FT of a Single Pulse

Let us find the Fourier transform of a single pulse as shown in Fig. 2.30. The single pulse is related to the rect(.) function by

$$p_s(t) = \text{rect}\left(\frac{t}{2\tau}\right) \tag{2.100}$$

The Fourier transform is given by

$$\begin{aligned}
\mathcal{F}\{p_s(t)\} &= \int_{-\infty}^{\infty} p_s(t)e^{-j\omega t}\,dt \\
&= \int_{-\tau}^{\tau} e^{-j\omega t}\,dt \\
&= -\frac{e^{-j\omega\tau}}{j\omega} + \frac{e^{j\omega\tau}}{j\omega} \\
&= 2\tau\frac{\sin(\omega\tau)}{\omega\tau} \tag{2.101}
\end{aligned}$$

The Fourier transform of a single pulse occupies an approximate bandwidth of

**Fig. 2.30**  A single pulse

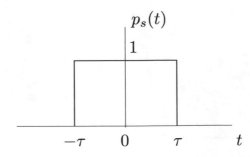

$$\omega_b \tau \cong \pi$$

$$f_b \cong \frac{1}{2\tau} \tag{2.102}$$

where $\omega_b$ and $f_b$ are the approximate radian frequency and frequency bandwidths, respectively.

### 2.11.4 FTs and IFTs of Time- and Frequency-Shifted Signals

**Example 2.19** Find the FT of the time-shifted impulse function, $\delta(t - t_0)$. (The impulse is at $t = t_0$.)

In this case

$$\mathcal{F}\{\delta(t - t_0)\} = \int_{-\infty}^{\infty} \delta(t - t_0)e^{-j\omega t}\, dt$$

$$= e^{-j\omega t_0} \int_{-\infty}^{\infty} \delta(t - t_0)\, dt$$

$$= 1e^{-j\omega t_0}$$

we notice that the time-shifted FT of $\delta(t - t_0)$ is a complex function with magnitude 1 and phase $-\omega t_0$. The phase is linear with frequency. This is shown in Fig. 2.31. The FT of the time-shifted impulse function contains all frequencies equally but each having a phase equal to $-\omega t_0$.

$$\boxed{\mathcal{F}\{\delta(t - t_0)\} = 1e^{-j\omega t_0}} \tag{2.103}$$

When we shift a function $f(t)$, in the time domain by $t_0$, then in the frequency domain there is the inclusion of a phase term, $e^{-j\omega t_0}$, in the FT.

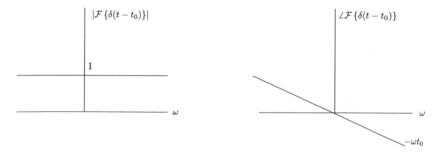

**Fig. 2.31** Magnitude and phase of $\mathcal{F}\{\delta(t - t_0)\}$

**Example 2.20** Find the inverse FT of the frequency-shifted impulse function, $\delta(\omega - \omega_0)$.

$$
\begin{aligned}
\mathcal{F}^{-1}\{\delta(\omega - \omega_0)\} &= \frac{1}{2\pi} \int_{-\infty}^{\infty} \delta(\omega - \omega_0) e^{j\omega t} d\omega \\
&= \frac{e^{j\omega_0 t}}{2\pi} \int_{-\infty}^{\infty} \delta(\omega - \omega_0) d\omega \\
&= \frac{1}{2\pi} e^{j\omega_0 t} \\
&= \frac{1}{2\pi} [\cos(\omega_0 t) + j \sin(\omega_0 t)] \quad\quad (2.104)
\end{aligned}
$$

This is a very important result. Sines and cosines of frequency $\omega_0$ have frequency-shifted delta functions!

$$
e^{j\omega_0 t} \Leftrightarrow 2\pi \delta(\omega - \omega_0) \quad\quad (2.105)
$$

This leads to the result that when we shift a function $F(\omega)$, in the frequency domain by $\omega_0$, then in the time domain there is the inclusion of a phase term, $e^{j\omega_0 t}$ in the IFT.

### 2.11.5 FT of the Cosine Signal

**Example 2.21** Looking at the FT pair (2.105), let us calculate the inverse FT of $\delta(\omega - \omega_0) + \delta(\omega + \omega_0)$.

From the definition,

$$
\begin{aligned}
\mathcal{F}^{-1}\{\delta(\omega - \omega_0) + \delta(\omega + \omega_0)\} &= \frac{1}{2\pi} \int_{-\infty}^{\infty} \{\delta(\omega - \omega_0) + \delta(\omega + \omega_0)\} e^{j\omega t} d\omega \\
&= \frac{1}{2\pi} e^{j\omega_0 t} \int_{-\infty}^{\infty} \delta(\omega - \omega_0) d\omega + \frac{1}{2\pi} e^{-j\omega_0 t} \int_{-\infty}^{\infty} \delta(\omega + \omega_0) d\omega \\
&= \frac{1}{2\pi} \left( e^{j\omega_0 t} + e^{-j\omega_0 t} \right) \\
&= \frac{1}{\pi} \cos \omega_0 t
\end{aligned}
$$

A very important result.

$$
\boxed{\cos \omega_0 t \Leftrightarrow \pi \{\delta(\omega - \omega_0) + \delta(\omega + \omega_0)\}} \quad\quad (2.106)
$$

## *2.11.6  FTs of Exponential-Related Signals*

**Example 2.22** Find the FT of the one-sided function: $e^{-at}u(t)$.

The FT is given by

$$
\mathcal{F}\{e^{-at}u(t)\} = \int_{-\infty}^{\infty} e^{-at}u(t)e^{-j\omega t}\,dt
$$

$$
= \int_{0}^{\infty} e^{-(a+j\omega)t}\,dt \quad (\text{since } u(t) = 0 \text{ for } t < 0)
$$

$$
= \left[ -\frac{e^{-(a+j\omega)t}}{(a+j\omega)} \right]_{t=0}^{\infty}
$$

$$
= \frac{1}{(a+j\omega)}
$$

$$
\boxed{e^{-at}u(t) \Leftrightarrow \frac{1}{(a+j\omega)}} \tag{2.107}
$$

The magnitude and phase of this function are shown in Fig. 2.32a,b. The magnitude of this function is

$$
\left| \frac{1}{(a+j\omega)} \right| = \frac{|1|}{|(a+j\omega)|}
$$

$$
= \frac{1}{\sqrt{a^2 + \omega^2}} \tag{2.108}
$$

and the phase

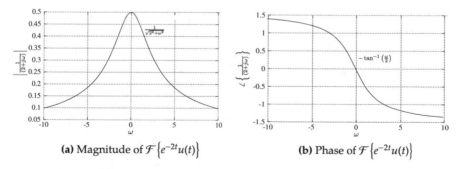

**(a)** Magnitude of $\mathcal{F}\{e^{-2t}u(t)\}$                 **(b)** Phase of $\mathcal{F}\{e^{-2t}u(t)\}$

**Fig. 2.32** FT details of $e^{-at}u(t)$

$$\angle \left\{ \frac{1}{(a + j\omega)} \right\} = \angle 1 - \angle (a + j\omega)$$

$$= 0 - \tan^{-1} \omega/a \qquad (2.109)$$

**Example 2.23** Find the FT of the double-sided exponential function

$$f(t) = e^{-a|t|}$$

The FT may be obtained from

$$\mathcal{F} \left\{ e^{-a|t|} \right\} = \int_{-\infty}^{0} e^{at} e^{-j\omega t} dt + \int_{0}^{\infty} e^{-at} e^{-j\omega t} dt$$

$$= \left[ \frac{e^{at} e^{-j\omega t}}{a - j\omega} \right]_{-\infty}^{0} + \left[ -\frac{e^{-at} e^{-j\omega t}}{a + j\omega} \right]_{0}^{\infty}$$

$$= \frac{1}{a - j\omega} + \frac{1}{a + j\omega}$$

$$= \frac{2a}{a^2 + \omega^2}$$

A plot is shown in Fig. 2.33.

**Example 2.24** If a function $f(t)$ is zero for $0 < t$ and its FT is $F(j\omega)$, show that the FT of $e^{-at} f(t)$ is $F(j\omega + a)$.

***Proof*** The FT of $f(t)$ is $F(j\omega)$, that is

$$F(j\omega) = \int_{-\infty}^{\infty} f(t) e^{-j\omega t} dt$$

$$= \int_{0}^{\infty} f(t) e^{-j\omega t} dt \qquad (\because \ f(t) = 0 \text{ for } t < 0)$$

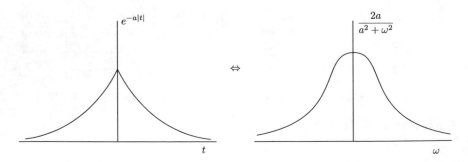

**Fig. 2.33** FT pair of $e^{-a|t|}$

and the FT of $e^{-at} f(t)$ is

$$
\begin{aligned}
F_1(j\omega) &= \int_{-\infty}^{\infty} f(t)e^{-at}e^{-j\omega t}dt \\
&= \int_{0}^{\infty} f(t)e^{-at-j\omega t}dt \qquad (\because \ f(t) = 0 \text{ for } t < 0) \\
&= \int_{0}^{\infty} f(t)e^{-(a+j\omega)t}dt \\
&= F(a + j\omega)
\end{aligned}
$$

### 2.11.7  FTs of the Signum and Unit Step Functions

**Example 2.25**  Find the FT of the sgn($t$) function.

Observe the function,

$$
f(t) = e^{-at}u(t) - e^{\,|\,at}u(-t) = \begin{cases} e^{-at} & \text{for } t > 0 \\ -e^{at} & \text{for } t < 0 \end{cases}
$$

which is shown in Fig. 2.34. Then by straightforward integration

$$
\begin{aligned}
\mathcal{F}\{f(t)\} &= \frac{1}{a + j\omega} - \frac{1}{a - j\omega} \\
&= \frac{-2j\omega}{a^2 + \omega^2}
\end{aligned}
$$

now

$$
\text{sgn}(t) = \lim_{a \to 0} f(t)
$$

therefore

$$
\mathcal{F}\{\text{sgn}(t)\} = \lim_{a \to 0} \left[ \frac{-2j\omega}{a^2 + \omega^2} \right]
$$

$$
\text{sgn}(t) \Leftrightarrow \frac{2}{j\omega}
$$

**Example 2.26**  Find the FT of the unit step function.

The unit step function can be written in terms of the sgn($t$) function as

$$
u(t) = 0.5\text{sgn}(t) + 0.5
$$

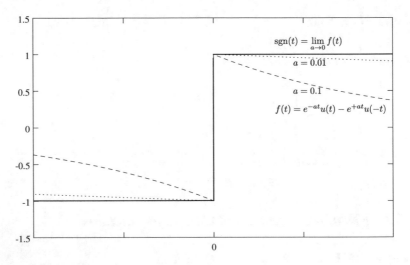

**Fig. 2.34** The signum function shown as a limit

Therefore, using the results of Example 2.18, we have

$$\mathcal{F}\{u(t)\} = 0.5\mathcal{F}\{\mathrm{sgn}(t)\} + \pi\delta(\omega)$$

$$= \frac{1}{j\omega} + \pi\delta(\omega)$$

**Example 2.27**  Find the inverse Fourier transform of $\mathrm{sgn}(\omega)$

As earlier, consider

$$\mathrm{sgn}(\omega) = \lim_{a\to 0} \begin{cases} e^{-a\omega} & \omega > 0 \\ -e^{a\omega} & \omega < 0 \end{cases} \text{ for } a > 0$$

then

$$\mathcal{F}^{-1}\{\mathrm{sgn}(\omega)\} = \lim_{a\to 0}\left[\frac{1}{2\pi}\left\{\int_{-\infty}^{0} -e^{a\omega}e^{j\omega t}\,d\omega + \int_{0}^{\infty} e^{-a\omega}e^{j\omega t}\,d\omega\right\}\right]$$

$$= \lim_{a\to 0}\left[\frac{1}{2\pi}\left\{\int_{-\infty}^{0} -e^{a\omega+j\omega t}\,d\omega + \int_{0}^{\infty} e^{-a\omega+j\omega t}\,d\omega\right\}\right]$$

$$= \lim_{a\to 0}\left[\frac{1}{2\pi}\left\{-\frac{1}{a+jt} - \frac{1}{jt-a}\right\}\right]$$

$$= -\frac{1}{j\pi t}$$

## 2.12   Some Properties of the Fourier Transform

The Fourier transform has many properties which provide perception into the nature of the transform. Most of these properties are used to derive other properties of communication systems and extend theory. Thus, if a time signal is changed in a particular way, what happens to its Fourier transform? And so a fundamental change in the way of thinking is required by the communication engineer in which a Fourier transform pair $x(t) \Leftrightarrow X(\omega)$ plays an important role.

### 2.12.1   Linearity

The FT of a linear combination of functions is the same linear combination of the individual FTs. That is (for two functions) if

$$x(t) \Leftrightarrow X(\omega)$$

and

$$y(t) \Leftrightarrow Y(\omega)$$

then

$$Ax(t) + By(t) \Leftrightarrow AX(\omega) + BY(\omega)$$

where $A$ and $B$ are two complex constants.

**Proof**

$$
\begin{aligned}
\mathcal{F}\{Ax(t) + By(t)\} &= \int_{-\infty}^{\infty} \{Ax(t) + By(t)\} e^{-j\omega t} dt \\
&= \int_{-\infty}^{\infty} Ax(t)e^{-j\omega t} dt + \int_{-\infty}^{\infty} By(t)e^{-j\omega t} dt \\
&= A\int_{-\infty}^{\infty} x(t)e^{-j\omega t} dt + B\int_{-\infty}^{\infty} y(t)e^{-j\omega t} dt \\
&= AX(\omega) + BY(\omega)
\end{aligned}
$$

$$\boxed{\mathcal{F}\{Ax(t) + By(t)\} \Leftrightarrow AX(\omega) + BY(\omega)}  \qquad (2.110)$$

### 2.12.1.1　FT of a Periodic Function

We now apply the linearity property of the Fourier transform to a specific case.

**Example 2.28**　Find the FT of a periodic function.

A Fourier series of a periodic function is represented by Eq. (2.77),

$$f(t) = \sum_{k=-\infty}^{k=\infty} F_k e^{jk\omega_0 t} \tag{2.111}$$

where $f(t)$ is a periodic function with a fundamental period $T$, and $\omega_0 = 2\pi/T$ is the fundamental frequency. From Example 2.20, we know that

$$e^{j\omega_0 t} \Leftrightarrow 2\pi \delta(\omega - \omega_0) \tag{2.112}$$

or

$$\mathcal{F}\{f(t)\} = \mathcal{F}\left\{ \sum_{k=-\infty}^{k=\infty} F_k e^{jk\omega_0 t} \right\}$$

$$= \sum_{k=-\infty}^{k=\infty} F_k \mathcal{F}\left\{ e^{jk\omega_0 t} \right\} \quad \text{(using linearity property)}$$

$$= 2\pi \sum_{k=-\infty}^{k=\infty} F_k \delta(\omega - k\omega_0) \tag{2.113}$$

**Example 2.29**　Apply the results of Example 2.28 to the case of the pulse train, $p_T(t)$, of Example 2.13.

From Example 2.15,

$$p_T(t) = \sum_{k=-\infty}^{\infty} F_k e^{jk\omega_0 t} \quad k = 0, \pm 1, \pm 2, \ldots$$

$$F_k = \frac{2\tau \sin(k\omega_0 \tau)}{T(k\omega_0 \tau)} \quad k = \ldots, -1, 0, 1, \ldots$$

So

$$\mathcal{F}\{p_T(t)\} = \sum_{k=-\infty}^{\infty} F_k \mathcal{F}\left\{ e^{jk\omega_0 t} \right\}$$

$$= \sum_{k=-\infty}^{\infty} \frac{2\tau \sin(k\omega_0 \tau)}{T(k\omega_0 \tau)} \delta(\omega - k\omega_0) \tag{2.114}$$

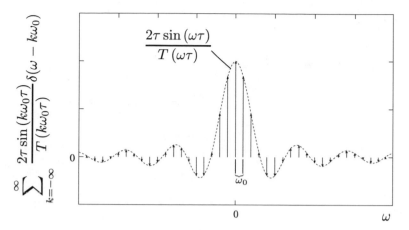

**Fig. 2.35**  Fourier transform of a pulse train shown in Fig. 2.21

The figure depicting the Fourier of the pulse train is shown in Fig. 2.35. Impulses are placed $\omega_0$ apart and have an envelope of

$$\frac{2\tau \sin (\omega\tau)}{T (\omega\tau)}$$

and the impulses occur at

$$\omega = k\omega_0 \quad k = 0, \pm 1, \ldots \tag{2.115}$$

and at $\omega = 0$ the value of the envelope is $2\tau/T$. The lower the value of $\tau$ (which is half of the pulse width), the lower the value of the envelope.

### 2.12.2   Time Scaling

If

$$x(t) \Leftrightarrow X(\omega)$$
$$x(at) \Leftrightarrow \frac{1}{|a|} X\left(\frac{\omega}{a}\right)$$

*Proof.* for $a > 0$

$$\mathcal{F}\{x(at)\} = \int_{-\infty}^{\infty} x(at)e^{-j\omega t}\, dt$$

$$= \int_{-\infty}^{\infty} x(\tau)e^{-j\omega \tau/a}\frac{d\tau}{a} \qquad (\text{subst. } \tau = at)$$

$$= \frac{1}{a}\int_{-\infty}^{\infty} x(\tau)e^{-j\omega \tau/a}\, d\tau$$

$$= \frac{X(\omega/a)}{a}$$

for $a < 0$

$$\mathcal{F}\{x(at)\} = \int_{-\infty}^{\infty} x(at)e^{-j\omega t}\, dt$$

$$= \int_{\infty}^{-\infty} x(\tau)e^{-j\omega \tau/a}\left(-\frac{d\tau}{|a|}\right) \qquad (\text{subst. } \tau = at)$$

$$= \frac{1}{|a|}\int_{-\infty}^{\infty} x(\tau)e^{-j\omega \tau/a}\, d\tau$$

$$= \frac{X(\omega/a)}{|a|}$$

so

$$\boxed{x(at) \Leftrightarrow \frac{1}{|a|}X\left(\frac{\omega}{a}\right)} \tag{2.116}$$

**Example 2.30** If $x(t) \Leftrightarrow X(\omega)$, then $x(-t) \Leftrightarrow$?

Applying the result of Sect. 2.12.2 for $a = -1$,

$$\boxed{x(-t) \Leftrightarrow X(-\omega)} \tag{2.117}$$

### 2.12.3  Frequency Scaling

If $x(t) \Leftrightarrow X(\omega)$, then

$$\boxed{\frac{1}{|a|}x\left(\frac{t}{a}\right) \Leftrightarrow X(a\omega)}$$

**Proof** for $a > 0$

$$\mathcal{F}^{-1}\{X(a\omega)\} = \frac{1}{2\pi}\left[\int_{-\infty}^{\infty} X(a\omega)e^{j\omega t}\,d\omega\right]$$

$$= \frac{1}{2\pi}\int_{-\infty}^{\infty} X(\Omega)e^{jt\Omega/a}\,\frac{d\Omega}{a} \qquad \text{(subst. } \Omega = a\omega\text{)}$$

$$= \frac{1}{a}\left[\frac{1}{2\pi}\int_{-\infty}^{\infty} X(\omega)e^{jt\omega/a}\,d\omega\right]$$

$$= \frac{x(t/a)}{a}$$

for $a < 0$

$$\mathcal{F}^{-1}\{X(a\omega)\} = \frac{1}{2\pi}\left[\int_{-\infty}^{\infty} X(a\omega)e^{j\omega t}\,d\omega\right]$$

$$= \frac{1}{2\pi}\int_{\infty}^{-\infty} X(\Omega)e^{jt\Omega/a}(-\frac{d\Omega}{a}) \qquad \text{(subst. } \Omega = a\omega\text{)}$$

$$= \frac{1}{|a|}\left[\frac{1}{2\pi}\int_{-\infty}^{\infty} X(\omega)e^{jt\omega/a}\,d\omega\right]$$

$$= \frac{x(t/a)}{|a|}$$

## 2.12.4   Time Shift

If $x(t) \Leftrightarrow X(\omega)$, then $x(t - t_0) \Leftrightarrow X(\omega)e^{-j\omega t_0}$. To prove this property,

$$\mathcal{F}[x(t - t_0)] = \int_{-\infty}^{\infty} x(t - t_0)e^{-j\omega t}\,dt$$

$$= \int_{-\infty}^{\infty} x(\tau)e^{-j\omega(\tau + t_0)}\,d\tau \qquad \text{(setting } \tau = t - t_0\text{)}$$

$$= e^{-j\omega t_0}\int_{-\infty}^{\infty} x(\tau)e^{-j\omega\tau}\,d\tau$$

$$= e^{-j\omega t_0} X(\omega)$$

therefore

$$\boxed{x(t - t_0) \Leftrightarrow X(\omega)e^{-j\omega t_0}} \qquad (2.118)$$

## 2.12.5   Frequency Shift

If $x(t) \Leftrightarrow X(\omega)$, then

$$\boxed{x(t)e^{j\omega_0 t} \Leftrightarrow X(\omega - \omega_0)} \qquad (2.119)$$

The proof along the same lines as before is

$$\mathcal{F}^{-1}[X(\omega - \omega_0)] = \frac{1}{2\pi} \int_{-\infty}^{\infty} X(\omega - \omega_0)e^{j\omega t} d\omega$$

$$= \frac{1}{2\pi} \int_{-\infty}^{\infty} X(\Omega)e^{j(\Omega + \omega_0)t} d\Omega \qquad (\text{setting } \Omega = \omega - \omega_0)$$

$$= \frac{e^{j\omega_0 t}}{2\pi} \int_{-\infty}^{\infty} X(\Omega)e^{j\Omega t} d\Omega$$

$$= e^{j\omega_0 t} x(t)$$

## 2.12.6   Duality

1. Time Duality: If

$$x(t) \Leftrightarrow X(\omega)$$

then

$$x(t) = \frac{1}{2\pi} \int_{-\infty}^{\infty} X(\omega)e^{j\omega t} d\omega$$

$$x(-t) = \frac{1}{2\pi} \int_{-\infty}^{\infty} X(\omega)e^{-j\omega t} dt \qquad (\text{subst. } -t \text{ for } t)$$

$$2\pi x(-\omega) = \int_{-\infty}^{\infty} X(t)e^{-j\omega t} d\omega \qquad (\text{interchanging } t \text{ and } \omega) \qquad (2.120)$$

so

$$\boxed{X(t) \Leftrightarrow 2\pi x(-\omega)} \qquad (2.121)$$

2. Frequency Duality: From equation set (2.120), interchanging $t$ and $\omega$,

$$x(\omega) = \frac{1}{2\pi} \int_{-\infty}^{\infty} X(\tau)e^{j\omega\tau} d\tau$$

Substitute $t = -\tau$

$$x(\omega) = \frac{1}{2\pi} \int_{\infty}^{-\infty} X(-t)e^{-j\omega t} d(-t)$$

$$2\pi x(\omega) = \int_{-\infty}^{\infty} X(-t)e^{-j\omega t} dt \qquad (2.122)$$

so

$$\boxed{\frac{X(-t)}{2\pi} \Leftrightarrow x(\omega)} \qquad (2.123)$$

The two equations, (2.121) and (2.123), are the two relations which may be used when the Fourier transform or inverse Fourier transform are known, and the other transform can be calculated.

### 2.12.6.1   IFT of a Periodic Function in Frequency

**Example 2.31**  Apply duality to a periodic function as in Example 2.28.

Let

$$f(t) = \sum_{k=-\infty}^{k=\infty} F_k e^{jk\omega_0 t}$$

be a periodic function. Here, $F_k$ are the exponential Fourier series coefficients of the periodic function. Then from Example 2.28, Eq. (2.113)

$$\underbrace{\sum_{k=-\infty}^{k=\infty} F_k e^{jk\omega_0 t}}_{f(t)} \Leftrightarrow \underbrace{2\pi \sum_{k=-\infty}^{k=\infty} F_k \delta(\omega - k\omega_0)}_{F(\omega)}$$

Now applying duality,

$$\underbrace{2\pi \sum_{k=-\infty}^{k=\infty} F_k \delta(t - kT)}_{F(t)} \Leftrightarrow \underbrace{2\pi \sum_{k=-\infty}^{k=\infty} F_k e^{-jkT\omega}}_{f(-\omega)} \qquad (2.124)$$

where we have replaced $\omega_0$ and $t$ by $T$ and $-\omega$, respectively. What does the relation in (2.124) say? The relation says that when we have a periodic function in the frequency domain then the function in the time domain appears as as a delta function train multiplied by the Fourier series coefficients.

## 2.12.7 Symmetry Property

If

$$x(t) \Leftrightarrow X(\omega)$$

and $x(t)$ is real, then

$$X^*(\omega) = X(-\omega)$$

**Proof**

$$X(\omega) = \int_{-\infty}^{\infty} x(t)e^{-j\omega t} dt$$

$$X^*(\omega) = \left[\int_{-\infty}^{\infty} x(t)e^{-j\omega t} dt\right]^*$$

$$= \int_{-\infty}^{\infty} x^*(t)\left(e^{-j\omega t}\right)^* dt$$

$$= \int_{-\infty}^{\infty} x(t)e^{j\omega t} dt \quad (x(t) \text{ is real})$$

$$= \int_{-\infty}^{\infty} x(t)e^{-j(-\omega)t} dt$$

$$= X(-\omega)$$

so

$$\boxed{X^*(\omega) = X(-\omega)} \tag{2.125}$$

basically this Eq. (2.125) says that if

$$X(\omega) = A(\omega) + jB(\omega)$$

then

$$X^*(\omega) = A(\omega) - jB(\omega)$$
$$= X(-\omega)$$
$$= A(-\omega) + jB(-\omega)$$

or

$$\boxed{\begin{array}{c} A(\omega) = A(-\omega) \\ B(\omega) = -B(-\omega) \end{array}}$$

The real part is an even function of $\omega$ while the imaginary part is an odd function of $\omega$. It also follows that the magnitude and phase are even and odd functions of $\omega$, respectively.

$$\boxed{\begin{array}{l} |X(\omega)| = |X(-\omega)| \\ \angle X(\omega) = -\angle X(-\omega) \end{array}}$$

## 2.12.8 Properties Based on the Properties of $x(t)$

1. If $x(t)$ is real, and $X(\omega) = \Re\{X(\omega)\} + j\Im\{X(\omega)\}$ ($\equiv X_r(\omega) + jX_i(\omega)$) (which is complex), then $X_r(\omega)$ is even and $X_i(\omega)$ is odd.
2. If $x(t)$ is imaginary, and $X(\omega) = X_r(\omega) + jX_i(\omega)$, then show that $X_r(\omega)$ is odd and $X_i(\omega)$ is even.
3. If $x(t)$ is real and even, then if $X(\omega) = X_r(\omega) + jX_i(\omega)$, show that $X_r(\omega)$ is real and even and $X_i(\omega)$ is zero.
4. If $x(t)$ is real and odd, then show that $X_r(\omega)$ is zero and $X_i(\omega)$ is odd.
5. If $x(t)$ is imaginary and even, then show that $X_r(\omega)$ is zero and $X_i(\omega)$ is even.
6. If $x(t)$ is imaginary and odd, then show that $X_r(\omega)$ is odd and $X_i(\omega)$ is zero.

**Exercise 2.3** Prove the results given in Properties 2.12.8, given above.

## 2.12.9 Time Differentiation

If $x(t) \Leftrightarrow X(\omega)$, then $d/dt\,[x(t)] \Leftrightarrow j\omega X(\omega)$.

**Proof**

$$x(t) = \frac{1}{2\pi}\left[\int_{-\infty}^{\infty} X(\omega)e^{j\omega t}\,d\omega\right]$$

$$\frac{d}{dt}x(t) = \frac{d}{dt}\left[\frac{1}{2\pi}\int_{-\infty}^{\infty} X(\omega)e^{j\omega t}\,d\omega\right]$$

$$= \frac{1}{2\pi}\int_{-\infty}^{\infty} [j\omega X(\omega)]\,e^{j\omega t}\,d\omega$$

therefore

$$\boxed{\frac{d}{dt}x(t) \Leftrightarrow j\omega X(\omega)}$$

**Exercise 2.4** Show that

$$\frac{d^n}{dt^n}x(t) \Leftrightarrow (j\omega)^n X(\omega)$$

## 2.12.10  *Frequency Differentiation*

If $x(t) \Leftrightarrow X(\omega)$, then

$$-jtx(t) \Leftrightarrow \frac{d}{d\omega}X(\omega)$$

*Proof.* To prove this theorem,

$$X(\omega) = \int_{-\infty}^{\infty} x(t)e^{-j\omega t}dt$$

then

$$\frac{d}{d\omega}X(\omega) = \frac{d}{d\omega}\left[\int_{-\infty}^{\infty} x(t)e^{-j\omega t}dt\right]$$

$$= \int_{-\infty}^{\infty} \frac{d}{d\omega}\left[x(t)e^{-j\omega t}\right]dt$$

$$= \int_{-\infty}^{\infty} \left[-jtx(t)\right]e^{-j\omega t}dt$$

$$= \mathcal{F}\{-jtx(t)\}$$

or

$$\boxed{-jtx(t) \Leftrightarrow \frac{d}{d\omega}X(\omega)}$$

**Exercise 2.5** Show that if $x(t)$ is real, then the magnitude of the FT is even and the phase is[3] odd.

---

[3] A very important result.

## 2.12.11  Convolution Theorem

Let us find the Fourier transform of a convolution:

$$\mathcal{F}\{x(t) * y(t)\} = \mathcal{F}\left\{\int_{-\infty}^{\infty} x(t-\tau)y(\tau)d\tau\right\}$$

$$= \int_{-\infty}^{\infty}\left\{\int_{-\infty}^{\infty} x(t-\tau)y(\tau)d\tau\right\}e^{-j\omega t}dt$$

$$= \int_{-\infty}^{\infty}\left\{\int_{-\infty}^{\infty} x(t-\tau)e^{-j\omega t}dt\right\}y(\tau)d\tau$$

$$= \int_{-\infty}^{\infty}\left\{X(\omega)e^{-j\omega\tau}\right\}y(\tau)d\tau \qquad (\because x(t-\tau) \Leftrightarrow X(\omega)e^{-j\omega\tau})$$

$$= X(\omega)\int_{-\infty}^{\infty} e^{-j\omega\tau}y(\tau)d\tau$$

$$= X(\omega)Y(\omega) \tag{2.126}$$

Therefore

$$\boxed{\mathcal{F}\{x(t) * y(t)\} = X(\omega)Y(\omega)} \tag{2.127}$$

Similarly, let us find the inverse FT of the convolution $F(\omega) * G(\omega)$:

$$\mathcal{F}^{-1}\{X(\omega) * Y(\omega)\} = \frac{1}{2\pi}\int_{-\infty}^{\infty}\{X(\omega) * Y(\omega)\}e^{j\omega t}d\omega$$

$$= \frac{1}{2\pi}\int_{-\infty}^{\infty}\left\{\int_{-\infty}^{\infty} X(\Omega)Y(\omega-\Omega)d\Omega\right\}e^{j\omega t}d\omega$$

$$= \int_{-\infty}^{\infty} X(\Omega)\left\{\frac{1}{2\pi}\int_{-\infty}^{\infty} Y(\omega-\Omega)e^{j\omega t}d\omega\right\}d\Omega$$

$$= \int_{-\infty}^{\infty} X(\Omega)\left\{y(t)e^{j\Omega t}\right\}d\Omega \qquad (\because y(t)e^{j\Omega t} \Leftrightarrow Y(\omega-\Omega))$$

$$= 2\pi y(t)\left\{\frac{1}{2\pi}\int_{-\infty}^{\infty} X(\Omega)e^{j\Omega t}d\Omega\right\}$$

$$= 2\pi y(t)x(t)$$

therefore we have

$$\boxed{\mathcal{F}^{-1}\{X(\omega) * Y(\omega)\} = 2\pi y(t)x(t)} \tag{2.128}$$

## *2.12.12   Time Integration*

We would like to consider the Fourier transform of the integral

$$\int_{-\infty}^{t} x(\tau)d\tau$$

To this end, consider Fig. 2.36. We are looking at the convolution of $u(t)$ with $x(t)$. In the figure, $x(\tau)$ and $u(t - \tau)$ are both shown with $\tau$ as the independent variable and $t$ is the point where the transition of the step function is shown. From the figure, it is obvious that

$$\int_{-\infty}^{t} x(\tau)d\tau = \int_{-\infty}^{\infty} u(t - \tau)x(\tau)d\tau$$

$$= u(t) * x(t) \tag{2.129}$$

so

$$\mathcal{F}\left\{\int_{-\infty}^{t} x(\tau)d\tau\right\} = \mathcal{F}\{u(t) * x(t)\}$$

$$= U(\omega)X(\omega)$$

$$= X(\omega)\left[\frac{1}{j\omega} + \pi\delta(\omega)\right]$$

$$= \frac{X(\omega)}{j\omega} + \pi X(\omega)\delta(\omega) \tag{2.130}$$

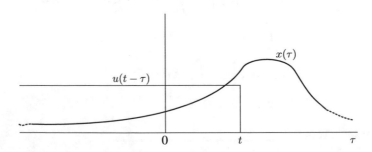

**Fig. 2.36**  Convolution of $x(t)$ with $u(t)$

## 2.12.13   Multiplication and Parseval's Theorems

If we consider the integral,

$$\int_{-\infty}^{\infty} x(t)y^*(t)dt$$

where $y^*(t)$ is the complex conjugate of $y(t)$, then what does it appear as in the frequency domain? To understand, let us look at

$$\int_{-\infty}^{\infty} x(t)y^*(t)dt = \int_{-\infty}^{\infty} \left[ \frac{1}{2\pi} \int_{-\infty}^{\infty} X(\omega)e^{j\omega t}d\omega \right] y^*(t)dt$$

$$= \int_{-\infty}^{\infty} X(\omega) \left[ \frac{1}{2\pi} \int_{-\infty}^{\infty} y^*(t)e^{j\omega t}dt \right] d\omega$$

$$= \int_{-\infty}^{\infty} X(\omega) \left[ \frac{1}{2\pi} \left\{ \int_{-\infty}^{\infty} y(t)e^{-j\omega t}dt \right\}^* \right] d\omega$$

$$= \frac{1}{2\pi} \int_{-\infty}^{\infty} X(\omega)Y^*(\omega)d\omega \qquad (2.131)$$

In this equation, (2.131), we replace $y^*(t)$ by $x^*(t)$; then we get *Parseval's theorem*:

$$\boxed{\int_{-\infty}^{\infty} x(t)x^*(t)dt = \frac{1}{2\pi} \int_{-\infty}^{\infty} X(\omega)X^*(\omega)d\omega} \qquad (2.132)$$

**Example 2.32** Apply Parseval's theorem on the time function $f(t) = e^{-at}u(t)$ where $a > 0$.

From Example 2.22,

$$e^{-at}u(t) \Leftrightarrow \frac{1}{(a+j\omega)}$$

then from Eq. (2.132), the left side of the equation is

$$\int_{-\infty}^{\infty} e^{-2at}u(t)dt = \int_{0}^{\infty} e^{-2at}dt$$

$$= \frac{1}{2a}$$

and the right side of the equation is

$$\frac{1}{2\pi} \int_{-\infty}^{\infty} \left[ \frac{1}{(a+j\omega)} \right] \left[ \frac{1}{(a-j\omega)} \right] d\omega = \frac{1}{2\pi} \int_{-\infty}^{\infty} \left[ \frac{1}{(a^2+\omega^2)} \right] d\omega$$

$$= \frac{1}{2\pi a} \left[ \tan^{-1} \left( \frac{\omega}{a} \right) \right]_{-\infty}^{\infty}$$

$$= \frac{1}{2\pi a} \left[ \frac{\pi}{2} - \left( -\frac{\pi}{2} \right) \right]$$

$$= \frac{1}{2a}$$

### 2.12.14  Energy Spectral Density

From Parseval's theorem, we conclude that the energy in a signal in the time domain is related to the integral of the FT of the signal. Thus[4]

$$\mathbb{E} = \int_{-\infty}^{\infty} x(t) x^*(t) dt = \frac{1}{2\pi} \int_{-\infty}^{\infty} X(\omega) X^*(\omega) d\omega$$

$$= \int_{-\infty}^{\infty} X(f) X^*(f) df$$

where $f$ is the frequency in hertz. Therefore, we call

$$\Psi(\omega) = \frac{X(\omega) X^*(\omega)}{2\pi} = \frac{|X(\omega)|^2}{2\pi} \tag{2.133}$$

$$\Psi(f) = |X(f)|^2 \tag{2.134}$$

as the energy spectral density. Thus, the energy in a signal in the time domain is linked to the energy of the signal in the frequency domain, and we can then say that the energy of the signal in the frequency range, $\omega_0 \leq \omega \leq \omega_1$, is given by

$$\frac{1}{2\pi} \int_{\omega_0}^{\omega_1} |X(\omega)|^2 \, d\omega + \frac{1}{2\pi} \int_{-\omega_1}^{-\omega_0} |X(\omega)|^2 \, d\omega = \int_{\omega_0}^{\omega_1} \Psi(\omega) d\omega \tag{2.135}$$

for a real time signal (Table 2.2).

---

[4] In the $f$ domain, the energy equation is $\int_{-\infty}^{\infty} x(t) x^*(t) dt = \int_{-\infty}^{\infty} X(f) X^*(f) df$.

**Table 2.2**  Properties of the Fourier transform

| Property | $x(t)$ | $X(\omega)$ |
|---|---|---|
| Definition | $x(t) = \frac{1}{2\pi} \int_{-\infty}^{\infty} X(\omega)d\omega$ | $X(\omega) = \int_{-\infty}^{\infty} x(t)dt$ |
| Time shifting | $x(t - t_0)$ | $X(\omega)e^{-j\omega t_0}$ |
| Frequency shifting | $x(t)e^{j\omega_0 t}$ | $X(\omega - \omega_0)$ |
| Time multiplication | $x(at)$ | $\frac{1}{|a|} X\left(\frac{\omega}{a}\right)$ |
| Duality | $X(t)$ | $2\pi x(-\omega)$ |
| Differentiation | $dx(t)/dt$ | $j\omega X(\omega)$ |
| $d^n/dt^n$ | $d^n x(t)/dt^n$ | $(j\omega)^n X(\omega)$ |
| Frequency differentiation | $(-jt)x(t)$ | $dX(\omega)/d\omega$ |
| $d^n/dt^n$ | $(-jt)^n x(t)$ | $d^n X(\omega)/d\omega^n$ |
| Time integration | $\int_{-\infty}^{t} x(\tau)d\tau$ | $\frac{1}{j\omega} X(\omega) + \pi X(0)\delta(\omega)$ |
| Parseval's theorem | $\int_{-\infty}^{\infty} x(t)x^*(t)dt = \frac{1}{2\pi} \int_{-\infty}^{\infty} X(\omega)X^*(\omega)d\omega$ | |

## 2.13  Signal Transmission Through Linear Systems

Linear time-invariant (LTI) systems are those systems which satisfy the following three criteria mathematically:

1. If the system has as an input $x(t)$ and its output $y(t)$

$$x(t) \Rightarrow y(t) \tag{2.136}$$

then when we use an input $ax(t)$, the output is $ay(t)$

$$ax(t) \Rightarrow ay(t) \tag{2.137}$$

Here, the arrow $\Rightarrow$ is used to designate the output of the system. This property is called the scaling property.
2. If the system has two separate inputs $x_1(t)$ and $x_2(t)$ so that the outputs to these two separate signals individually are $y_1(t)$ and $y_2(t)$, then if the input is $x_1(t) + x_2(t)$ the output is $y_1(t) + y_2(t)$,

$$x_1(t) \Rightarrow y_1(t) \quad \text{and}$$
$$x_2(t) \Rightarrow y_2(t) \quad \text{then}$$
$$x_1(t) + x_2(t) \Rightarrow y_1(t) + y_2(t) \tag{2.138}$$

This property is called the additive property. Both these properties together are called the superposition properties.
3. If the system has an input $x(t)$ which gives an output $y(t)$ then if the input is given at a later time $t_0$, then the output is also delayed by time $t_0$. Or more formally,

**Fig. 2.37** Impulse response
of an LTI system

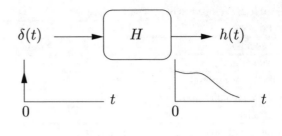

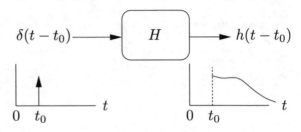

$$x(t) \Rightarrow y(t)$$
$$x(t - t_0) \Rightarrow y(t - t_0) \tag{2.139}$$

LTI systems have a property which involves the impulse response of a system. The impulse response of a system is the output of the system when the input is a unit impulse. This is shown in Fig. 2.37. The figure shows that when the input is $\delta(t)$, the output is the impulse response, $h(t)$, and when the input is $\delta(t - t_0)$, the time-shifted delta function, then the output is also time-shifted by the same amount due to the property of time invariance. That is

$$\delta(t) \Rightarrow h(t)$$
$$\delta(t - t_0) \Rightarrow h(t - t_0) \tag{2.140}$$

Now we take a look at another interesting property of the delta function:

$$\int_{-\infty}^{\infty} x(\tau)\delta(\tau)d\tau = \int_{-\infty}^{\infty} x(0)\delta(\tau)d\tau \quad \text{(by Eqn (2.16))}$$

$$= x(0) \int_{-\infty}^{\infty} \delta(\tau)d\tau$$

$$= x(0) \quad \text{(the integral of } \delta(\tau) \text{ is 1)} \tag{2.141}$$

And similarly,

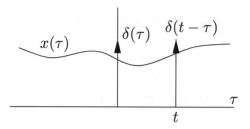

**Fig. 2.38**  $x(t)$ and $\delta(t)$

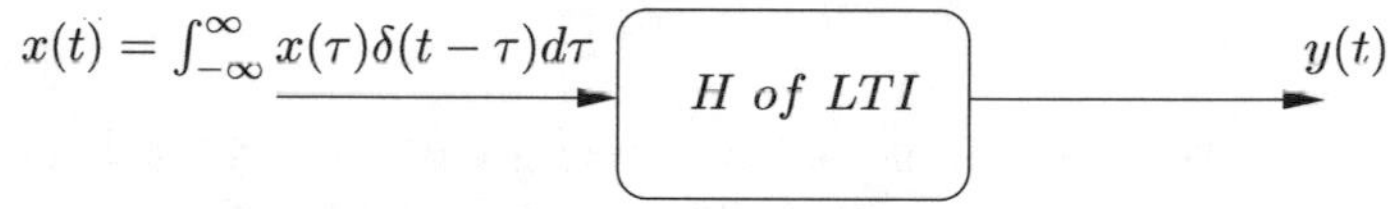

$$x(t) = \int_{-\infty}^{\infty} x(\tau)\delta(t-\tau)d\tau \longrightarrow \boxed{H \ of \ LTI} \longrightarrow y(t)$$

**Fig. 2.39**  Analysis of an LTI system using the impulse response

$$\int_{-\infty}^{\infty} x(\tau)\delta(t-\tau)d\tau = \int_{-\infty}^{\infty} x(t)\delta(t-\tau)d\tau \quad \text{(by Eqn (2.16))}$$

$$= x(t) \int_{-\infty}^{\infty} \delta(t-\tau)d\tau$$

$$= x(t) \quad \text{(the integral of } \delta(t-\tau) \text{ is 1)} \qquad (2.142)$$

Or in other words, the convolution of $x(t)$ with $\delta(t)$ is $x(t)$! This is shown in Fig. 2.38.
Now let us see how the linear LTI operator operates on an input signal $x(t)$:

$$y(t) = H\{x(t)\} \qquad (2.143)$$

Similarly, for an input $ax(t)$ as in Eq. (2.137),

$$H\{ax(t)\} = aH\{x(t)\} = ay(t) \qquad (2.144)$$

and for an input $x_1(t) + x_2(t)$ as per Eq. set (2.138),

$$H\{x_1(t) + x_2(t)\} = H\{x_1(t) + x_2(t)\}$$

$$= H\{x_1(t)\} + H\{x_2(t)\}$$

$$= y_1(t) + y_2(t) \qquad (2.145)$$

Concentrating on an LTI system as shown in Fig. 2.39 where an input $x(t)$ is
shown, we know that the function can be written as a convolution integral as given
in Eq. (2.142). The signal is fed into an LTI system as the integral 'sum'

$$x(t) = \int_{-\infty}^{\infty} x(\tau)\delta(t-\tau)d\tau$$

The integral says that we feed into the system a sum of terms

$$\lim_{i \to \infty} \sum_i x(\tau_i)\delta(t - \tau_i)\Delta\tau_i \tag{2.146}$$

where the coefficient of each $\delta(t - \tau)$ is $x(\tau_i)\Delta\tau_i$. When the system operator, $H_{(t)}$, operates on this sum

$$H_{(t)}\left\{\lim_{i \to \infty} \sum_i x(\tau_i)\delta(t - \tau_i)\Delta\tau_i\right\} = \lim_{i \to \infty} \sum_i x(\tau_i)H_{(t)}\left\{\delta(t - \tau_i)\right\}\Delta\tau_i \tag{2.147}$$

the subscript $(t)$ implies that the operator operates on the $t$ variable and *not* on the $\tau$ variable. Equation (2.147) is possible due to the superposition properties: the properties of superposition and scaling. But we know that

$$H_{(t)}\left\{\delta(t - \tau_i)\right\} = h(t - \tau_i)$$

is true from Eq. (2.140) (where $h(t - \tau)$ is the shifted impulse response). Therefore

$$H_{(t)}\left\{\lim_{i \to \infty} \sum_i x(\tau_i)\delta(t - \tau_i)\Delta\tau_i\right\} = \lim_{i \to \infty} \sum_i x(\tau_i)H_{(t)}\left\{\delta(t - \tau_i)\right\}\Delta\tau_i$$

$$= \lim_{i \to \infty} \sum_i x(\tau_i)h(t - \tau_i)\Delta\tau_i$$

or going back to the integral

$$H\{x(t)\} = \int_{-\infty}^{\infty} x(\tau)h(t - \tau)d\tau$$

$$= y(t)$$

therefore we have the important result that the output $y(t)$ is the convolution of the input, $x(t)$ with the impulse response $h(t)$:

$$\boxed{y(t) = \int_{-\infty}^{\infty} x(\tau)h(t - \tau)d\tau} \tag{2.148}$$

The importance of these results are highlighted when we go over to the transform domain. If the Fourier transform of the impulse response is $H(\omega)$

$$h(t) \Leftrightarrow H(\omega) \tag{2.149}$$

and the FT of the input signal is $X(\omega)$

$$x(t) \Leftrightarrow X(\omega) \tag{2.150}$$

then the FT of the output is given by

$$\boxed{Y(\omega) = H(\omega)X(\omega)} \tag{2.151}$$

which is true by Eq. (2.127). $H(\omega)$ is also called the frequency response of the filter.

### 2.13.1  Paley–Wiener Criterion

The necessary and sufficient condition that $H(\omega)$ is the frequency response of a filter is

$$\int_{-\infty}^{\infty} \frac{\ln |H(\omega)|}{1 + \omega^2} d\omega < \infty$$

Thus, for example, the gain $|H(\omega)|$ may be zero only for a finite number of frequencies, but it cannot be zero for a band of frequencies. By this criterion, the gain may *approach* zero over a band of frequencies but not be zero over that band. By this criterion, it is impossible to realise a filter such as

$$H(\omega) = \begin{cases} 1 & -2\pi B \le \omega \le 2\pi B \\ 0 & \text{elsewhere} \end{cases}$$

since, then $\ln |H(\omega)|$ will be $-\infty$ over those portions of the frequency where $|H(\omega)|$ will be zero.

However, the filter

$$H(\omega) = \begin{cases} 1 & -2\pi B \le \omega \le 2\pi B \\ > \varepsilon & \text{elsewhere} \end{cases}$$

where $\varepsilon$ is a small positive value, is a realisable filter, $\ln(\varepsilon) \ne -\infty$.

## 2.14  Distortionless LTI Systems

Let us take sinusoidal radiated wave at frequency $\omega$ being received at a distance $d$ from the transmitter. The electric field function between the two transmitting and receiving ends will be of the form

$$E_T = (A/z) \sin(\omega t - \beta z) \tag{2.152}$$

where $A$ is a constant, $\beta = 2\pi/\lambda$ and $z$ is the direction of propagation. The $A/z$ term says that the field decays as $1/z$ as the wave moves away from the transmitting station, while the argument $\omega t - \beta z$ ensures that the wave is a travelling wave.

On receiving the signal, at the receiving end, the electric field (and therefore the voltage received) is of the form

$$E_R = B \sin(\omega t - \theta) \tag{2.153}$$

where $B = A/d$ and $\theta = \beta d$. We notice that the amplitude is not a function of frequency but let us examine the nature of $\theta$ (let $f$ be the frequency in Hertz):

$$
\begin{aligned}
\theta &= \beta d \\
&= (2\pi/\lambda)d \\
&= (2\pi f/c)d \qquad (c \text{ is the velocity of the wave; } c = f\lambda) \\
&= (\omega/c)d \\
&= \omega(d/c) \tag{2.154} \\
&= \omega t_0 \qquad (\text{where } t_0 = d/c) \tag{2.155}
\end{aligned}
$$

And we know that the wave is undistorted. So what are the conditions of distortionless transmission?

1. The amplitude of the wave is not a function of frequency. In this case, it is $A/d$.
2. The phase of the wave is a *linear* function of frequency. In this case, it is $\omega(d/c)$.

These results are shown in Fig. 2.40. The figure shows the amplitude and phase characteristics of the signal with frequency $\omega$. In fact, this is the FT of the impulse response of the ideal distortionless channel. If we compare with Fig. 2.31, then we find that both figures are identical and that Fig. 2.40 represents the Fourier transform of

$$B\delta(t - t_0)$$

where $B$ is a constant. If into this channel we input $A\delta(t)$, we get on the output $(A/d)\delta(t - t_0)$ where $t_0 = d/c$.

Another way to look at this problem is that if $x(t)$ is the input to a system, and $y(t)$ is the output, the signal is not distorted if

**Fig. 2.40** Amplitude and phase characteristic of Distortionless transmission

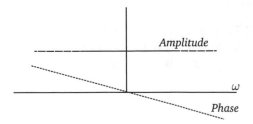

$$y(t) = kx(t - t_0) \tag{2.156}$$

where $k$ and $t_0$ are constants. This equation, (2.156), says that the signal is not distorted if

1. The signal is uniformly amplified or attenuated by a factor $k$, and
2. The signal is delayed by a time $t_0$.

We now take the FT of Eq. (2.156). Then

$$
\begin{aligned}
Y(\omega) &= kX(\omega)e^{-j\omega t_0} \\
&= \underbrace{\left[ke^{-j\omega t_0}\right]}_{H(\omega)} X(\omega) \\
&= H(\omega)X(\omega)
\end{aligned}
\tag{2.157}
$$

where $H(\omega)$ is the frequency response of the channel or system (if the channel is considered a system). To find the impulse response of such a system,

$$h(t) = k\delta(t - t_0) \tag{2.158}$$

Let us for a moment reflect upon this result. Suppose we want a distortionless *system* such as a low-pass filter or a bandpass filter, what should be their characteristics? It should be the same as points 1 and 2 given above! These results are shown diagrammatically in Fig. 2.41.

**Fig. 2.41** Ideal low- and bandpass filters

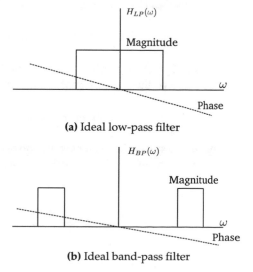

**(a)** Ideal low-pass filter

**(b)** Ideal band-pass filter

## 2.15  Signal Distortion

As signals pass through non-ideal linear systems or channels, we can model them through their impulse responses or equivalently their frequency responses. The two of them are related by

$$h(t) \Leftrightarrow H(\omega) \tag{2.159}$$

where $h(t)$ is the impulse response and $H(\omega)$ is the frequency response of the system or channel.

If we observe the output of such a system, then

$$y(t) = h(t) * x(t) \tag{2.160}$$

or

$$Y(\omega) = H(\omega)X(\omega) \tag{2.161}$$

which implies that since $X(\omega)$ is multiplied by $H(\omega)$, the magnitude of and phase of $Y$ is changed from that of $X$ by multiplication of the factor $H(\omega)$. That is

$$|Y(\omega)| = |H(\omega)|\,|X(\omega)| \tag{2.162}$$

We can see from this equation, (2.162), that the system introduces *no distortion* of the magnitude spectrum if only

$$|H(\omega)| = k, \qquad \text{a constant}$$

Similarly, from Eq. (2.161), we can see that

$$\angle Y(\omega) = \angle H(\omega) + \angle X(\omega) \tag{2.163}$$

so the output phase too suffers a distortion. From our earlier discussion, it is clear that the phase of the output signal is *not distorted* if the phase

$$\angle H(\omega) = -k\omega, \qquad k, \text{ a constant}$$

In short, every system introduces some sort of distortion in the signal, and the question is whether the distortion is acceptable or unacceptable.

### 2.15.1  Equalisation

Channel distortion or system distortion may be corrected through what is known as an equaliser filter. Essentially, we use the frequency responses of the channel and the equaliser to remove the distortion introduced. Thus, we know that the undistorted frequency response is $H_{ud}(\omega) = ke^{-j\omega t_0}$ where $k$ and $t_0$ are constants. Therefore, we design an equaliser, $H_{eq}(\omega)$, which satisfies the equation,

$$H(\omega)H_{eq}(\omega) = ke^{-j\omega t_0} \qquad (2.164)$$

Figure 2.42 shows how the equaliser is used. From $\sum quation$ 2.164,

$$H_{eq}(\omega) = \frac{ke^{-j\omega t_0}}{H(\omega)} \qquad (2.165)$$

In the figure,

$$X(\omega) = H(\omega)M(\omega)$$
$$\text{and } \hat{M}(\omega) = kM(\omega)e^{-j\omega t_0} \qquad (2.166)$$

where

$$m(t) \Leftrightarrow M(\omega)$$

and

$$\hat{m}(t) \Leftrightarrow \hat{M}(\omega)$$

**Example 2.33** The frequency characteristic of a channel is proportional to $Ke^{-\alpha\omega}$ in the frequency range $0 \le \omega \le \omega_0$. Design an equaliser to correct this frequency response.

The equaliser characteristic must be the reciprocal of this response. Therefore

$$H_{eq}(\omega) \propto \frac{k}{H(\omega)}$$

These two frequency responses and their product are shown in Fig. 2.43.

**Fig. 2.42** Equaliser filter

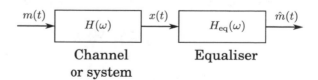

Channel
or system

Equaliser

**Fig. 2.43** Figure for
Example 2.33

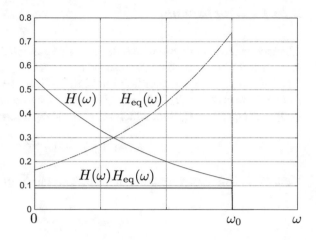

## 2.15.2   Nonlinear Channel Distortion

A channel may introduce nonlinearities in the signal. Such a channel may be modelled
by the equation,

$$y(t) = h_0 + h_1 x(t) + h_2 [x(t)]^2 + \ldots$$

$$= \sum_{i=0}^{\infty} h_i [x(t)]^i \qquad (2.167)$$

where some of the $h_i$ may be zero.

To take a concrete example, if the input to such a channel, characterised by

$$y(t) = x(t) + 0.01 [x(t)]^2$$

and the input is $A \cos(\omega_0 t)$, having a bandwidth of $\omega_0$, then

$$y(t) = 0.01 A^2 \cos(\omega_0 t)^2 + A\cos(\omega_0 t)$$

$$= 0.01 \left( \frac{\cos(2\omega_0 t) A^2}{2} + \frac{A^2}{2} \right) + \cos(\omega_0 t) A \qquad (2.168)$$

We observe that two new (small) terms have appeared, namely

$$0.01 \left( \frac{\cos(2\omega_0 t) A^2}{2} + \frac{A^2}{2} \right)$$

one at $\omega = 0$ and the other at $\omega = 2\omega_0$.

## 2.16  Time-Limited and Band-Limited Signals

A time-limited signal is one which *does not* tend to infinity in any direction. For example, the signal

$$f(t) = \begin{cases} e^{-a|t|} & -t_0 \le t \le t_0 \\ 0 & \text{otherwise} \end{cases}$$

is a time-limited signal, since it lies in the interval $-t_0 \le t \le t_0$ and zero elsewhere. The energy in a time-limited signal may be made larger by increasing the value of $t_0$, and it can be made comparable to an infinite time signal by such means. For example, if we were to consider the equivalent infinite time signal,

$$g(t) = e^{-a|t|} \qquad -\infty \le t \le \infty$$

then its energy is given by

$$\mathbb{E} = \int_{-\infty}^{\infty} \left[ e^{-2at} u(t) + e^{2at} u(-t) \right] dt = \frac{1}{a}$$

The energy in the frequency domain, of the FT, of such a signal would also be

$$\frac{1}{2\pi} \int_{-\infty}^{\infty} G(\omega) G^*(\omega) d\omega = \frac{1}{a}$$

by Parseval's theorem (or Rayleigh's energy theorem). The energy of the finite time signal would be, on the other hand,

$$\mathbb{E}_{ft} = \int_{-t_0}^{t_0} \left[ e^{-2at} u(t) + e^{2at} u(-t) \right] dt = \frac{1}{a} - \frac{e^{2at_0}}{a}$$

If we let

$$t_0 = \frac{\ln(10)}{a} = 2.303$$

then

$$\mathbb{E}_{ft} = \frac{1 - e^{-\ln(100)}}{a}$$
$$= \frac{1 - (1/e^{\ln(100)})}{a}$$
$$= \frac{0.99}{a}$$
$$= 0.99\mathbb{E}$$

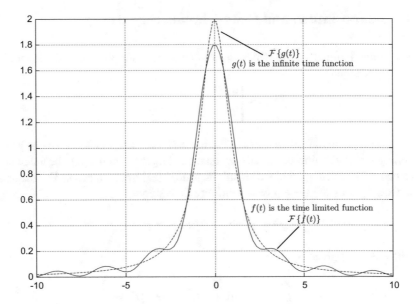

**Fig. 2.44**  FTs of $f(t)$ and $g(t)$

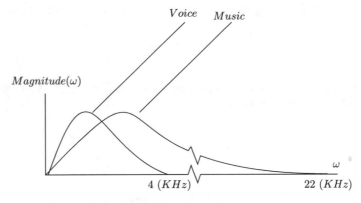

**Fig. 2.45**  Approximate FTs of two band-limited signals

that is, the energy in the time-limited signal is 0.99 times the energy in the infinite time signal, and therefore the energy in the FT will accordingly be only 0.99 times the energy of the FT of the infinite time signal. From this, we may conclude that the two FTs (the infinite time, and the finite time signals) would differ from each other only slightly. The two FTs are shown in Fig. 2.44.

Similar remarks apply to signals in the frequency domain. Theoretically, most signals have infinite bandwidth, but if we were to limit the bandwidth to a finite value, such that the bulk of the energy is retained, then in the time domain there will only be a slight difference between the two signals. The approximate Fourier transforms for two band-limited signals, namely voice and music, are shown in Fig. 2.45.

## 2.17 The Hilbert Transform[5]

The Hilbert transform of a function is by definition,

$$\mathcal{H}\{x(t)\} = x_h(t) = \frac{1}{\pi} \int_{-\infty}^{\infty} \frac{x(\tau)}{t - \tau} d\tau \tag{2.169}$$

which is the convolution of $x(t)$ with $1/\pi t$,

$$\mathcal{H}\{x(t)\} = x_h(t) = x(t) * \left[\frac{1}{\pi t}\right] \tag{2.170}$$

If we take the FT of this convolution,

$$X_h(\omega) = X(\omega) \times \mathcal{F}\left\{\frac{1}{\pi t}\right\} \tag{2.171}$$

then from Example 2.25,

$$\mathcal{F}\{\text{sgn}(t)\} = \frac{2}{j\omega} \tag{2.172}$$

and using duality from Eq. (2.123) and Example 2.27,

$$\frac{j}{\pi t} \Leftrightarrow \text{sgn}(\omega) \tag{2.173}$$

therefore

$$\mathcal{F}\left\{\frac{1}{\pi t}\right\} = -j\,\text{sgn}(\omega) \tag{2.174}$$

therefore

$$X_h(\omega) = -jX(\omega)\text{sgn}(\omega)$$
$$jX_h(\omega) = X(\omega)\text{sgn}(\omega) \tag{2.175}$$

From the definition of the Hilbert transform and its FT, we may infer the following properties.

1. The function and its Hilbert transform have the same magnitude spectrum. This property may be inferred from the following facts.

$$x(t) \Leftrightarrow X(\omega)$$
$$x_h(t) \Leftrightarrow -j\,\text{sgn}(\omega)X(\omega) \tag{2.176}$$

---

[5] This section may be read later when required.

**Fig. 2.46**  $-90°$ Phase
shifter

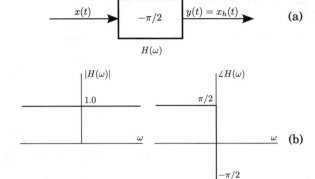

therefore,

$$|X_h(\omega)| = |-j\mathrm{sgn}(\omega)X(\omega)|$$
$$= |-j||\mathrm{sgn}(\omega)||X(\omega)|$$
$$= |X(\omega)|$$

2. For a real-time function, if we pass it through a $-\pi/2$ phase shifter, then the
   output of the filter is the Hilbert transform of the input. Referring to Fig. 2.46a,
   $x(t)$ is the input to the phase shifter whose output is $y(t)$. The phase shifter only
   has an effect on the phase, while the magnitude is not changed. We know that if
   we give all the positive frequencies a phase shift of $-\pi/2$, for a realisable filter,
   then the negative frequencies will suffer a phase shift of $+\pi/2$. Based on these
   ideas, the frequency response of the phase shifter is shown in the (b) part of the
   figure. If we examine the filter in more detail, then the filter response shows that
   this is the same as the operation of the Hilbert transform on the input function.
3. The Hilbert transform applied twice gives us the same function with a negative
   sign. The Hilbert transformer then can be considered to be a linear filter with a
   frequency response,

$$H_h(\omega) = \begin{cases} e^{-j\pi/2} & \omega > 0 \\ e^{j\pi/2} & \omega < 0 \end{cases} \tag{2.177}$$

therefore when we pass a function through two Hilbert transform filters, then in
the frequency domain,

$$\mathcal{H}\{\mathcal{H}[x(t)]\} \Leftrightarrow \left[-j\mathrm{sgn}(\omega)\right]\left[-j\mathrm{sgn}(\omega)X(\omega)\right]$$
$$= -X(\omega) \tag{2.178}$$

and $-X(\omega)$ has an IFT of $-x(t)$.

4. The function $x(t)$ and its Hilbert transform are orthogonal. To prove this property, consider Eq. (2.131),

$$\int_{-\infty}^{\infty} x(t)y^*(t)dt = \frac{1}{2\pi} \int_{-\infty}^{\infty} X(\omega)Y^*(\omega)d\omega$$

where $x(t)$ and $y(t)$ are any two functions. We replace $y(t)$ by $x_h(t)$, so

$$\int_{-\infty}^{\infty} x(t)x_h^*(t)dt = \frac{1}{2\pi} \int_{-\infty}^{\infty} X(\omega)X_h^*(\omega)d\omega$$

But $x_h(t)$ is a real function, therefore $x_h(t) = x_h^*(t)$, and hence

$$\begin{aligned}
\int_{-\infty}^{\infty} x(t)x_h(t)dt &= \frac{1}{2\pi} \int_{-\infty}^{\infty} X(\omega)X_h^*(\omega)d\omega \\
&= \frac{1}{2\pi} \int_{-\infty}^{\infty} X(\omega)\left[-j\,\mathrm{sgn}(\omega)X(\omega)\right]^* d\omega \\
&= \frac{1}{2\pi} \int_{-\infty}^{\infty} j\,\mathrm{sgn}(\omega)X(\omega)\left[X(\omega)\right]^* d\omega \\
&= \frac{1}{2\pi} \int_{-\infty}^{\infty} j\,\mathrm{sgn}(\omega)|X(\omega)|^2 d\omega \\
&= 0
\end{aligned}$$

In communication systems where Hilbert transforms are used, the definition of what is known as the pre-envelope for a function, for positive frequencies, is

$$x_1(t) = x(t) + jx_h(t) \tag{2.179}$$

If we take the FT of this complex function,

$$\begin{aligned}
\mathcal{F}\{x_+(t)\} &= \mathcal{F}\{x(t) + jx_h(t)\} \\
&= X(\omega) + j\left[-j\,\mathrm{sgn}(\omega)X(\omega)\right] \\
&= \left[1 + \mathrm{sgn}(\omega)\right] X(\omega) \\
&= 2u(\omega)X(\omega) \tag{2.180}
\end{aligned}$$

then the function $X_+(\omega)$ contains all the positive frequencies in $X(\omega)$ but multiplied by two. Similarly, the pre-envelope for negative frequencies is given by

$$x_-(t) = x(t) - jx_h(t) \tag{2.181}$$

and the function $X_-(\omega)$ contains all the negative frequencies in $X(\omega)$ but multiplied by two. These two functions are shown symbolically in Fig. 2.47.

**Fig. 2.47** Pre-envelope
definitions

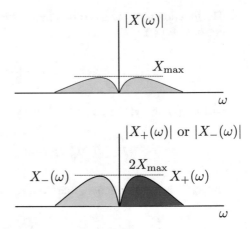

## 2.18   Important Concepts and Formulae

- Most analogue signals which occur in nature are highly oscillatory, and which are generally restricted in bandwidth. A good example is speech.
- Discrete or digital signals are derived from analogue signals by *sampling*. The equation of a typical sampled signal is given in Eq. (2.1):

$$f[n] = \sin (\omega_0 t)|_{t=nT}$$

- Signal transformation are of two types: transformations of the dependent variable, and transformations of the independent variable
- Transformations of the dependent variable are

  - Multiplication by a constant, $f(t) \rightarrow af(t)$.
  - Addition of a constant, $f(t) \rightarrow a + f(t)$.
  - Two signals may be multiplied together, $f_1(t)f_2(t)$.
  - A signal may be squared, $f(t)^2$.

- Transformations of the independent variable are

  - The transformation in $f(t)$ where $t$ is replaced by $at$ where $a$ is a real number.
  - The second case which is of importance is the case of $f(t)$ when $t$ is replaced by $-t$.
  - The third case of interest is when the variable $t$ is shifted by $\tau$. In this case, the two functions are $f(t)$ and $f(t - \tau)$.

- The energy of a signal is formally defined as

$$\mathbb{E}\{f(t)\} \triangleq \int_{-\infty}^{\infty} [f(t)]^2\, dt$$

where $\mathbb{E}$ is the energy of the signal $f(t)$.
- For a complex signal, the energy is

$$\mathbb{E}\{f(t)\} \triangleq \int_{-\infty}^{\infty} f(t) f(t)^*\, dt = \int_{-\infty}^{\infty} |f(t)|^2\, dt$$

- The power in a signal is

$$\mathbb{P}\{f(t)\} \triangleq \lim_{T \to \infty} \left\{ \frac{1}{T} \int_{-T/2}^{T/2} [f(t)]^2\, dt \right\}$$

- The impulse function, $\delta(t)$, is defined as

$$\int_{-\infty}^{\infty} \delta(t)\, dt = 1$$

where the function

$$\delta(t) = \begin{cases} \infty & \text{at } t = 0 \\ 0 & \text{elsewhere} \end{cases}$$

- The unit step function is defined by

$$u(t) = \begin{cases} 1 & t > 0 \\ 0 & t < 0 \\ 0.5 & t = 0 \end{cases}$$

- The signum function is

$$\text{sgn}(t) = \begin{cases} 1 & t > 0 \\ 0 & t = 0 \\ -1 & t < 0 \end{cases}$$

- The sinc function is

$$\text{sinc}(\omega) = \begin{cases} 1 & \omega = 0 \\ \sin(\omega)/\omega & \omega \neq 0 \end{cases}$$

- The rect function,

$$\text{rect}(t) = \begin{cases} 1 & |t| < 0.5 \\ 0.5 & |t| = 0.5 \\ 0 & |t| > 0.5 \end{cases}$$

- Convolution is defined by the integral

$$f(t) * g(t) = h(t) = \int_{-\infty}^{\infty} f(t - \tau)g(\tau)d\tau \quad \text{where} \ -\infty < t < \infty$$

- A trigonometric Fourier series is given by

$$x_T(t) = a_0 + \sum_{k=1}^{\infty} a_k \cos(k\omega_0 t) + \sum_{k=1}^{\infty} b_k \sin(k\omega_0 t)$$

where

$$a_0 = \frac{1}{T} \int_{t_0}^{t_0+T} x_T(t)dt$$

$$a_1 = \frac{2}{T} \int_{t_0}^{t_0+T} x_T(t) \cos(\omega_0 t)dt$$

$$\dots = \dots$$

$$a_m = \frac{2}{T} \int_{t_0}^{t_0+T} x_T(t) \cos(m\omega_0 t)dt$$

and

$$b_m = \frac{2}{T} \int_{t_0}^{t_0+T} x_T(t) \sin(m\omega_0 t)dt$$

- The Dirichlet conditions are *sufficient* conditions for a *real-valued, periodic function* $f_T(x)$ to be equal to the sum of its Fourier series at each point where $f$ is *continuous:*

  1. $f_T(x)$ must have a finite number of extrema in any given interval.
  2. $f_T(x)$ must have a finite number of discontinuities in any given interval.
  3. $f_T(x)$ must be absolutely integrable over a period.
  4. $f_T(x)$ must be bounded.

- Two real functions $g_1(t)$ and $g_2(t)$ are orthogonal on the interval $t_0 \le t \le t_0 + T$ if

$$\int_{t_0}^{t_0+T} g_1(t)g_2(t)dt = 0$$

- We may write an exponential Fourier series in the form

$$f_T(t) = \sum_{k=-\infty}^{k=\infty} F_k e^{jk\omega_0 t}$$

where

$$F_k = \frac{1}{T} \int_{t_0}^{t_0+T} f_T(t)e^{-jk\omega_0 t}\, dt$$

- Parseval's theorem for Fourier series. If

$$f_T(t) = a_0 + \sum_{k=1}^{\infty} a_k \cos(k\omega_0 t) + \sum_{k=1}^{\infty} b_k \sin(k\omega_0 t)$$

then the power of the periodic function is

$$\mathbb{P} = \int_T f_T^2(t)\, dt = a_0^2 + \frac{1}{2}\left[a_1^2 + a_2^2 + \cdots\right] + \frac{1}{2}\left[b_1^2 + b_2^2 + \cdots\right]$$

- Similarly, if

$$f_T(t) = \sum_{k=-\infty}^{\infty} F_k e^{jk\omega_0 t}$$

then

$$\mathbb{P} = \frac{1}{T} \int_{-T/2}^{T/2} f_T(t) f_T^*(t)\, dt - \sum_{k=-\infty}^{\infty} |F_k|^2$$

- The FT and IFT are related by the equations

$$F(\omega) = \int_{-\infty}^{\infty} f(t)e^{-j\omega t}\, dt$$
$$f(t) = \frac{1}{2\pi} \int_{-\infty}^{\infty} F(\omega)e^{j\omega t}\, d\omega$$

- The following FTs exist for special functions
  - $\delta(t)$,

$$\delta(t) \Leftrightarrow 1$$

  - $\delta(\omega)$

$$\frac{1}{2\pi} \Leftrightarrow \delta(\omega)$$

  - For a pulse

$$p_s(t) = \begin{cases} 1 & -\tau \le t \le \tau \\ 0 & \text{elsewhere} \end{cases}$$

$$p_s(t) \Leftrightarrow 2\tau \frac{\sin \omega\tau}{\omega\tau}$$

  - $f(t - t_0)$

$$f(t - t_0) \Leftrightarrow F(\omega)e^{-j\omega t_0}$$

– $F(\omega - \omega_0)$

$$\frac{1}{2\pi} f(t)e^{j\omega_0 t} \Leftrightarrow F(\omega - \omega_0)$$

– Cosine function

$$\cos \omega_0 t \Leftrightarrow \pi \left[\delta(\omega - \omega_0) + \delta(\omega + \omega_0)\right]$$

– sgn($t$) function

$$\text{sgn}(t) \Leftrightarrow \frac{2}{j\omega}$$

– $u(t)$

$$u(t) \Leftrightarrow \frac{1}{j\omega} + \pi\delta(\omega)$$

- Properties of the Fourier transform

  1. Linearity

  $$x(t) \Leftrightarrow X(\omega)$$
  $$y(t) \Leftrightarrow Y(\omega) \quad \text{then}$$
  $$Ax(t) + By(t) \Leftrightarrow AX(\omega) + BY(\omega)$$

  2. Time scaling

  $$x(t) \Leftrightarrow X(\omega)$$
  $$x(at) \Leftrightarrow \frac{1}{|a|} X\left(\frac{\omega}{a}\right)$$

  3. Frequency scaling

  $$\frac{1}{|a|} x\left(\frac{t}{a}\right) \Leftrightarrow X(a\omega)$$
  $$\text{if } x(t) \Leftrightarrow X(\omega)$$

  4. Time shift

  $$x(t) \Leftrightarrow X(\omega)$$
  $$x(t - t_0) \Leftrightarrow X(\omega)e^{-j\omega t_0}$$

5. Frequency shift

$$x(t) \Leftrightarrow X(\omega)$$

$$x(t)e^{j\omega_0 t} \Leftrightarrow X(\omega - \omega_0)$$

6. Duality

$$x(t) \Leftrightarrow X(\omega)$$

$$X(t) \Leftrightarrow 2\pi x(-\omega)$$

and

$$x(t) \Leftrightarrow X(\omega)$$

$$\frac{X(-t)}{2\pi} \Leftrightarrow x(\omega)$$

7. Time differentiation

$$x(t) \Leftrightarrow X(\omega)$$

$$\frac{d}{dt}x(t) \Leftrightarrow j\omega X(\omega)$$

8. Frequency differentiation

$$x(t) \Leftrightarrow X(\omega)$$

$$-jtx(t) \Leftrightarrow \frac{d}{d\omega}X(\omega)$$

9. Convolution theorem

$$x_1(t) \Leftrightarrow X_1(\omega)$$

$$x_2(t) \Leftrightarrow X_2(\omega)$$

$$x_1(t) * x_2(t) \Leftrightarrow X_1(\omega)X_2(\omega)$$

and

$$x_1(t) \Leftrightarrow X_1(\omega)$$

$$x_2(t) \Leftrightarrow X_2(\omega)$$

$$2\pi x_1(t)x_2(t) \Leftrightarrow X_1(\omega) * X_2(\omega)$$

• FT of a periodic function:

$$\mathcal{F}\{f(t)\} = \mathcal{F}\left\{\sum_{k=-\infty}^{k=\infty} F_k e^{jk\omega_0 t}\right\}$$

$$= \sum_{k=-\infty}^{k=\infty} F_k \mathcal{F}\left\{e^{jk\omega_0 t}\right\}$$

$$= 2\pi \sum_{k=-\infty}^{k=\infty} F_k \delta(\omega - k\omega_0)$$

where $F_k$ are the Fourier series coefficients of the periodic function.
- IFT of a periodic function in the frequency domain

$$\sum_{k=-\infty}^{k=\infty} F_k \delta(t - kT) \Leftrightarrow \sum_{k=-\infty}^{k=\infty} F_k e^{-jkT\omega}$$

where

$$F(\omega) = \sum_{k=-\infty}^{k=\infty} F_k e^{jkT\omega}$$

- Rayleigh's energy theorem

$$\int_{-\infty}^{\infty} x(t)x^*(t)dt = \frac{1}{2\pi} \int_{-\infty}^{\infty} X(\omega)X^*(\omega)d\omega$$

- Energy spectral density

$$\Psi(\omega) = \frac{X(\omega)X^*(\omega)}{2\pi} = \frac{|X(\omega)|^2}{2\pi}$$

- Transmission through linear systems

$$y(t) = \int_{-\infty}^{\infty} x(\tau)h(t - \tau)d\tau$$

and

$$Y(\omega) = H(\omega)X(\omega)$$

- Distortionless transmission. The frequency response is

$$H(\omega) = ke^{-j\omega t_0}$$

in the band of frequencies considered.
- Equalisation

$$H_{eq}(\omega) \propto \frac{k}{H(\omega)}$$

- Hilbert transform

$$\mathcal{H}\{x(t)\} = x_h(t) = \frac{1}{\pi} \int_{-\infty}^{\infty} \frac{x(\tau)}{t - \tau} d\tau$$

and

$$X_h(\omega) = -jX(\omega)\mathrm{sgn}(\omega)$$

- And

$$H_h(\omega) = \begin{cases} e^{-j\pi/2} & \omega > 0 \\ e^{j\pi/2} & \omega < 0 \end{cases}$$

- Pre-envelope for positive frequencies

$$x_+(t) = x(t) + jx_h(t)$$

and

$$\mathcal{F}\{x_+(t)\} = 2u(\omega)X(\omega)$$

## 2.19  Self-Assessment

### 2.19.1  Review Questions

1. Where would signal transformation be useful when we consider communication systems?
2. Is the energy of $\delta(t)$ finite or infinite?. Explain.
3. Explain why, for a time-limited signal, the energies in the signals $f(t)$, $f(-t)$ and $f(t - \tau)$ are identical.
4. In some cases, the energies of a signal are infinite but they are not periodic functions. Give examples.
5. Show that

$$u(t) = \int_{\infty}^{t} \delta(\tau)d\tau$$

with an explanation.
6. Why is

$$\int_{-\infty}^{\infty} \delta(\tau)x(t - \tau)d\tau = x(t)$$

Explain in your own words.

7. Except for sines, cosines and complex exponentials, give an example of two functions which are orthonormal.
8. Find the IFT of

$$F_1(\omega) = F_2(\omega) + c$$

where $c$ is a constant.
9. If the signal $x_1(t)$ occupies a bandwidth of $\omega_1$ and $x_2(t)$ occupies a bandwidth of $\omega_2$, then show that $x_1(t)x_2(t)$ occupies an approximate bandwidth of $\omega_1 + \omega_2$. (Note that $x_1(t)x_2(t)$ must not be equal to $x_1(t)$ or $x_2(t)$ as in the case of the rect function; that is, rect$(t)$rect$(t)$ =rect$(t)$.)

## 2.19.2  Numerical Problems

1. For $p_s(t)$ of Fig. 2.48, find $x(t) = 4 + 5p_s[7(t - 2\tau)]$.
2. Find the energy of the signal

$$Ae^{-a|t|}$$

where $a$, $A$ are both greater than zero and $-\infty < t < \infty$.
3. How are the energies in the signals $f(t)$ and $f(at)$ related for both of the cases, $a < 0$ and $a > 0$?.
4. How are the energies in the signals $f(t)$ and $f(-t)$ related?
5. Find the energy in the signal

$$f(t) = \sum_{i=1}^{N} f_i(t)$$

when

(a) all the $f_i(t)$ are all orthogonal to each other, and
(b) when they are not orthogonal to each other.
     Using this result, find the power in the signal

**Fig. 2.48**  Pulse for Example 9

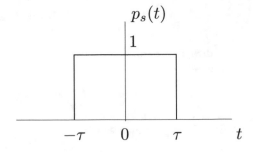

$p_s(t)$

$$10\cos(3t + \pi/3) + 5\sin(3t + \pi/3)$$

6. Show that if the energy of a signal is finite, then the power of that signal is zero.
7. Show that the integral of

(a)

$$\int_{-5}^{5} \left( \frac{\sin t \cos t}{t^2 + 5} \right) \delta(t) dt = 0$$

(b)

$$\int_{-1}^{5} \left( \frac{1}{jt} \right) \delta(t + 5) dt = 0$$

$$\int_{-10}^{5} \left( \frac{1}{jt} \right) \delta(t + 5) dt = -\frac{1}{j5}$$

8. Prove that

$$\int_{-\infty}^{\infty} \delta(at) dt = \frac{1}{|a|}$$

and therefore show that

$$\delta(at) = \frac{1}{|a|} \delta(t).$$

What is the relation between $\delta(\omega)$ and $\delta(f)$?
9. Find the convolution of the function $p_s(t)$ shown in Fig. 2.48 with itself.
10. Find the trigonometric Fourier series of

$$f_T(t) = 2 + 3\cos(t) - 4\sin(5t - \frac{\pi}{4})$$

11. Find the trigonometric Fourier series of the periodic signals given in Fig. 2.49.
12. Find the exponential Fourier series for the periodic signals given in Fig. 2.49.
13. Work out Eqs. (2.77),

$$f_T(t) = \sum_{k=-\infty}^{k=\infty} F_k e^{jk\omega_0 t}$$

and (2.78),

$$F_k = \frac{1}{T} \int_{t_0}^{t_0+T} f_T(t) e^{-jk\omega_0 t} dt$$

in detail.
14. If

$$x_T(t) = a_0 + \sum_{k=1}^{\infty} a_k \cos(k\omega_0 t) + \sum_{k=1}^{\infty} b_k \sin(k\omega_0 t)$$

**Fig. 2.49**  Various periodic
signals

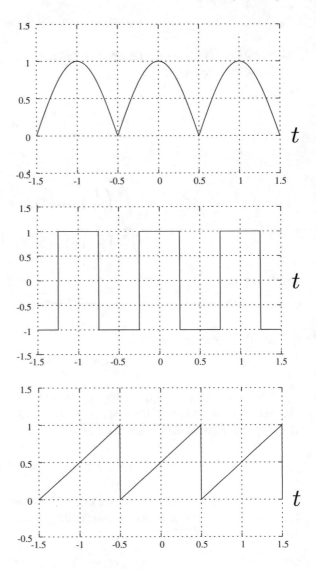

and the same signal has the expansion

$$x_T(t) = \sum_{k=-\infty}^{k=\infty} F_k e^{jk\omega_0 t}$$

then find the relation between the coefficients $F_k$, $a_k$ and $b_k$.

**Hint.** $\cos(k\omega_0 t) = (1/2)\left(e^{jk\omega_0 t} + e^{-jk\omega_0 t}\right)$ and $\sin(k\omega_0 t) = (1/2j)\left(e^{jk\omega_0 t} - e^{-jk\omega_0 t}\right)$

15. If a periodic function $x_T(t)$ is an even function, then all the sine term of the Fourier series vanish, and if it is an odd function then all the cosine terms and the constant term vanish. Prove this.
16. Prove that the Fourier coefficients of an exponential Fourier series for an even function are real and for an odd function are imaginary.
17. Prove that the energy per cycle of a periodic function is equal to the power of the signal.
18. If

$$x(t) \Leftrightarrow X(\omega)$$

then show that

$$\int_{-\infty}^{\infty} X(\omega)d\omega = 2\pi x(0)$$

$$\text{and } \int_{-\infty}^{\infty} x(t)dt = X(0)$$

**Hint:** Use $X(\omega) = \int_{-\infty}^{\infty} x(t)e^{-j\omega t}dt$, etc.

19. Prove all the properties given in Sect. 2.12.8.
20. Find the integration of the sinc(.) function using Fourier transforms

$$\int_{-\infty}^{\infty} \frac{\sin(\omega\tau)}{\omega\tau}d\omega$$

21. First plot and then find the Fourier transform of

$$x(t) = \begin{cases} 1 & \text{for } -\tau < t < 0 \\ -1 & \text{for } 0 < t < \tau \\ 0 & \text{elsewhere} \end{cases}$$

using the properties of the Fourier transform.
22. First plot and then find the Fourier transform of

$$x(t) = \begin{cases} 1 + t/\tau & \text{for } -\tau < t < 0 \\ 1 - t/\tau & \text{for } 0 < t < \tau \\ 0 & \text{elsewhere} \end{cases}$$

using the properties of the Fourier transform.
23. Find the energy of the signal

$$\text{sinc}(at)$$

where $a > 0$ and $-\infty < t < \infty$ by using properties of the Fourier transform.

24. Consider two pulses

$$p_s(t) = \begin{cases} 1 & \text{for } |t|, < \tau \\ 0 & \text{elsewhere} \end{cases}$$

and

$$p_s(t - 2\tau)$$

then based on these two pulses, show that

$$\int_{-\infty}^{\infty} \left[ \frac{\sin \omega\tau}{\omega\tau} \right]^2 e^{2j\omega\tau} d\omega = 0$$

25. Show that

$$x(t) * \delta(t - t_0) = x(t - t_0)$$

26. Show that the convolution of the sinc(x) with itself is a sinc function.

$$\text{sinc}(x) * \text{sinc}(x) = \text{sinc}(x)$$

**Hint:** Consider multiplication in the time domain.

27. Find the FT of the function

$$f(t) = A \sin(\omega_0 t)$$

28. If a function is zero for $t > 0$ and its FT is $F(j\omega)$, find the FT of $e^{at} f(t)$ where $a > 0$.

29. Find the FT of the function $f(t) = p_s(t) * p_s(t)$ where $p_s(t)$ is shown in Fig. 2.48.

30. If we consider the pulse train shown in Fig. 2.21, find the FT of this function.

31. Show that finite bandwidth pulse

$$F(j\omega) = \begin{cases} A & -2\pi B \leq \omega \leq 2\pi B \\ 0 & \text{elsewhere} \end{cases}$$

requires an infinite time signal as its IFT.

32. If a signal of finite energy is passed through a realisable filter, then show that the output energy is also finite.

33. If the *root mean square bandwidth* is given by

$$W_{rms} = \left( \frac{\int_{-\infty}^{\infty} f^2 |G(f)|^2 df}{\int_{-\infty}^{\infty} |G(f)|^2 df} \right)^{1/2} \tag{2.182}$$

find the rms bandwidth for

$$G(f) = \begin{cases} 1 & -B \leq f \leq B \\ 0 & \text{elsewhere} \end{cases}$$

34. Find rms (Eq. (2.182)) bandwidth for the signal

$$g(t) = \begin{cases} A & -\tau \leq t \leq \tau \\ 0 & \text{elsewhere} \end{cases}$$

35. If the *root mean square time duration* of a signal is given by

$$T_{rms} = \left( \frac{\int_{-\infty}^{\infty} t^2 |f(t)|^2 \, dt}{\int_{-\infty}^{\infty} |f(t)|^2 \, dt} \right)^{1/2}$$

   find the rms time duration for

$$f(t) = e^{-at} u(t), \ a > 0$$

36. Find the Hilbert transform of the functions (a) $\sin(\omega_0 t)$, (b) $\cos(\omega_0 t)$ and (c) $e^{j\omega_0 t}$.

37. Also find the pre-envelope for positive frequencies of the functions (a) $\sin(\omega_0 t)$, (b) $\cos(\omega_0 t)$ and (c) $e^{j\omega_0 t}$

38. From the convolution definition of Hilbert transforms,

$$\mathcal{H}[f_1(t) * f_2(t)] = f_{1h}(t) * f_2(t) = f_1(t) * f_{2h}(t)$$

# Chapter 3
# Probability, Random Variables and Random Processes

Concepts of probability are used to describe the various fluctuations in the communication process. When we send a message its content is known to us as something which we call 'deterministic', in the sense it *is* what we want to send or transmit, and deterministic phenomena are those which we can term as 'known beforehand'. The randomness in this deterministic signal is the fluctuation produced in it by the addition of noise to the signal as considered in Fig. 1.5. To consider the random fluctuations, we are compelled to use applications of the theory of probability.

The requirement of the theory of probability in communication is that noise plays an important role in the result of a communication process, and noise is a random signal. Thus, we find that a signal is corrupted by noise that noise power can be generally characterised and we can minimise the signal distortion by increasing the signal power. Since we are dealing with a phenomenon which is not directly under our control, the noise may sometimes overwhelm the signal power.

One of the most important points in considering phenomenon which are not under our control is to characterise them. In the case of a random phenomenon, if we perform an experiment whose outcomes are random in nature, the number of the result of the outcomes of the experiment may sometimes be finite, and sometimes infinite. According to the theory of probability the random outcome, called an *event*, of some *process* or *experiment* is not predictable. However what we can say about the experiment or process is what we call the *probability of occurrence* of the outcome. The word 'probability' refers in an approximate sense to the relative frequency of occurrence of the outcome with respect to all the outcomes. As an example if we toss a die, the number of outcomes is finite, (six outcomes) which is considered to be technically called *the experiment*. In this case, we say that the probability of occurrence of an event, in particular, the number '1' is 1/6. The following two points are to be always kept in mind

- What we mean by this is that if we toss a die a very large number of times then the number '1' will appear *approximately* (but not *exactly*) 1/6th of the total number of tosses. We have used the word approximately to indicate that the relative frequency is *not* an exact number, but the probability of the occurrence of an event is.

© The Author(s), under exclusive license to Springer Nature Singapore Pte Ltd. 2022     109
S. Bhooshan, *Fundamentals of Analogue and Digital Communication Systems*,
Lecture Notes in Electrical Engineering 785,
https://doi.org/10.1007/978-981-16-4277-7_3

- Also, and this is important, *if we toss a large number of dice only once, then the number '1' will appear approximately* (but not *exactly*) on 1/6th of the total number of dice.

These two methods of equating probability with our everyday experiences are very effective.

Writing this in mathematical language, if $N$ is the total number of tosses, (or the total number of dice being tossed) and $n_1$ is the number of occurrences of the number '1', then

$$n_1 \cong \underbrace{\frac{1}{6}}_{\text{probability}} \times N \quad \text{(for a large value of } N) \tag{3.1}$$

Similarly, the number '2' will occur $n_2$ times which also will occur approximately 1/6th of the total number of outcomes. And this applies to all the outcomes (the numbers '1' to '6'), so for $i = 1 \ldots 6$,

$$n_i \cong \frac{1}{6}N \quad \text{(for a large value of } N) \tag{3.2}$$

$$\text{with } \sum_{i=i}^{6} n_i = N \tag{3.3}$$

which makes us say that all these events are *equally probable*. Having expressed the concept of probability in layman terms, we now go over to the *mathematical* concepts involved always *keeping in mind the relative frequency of occurrence of an event at the back of our minds to interpret probabilities.*

## 3.1 Basic Concepts

We start the mathematical formulation on the basis of three definitions:

1. The *sample space*, $S$ is the set of all possible outcomes of an experiment

$$S = \{\text{all possible outcomes}\}$$

which for our die is the set $\{1, 2, 3, 4, 5, 6\}$.
2. The set, $S$, may have

   (a) A finite number of (elementary) outcomes.
   (b) or an infinite number of (elementary) outcomes.

3. The *sample points*, are all possible outcomes $x$, such that $x \in S$ are the *elementary outcomes*.

4. An event, $E$, is a subset of $S$: $E \subset S$. For example, $E = \{1, 2, 3\}$ for our die if we want to consider the outcomes as any of the numbers 1, 2 or 3.

For basic set theory definitions, see Appendix B.

**Example 3.1**  Define the sample space of the tossing of two dice.

If we label the dice '1' an '2', then, the first die can show the six numbers, $(1 \ldots 6)$ and for each of these six numbers the second die will show six numbers. Therefore there are thirty-six outcomes:

$$S = \{\text{all combinations of } (i, j), \ i = 1 \ldots 6; \ j = 1 \ldots 6\}$$
$$= \{(1, 1), (1, 2) \ldots (1, 6), (2, 1) \ldots (2, 6), \ldots (6, 1) \ldots (6, 6)\} \qquad (3.4)$$

**Example 3.2**  Define the event of getting a '1' on one die and '2' on the other when we toss two dice.

Here the event is the set
$$E = \{(1, 2), (2, 1)\}$$

**Example 3.3**  Define the event of getting at least a '1' on tossing two dice.

The event is

$$E = \left\{ \underbrace{(1, 1), (1, 2) \ldots (1, 6)}_{6 \text{ outcomes}}, \underbrace{(2, 1), (3, 1) \ldots (6, 1)}_{5 \text{ outcomes}} \right\}$$

## 3.2  Axioms of Probability

The following is an exposition about modern probability theory.[1] If $S$ be the sample space of a random experiment, and

$$E \subset S \qquad (3.5)$$

where $E$ is any event which satisfies Eq. (3.5). We then denote a number $P(E)$ in a uniform manner[2] called the probability of the event $E$ which satisfies

1. $1 \geq P(E) \geq 0$ and which is true for all events, $E \subset S$. For example for the toss of two dice, if the event $E$ is "getting a '1' on one die and '2' on the other", then from Example 3.2, $P(E) = 1/18$. Clearly, $0 \leq P(E) \leq 1$ and if we were to toss

---

[1] Modern probability theory is a mathematical theory and is not concerned with how the probability of an event is obtained. Classical probability theory, or the relative frequency approach tries to answer that question.

[2] This should be consistent with the relative frequency definition given earlier to reflect reality.

two dice 'a very large number of times' the relative frequency of this event would be about 1/18.

2. $P(S) = 1$. Here the example is the event "getting any number on die '1' and any number on die '2'" is the sample space itself: it will always happen, therefore probability of this event is $P(S) = 1$.

3. If two events are mutually exclusive,

$$E_1 \cap E_2 = \phi \qquad (3.6)$$

where $\phi$ is the empty set, (which in reality means that the probability of occurrence an event $E_1$ *and* $E_2$ is zero) then the probability of occurrence of $E_1$ *or* $E_2$ is given by

$$P(E_1 \cup E_2) = P(E_1) + P(E_2)$$

**Example 3.4** To illustrate this example we consider two events, when we toss a single die:

$$E_1 = \{1\}$$

and

$$E_2 = \{2\}$$

That is we are considering the event we get both '1' and '2'. Obviously this is impossible, so

$$E_1 \cap E_2 = \phi$$

Also

$$P(E_1 \cap E_2) = 0$$

Then the probability of occurrence of the event getting a '1' *or* '2' is

$$P(E_1 \cup E_2) = P(E_1) + P(E_2) = \frac{1}{6} + \frac{1}{6}$$

Considering the following situation as an application, if

$$E_i, \ i = 1, 2, ...n$$

is a set of mutually exclusive events

$$E_i \cap E_j = \phi, \ \forall \, i, \ j; \ i \neq j$$

and satisfying the two conditions,

**Fig. 3.1** Diagram showing mutually exclusive and exhaustive events

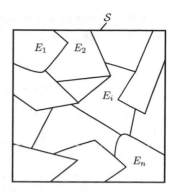

$$E_i \subset S, \; \forall \, i \qquad (3.7)$$

$$\text{and} \bigcup_{i=1}^{n} E_i = S \qquad (3.8)$$

then

$$P\left[\bigcup_{i=1}^{n} E_i\right] = \sum_{i=1}^{n} P(E_i) = 1 \qquad (3.9)$$

this situation is depicted in Fig. 3.1, through a Venn diagram.

For this case, using the example of the toss of two dice, we use two dice which are coloured as green and blue. The mutually exclusive events, $E_i$ are described as "the number $i \, (= 1, \ldots 6)$ on the green die and any number on the blue die:"

$$E_i = \{(i, 1), \ldots (i, 6)\} \; i = 1 \ldots 6$$

we can see that these events are mutually exclusive, and exhaustive[3] as given in (3.8). Then

$$P(E_i) = \frac{1}{6} \; \forall \, i = 1, \ldots 6$$

and

$$\sum_{i=1}^{6} P(E_i) = 1$$

### 3.2.1 Equally Probable Events and Union of Two Events

**Example 3.5** Definition of equally probable events.

---

[3] Exhaustive would mean that $\bigcup_{i=1}^{n} E_i = S$.

Taking a very simple example, if we toss a die we get any of the numbers, '1' to '6.' It is obvious from our earlier discussion that the following six events

$$E_1 = \{\text{getting a '1' on the toss of a die}\}$$
$$E_2 = \{\text{getting a '2' on the toss of a die}\}$$
$$\vdots = \vdots$$
$$E_6 = \{\text{getting a '6' on the toss of a die}\}$$

are all mutually exclusive and exhaustive. Also common sense tells us that each of these events is equally probable since when we toss the die, any number will turn up with the *same* frequency. Therefore if we call the probability of occurrence of one such event as $p$ then

$$P\left[\bigcup_{i=1}^{6} E_i\right] = \sum_{i=1}^{6} P(E_i) = 6p = 1$$

or

$$p = \frac{1}{6}$$

similarly for the toss of two dice, if we define the event

$$E_1 = \{\text{'1' on die one and '1' on die two}\}$$

then there are thirty-six such outcomes,

$$E_k = \{\text{'i' on die one and 'j' on die two}\} \qquad i, j = 1, \ldots 6$$

which are mutually exclusive, exhaustive and equally probable, therefore the probability of occurrence of only one such event is $1/36$.

**Example 3.6** Definition of events consisting of equally probable events.

Let us take the example of the probability of tossing two dice and obtaining '1' on one die, and '2' on the other, *or*, '1' on one die, and '3' on the other. Let us look at the problem carefully and we realise that we are to consider two events: event 1 *or* event 2. When we have an 'or' between the two events implies that we are essentially talking about the union of those events.

$$E_1 = \{(1, 2), (2, 1)\}$$
$$E_2 = \{(1, 3), (3, 1)\}$$

then the event of interest is

$$E = E_1 \cup E_2 = \{(1, 2), (2, 1), (1, 3), (3, 1)\}$$

and $E_1$ and $E_2$ are mutually exclusive and

$$P(E_1) = \frac{1}{18}$$
$$P(E_2) = \frac{1}{18}$$

therefore $P(E) = P(E_1) + P(E_2) = \frac{1}{9}$

## 3.2.2 Complement of an Event

(see Appendix B) If we consider the event $E$ and its complement, $E^c$

$$E \cup E^c = S$$
$$\text{and } E \cap E^c = \phi$$
$$\text{therefore } P(E \cup E^c) = P(S)$$

but from the earlier discussion $P(E) + P(E^c) = 1$      (mutually exclusive)

therefore the probability of occurrence of $E^c$ is given by

$$P(E^c) = 1 - P(E) \tag{3.10}$$

this equation tells the probability of occurrence of the event that event $E$ *does not occur* is $1 - P(E)$.

**Example 3.7** If we toss two dice what is the probability of not getting doubles?

The probability of occurrence of the event of getting doubles is $P(E) = P\{(1, 1), (2, 2), \ldots (6, 6)\} = 6/36$. Therefore the probability of occurrence of the event of not getting doubles is $P(E^c)$

$$P(E^c) = 1 - \frac{6}{36} = \frac{30}{36}$$

## 3.2.3 Probability of the Union of Events

If there are two events, $E_1$ and $E_2$ which are not mutually exclusive, then

$$P(E_1 \cup E_2) = P(E_1) + P(E_2) - P(E_1 \cap E_2) \tag{3.11}$$

**Fig. 3.2** Union of two
events, $E_1$ and $E_2$

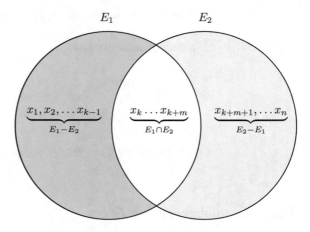

The reader should remember that $P(E_1 \cup E_2)$ is the same as saying "The probability of occurrence of events $E_1$ *or* $E_2$ taking place is...". What follows is not a definitive proof but rather an explanation. Let us suppose that there are elementary events $x_1, x_2, \ldots x_n$ which are mutually exclusive events and which comprise the event $E_1 \cup E_2$ (see Fig. 3.2). Then

$$E_1 \cup E_2 = \underbrace{x_1, x_2, \ldots x_{k-1}}_{E_1 - E_2}, \underbrace{x_k \ldots x_{k+m}}_{E_1 \cap E_2}, \underbrace{x_{k+m+1}, \ldots x_n}_{E_2 - E_1}$$

so

$$E_1 \cup E_2 = (E_1 - E_2) \cup (E_1 \cap E_2) \cup (E_2 - E_1) \tag{3.12}$$

these three sets are mutually exclusive. But

$$E_1 = (E_1 - E_2) \cup (E_1 \cap E_2)$$
$$P(E_1) = P(E_1 - E_2) + P(E_1 \cap E_2) \quad \text{(mutually exclusive events)}$$
$$\text{or } P(E_1 - E_2) = P(E_1) - P(E_1 \cap E_2) \tag{3.13}$$

similarly

$$P(E_2 - E_1) = P(E_2) - P(E_1 \cap E_2) \tag{3.14}$$

so from Eq. (3.12)

$$P(E_1 \cup E_2) = P(E_1 - E_2) + P(E_1 \cap E_2) + P(E_2 - E_1)$$
$$= P(E_1) - P(E_1 \cap E_2) + P(E_1 \cap E_2) + P(E_2) - P(E_1 \cap E_2) \text{ (using earlier Eqns)}$$
$$= P(E_1) + P(E_2) - P(E_1 \cap E_2) \tag{3.15}$$

## 3.3 Conditional Probability and Independent Events

Suppose we have a sample space, $S$, which contains two events, $A$ and $B$. The probability of occurrence of the event $A$ is $P(A)$, and that of $B$ is $P(B)$. We want to find the probability of occurrence of the joint event $A \cap B$, which is the same as saying, "the probability of occurrence of events $A$ *and* $B$ taking place ..." Let us look at the following experiment: two dice are tossed, a green one and a blue one. Two people are involved, you and another person. The other person can see result of the result of the fall of the dice, but you cannot. The event $A$ is: a '1' appears on the green die, while the event $B$ is a '2' appears on the blue die. The two sample spaces are (*(green outcome, blue outcome)*)

$$A = \{(1, 1) \ldots (1, 6)\} \quad \text{(6 outcomes)}$$
$$B = \{(1, 2), (2, 2), \ldots (6, 2)\} \quad \text{(6 outcomes)}$$

and

$$A \cap B = \{(1, 2)\}$$
$$P(A \cap B) = \frac{1}{36}$$

where the event $A \cap B$ is '1' appears on the green die and '2' appears on the blue die. Now the two dice are tossed. There will be many outcomes when neither a '1' or '2' appear. However, when a '2' appears on the blue die the other person announces it: "a '2' has appeared on the blue die". Now the probability of occurrence of '1' appearing on the green die has increased to 1/6: the probability of occurrence of the conditional event $P(A|B) = 1/6$, (read as "probability of occurrence of the event $A$ given $B$ has occurred) since the sample space

$$A|B = \{(1, 2), (2, 2) \ldots (6, 2)\} \quad \text{(6 outcomes)} \tag{3.16}$$

Or

$$P(A \cap B) = P(A|B)P(B) \tag{3.17}$$

To check this equation, substitute $B$ for $A$, which gives

$$P(B) = P(B|B)P(B) = P(B)$$

or $A$ for $B$

$$P(A) = P(A|A)P(A) = P(A)$$

If we study Eq. (3.17) in a little more detail for the two die problem we know that the outcome of event $A$ does not depend on the event $B$, so

**Fig. 3.3** Binary symmetric
channelBinary symmetric
channel

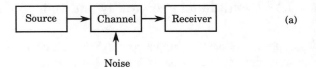

(a)

$X_0, 0$ transmitted     $1 - p$     $Y_0, 0$ received
$p_0$                                                                   (b)

$X_1, 1$ transmitted     $1 - p$     $Y_1, 1$ received
$p_1$

$$P(A|B) = P(A) \quad \text{(for independent events)} \tag{3.18}$$

and Eq. (3.17) becomes

$$P(A \cap B) = P(A)P(B) \tag{3.19}$$

So whenever we want to find the joint probability of occurrence of two events we write Eq. (3.17), then decide whether the events are independent or not, and if they are, use Eq. (3.18).

We may also write Eq. (3.17) as

$$P(A|B) = \frac{P(A \cap B)}{P(B)}$$
$$= \frac{P(B|A)P(A)}{P(B)} \tag{3.20}$$

which is called Baye's theorem or rule.

**Example 3.8** Binary symmetric channel

Referring to Fig. 3.3, when we transmit symbols 0s and 1s through a noisy channel, and noise is added to each symbol as shown in Fig. 3.3a. If 1 is received given that a 0 is transmitted, the probability of this happening is $p$, while if a 0 was received given that 1 was transmitted, then also the crossover probability is also $p$. The channel which behaves in this way is referred to a binary symmetric channel as shown in Fig. 3.3b. Or

$$P_{y|x}(1|0) = P_{y|x}(0|1) = p \tag{3.21}$$

The probability that the source transmits a 0, event $x_0$, is $p_0$ while that the source transmits a 1, event $x_1$, is $p_1$.

$$P_x(0) = p_0$$
$$P_y(1) = p_1 \qquad (3.22)$$

Other possibilities are: a 0 is received, event $y_0$, or a 1 is received, event $y_1$. Similarly the conditional probability that a 1 was received given that 1 was transmitted, or that a 0 was received given that a 0 was transmitted is given by

$$P_{y|x}(0|1) = P_{y|x}(1|1) = 1 - p$$

## 3.4 Random Variables

When we toss a die the outcome is one of six numbers, '1', ..., '6'. We now define these numbers as a variable $X$ (by convention, an uppercase letter), in the sense that the variable $X$ *will take on* one of these six numbers. However we cannot definitely say which of these numbers will appear. Therefore $X$ is given the name 'random variable'.

The notation which will be adopted is that the random $X$ takes on the value of $x \in \{1, 2 \ldots 6\}$, which is the sample space $S$. In this case we are dealing with a *discrete* random variable. A second example is the length of screws produced in a factory where the nominal length of the screw may be, for example, 1 cm, and the random variable $L$ (the length of the screw) will take on values of $l \in S = \{\text{interval}, (0, L_{max})\}$ where $L_{max}$ is a large number which is envisaged to be the maximum value which $L$ may take, though $L$ will almost always take on values around 1 cm. In this case the random variable $L$ will take on continuous values. These two sample spaces are shown in Fig. 3.4.

Outcomes of a die

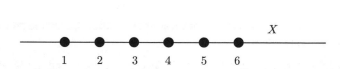

Outcomes of the length of a factory produced screw

**Fig. 3.4** Sample spaces of the examples

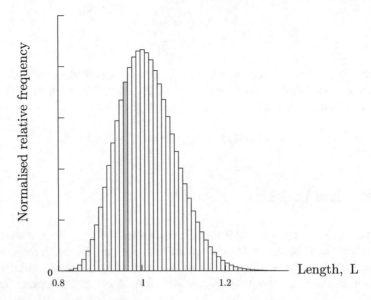

**Fig. 3.5** Relative frequency approach to continuous random variables

Each random variable is therefore defined on its own sample space, and we can ask the question that what is the probability of occurrence of the random variable $2 \leq X \leq 4$ for the toss of a die? The answer to this question is

$$P(2 \leq X \leq 4) = \sum_{i=2}^{i=4} P(X = i)$$

$$= \frac{3}{6} \tag{3.23}$$

However for continuous random variables we can approach the problem from the relative frequency approach.

1. Supposing we produced $N = 10,000$ screws a day (note, by hand and not by machine in order to obtain a wider variation)
2. Divide the interval $[0, L_{\max}]$ into intervals of 0.05 $(= \Delta l)$ cm intervals,
3. Plot the outcome of the measured screws.
4. We would then have a plot of the number of screws in that interval. For example in the range 0 to 0.05 we had no screws $(= n_1)$, in the interval 0.05 to 0.1 we had again no screws, $(= n_2)$ and so on. We could then plot $n_i/N$, the number of screws in the interval $[0.05(i-1), 0.05i]$ and we would get a plot similar to the plot shown in Fig. 3.5.
5. Each pulse of the graph would be of width 0.05 and value $n_i/N$, with

$$\sum_{i=1} n_i = N \tag{3.24}$$

$$\text{or} \sum_{i=1} \frac{n_i}{N} = 1 \tag{3.25}$$

so that $n_i/N$ could be interpreted as the approximate probability of having the screw in interval $i$.

6. Now in the next step we allow each $n_i/N$ to divided by the interval duration, 0.05, $y(n_i) = n_i/(0.05N)$, (which is $y(n_i) = n_i/(\Delta l N)$) and then make a new plot which we call the 'Normalised relative frequency'.
7. In this plot *the area* (one pulse shown shaded) of each pulse represents the approximate probability of the screw being in the $i$th interval.
8. We now define a smooth function, $f_L(l)$ which passes through the centre at the maximum of all the pulses and which may be used (in this case, approximately) to calculate the probability, *by area*

$$P(l_1 \le l \le l_2) = \int_{l_1}^{l_2} f_L(l)dl \tag{3.26}$$

where $P(l_1 \le l \le l_2)$ is the probability that the screw lies in the interval $l_1 \le l \le l_2$.

The (exact *not* the approximate) function $f_L(l)$ is called the *probability density function* (PDF) the random variable, $L$.

### 3.4.1 Properties of the Probability Density Function

From our discussion that the probability density function of a continuous random variable has some important properties:

1. The function, $f_X(x)$, (the PDF) takes real values and is defined on the real line, $x \in \mathbb{R}$, generally $-\infty < X < \infty$.
2. On some parts of the line the function $f_X(x)$ may be zero. For example in the screw example $f_L(l) = 0$ for $-\infty < L < 0$.
3. $f_X(x) \ge 0$ everywhere.
4. The integral

$$P(-\infty < X < \infty) = \int_{-\infty}^{\infty} f_X(\sigma)d\sigma = 1 \tag{3.27}$$

since the probability that $x$ lies in the range, $-\infty$ to $\infty$ is unity.

## 3.5  Cumulative Distribution Function

The cumulative distribution function (CDF) is defined as

$$F_X(x) = \int_{-\infty}^{x} f_X(\sigma)d\sigma \tag{3.28}$$

it is the probability

$$P(-\infty < X < x) = F_X(x) \tag{3.29}$$

therefore the probability that $x_1 \leq X \leq x_2$ is

$$F_X(x_2) - F_X(x_1) \tag{3.30}$$

the CDF has the following two important properties

1.  $F_X(x)$ is bounded by zero and one.

$$0 \leq F_X(x) \leq 1 \tag{3.31}$$

   which is proved using Eq. (3.28).
2.  The CDF is monotonically increasing

$$F_X(x_2) > F_X(x_1) \quad \text{when } x_2 > x_1 \tag{3.32}$$

**Example 3.9**  For a (hypothetical) triangular PDF, the CDF is shown in Fig. 3.6. The function is defined as

$$f_X(x) = \begin{cases} 0 & -\infty < X < -1 \\ x+1 & -1 < X < 0 \\ 1-x & 0 < X < 1 \\ 0 & 1 < X < \infty \end{cases}$$

Let us see whether all the properties given above are satisfied.

**PDF properties**

1.  The PDF $f_X(x)$ takes on real values in the interval $-\infty < X < \infty$.
2.  The function $f_X(x)$ is zero in the intervals $-\infty < X < -1$ and $1 < X < \infty$.
3.  The function $f_X(x) \geq 0$ everywhere.
4.  The integral

$$\int_{-\infty}^{\infty} f_X(x)dx = 1$$

   since the base length of the triangle is 2 and the height is 1.

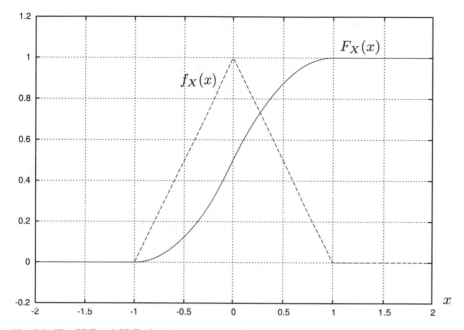

**Fig. 3.6** The PDF and CDF of a triangular probability density function

**CDF properties**
The function

$$F_X(x) = \int_{-\infty}^{x} f_X(x)dx$$

1. $0 \le F_X(x) \le 1$
2. $F_X(x)$ is monotonically increasing

## 3.6 Examples of Random Variables

### 3.6.1 Discrete Random Variables

Discrete random variable occurs when the random variable only takes on discrete values, as the following example suggests.

**Example 3.10** Find the probability density function and cumulative distribution function of the random variable corresponding to the toss of a dice.

The toss of a dice is a discrete random variable. From the discussion given, the PDF generally applies to a continuous random variable. However we may use a discrete approach to the problem. The discrete approach implies that we use summation rather than integration, and the notation is the $f_X[x]$ where the square brackets implies a discrete random variable. Then

$$f_X[x] = \begin{cases} \frac{1}{6} & \text{for } 1 \leq x \leq 6, \ x = 1, \ldots 6 \\ 0 & 0 \text{ elsewhere} \end{cases} \tag{3.33}$$

using this notation

$$F_X[x] = \sum_{n=1}^{x} f_X[x] = \frac{x}{6} \text{ where } x = 1, \ldots 6 \tag{3.34}$$

The problem which we can envisage is that in the most general case some particular random variable may take continuous values in some intervals and discrete values at some points. So to make Eqs. (3.33) and (3.34), 'continuous' we use the property of a unit impulse:

$$\int_{-\infty}^{\infty} \delta(\sigma)d\sigma = 1 \tag{3.35}$$

which gives us

$$f_X(x) = \begin{cases} \sum_{x_0=1}^{6} \frac{1}{6}\delta(x - x_0) & \text{for } 1 \leq x_0 \leq 6, \ x_0 = 1, \ldots 6 \\ 0 & 0 \text{ elsewhere} \end{cases} \tag{3.36}$$

and

$$F_X(x) = \int_{\sigma=-\infty}^{x} f_X(\sigma)d\sigma \tag{3.37}$$

the PDF and CDF of this random variable are shown in Fig. 3.7.

### 3.6.2  Uniformly Distributed Random Variable

Uniformly distributed random variables occur in those cases where we expect the outcome to occur with equal probability in any of a band of outcomes. This distribution is the continuous analogue of the toss of a die. Thus, for example, when we toss a die the outcomes are all equally probable, though the outcomes are discrete. In the same manner we may visualise a situation where the outcomes of a random variable are continuous but 'equally probable'

**Example 3.11** Consider the case of a random variable whose PDF is constant over a region $x_0 \leq X \leq x_1$ as shown in Fig. 3.8. We have to find the value of $A$.

To find the value of $A$,

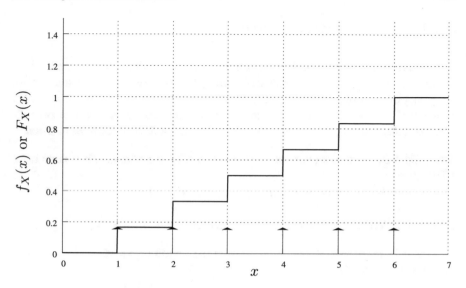

**Fig. 3.7** PDF and CDF of the toss of a die

**Fig. 3.8** Evenly distributed random variable

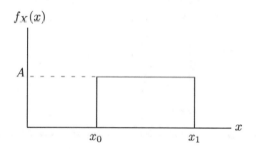

$$P(-\infty < X < \infty) = \int_{-\infty}^{\infty} A\,dx = 1 \tag{3.38}$$

since the probability that $x$ lies somewhere in the infinite interval is 1. Or

$$A \int_{x_0}^{x_1} dx = 1$$

$$\text{or } A(x_1 - x_0) = 1$$

$$\text{therefore } A = \frac{1}{x_1 - x_0} \tag{3.39}$$

### 3.6.3   *Gaussian Random Variable*

The Gaussian random variable occurs most often in practice. For example, if we conduct an exam where the number of participants is very large, then the relative frequency of the results will be close to a Gaussian distribution. The case of a number of identical random variables *added together has a distribution which is close to the Gaussian distribution*, a result which is formally called the *central limit theorem.*

**Example 3.12**   A random variable which is Gaussian (or normal) distributed has a PDF given by

$$f_X(x) = Ae^{-Bx^2} \qquad -\infty < X < \infty$$

find the relation between $A$ and $B$. $A$, $B$ are both greater than zero (see Fig. 3.9).

The probability that $X$ lies somewhere in the region $(-\infty, \infty)$ $(P(-\infty < x < \infty))$ is one. Therefore

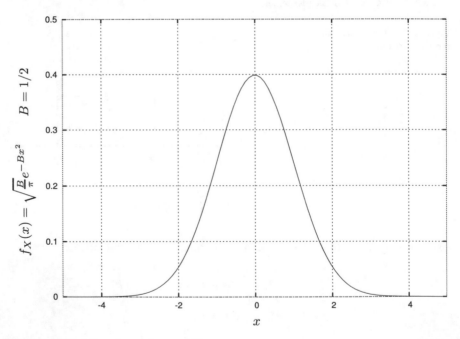

**Fig. 3.9**   PDF of Gaussian random variable

$$\int_{-\infty}^{\infty} Ae^{-Bx^2} dx = 1$$

$$\text{or } \frac{\sqrt{\pi} A}{\sqrt{B}} = 1$$

$$A = \sqrt{\frac{B}{\pi}} \tag{3.40}$$

To integrate

$$I = \int_{-\infty}^{\infty} e^{-Bx^2} dx$$

we consider the double integral

$$I^2 = \int_{-\infty}^{\infty} \int_{-\infty}^{\infty} e^{-B(x^2+y^2)} dx dy$$

and going over to cylindrical coordinates,

$$I^2 = \int_0^{2\pi} \int_0^{\infty} e^{-B\rho^2} \rho d\rho d\phi$$

$$= 2\pi \int_0^{\infty} e^{-B\rho^2} \rho d\rho$$

set $\rho^2 = x, 2\rho d\rho = dx$

$$I^2 = \pi \int_0^{\infty} e^{-Bx} dx$$

$$= \frac{\pi}{B}$$

$$I = \sqrt{\frac{\pi}{B}}$$

therefore using Eq. (3.40),

$$f_X(x) = \sqrt{\frac{B}{\pi}} e^{-Bx^2} \tag{3.41}$$

### 3.6.4 Poisson Process—Arrival Time

Take a specific process which consists of the time of occurrence a single event, since its last occurrence, such as the emission of an electron from a source emitting these particles at random times. Generally the particle will be emitted randomly, and the Poisson arrival time can be considered to be a random variable. This Poisson process describes the birth–death process, and the arrival of packets at a server in digital communication.

**Example 3.13** A random variable, $T$, is the time taken for a particular event to take place (e.g. the arrival of a particle from a radioactive substance.) The PDF is

$$f_T(t) = Ae^{-t/\tau} \quad 0 \le T < \infty$$

find the relation between $A$ and $\tau$, and find the probability that the event takes place in time $0 \le T \le 3\tau$

As considered earlier, the probability that $T$ lies in the interval $[0, \infty)$ is one. Therefore

$$\int_0^\infty Ae^{-t/\tau} dt = 1$$

$$\text{or } A\tau = 1$$

$$A = \frac{1}{\tau}$$

therefore

$$f_T(t) = \frac{1}{\tau} e^{-t/\tau} \quad 0 \le T < \infty \tag{3.42}$$

the probability that the event happens in time $[0, 3\tau]$ is

$$\frac{1}{\tau} \int_0^{3\tau} e^{-t/\tau} dt = 1 - e^{-3}$$

$$= 0.9502$$

so the probability that the event will happen in time $3\tau$ is 0.9502. The meaning of this result is that we can say with great confidence that around 95% of the time a particle will arrive in the interval $[0, 3\tau]$ since the arrival of the last particle.

**Example 3.14** For the PDF of Eq. (3.42), find the probability of arrival in the time $[\tau, 2\tau]$.

The required probability is the area of the shaded region shown in Fig. 3.10. The required probability is 0.2325.

The arrival of exactly $k$ electrons in the time interval $[0, t]$ is given by

$$P(X = k) = \frac{1}{k!} \left(\frac{t}{\tau}\right)^k e^{-t/\tau} \quad k = 0, 1, \dots \tag{3.43}$$

A plot of this probability is shown in Fig. 3.11. The figure shows the probability of arrival of exactly 5 electrons in time $t$. Since $\lambda = 1/\tau$ is the arrival rate in arrivals

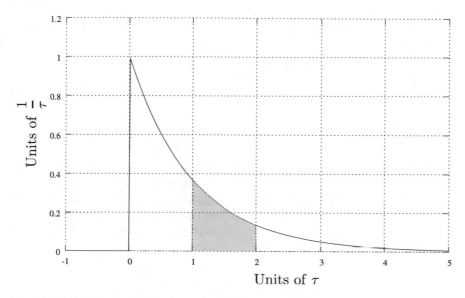

**Fig. 3.10** Probability density function and probability of arrival of $T \in [\tau, 2\tau]$

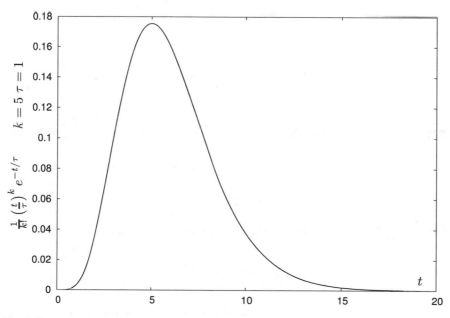

**Fig. 3.11** PDF of the Poisson process for $k = 5$ and $\tau = 1$

per second, the highest probability is at $t = 5$ and after that, the probability of arrival of more than 5 electrons increases. For $t < 5$ the probability of arrival of exactly 5 electrons is small.

The PDF for the occurrence time of the random variable $T$ for exactly $k$ events can be shown to be

$$f_T(k, t) = \frac{t^{k-1}}{(k-1)!\tau^k} e^{-(t/\tau)} \tag{3.44}$$

## 3.7   Statistical Averages

We know from simple reasoning that in a board examination of national or international importance, where $N$ ($N$ very large) students appear, then we may regard the number of students who obtain $i$ marks as a random variable. If the marks obtained are rounded off to the nearest whole number, and if there are $n_0$ students with 0 marks, $n_1$ students with 1 mark, ..., $n_i$ students with $i$ marks, and so on, then the average in the board examination is

$$\text{Average marks} = \frac{\sum_{i=0}^{i=100} i n_i}{\sum_{i=0}^{i=100} n_i} = \frac{\sum_{i=0}^{i=100} i n_i}{N}$$

If we regard $n_i/N$ as the relative frequency of occurrence, (the probability) of obtaining $i$ marks, then

$$\text{Average marks} = \sum_{i=0}^{i=100} \underbrace{i}_{"rv"} \underbrace{\frac{n_i}{N}}_{"Probability"}$$

from this result, we can proceed to calculate the average value of a random variable, $x$.

The probability of occurrence of an event, $X$, in the interval $dx$, $(x \leq X \leq x + dx)$ is

$$f_X(x)dx$$

then (using the earlier reasoning) the first step in finding the mean is

$$\underbrace{x}_{"Marks"} \underbrace{f_X(x)dx}_{"Probability"}$$

therefore we 'sum' (integrate) this expression to give the statistical average of the event

$$\text{Mean } x = E[X] = m_X = \int_{-\infty}^{\infty} x f_X(x)dx \tag{3.45}$$

The average of a random variable gives us an important indicator about the random variable. For any distribution, the average is an indicator about the value of the outcomes. These are

1. If the PDF is symmetrical about the mean then *approximately* half the outcomes will be greater than the average, and about half less than the average.
2. For PDFs which have a single, sharp, hump, then the average will generally be close to that hump, and we can say that the '*approximate*' value of most outcomes of the random variable will close to the average value of the random variable.
3. We can compare the actual output with the average and realise the difference. For an example of the board examination, has a student done much better than the average or much worse? In this case it is an indicator of the performance of the student.

These conclusions are extremely general in nature and help us have a better understanding about an outcome. Let us concentrate on various examples.

**Example 3.15** Find the average of the random variable which is uniformly distributed.

The PDF for a uniformly distributed random variable is

$$f_X(x) = \frac{1}{x_1 - x_0}$$

(see Sect. 3.6.2) so the average is

$$m_X = \int_{x_0}^{x_1} \frac{x}{x_1 - x_0} dx = \frac{x_1 + x_0}{2}$$

we can say from this result that the outcomes of the random variable $x$ will fall into two categories: half the times be in the range $x_0 \leq X \leq m_X$ and half the other times it will be in the range $m_X \leq X \leq x_1$.

**Example 3.16** Find the average of the PDF which has a Gaussian distribution (see Sect. 3.6.3).

In this case $f_X(x)$ is given by

$$f_X(x) = \sqrt{\frac{B}{\pi}} e^{-Bx^2}$$

so the average is

$$E[X] = m_X = \sqrt{\frac{B}{\pi}} \int_{-\infty}^{\infty} x e^{-Bx^2} dx = 0$$

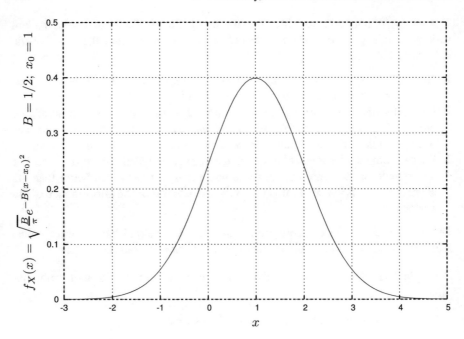

**Fig. 3.12** Gaussian distribution with nonzero mean

**Example 3.17** Find the average of the PDF which has a shifted Gaussian distribution (see Fig. 3.12).

The PDF in this case is

$$f_X(x) = \sqrt{\frac{B}{\pi}} e^{-B(x-x_0)^2}$$

and the average is then

$$m_X = x_0$$

this left as an exercise for the student to do.

**Example 3.18** Find the average time of arrival of a particle which follows the Poisson process, whose PDF is

$$f_T(t) = \frac{1}{\tau} e^{-t/\tau} \qquad 0 \leq T < \infty$$

To find the average arrival time, $m_T$, we have to calculate

$$m_T = \int_0^\infty \frac{t}{\tau} e^{-t/\tau} dt = \tau$$

which means that 'on the average' the particle will arrive in time $\tau$. This does not imply that particles will arrive at time $\tau$, but that we can 'expect' them to arrive around $t = \tau$. It is because of this reason that the mean is also called the 'expectation'.

### 3.7.1 Moments

Another set of averages which are often used are the $n$th moments and $n$th central moments of a random variable. In general, the $n$th moment of a random variable, $X$ is given by

$$E[X^n] = \int_{-\infty}^{\infty} x^n f_X(x)dx \tag{3.46}$$

where $E[\cdots]$ is the mean or expectation. The $n$th central moment of a random variable is

$$E[(X - m_X)^n] = \int_{-\infty}^{\infty} (x - m_X)^n f_X(x)dx \tag{3.47}$$

the most used moments are the ones with $n = 2$. For example in Eq. (3.46), for $n = 2$,

$$E[X^2] = \int_{-\infty}^{\infty} x^2 f_X(x)dx \tag{3.48}$$

gives us the mean of the square of the random variable, which may be connected to the 'power' of the random variable. In Eq. (3.47), when $n = 2$,

$$\sigma_X^2 = E[(X - m_X)^2] \tag{3.49}$$

is called the *variance* of the random variable. Looking at the variance in more detail,

$$\begin{aligned}
\sigma_X^2 &= E[(X - m_X)^2] \\
&= \int_{-\infty}^{\infty} (x - m_X)^2 f_X(x)dx \\
&= \int_{-\infty}^{\infty} [x^2 - 2xm_X + m_X^2] f_X(x)dx \\
&= E[x^2] - m_X^2 \tag{3.50}
\end{aligned}$$

$\sigma_X$ is also called the *standard deviation* of the random variable. In this way we can apply the average to any function $g(X)$,

$$E[g(X)] = \int_{-\infty}^{\infty} g(x) f_X(x)dx \tag{3.51}$$

**Example 3.19** Find the variance of the random variable which is uniformly distributed, as given in Example 3.15.

To find the variance, we first find the second moment

$$E[x^2] = \frac{1}{x_1 - x_0} \int_{x_0}^{x_1} x^2 dx = \frac{x_1^2 + x_0 x_1 + x_0^2}{3} \tag{3.52}$$

and

$$
\begin{aligned}
\sigma_X^2 &= E[x^2] - m_X^2 \\
&= \frac{x_1^2 + x_0 x_1 + x_0^2}{3} - \left(\frac{x_0 + x_1}{2}\right)^2 \\
&= \frac{(x_1 - x_0)^2}{12}
\end{aligned}
$$

in many cases, $x_0 = -x_1 = l$, (that is the random variable lies in the interval $[-l, l]$, where $l > 0$) then

$$\sigma_X^2 = \frac{l^2}{3}$$

**Example 3.20** Find the variance of the PDF which has a Gaussian distribution (see Sect. 3.6.3).

In this case $f_X(x)$ is given by

$$f_X(x) = \sqrt{\frac{B}{\pi}} e^{-Bx^2} \tag{3.53}$$

since the mean, $m_X = 0$,

$$
\begin{aligned}
E[(X - m_X)^2] &= E[X^2] \\
&= \sqrt{\frac{B}{\pi}} \int_{-\infty}^{\infty} x^2 e^{-Bx^2} dx \\
&= \sqrt{\frac{B}{\pi}} \times \frac{\sqrt{\pi}}{2 B^{\frac{3}{2}}} \\
\sigma_X^2 &= \frac{1}{2B}
\end{aligned}
$$

so

$$B = \frac{1}{2\sigma_X^2}$$

so the Gaussian random variable with zero mean has the PDF

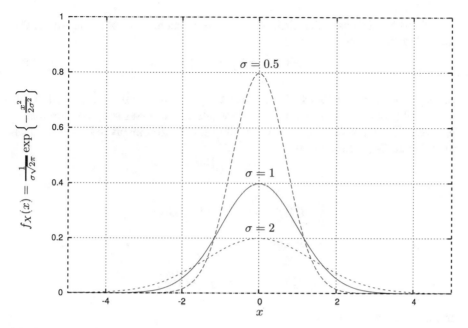

**Fig. 3.13** Effect of $\sigma$ on the Gaussian distribution

$$f_X(x) = \frac{1}{\sigma_X \sqrt{2\pi}} \exp\left\{-\frac{x^2}{2\sigma_X^2}\right\} \tag{3.54}$$

where we have used Eq. (3.53). Similarly, the Gaussian random variable with mean $m_X$ is

$$f_X(x) = \frac{1}{\sigma_X \sqrt{2\pi}} \exp\left\{-\frac{(x - m_X)^2}{2\sigma_X^2}\right\} \tag{3.55}$$

In the future, references to the mean and standard deviation may be written as just $m$ and $\sigma$ where the meaning is clear.

**Example 3.21** Show how the graph of the normal distribution is affected by the parameter $\sigma$.

Figure 3.13 shows the effect of $\sigma$ on the Gaussian distribution. We can see from the graph that three distributions have been drawn. The distribution $\sigma = 1$ is called the normalised distribution, while $\sigma = 0.5$ and $\sigma = 2$ are also drawn. As $\sigma$ increases, the distribution has more spread. Therefore we can say about the standard deviation, that as $\sigma$ increases, the distribution is "more spread out". This is also seen in Example 3.19, where

$$\sigma = \frac{(x_1 - x_0)}{\sqrt{12}}$$

and the greater the value of the spread, $x_1 - x_0$, the greater the value of $\sigma$.

This aspect reflects in the Chebyshev inequality but in terms of probabilities. The inequality states that

$$P[|X - m| \geq \epsilon] \leq \left(\frac{\sigma}{\epsilon}\right)^2 \qquad (3.56)$$

for any continuous random variable. The Chebyshev inequality may be used to calculate tail probabilities. For example, we would like to find the probability that $X$ is a very large number where the probability distribution involved is the Gaussian distribution. So in this case, if

$$X - m \geq 10\sigma$$

then

$$P[|X - m| \geq 10\sigma] \leq 0.01$$

Another way of writing this inequality is

$$P[|X - m| \leq \epsilon] \geq 1 - \left(\frac{\sigma}{\epsilon}\right)^2 \qquad (3.57)$$

setting $\epsilon = 1.5\sigma$, then

$$P[|X - m| \leq 1.5\sigma] \geq 1 - \frac{1}{1.5^2} = 0.56 \qquad (3.58)$$

or

$$P(m - 1.5\sigma \leq X \leq m + 1.5\sigma) \geq 0.56 \qquad (3.59)$$

applying this to the uniform distribution, $(m = (x_1 + x_0)/2$, and $\sigma = (x_1 - x_0)/\sqrt{12})$ we get this probability to be 0.8662.

**Example 3.22** Investigate the Gaussian random variable and the probabilities associated with the random variable falling in various intervals.

If the random variable $X$ is Gaussian, then

$$Pr\{x_0 \leq X < \infty\} = \frac{1}{\sigma_X\sqrt{2\pi}} \int_{x_0}^{\infty} \exp\left\{-\frac{(x - m_X)^2}{2\sigma_X^2}\right\} dx$$

is of great importance in communication systems. In this equation setting

$$x = \sqrt{2}\sigma_X y + m_X$$
$$dx = \sqrt{2}\sigma_X dy$$

or

$$\Pr\{x_0 < X \le x_0\} = \frac{1}{\sqrt{\pi}} \int_{\frac{x_0-m_x}{\sigma_X\sqrt{2}}}^{\infty} \exp\{-y^2\} \, dy$$

$$= \frac{1 - \mathrm{erf}\left(\frac{x_0-m_x}{\sigma_X\sqrt{2}}\right)}{2} \tag{3.60}$$

where

$$\mathrm{erf}(x) = \frac{2}{\sqrt{\pi}} \int_0^x e^{-t^2} \, dt \tag{3.61}$$

## 3.8  Characteristic Functions

The Fourier transform of the PDF of a random variable is called the characteristic function of a random variable. The characteristic function can also be viewed as an expectation. Thus

$$\Phi_X(v) = E[e^{jvx}] = \int_{-\infty}^{\infty} f_X(x) e^{jvx} \, dx \tag{3.62}$$

(we use $\Phi_X$ since $F_X$ is used to represent the CDF of $f_X$). And

$$f_X(x) = \frac{1}{2\pi} \int_{-\infty}^{\infty} \Phi_X(v) e^{-jvx} \, dx \tag{3.63}$$

The characteristic function is useful in many ways. For example,

$$\left.\frac{d\Phi_X(v)}{dv}\right|_{v=0} = j \int_{-\infty}^{\infty} x f_X(x) e^{jvx} \, dx \Big|_{v=0}$$

$$= j \int_{-\infty}^{\infty} x f_X(x) \, dx$$

$$= jE[x]$$

$$= jm_X \tag{3.64}$$

or

$$\left.\frac{d^2\Phi_X(v)}{dv^2}\right|_{v=0} = - \int_{-\infty}^{\infty} x^2 f_X(x) e^{jvx} \, dx \Big|_{v=0}$$

$$= - \int_{-\infty}^{\infty} x^2 f_X(x) \, dx$$

$$= -E[x^2] \tag{3.65}$$

which is proportional to the second moment of the random variable, and so on.

Another application is to calculate the sums of random variables. For example, if there are two independent random variable, $X_1$ and $X_2$, then

$$E[e^{jv(X_1+X_2)}] = E[e^{jvX_1}]E[e^{jvX_2}]$$

that is their characteristic functions get multiplied. Similarly, for many independent RVs,

$$E[e^{jv(\sum X_i)}] = \prod E[e^{jvX_i}] \tag{3.66}$$

Since the characteristic functions get multiplied in the 'frequency' domain, in the 'time' domain, we must convolve the PDFs. For example, if the random variable $X$ is the sum of many independent random variables,

$$X = \sum X_i \tag{3.67}$$

then

$$f_X(x) = f_{X_1}(x_1) * f_{X_2}(x_2) * \cdots \tag{3.68}$$

**Example 3.23** If several $(N)$ uniformly distributed random variable $X_i$ are added, find the characteristic function of their sum.

Let these random variables be identical with mean zero and their variance, $\sigma = 1$. For this case,

$$f_{X_i}(x_i) = \begin{cases} 0 & x_i < \sqrt{3} \\ \frac{1}{2\sqrt{3}} & -\sqrt{3} < x_i < \sqrt{3} \\ 0 & x_i > \sqrt{3} \end{cases} \tag{3.69}$$

then the characteristic functions are calculated as follows.

$$\Phi_{X_i}(v) = \frac{1}{2\sqrt{3}} \int_{-\sqrt{3}}^{\sqrt{3}} e^{jvx_i} dx_i = \frac{\sin\left(\sqrt{3}v\right)}{\sqrt{3}v} \tag{3.70}$$

then the characteristic function of their sum is

$$\Phi_X(v) = \left[ \frac{\sin\left(\sqrt{3}v\right)}{v\sqrt{3}} \right]^N \tag{3.71}$$

For the Gaussian random variable,

$$f_X(x) = \frac{1}{\sigma_X\sqrt{2\pi}} \exp\left[ -\frac{(x-m_x)^2}{2\sigma_X^2} \right] \tag{3.72}$$

the characteristic function is

$$\Phi_X(v) = \exp\left[ jvm_x - \frac{1}{2}v^2\sigma_X^2 \right]$$ (3.73)

## 3.9 Transformations or Functions of a Random Variable

Very often, we may have characterised a random variable $x$ in terms of its PDF $f_X(x)$ but we may require the PDF of another random variable, $y$ which is the transformation or function of the random variable $x$. Namely,

$$Y = g(X)$$ (3.74)

an example of which is shown in Fig. 3.14. If we take a look at the CDF

$$F_Y(y) = P(Y \le y) = P(g(X) \le y)$$ (3.75)

then we have to calculate and implement all the intervals where $g(X) \le y$. To do this we have to focus on the expression

$$f_Y(y)dy$$

where $dy$ is positive. We know that this is the tiny probability

$$dP = P(y \le Y \le y + dy)$$

Concentrating, next on the expression,

$$y = g(x)$$

which from observing the figure we can clearly see that the line $Y = y$ cuts the function at various points, $x = x_i \ i = 1, \ldots 4$. These are the values of $x$ corresponding to the value of $y$. Then

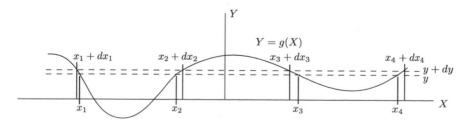

**Fig. 3.14** Example transformation $g(X)$

$$dP = f_Y(y)dy = f_X(x_1)|dx_1| + f_X(x_2)|dx_2| + f_X(x_3)|dx_3| + \cdots \qquad (3.76)$$

When we consider this equation, we have taken the magnitudes of the $dx$s. Why is this? Observing the figure we notice that at $x_1$ and $x_3$ the $dx_1$ and $dx_3$ are of negative values, while at $x_2$ and $x_4$ the $dx$s are positive. Clearly, each of the expressions

$$f_X(x_i)dx_i$$

should be tiny probability, and this *must be* positive! Therefore rewriting Eq. (3.76), we may write instead,

$$dP = f_Y(y)|dy| = f_X(x_1)|dx_1| + f_X(x_2)|dx_2| + f_X(x_3)|dx_3| + \cdots \qquad (3.77)$$

or

$$
\begin{aligned}
f_Y(y) &= f_X(x_1)\frac{|dx_1|}{|dy|} + f_X(x_2)\frac{|dx_2|}{|dy|} + f_X(x_3)\frac{|dx_3|}{|dy|} + \cdots \\
&= \frac{f_X(x_1)}{\left|\frac{dy}{dx}\right|_{x=x_1}} + \frac{f_X(x_2)}{\left|\frac{dy}{dx}\right|_{x=x_2}} + \frac{f_X(x_3)}{\left|\frac{dy}{dx}\right|_{x=x_3}} + \cdots
\end{aligned}
\qquad (3.78)
$$

In this expression,

$$x_i = g^{-1}(y) \qquad (3.79)$$

and furthermore, each of the $x_i$s must be written in terms of $y$ to make it consistent with the definition of $f_Y(y)$.

**Example 3.24** $X$ is a uniformly distributed random variable, where

$$
f_X(x) = \begin{cases} 1 & -0.5 \le x \le 0.5 \\ 0 & \text{otherwise} \end{cases}
$$

while $Y$ is another random variable related to $X$ by

$$Y = 4X + 1 \qquad (3.80)$$

Find $f_Y(y)$.

First we investigate the limits of $Y$. When $x = -0.5$, $y = -1$ and when $x = 0.5$, $y = 3$. Therefore $Y$ lies in the range $[-1, 3]$. From the given Eq. (3.80),

$$X = \frac{Y - 1}{4}$$

or

$$\left|\frac{dX}{dY}\right| = \frac{1}{4}; \qquad \left|\frac{dY}{dX}\right| = 4$$

Next,

$$f_Y(y) = \frac{f_X(x)}{\left|\frac{dy}{dx}\right|}$$

$$= \begin{cases} \frac{1}{4} & \text{for } -1 \le y \le 3 \\ 0 & \text{otherwise} \end{cases} \tag{3.81}$$

which is a perfectly acceptable PDF.

**Example 3.25** $X$ is a uniformly distributed random variable, where

$$f_X(x) = \begin{cases} 1 & -0.5 \le x \le 0.5 \\ 0 & \text{otherwise} \end{cases}$$

while $Y$ is another random variable related to $X$ by

$$Y = X^2 \tag{3.82}$$

Find $f_Y(y)$.

As in the earlier example, first we investigate the limits of $Y$. When $x = -0.5$, $y = 0.25$ and when $x = 0.5$, $y = 0.25$. Therefore $Y$ lies in the range $[0, 0.25]$. For each value of $y$ there are two values of $x$, $x = +\left|\sqrt{y}\right|$ and $x = -\left|\sqrt{y}\right|$. From the given Eq. (3.82),

$$X = \sqrt{Y}$$

$$\left|\frac{dX}{dY}\right| = \left|\frac{1}{2\sqrt{Y}}\right|$$

$$\left|\frac{dY}{dX}\right| = +2\sqrt{Y}$$

therefore

$$f_Y(y) = \frac{1}{2\sqrt{y}} + \frac{1}{2\sqrt{y}}$$

$$= \begin{cases} \frac{1}{\sqrt{y}} & 0 \le y \le 0.25 \\ 0 & \text{otherwise} \end{cases}$$

(Check: is this a valid PDF?

$$\int_0^{0.25} \frac{dy}{\sqrt{y}} = 1$$

other criteria are also satisfied,

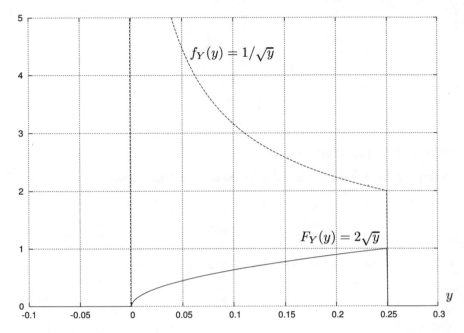

**Fig. 3.15**  Figure for Example 3.25

$$F_Y(y) = 2\sqrt{y}$$

which is a monotonically increasing function. And

$$F_Y(0) = 0$$
$$F_Y(0.25) = 1$$

etc.)

Both these functions are shown in Fig. 3.15.

## 3.10  Several Random Variables

We now consider the probability density and distribution functions of several random variables. Two or more than two random variables would occur when we have to consider experiments where there are more than one parameters which are random in nature. For example, in the discrete case if two dice are thrown (which are distinguishable) and output are $X =$ one of $\{x_1 = 1, \ldots x_6 = 6\}$ and $Y = 1, \ldots 6$. If consider another case, where the random variables are continuous in nature,

$$g(A, \Phi) = a \cos(\omega_0 t + \phi)$$

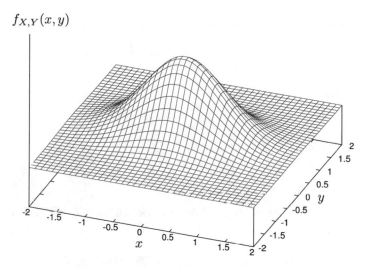

$f_{X,Y}(x,y)$

**Fig. 3.16** Typical joint probability density function

is a signal where $A$ and $\Phi$ are both random variables with

$$0 \leq A \leq \infty$$
$$-\pi \leq \Phi \leq \pi$$

with a joint PDF

$$f_{A,\Phi}(a,\phi)$$

such that the cumulative distribution function is

$$F_{A,\Phi}(a,\phi) = P(0 \leq A \leq a, \text{ and } -\pi \leq \Phi \leq \pi) = \int_0^a \int_{-\pi}^{\phi} f_{A,\Phi}(\alpha,\beta)d\alpha d\beta$$

$$(3.83)$$

In the general case, if $X$ and $Y$ are two random variables, then their joint probability density function is given by $f_{X,Y}(x, y)$ and the cumulative distribution function is

$$F_{X,Y}(x,y) = P(-\infty \leq X \leq x, \text{ and } -\infty \leq Y \leq y) = \int_{-\infty}^{x} \int_{-\infty}^{y} f_{X,Y}(\alpha,\beta)d\alpha d\beta$$

$$(3.84)$$

with,

$$\int_{-\infty}^{\infty} \int_{-\infty}^{\infty} f_{X,Y}(\alpha,\beta)d\alpha d\beta = 1 \qquad (3.85)$$

An example of a joint PDF is shown in Fig. 3.16. If we look at the probability,

$$P(x_1 \le x \le x_2, \text{ and } y = \text{anything}) = P(x_1 \le x \le x_2, \text{ and } -\infty \le y \le \infty)$$

$$= \int_{x_1}^{x_2} dx \int_{-\infty}^{\infty} f_{X,Y}(x, y)dy$$

then we realise that

$$f_X(x) = \int_{-\infty}^{\infty} f_{X,Y}(x, y)dy \tag{3.86}$$

and

$$f_Y(y) = \int_{-\infty}^{\infty} f_{X,Y}(x, y)dx \tag{3.87}$$

Next we consider the case where we define the conditional PDF of $X$ and $Y$. For this case,

$$f_{X,Y}(x, y) = f_{X|Y}(x|y) f_Y(y) = f_{Y|X}(y|x) f_X(x)$$

$$\text{or } f_{X|Y}(x|y) = \frac{f_{X,Y}(x, y)}{f_Y(y)}; \text{ and } f_{Y|X}(y|x) = \frac{f_{X,Y}(x, y)}{f_X(y)}$$

and further, the integral

$$\int_{-\infty}^{\infty} f_{X|Y}(x|y)dx \overset{?}{=} 1$$

should be unity, if $f_{X|Y}(x|y)$ is a valid PDF. To corroborate this,

$$\int_{-\infty}^{\infty} \frac{f_{X,Y}(x, y)}{f_Y(y)} dx = \frac{f_Y(y)}{f_Y(y)} = 1$$

As we know that in the case of *independent random variables*,

$$f_{X|Y}(x|y) = f_X(x)$$

therefore

$$f_{X,Y}(x, y) = f_X(x) f_Y(y)$$

in this way we may define joint probability density functions and cumulative distribution functions of more than two random variables.

## 3.11 Random Processes[4]

A random process is of a random function of time. This applies specifically to noise as whose concept was introduced in Chap. 1, in Fig. 1.5. We can see by perusal of the figure that the noise consists of random fluctuations of voltage or current (as the case may be) as a function of time and can be denoted by $n(t)$. Since noise is time variant, it cannot be characterised by a single random variable. Noise is the outcome of some electronic or electrical process which is proceeding in the background and creating unwanted random signals. A sample of a random signal is shown in Fig. 3.17.

To characterise a random process, we have to consider a sample space $S$, where the sample points denote the individual random signals as shown in Fig. 3.18. In this figure only a few sample points are shown, specifically $n_1(t), \ldots n_4(t)$. Each of these sample points has a noise trace which is shown in Fig. 3.19. The sample space, $S$, has a continuously infinite number of such points, since there are an infinite number of noise signals, which differ from each other. A collection of these random signals is called an *ensemble*.

If we have to define a random process, we may write it mathematically as

$$N(t, s_i) \quad -\infty < t < \infty; \ i = 1, \ldots, \infty \tag{3.88}$$

which takes on values

$$n_i(t) = N(t, s_i) \tag{3.89}$$

and it is customary to write

$$N(t) \equiv N(t, s_i)$$

as the random process.

The next question we ask ourselves is that from a random process, how are we to define a random variable? A very simple method is to observe a fixed point $t = t_0$ and take a look at the functions

$$\{n_1(t_0), n_2(t_0), n_3(t_0), \ldots, n_j(t_0), \ldots\}$$

as the number of such 'sample points' tends to infinity, it defines an random variable, $N(t_0)$ or $N_0$. For four such functions, $t_0$ is shown in Fig. 3.19. Similarly we can define another random variable, by looking at the functions at time $t = t_1$ (see Fig. 3.19),

$$\{n_1(t_1), n_2(t_1), n_3(t_1), \ldots, n_j(t_1), \ldots\}$$

and define the random variable $N_1$, and so on. A special case which is of importance is that the density functions, $f_{N_0}(n_0)$ and $f_{N_1}(n_1)$ satisfy, (for all values of $t_0$ and $t_1$)

---

[4] This and the following sections may be read later.

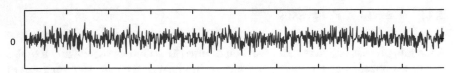

0

**Fig. 3.17**  Sample random signal

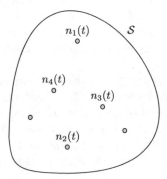

**Fig. 3.18**  Sample space of a random process which is also called an ensemble

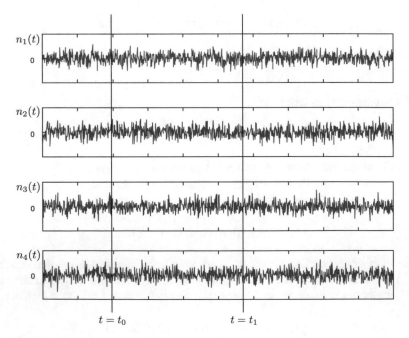

**Fig. 3.19**  Random noise samples, $n_1(t), \dots n_4(t)$

$$f_{N_0}(n_0)_{\text{at } t_0} = f_{N_1}(n_1)_{\text{at } t_1} = f_N(n) \tag{3.90}$$

in this equation, time does not play a role in the probability density function of the random process. Also if we find the joint probability density functions of the two random variables, $N_1$, and $N_2$, then

$$f_{N_1,N_2}(n_1, n_2)_{t_1,t_2} = f_{N_1 N_2}(n_1, n_2)_{t_2-t_1} \tag{3.91}$$

that is the joint probability density function of $N_1$, and $N_2$ is the same as the joint probability density function of the random variables $N(t_1)$ and $N(t_2)$ for all values of $t_1$ and $t_2$ *depends only on the difference,*

$$\tau = t_2 - t_1$$

When Eqs. (3.90) and (3.91) are satisfied, these are the *necessary conditions* for a random process to be called *a stationary random process*, but these *are not the sufficient conditions.* However when Eqs. (3.90) and (3.91) are satisfied, we call the process a *wide sense stationary process*, (or *stationary up to order 2*) and for all practical purposes, the process is considered stationary.[5]

## 3.11.1 Mean of a Stationary Process

If we take a look at a stationary process, $N(t)$ then the *ensemble* mean for any random stationary process is

$$
\begin{aligned}
m_{N(t)} &= m_N(t) \\
&= E[N(t)] \\
&= \int_{-\infty}^{\infty} n f_N(n) dn \quad \text{(since } f_N(n) \text{ is time } - \text{ independant)} \\
&= m_N
\end{aligned}
$$

so we come to the conclusion that the mean of a stationary process is a constant. The meaning of ensemble mean means that we look at *all the noise samples in the random process* and take the mean at any time $t_0$ across the ensemble.

---

[5] A random process is *strict sense stationary*, when the following condition outlined below has is met. Let $N(t_1)$, $N(t_2)$, $\cdots$ $N(t_j)$ be the random variables at times $t = t_1, t_2, \ldots t_j$, when $j$ can take any value. Then

$$f_{N_1,N_2,...N_j}(n_1, n_2, \ldots n_j) = f_{N(t_1+\tau),N(t_2+\tau),...N(t_j+\tau)}(n_1, n_2, \ldots n_j) \tag{3.92}$$

for the case when each of the random variables, $N(t_1 + \tau)$, $N(t_2 + \tau)$, $\ldots N(t_j + \tau)$ are evaluated at each corresponding times plus $\tau$. The condition given above is the sufficient condition. If this equation is satisfied for $j = j_0$ but not for $j = j_0 + 1$, then the process is stationary up to order $j_0$.

## 3.11.2   Autocorrelation and Covariance Functions

The auto correlation function of random stationary process, $N(t)$ is defined as

$$R_N(t_1, t_2) = E[N(t_1)N(t_2)] \tag{3.93}$$

$$= \int_{-\infty}^{\infty} \int_{-\infty}^{\infty} n_1 n_2 f_{N_1 N_2}(n_1, n_2) dn_1 dn_2$$

$$= R_N(t_2 - t_1)$$

where $N(t_1) \equiv N_1$ and $N(t_2) \equiv N_2$ are the random variables at times $t_1$ and $t_2$ respectively, and $f_{N_1 N_2}(n_1, n_2)$ is the joint probability density function of $N_1$ and $N_2$.

If we observe the autocorrelation function, $R_N(t_1, t_2)$, a little more closely, we expect on the basis of Eq. (3.91) that

$$R_N(t_1, t_2) = R_N(t_2 - t_1) = R(t_1 - t_2) \tag{3.94}$$

Some significance of what the general shape of the autocorrelation function is, may be deduced from Eq. (3.93) and Fig. 3.20. Three noise traces are shown in the figure; $n_1(t)$, the solid curve (—), $n_2(t)$, the dashed curve (- - -), and $n_3(t)$, the dot dash curve (· — ·). We are considering

$$E[N(t_1)N(t_2)] \tag{3.95}$$

and if we observe Fig. 3.20, we can see that as $t_2$ is far from $t_1$ ($t = t_2'$ of the figure) we expect (3.95) to be close to zero since the product $N(t_1)N(t_2)$ will "'on the average' will be positive and negative equal number of times. However as the difference $\tau = t_2 - t_1$ ($t = t_2''$ of the figure) becomes small, the expected value will become positive, *though small since the noise is not able to do a zero crossing*. However the value of (3.95) will be the greatest when $\tau = 0$ (or when $t_2 = t_1$). This implies that when $\tau$ is small, there is good correlation in a noise signal, and poor correlation when $\tau$ is large.

$$R_N(\tau)|_{\tau=0}$$

**Fig. 3.20**  Magnified noise trace

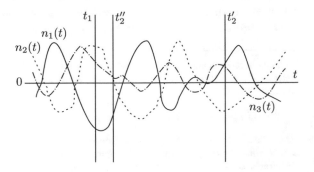

is maximum, and the value of $R_N(\tau)$ decreases rapidly as $\tau$ moves away from zero.

These remarks (the italics in the above sentences) has implications on the maximum frequency of the spectrum of $R_N(\tau)$. In a similar manner we can compute the covariance of a random process.

$$
\begin{aligned}
C_N(t_1, t_2) &= E\left[\{N(t_1) - m_N\}\{N(t_2) - m_N\}\right] \\
&= E\left[N(t_1)N(t_2) - m_N N(t_1) - m_N N(t_1) + m_N^2\right] \\
&= R_N(t_2 - t_1) - m_N^2 \tag{3.96}
\end{aligned}
$$

In actual practice we may *never* have to calculate $f_{N_1 N_2}(n_1, n_2)$ but usually we calculate directly the mean and autocorrelation function as seen in examples (Examples 3.26, 3.27, and 3.28) appearing later in this section.

### 3.11.2.1 Properties of the Autocorrelation Function

1. The autocorrelation function is an even function of $\tau$, which means that

$$
\begin{aligned}
R_N(t_1 - t_2) &= R_N(t_2 - t_1) \\
\Rightarrow R_N(\tau) &= R_N(-\tau) \tag{3.97}
\end{aligned}
$$

2. Since

$$
R_N(0) = E[N(t)^2]
$$

therefore

$$
R_N(0) > 0
$$

which is mean square value of the process.

3.

$$
R_N(0) \geq |R_N(\tau)|
$$

for all values of $\tau$. Consider

$$
E[\{N(t + \tau) \pm N(t)\}^2] = E[N(t + \tau)^2 \pm 2N(t + \tau)N(t) + N(t)^2]
$$
$$
\geq 0
$$

For a stationary process,

$$
E[N(t + \tau)^2] = E[N(t)^2] = R_N(0)
$$

and

$$
R_N(\tau) = E[N(t + \tau)N(t)]
$$

therefore

$$2R_N(0) \pm 2R_N(\tau) \geq 0$$

or

$$|R_N(\tau)| \leq R_N(0)$$

**Example 3.26**  A generator produces a waveform,

$$N(t) = A\cos(\omega t + \Phi)$$

where $\Phi$ is randomly distributed in the interval, $0 \leq \Phi \leq 2\pi$. $A$ and $\omega$ are constants. The probability density function of $\Phi$ is

$$f_\Phi(\phi) = \begin{cases} \frac{1}{2\pi} & 0 \leq \phi \leq 2\pi \\ 0 & \text{otherwise} \end{cases}$$

Investigate the properties of this random process. Is it wide-sense stationary?

To investigate the properties of the random process,

$$\begin{aligned} E[N(t)] &= \int_0^{2\pi} f_\Phi(\phi) N(t) d\phi \\ &= \frac{A}{2\pi} \int_0^{2\pi} \cos(\omega t + \phi) d\phi \end{aligned}$$

where $A\cos(\omega t + \phi)$ is a function of the random variable, $\phi$. If we evaluate this integral, we find that it is zero. That is, the ensemble mean $m_N$ is zero, which is independent of time. The process is first-order stationary.

Let now compute the autocorrelation function, $R_N(\tau)$.

$$\begin{aligned} R_N(t_1, t_2) &= E[A^2 \cos(\omega t_1 + \phi) \cos(\omega t_2 + \phi)] \\ &= \frac{A^2}{2\pi} \int_0^{2\pi} \cos(\omega t_1 + \phi) \cos(\omega t_2 + \phi) d\phi \\ &= \frac{A^2}{2\pi} \int_0^{2\pi} \left[ \frac{\cos(t_2\omega + t_1\omega + 2\phi)}{2} + \frac{\cos(t_2\omega - t_1\omega)}{2} \right] d\phi \\ &= \frac{A^2}{2\pi} \times \pi \cos[\omega(t_2 - t_1)] \\ &= \frac{A^2}{2} \cos[\omega(t_2 - t_1)] \end{aligned}$$

since the autocorrelation function is a function of $t_2 - t_1$, the process is second-order stationary, and hence wide-sense stationary.

**Example 3.27**  A generator produces a waveform,

**Fig. 3.21** A random binary signal

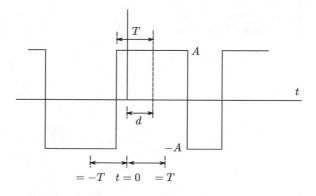

$$N(t) = A \cos(\omega t + \phi)$$

where $A$ is randomly distributed in the interval, $A_1 \leq A \leq A_2$, where $A_1$ and $A_2 > 0$. Here, $\omega$ and $\phi$ are considered constants. The probability density function is

$$f_A(a) = \begin{cases} \frac{1}{A_2 - A_1} & A_1 \leq a \leq A_2 \\ 0 & \text{otherwise} \end{cases}$$

Investigate the properties of this random process.

In this case,

$$E[A \cos(\omega t + \phi)] = \cos(\omega t + \phi) E[A]$$

since $m_N = \cos(\omega t + \phi) E[A]$ is a function of time, and $E[A] \neq 0$, the process is *not* a stationary process.

**Example 3.28** Investigate the parameters of the random binary process as shown in Fig. 3.21.

First of all we have to characterise the process. The process consists of pulses amplitude $A$ or $-A$ which come along in a random fashion. At any time $t$,

$$P(\text{positive pulse}, A) = \frac{1}{2}$$

$$P(\text{negative pulse}, -A) = \frac{1}{2} \tag{3.98}$$

so the mean at any time $t$ the ensemble mean $m_N$ is

$$E[N(t)] = A \times \frac{1}{2} + -A \times \frac{1}{2} = 0 \tag{3.99}$$

next we have to characterise the autocorrelation function, $R_N(t_1, t_2)$. Without loss of generality, we can probe the properties of the system looking at $t = t_1 = 0$ since

**Fig. 3.22** Autocorrelation
function of a random binary
signal

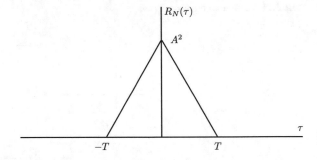

$t = 0$ has no special distinction. The autocorrelation function,

$$R_N(t_1, t_2) = E[N(t_1)N(t_2)]$$
$$= E[N(0)N(\tau)] \tag{3.100}$$

Where $\tau = t_2 - t_1$. The first point which is of importance is that

$$R_N(0) = \begin{cases} A \times A = A^2 & \text{for a positive pulse} \\ -A \times -A = A^2 & \text{for a negative pulse} \end{cases} \tag{3.101}$$

the second point which is of importance is that if $\tau > T$ or $\tau < -T$, the signal
ensemble is uncorrelated, and in that case $R_N(\tau) = 0$ (Fig. 3.22).

$$R_N(\tau) = \begin{cases} 0 & \tau > T \\ 0 & \tau < -T \end{cases} \tag{3.102}$$

we now look at when $-T < \tau < T$. The parameter $d$ shown in the figure has a
probability density function which is uniformly distributed and is given by

$$f_D(d) = \begin{cases} \frac{1}{T} & 0 \le d \le T \\ 0 & \text{elsewhere} \end{cases}$$

so for $T > \tau > 0$ in this range, the mean is

$$R_N(\tau) = \int_0^{T-\tau} A^2 \times \frac{1}{T} d(d) = \frac{A^2}{T} d \Big|_0^{T-\tau}$$
$$= \frac{A^2}{T} \{T - \tau\}$$
$$= A^2 (1 - \frac{\tau}{T}) \tag{3.103}$$

The $A^2$ appears due to Eq. (3.101). Since there is no reason to believe that for $-T < \tau < 0$, the auto correlation function is not a mirror image of Eq. (3.103), for this range,

$$R_N(\tau) = A^2(1 + \frac{\tau}{T}) \tag{3.104}$$

therefore

$$R_N(\tau) = \begin{cases} 0 & \tau < -T \\ A^2(1 + \frac{\tau}{T}) & -T \le \tau \le 0 \\ A^2(1 - \frac{\tau}{T}) & 0 \le \tau \le T \\ 0 & \tau > T \end{cases} \tag{3.105}$$

substituting back, the value of $\tau$,

$$R_N(t_1, t_2) = \begin{cases} 0 & t_2 - t_1 < -T \\ A^2(1 + \frac{t_2 - t_1}{T}) & -T \le \tau \le 0 \\ A^2(1 - \frac{t_2 - t_1}{T}) & 0 \le \tau \le T \\ 0 & \tau > T \end{cases}$$

$$= R_N(t_2 - t_1)$$

$$= R_N(t_1 - t_2) \tag{3.106}$$

since $m_N$ is time-independent and $R_N(t_1, t_2) = R_N(t_2 - t_1) = R_N(t_1 - t_2)$, the process is wide-sense stationary.

## 3.12 Ergodicity

Ergodicity[6] is the property of an ensemble of random processes, where you look at the statistical properties of any one of the sample points of the sample space, as opposed to looking at the statistical proprieties of the full sample space. Taking a concrete example, if we look at a random process, our first instinct would be to take the mean of the process as the average of the process itself, *over time*. This would mean that in Fig. 3.17,

$$\mu_N(T) = \frac{1}{2T} \int_{-T}^{T} N(t)dt \tag{3.107}$$

thus we can see that

1. For different noise samples (and the same value of $T$) $\mu_N(T)$ would then be a random variable.
2. The value of any one mean, $\mu_N(T)$ as we let $T \to \infty$ would then be a constant.

---

[6] This section may be skipped on first reading.

A process is ergodic if three criteria are satisfied.

1. The average of the ergodic mean, $\mu_N(T)$ is equal to the ensemble mean, $\mu_N$

$$
\begin{aligned}
E[\mu_N(T)] &= \frac{1}{2T} \int_{-T}^{T} E[N(t)]dt \\
&= \frac{1}{2T} \int_{-T}^{T} \mu_N dt \\
&= \mu_N
\end{aligned}
\tag{3.108}
$$

where we have interchanged the expectation and integral operators.
2. The ergodic mean and ensemble mean are related by the equation,

$$
\lim_{T \to \infty} \mu_N(T) = \mu_N
\tag{3.109}
$$

3. The variance of the ergodic mean in the limit as $T \to \infty$ is zero

$$
\lim_{T \to \infty} [\text{var}(\mu_N(T))] = 0
\tag{3.110}
$$

or that

$$
\lim_{T \to \infty} E[\{\mu_N(T) - \mu_N\}^2] = \lim_{T \to \infty} E[\mu_N^2(T) - 2\mu_N(T)\mu_N + \mu_N^2]
$$
$$
= 0
$$

We may now proceed to look at the ergodic statistics of the autocorrelation function. The ergodic autocorrelation function is

$$
R_N(\tau, T) = \frac{1}{2T} \int_{-T}^{T} N(t + \tau)N(t)dt
\tag{3.111}
$$

the meaning of an autocorrelation function *which is ergodic* is when

1. The relation,
$$
\lim_{T \to \infty} R_N(\tau, T) = R_N(\tau) \qquad \forall \text{ values of } \tau
$$

2. And the relation,

$$
\lim_{T \to \infty} \{\text{var}[R_N(\tau, T)]\} = 0 \qquad \forall \text{ values of } \tau
$$

both hold.

We state here, *without proof, that for a process to be ergodic, it must be WSS.*

## 3.13 Power Spectral Density. Wiener–Khinchin Theorem

When we look any sample of a stationary random process, $x(t)$ and try to find the energy in that signal,

$$\mathbb{E}\{x(t)\} = \int_{-\infty}^{\infty} |x(t)|^2 dt$$

we find that this energy is infinite, and it is because of this that the FT of $x(t)$ may not exist.[7] The next point which is of importance is that the power will vary from sample point to sample point, and therefore the power in a particular sample will not be equal to the power in another sample. However we are interested in the *power spectral density* of a random process, which is a statistical characterisation of a random process in the frequency domain. Power in any general sample of a random process is

$$\mathbb{P}\{x(t)\} = \lim_{T \to \infty} \left\{ \frac{1}{T} \int_{-T/2}^{T/2} |x(t)|^2 dt \right\}$$

If we define

$$x_T(t) = \begin{cases} x(t) & -T/2 \le t \le T/2 \\ 0 & \text{otherwise} \end{cases}$$

then, we can define the FT of this function:

$$\mathcal{F}\{x_T(t)\} = X_T(\omega) = \int_{-T/2}^{T/2} x(t) e^{-j\omega t} dt$$

Next we determine the function,

$$|X_T(\omega)|^2 = X_T(\omega) * X_T(\omega)^*$$

$$= \left[ \int_{-T/2}^{T/2} x(t) e^{-j\omega t} dt \right] \left[ \int_{-T/2}^{T/2} x(t) e^{-j\omega t} dt \right]^*$$

$$= \left[ \int_{-T/2}^{T/2} x(t_1) e^{-j\omega t_1} dt_1 \right] \left[ \int_{-T/2}^{T/2} x(t_2) e^{j\omega t_2} dt_2 \right]$$

where in the last equation we have assumed that $x(t_2)$ is real. Or

$$|X_T(\omega)|^2 = \int_{-T/2}^{T/2} \int_{-T/2}^{T/2} x(t_1) x(t_2) e^{-j\omega t_1} e^{j\omega t_2} dt_1 dt_2$$

$$= \int_{-T/2}^{T/2} \int_{-T/2}^{T/2} x(t_1) x(t_2) e^{-j\omega(t_1 - t_2)} dt_1 dt_2$$

---

[7] One of the requirements that the FT of signal exists is that it be absolutely integrable.

Since we are working with random processes, we would like to look at the mean value of the FT transform, $|X_T(\omega)|$ at every frequency $\omega$:

$$
\begin{aligned}
E\left[|X_T(\omega)|^2\right] &= E\left[\int_{-T/2}^{T/2}\int_{-T/2}^{T/2} x(t_1)x(t_2)e^{-j\omega(t_1-t_2)}dt_1 dt_2\right] \\
&= \int_{-T/2}^{T/2}\int_{-T/2}^{T/2} E\left[x(t_1)x(t_2)e^{-j\omega(t_1-t_2)}\right]dt_1 dt_2 \\
&= \int_{-T/2}^{T/2}\int_{-T/2}^{T/2} E\left[x(t_1)x(t_2)\right]e^{-j\omega(t_1-t_2)}dt_1 dt_2
\end{aligned}
$$

In this last equation, for a wide-sense stationary process,

$$
E\left[x(t_1)x(t_2)\right] = R_X(t_1 - t_2) = R_X(\tau) \tag{3.112}
$$

So

$$
\begin{aligned}
E\left[|X_T(\omega)|^2\right] &= \int_{-T/2}^{T/2}\int_{-T/2}^{T/2} R_X(t_1 - t_2)e^{-j\omega(t_1-t_2)}dt_1 dt_2 \\
&= \int_{-T/2}^{T/2}\int_{-T/2}^{T/2} R_X(x - y)e^{-j\omega(x-y)}dx dy \tag{3.113}
\end{aligned}
$$

where we have substituted $x$ for $t_1$ and $y$ for $t_2$ to make these variables in keeping with existing mathematical texts where double integrals are considered. Without going into the mathematical details, we can show that

$$
S_X(\omega) \Leftrightarrow R_X(\tau) \tag{3.114}
$$

which is also called the Wiener–Khinchin theorem. In this equation, $S_X(\omega)$ is the *power spectral density* of the random process $X(t)$.

The mathematical details are given in Appendix G (Fig. 3.23).

The importance of the power spectral density of any WSS noise is the power in the *spectral (or frequency)* components of the noise in any frequency range. Thus, for example, the power in the frequency range $\omega_a$ to $\omega_b$ is given by

$$
P_{ab} = 2\int_{\omega_a}^{\omega_b} S_X(\omega)d\omega \tag{3.115}
$$

**Example 3.29** Find the power spectral density of the random binary wave

We know that

$$
S_X(\omega) \Leftrightarrow R_X(\tau)
$$

and for a random binary wave, (from Example 3.28, Sect. 3.11.2)

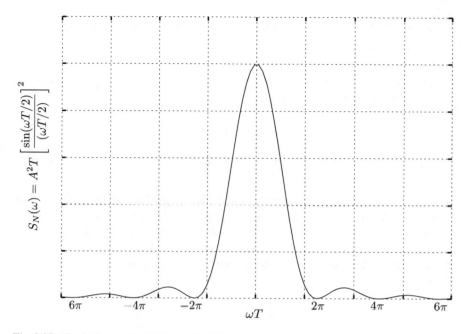

**Fig. 3.23** The PSD computed in Example 3.29

$$R_N(\tau) = \begin{cases} 0 & \tau < -T \\ A^2(1 + \frac{\tau}{T}) & -T \leq \tau \leq 0 \\ A^2(1 - \frac{\tau}{T}) & 0 \leq \tau \leq T \\ 0 & \tau > T \end{cases} \qquad (3.116)$$

therefore,

$$\begin{aligned} S_N(\omega) &= \int_{-T}^{0} A^2(1 + \frac{\tau}{T})e^{-j\omega\tau}d\tau + \int_{0}^{T} A^2(1 - \frac{\tau}{T})e^{-j\omega\tau}d\tau \\ &= A^2\left[\frac{j\omega T + 1}{\omega^2 T} - \frac{e^{j\omega T}}{\omega^2 T} - \frac{e^{-j\omega T}}{\omega^2 T} - \frac{j\omega T - 1}{\omega^2 T}\right] \\ &= -A^2\left[\frac{e^{-j\omega T}\left(e^{2j\omega T} - 2e^{j\omega T} + 1\right)}{\omega^2 T}\right] \\ &= A^2\left[\frac{2 - 2\cos(\omega T)}{\omega^2 T}\right] \\ &= A^2 T\left[\frac{\sin(\omega T/2)}{(\omega T/2)}\right]^2 \end{aligned}$$

## 3.14    Transmission of Random Processes Through Linear Systems

Next we look at what the output of a linear system is when the input is a random process as shown in Fig. 3.24. For any linear system,

$$y(t) = h(t) * x(t)$$
$$= \int_{-\infty}^{\infty} h(\tau)x(t-\tau)d\tau \tag{3.117}$$

where $x(t)$ is the input, $y(t)$ is the input and $h(t)$ is the impulse response of the system. When the input to a linear system is a random process, we are interested more in the statistical characterisation of the whole phenomenon rather than in details. Hence, the above will therefore apply to the case when the input is a random process, and by implication, the output too will be a random process. If we investigate the mean of the output,

$$E[y(t)] = E[h(t) * x(t)]$$
$$= E\left[\int_{-\infty}^{\infty} h(\tau)x(t-\tau)d\tau\right]$$
$$= \int_{-\infty}^{\infty} h(\tau)E[x(t-\tau)]d\tau \tag{3.118}$$

For WSS process,

$$E[x(t-\tau)] = \mu_X \tag{3.119}$$

which is a constant. Therefore,

$$E[y(t)] = \mu_Y = \mu_X \int_{-\infty}^{\infty} h(\tau)d\tau$$

Next we take a look at the output autocorrelation function of WSS (wide-sense stationary) process, then

$$S_Y(\omega) = |H(\omega)|^2 S_X(\omega) \tag{3.120}$$

the proof of which is given in Appendix H.

**Example 3.30** If we pass a WSS random process, $X(t)$ with autocorrelation function

$$R_X(\tau) = \delta(\tau)$$

**Fig. 3.24** Input and output of a linear system

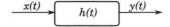

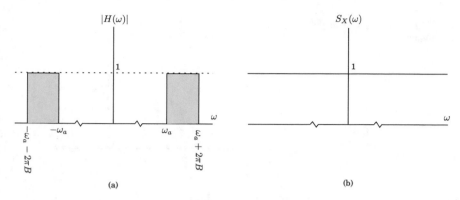

**Fig. 3.25**  PSD of random process and filter characteristics for Example 3.30

through a BPF filter with characteristics

$$H(\omega) = \begin{cases} 1e^{j\omega} & \omega_a \le \omega \le \omega_a + 2\pi B \\ 1e^{-j\omega} & -(\omega_a + 2\pi B) \le \omega \le -\omega_a \\ 0 & \text{otherwise} \end{cases}$$

find the average output power.

For the autocorrelation function given (see Fig. 3.25),

$$S_X(\omega) = 1$$

and,

$$|H(\omega)|^2 = 1$$

in the frequency ranges $\omega_a \le \omega \le \omega_a + 2\pi B$ and $-(\omega_a + 2\pi B) \le \omega \le -\omega_a$,

$$E[Y(t)^2] = \frac{1}{2\pi} \left[ \int_{-(\omega_a + 2\pi B)}^{-\omega_a} |H(\omega)|^2 S_X(\omega) d\omega + \int_{\omega_a}^{(\omega_a + 2\pi B)} |H(\omega)|^2 S_X(\omega) d\omega \right]$$

$$= \frac{1}{2\pi} [4\pi B]$$

$$= 2B$$

## 3.14.1  Noise Equivalent Bandwidth

Let us understand the concept of noise equivalent bandwidth with the help of an example. Suppose we pass white noise of magnitude $N_0/2$ through a lowpass RC

**Fig. 3.26** Noise equivalent bandwidth

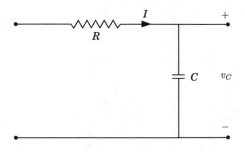

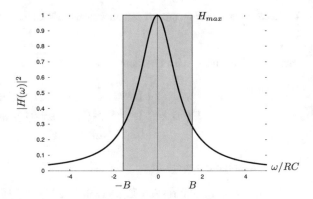

filter shown in Fig. 3.26 with the characteristics of the filter shown below. The output noise has a power spectral density given by

$$S_o(\omega) = |H(\omega)|^2 \frac{N_0}{2} \qquad (3.121)$$

and the average power output in the noise is

$$\mathbb{P}_{av} = \frac{N_0}{4\pi} \int_{-\infty}^{\infty} |H(\omega)|^2 d\omega \qquad (3.122)$$

This noise is contained in a band of frequencies (shown shaded in the figure) as if it was passed through an ideal LPF,

$$\mathbb{P}_{av} = \frac{N_0}{4\pi} \int_{-B}^{B} |H_{max}|^2 d\omega$$

$$= \frac{N_0 |H_{max}|^2 \times 2B}{4\pi} \qquad (3.123)$$

equating these two results,

$$B = \frac{\int_0^\infty |H(\omega)|^2 d\omega}{|H_{max}|^2} \qquad (3.124)$$

which is the noise equivalent bandwidth.

**Example 3.31** Calculate the noise equivalent bandwidth of the circuit shown in Fig. 3.26.

The transfer function of the circuit is

$$H(\omega) = \frac{1}{j\omega C R + 1}$$

and therefore

$$B = \frac{\int_{-\infty}^\infty |H(\omega)|^2 d\omega}{2 |H_{max}|^2} = \frac{1}{2} \int_{-\infty}^\infty \frac{d\omega}{\omega^2 C^2 R^2 + 1}$$
$$= \frac{\pi}{2RC} \text{ rad/s}$$

where $H_{max} - 1$.

The closer the given LPF is to the ideal LPF, the closer will $B$ be to one, as shown in the next example,

**Example 3.32** For the Butterworth filter with

$$|H(j\omega)|^2 = \frac{1}{1 + \omega^{2n}}$$

for $n = 5$,

$$B = \frac{\int_{-\infty}^\infty |H(\omega)|^2 d\omega}{2 |H_{max}|^2} = \frac{\pi}{10 \sin\left(\frac{\pi}{10}\right)} \approx \frac{\pi}{10 \times \left(\frac{\pi}{10}\right)} \approx 1$$

as filter approaches the ideal brick wall characteristic, the bandwidth, $B$ approaches 1 r/s.

## 3.15 Properties of the Power Spectral Density

The power spectral density has various properties. The relationship between the autocorrelation function and the power spectral density is connected by the two equations,

$$S_X(\omega) = \int_{-\infty}^{\infty} R_X(\tau)e^{-j\omega\tau}d\tau$$

$$R_X(\tau) = \frac{1}{2\pi}\int_{-\infty}^{\infty} S_X(\omega)e^{j\omega\tau}d\omega \qquad (3.125)$$

Some of these properties are

1. The first property is

$$S_X(0) = \int_{-\infty}^{\infty} R_X(\tau)d\tau \qquad (3.126)$$

which follows from setting $\omega = 0$ in the first relation of Eq. (3.125).

2. The second property is

$$E[X^2(t)] = R(0) = \frac{1}{2\pi}\int_{-\infty}^{\infty} S_X(\omega)d\omega \qquad (3.127)$$

which follows from the second of the relations in Eq. (3.125). This property is used in the proof given in Appendix H. Therefore if we want to find the power contributed by the spectral components where $\omega_a \leq \omega \leq \omega_b$, then

$$\mathbb{P}(\omega_a \leq \omega \leq \omega_b) = \frac{1}{2\pi}\left\{\int_{\omega_a}^{\omega_b} S_X(\omega)d\omega + \int_{-\omega_b}^{-\omega_a} S_X(\omega)d\omega\right\} \qquad (3.128)$$

3. The PSD of a random process is non-negative,

$$S_X(\omega) \geq 0 \qquad (3.129)$$

this property follows from Eq. (3.127), since $E[X^2(t)]$ is always greater than zero. Furthermore if, in some interval, $S_X(\omega)$ is negative, then Eq. (3.128) would give us a negative result and which would contradict the definition of power being non-negative.

4. The power spectral density is an even function of $\omega$ and is real. To prove this, consider

$$S_X(\omega) = \int_{-\infty}^{\infty} R(\tau)[\cos(\omega\tau) - j\sin(\omega\tau)]d\tau$$

$$= \int_{-\infty}^{\infty} R(\tau)\cos(\omega\tau)d\tau \qquad (3.130)$$

Since $R(\tau)$ is even and $\sin(\omega\tau)$ is odd. Observing this equation, we can see that $S_X(\omega)$ is real

$$\Im[S_X(\omega)] = 0$$

Also substituting $-\omega$ for $\omega$ we see that

$$S_X(\omega) = S_X(-\omega)$$

Furthermore, these properties follow from Sect. 2.12.8, Property 3, since $R(\tau)$ is real and even for a wide-sense stationary process.

If we use the power spectral density in terms of $f$, the frequency in hertz, then

$$E[X^2(t)] = \frac{1}{2\pi} \int_{-\infty}^{\infty} S_X(\omega)d\omega = \int_{-\infty}^{\infty} S_X(f)df$$

$$= \int_{-\infty}^{\infty} S_X(f)df$$

$$= 2 \int_{0}^{\infty} S_X(f)df \quad (S_X(f) \text{ is even}) \tag{3.131}$$

## 3.16 Noise

### 3.16.1 White Noise

White noise is a *theoretical random process*, which is considered to be an *extremely uncorrelated*.[8] Looking at various random processes, whose autocorrelation functions are shown in Fig. 3.27, we observe autocorrelation functions $a$, $b$, $c$ and $d$. We can clearly see that $a$ is more correlated than $b$ which is more correlated than $c$, and so on, with $d$ the most uncorrelated function among the four. If we continue this process we reach the logical conclusion that a process which is *completely uncorrelated* is the impulse function.[9] That is

$$R_W(\tau) \propto \delta(\tau)$$

White noise is exactly such an (imaginary) process. That is

$$R_W(\tau) = \frac{N_0}{2}\delta(\tau) \tag{3.132}$$

whose power spectral density is

$$S_W(\omega) = \frac{N_0}{2} \tag{3.133}$$

this last equation implies that the spectrum of white noise contains all frequencies in the same proportion, and therefore it is termed *white noise*. White noise is an

---

[8] When we say that the autocorrelation function is uncorrelated, we mean that for values of $\tau$ which are small, $R(\tau)$ is small and $\approx 0$.

[9] Something like how a Gaussian function tends to the impulse function as considered in Sect. 2.4.1.

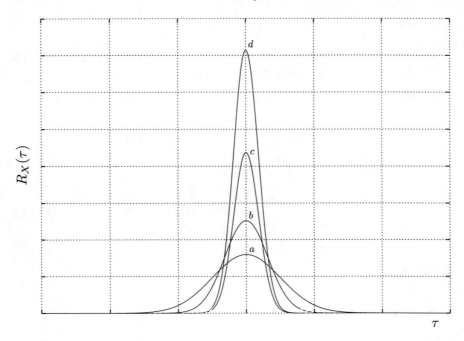

**Fig. 3.27**  Autocorrelation functions with different levels of correlations

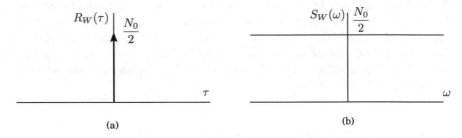

**Fig. 3.28**  White noise characteristics

idealisation, though in practice there are many PSDs which approach the concept of white noise, none can be termed as white noise *exactly*.

The reason that we have used $N_0/2$ is that in a band of frequency of $B$ Hz, the average power of white noise is

$$2 \int_{f_0}^{f_0+B} S_W(f) df = \frac{N_0}{2} \times 2B = N_0 B \tag{3.134}$$

(see Example 3.30 of Sect. 3.14). The characteristics of white noise are shown in Fig. 3.28.

## 3.16.2  Thermal Noise

Thermal noise, also called Johnson noise, or Nyquist noise, is the noise generated by electrons in their thermal agitation in the crystalline lattice in conductors. The names associated with thermal noise is due to the two papers by Johnson [1] and Nyquist [2]. This noise is present if we measure the average rms voltage, $V_{th}(t)$, across a resistor of resistance $R$ ($\Omega$), for a bandwidth of $B$ (Hz), the average value is

$$\sqrt{E[V_{th}^2(t)]} = \sqrt{4kTRB} \tag{3.135}$$

where the process is assumed to be WSS, $E[\cdots]$ is the expected value, $k$ is the Boltzmann's constant ($= 1.381 \times 10^{-23}$, MKS units), and $T$ is the absolute temperature in $^\circ$K.

Similarly we may define the rms of the noise current to be,

$$\sqrt{E[I_{th}^2(t)]} = \sqrt{4kTGB} \tag{3.136}$$

The power spectral density as estimated by researchers for a resistor of value $R$ ohms is given by

$$S_V(f) = \frac{2Rhf}{e^{\frac{hf}{kT}} - 1}$$

where $h$ is Planck's constant, ($= 6.626 \times 10^{-34}$ MKS units). If we expand the function

$$e^{\frac{hf}{kT}} - 1$$

in a Taylor series expansion we get

$$e^{\frac{hf}{kT}} - 1 \approx \frac{hf}{kT}$$

where we have assumed that

$$\frac{hf}{kT} \ll 1$$

$$f \ll \frac{kT}{h} = 6.25 \times 10^{12} \text{Hz} \ (T = 300^\circ\text{K})$$

This range of frequencies (0 to $6.25 \times 10^{12}$ Hz) is well within the range used for communication purposes, and therefore if we use the Taylor series expansion we get

$$S_V(f) = \frac{2Rhf}{\left[\frac{hf}{kT}\right]} = 2RkT \tag{3.137}$$

**Fig. 3.29** Thevenin and
Norton equivalents of a noisy
resistor

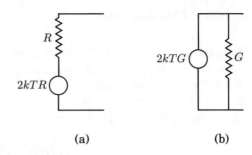

(a)                              (b)

where $f \ll 10^{12}$ Hz. Examining this expression, we know that the PSD of thermal
noise approximates that white noise, since the autocorrelation function

$$R_V(\tau) = \mathcal{F}^{-1}[S_V(f)] = 2kRT\delta(\tau) \qquad (3.138)$$

A practical resistor may be therefore modelled as an ideal resistor with a noisy voltage
source in series, or an ideal conductance with a noisy current source in parallel. These
equivalent circuits are shown in Fig. 3.29.

## 3.17  Sum of Two Random Processes. Cross Power Spectral Densities

Consider two random processes, $X(t)$ and $Y(t)$, and the sum of these two processes,

$$Z(t) = X(t) + Y(t) \qquad (3.139)$$

then,

$$E[Z^2(t)] = E[X^2(t) + Y^2(t) + 2X(t)Y(t)] \qquad (3.140)$$

we already know and have dealt with the functions

$$E[X^2(t)], \text{ and } E[Y^2(t)]$$

but to consider

$$E[X(t)Y(t)]$$

we have to define the cross correlation function,

$$R_{XY}(t_1, t_2) = E[X(t_1)Y(t_2)] \qquad (3.141)$$

the processes are called jointly wide-sense stationary if

**Fig. 3.30** Circuit for problem 3.33

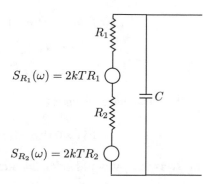

$$S_{R_1}(\omega) = 2kTR_1$$

$$S_{R_2}(\omega) = 2kTR_2$$

$$R_{XY}(t_1, t_2) = R_{XY}(t_1 - t_2) = R_{XY}(\tau) \tag{3.142}$$

similarly we may define an cross correlation function,

$$R_{YX}(t_1, t_2) = E[Y(t_1)X(t_2)] = R_{YX}(t_1, t_2) = R_{YX}(\tau) \tag{3.143}$$

Having defined these cross correlation functions, we may now define their spectral densities,

$$S_{XY}(\omega) \Leftrightarrow R_{XY}(\tau)$$
$$S_{YX}(\omega) \Leftrightarrow R_{YX}(\tau)$$

Also if

$$E[X(t_1)Y(t_2)] = E[X(t_1)]E[Y(t_2)] \tag{3.144}$$

then these processes are *uncorrelated* or *orthogonal*, and in many (if not the majority of cases) this is true.

**Example 3.33** Find the PSD of the noise at the output of the circuit shown in Fig. 3.30.

For the circuit shown,

$$
\begin{aligned}
H(\omega) &= \frac{1/j\omega C}{R_1 + R_2 + 1/j\omega C} \\
&= \frac{1}{j\omega C(R_1 + R_2) + 1}
\end{aligned}
\tag{3.145}
$$

Since the two noise voltages add,

$$Z(t) = V_{R_1}(t) + V_{R_2}(t) \tag{3.146}$$

and

$$E[Z(t)^2] = E\left[\left\{V_{R_1}(t) + V_{R_2}(t)\right\}^2\right]$$
$$= E[V_{R_1}(t)^2] + 2E[V_{R_1}(t)V_{R_1}(t)] + E[V_{R_2}(t)^2]$$
$$= E[V_{R_1}(t)^2] + E[V_{R_2}(t)^2] \tag{3.147}$$

Where we have assumed that

$$E[V_{R_1}(t)V_{R_1}(t)] = E[V_{R_1}(t)]E[V_{R_2}(t)] = 0$$

the two noise processes being uncorrelated, and since

$$E[V_{\text{th}}(t)] = 0 \tag{3.148}$$

for both the noise voltage of $R_1$ and $R_2$. The PSD is (using superposition of the noise in terms of $V^2$is)

$$S_{R_1+R_2}(\omega) = S_{R_1}(\omega) + S_{R_2}(\omega)$$
$$= 2kT(R_1 + R_2)$$

where $T$ is the ambient temperature. Setting

$$R_1 + R_2 = R$$

we have

$$|H(\omega)|^2 = \frac{1}{1 + \omega^2 C^2 R^2}$$

therefore the output PSD is

$$S_Z(\omega) = S_{R_1+R_2}(\omega)\,|H(\omega)|^2 = \frac{2kTR}{1 + \omega^2 C^2 R^2} \tag{3.149}$$

this result tells us, that whether we use a single resistor or two resistors, we get the same PSD on the output. If we work out the inverse FT of this expression,

$$R_Z(\tau) = \begin{cases} (2kT/C)e^{-\tau/RC} & \tau > 0 \\ (2kT/C)e^{\tau/RC} & \tau < 0 \end{cases}$$

Let us see what the average noise voltage is

$$E[Z(t)^2] = \frac{1}{2\pi} \int_{-\infty}^{\infty} S_Z(\omega)d\omega$$
$$= R_Z(0)$$
$$= 2kT/C$$

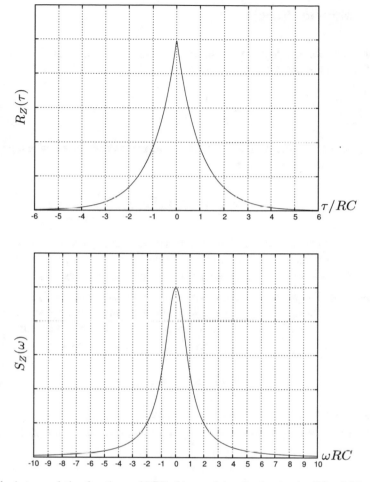

**Fig. 3.31** Autocorrelation function and PSD of two resistors for the circuit of Fig. 3.30

for $T = 300°$K, and $C = 1$pF

$$E[Z(t)^2] = 8.28 \times 10^{-9} \ (\text{V}^2)$$

The plots of $S_Z(\omega)$ and $R_Z(\tau)$ are shown in Fig. 3.31.

## 3.18  Gaussian Process

The Gaussian process is widespread in nature.[10] For a wide-sense stationary process, the PDF of a Gaussian process is given by

$$f_X(x) = \frac{1}{\sigma_X \sqrt{2\pi}} e^{-[(x-\mu_X)^2/(2\sigma_X^2)]} \tag{3.150}$$

where $\mu_X$ is the mean and $\sigma_X$ is the variance of the process. If we take a particular process, $X(t)$, and form the integral

$$Y = \int_0^T g(t)X(t)dt \tag{3.151}$$

where $T$ is called the observation interval, then this integral gives us a random variable $Y$. $g(t)$ is any function such that the mean square value of $Y$ is finite; if *for all such $g(t)$s Y is Gaussian distributed, then $X(t)$* is a Gaussian process.

**Example 3.34**  For example for the simple case of $g(t) = \delta(t - T/2)$ we get the case of the random variable $Y = X(T/2)$ (a single point). The mean square value of $Y$ is finite then for $X(t)$ is to be a Gaussian process, $Y$ should be Gaussian.

Furthermore, since $T$ is arbitrary, all the random variables defined by

$$Y = X(t = t_k) \qquad k = 0, 1, \ldots \tag{3.152}$$

must all be Gaussian. Similarly for $g(t) = 1$, and

$$Y = \int_0^T X(t)dt \tag{3.153}$$

should also be Gaussian.

Using this property of a random variable, we can show that when the input to a LTI filter is a Gaussian process, then the output is also a Gaussian process. To prove this property,

$$Y(t) = \int_{\tau=0}^{\infty} h(t - \tau)X(\tau)d\tau \tag{3.154}$$

we now formulate that the random variable,

$$Z = \int_{t=0}^T g(t)Y(t)dt$$

$$= \int_{t=0}^T g(t) \int_{\tau=0}^{\infty} h(t - \tau)X(\tau)d\tau dt$$

---

[10] Due to central limit theorem, considered in a later section.

should be a Gaussian random variable (for $g(t)$ as specified earlier) then $Y(t)$ is a Gaussian random process. Therefore,

$$g_1(\tau) = \int_{t=0}^{T} g(t)h(t - \tau)dt$$

is now the 'new' function which also belongs to the class specified, since $h(\tau)$ is stable, and linear function. Therefore formulating a new integral,

$$Z = \int_{\tau=0}^{\infty} g_1(\tau)X(\tau)d\tau$$

In this equation, the support of $g(\tau)$ will be the interval $[0, T]$, or more specifically, $g_1(\tau)$ will be zero outside this interval. Therefore

$$Z = \int_{\tau=0}^{\infty} g_1(\tau)X(\tau)d\tau$$
$$= \int_{\tau=0}^{T} g_1(\tau)X(\tau)d\tau$$

which on closer examination is a Gaussian random variable since $X(\tau)$ is a Gaussian random process.

### 3.18.1   Central Limit Theorem

For a random variable to be Gaussian, its PDF must be

$$f_X(x) = \frac{1}{\sigma_X\sqrt{2\pi}} \exp\left\{-\frac{(x - m_X)^2}{2\sigma_X^2}\right\} \qquad (3.155)$$

We define the $\mathcal{N}(0, 1)$ distribution as the Gaussian PDF with mean, $m_X = 0$ and $\sigma_X = 1$. Or

$$\mathcal{N}(0, 1) = \frac{1}{\sqrt{2\pi}} \exp\left\{-\frac{x^2}{2}\right\} \qquad (3.156)$$

the central limit theorem states, that if there are $N$ identical and independent random variables, $X_i$ with mean $m$ and variance $\sigma^2$ then we normalise each of these random variables as

$$Y_i = \frac{X_i - m}{\sigma} \qquad (3.157)$$

so that

$$E[Y_i] = E\left[\frac{X_i - m}{\sigma}\right] = 0$$

and

$$\text{var}[Y_i] = E[Y_i^2] = E\left[\left(\frac{X_i - m}{\sigma}\right)^2\right]$$
$$= \frac{\sigma^2}{\sigma^2} = 1$$

We now define a new random variable,

$$Z = \frac{1}{\sqrt{N}} \sum_1^N Y_i$$

then the central limit theorem says that the PDF of $Z$ approaches the Gaussian distribution, $\mathcal{N}(0, 1)$ in the limit as $N$ tends to infinity.

**Example 3.35** Apply the central limit theorem to $X_i$ which are uniformly distributed random variables, with

$$f_{X_i}(x_i) = \begin{cases} 0 & x_i < \sqrt{3} \\ \frac{1}{2\sqrt{3}} & -\sqrt{3} < x_i < \sqrt{3} \\ 0 & x_i > \sqrt{3} \end{cases} \tag{3.158}$$

which have $m = 0$, and $\sigma = 1$ as given in Example 3.23.

We consider the new random variable

$$Z = \frac{1}{\sqrt{N}} \sum_1^N X_i$$

To consider this random variable, we look at

$$Y_i = \frac{X_i}{\sqrt{N}}$$
$$\text{lower limit of } Y_i = -\sqrt{\frac{3}{N}}$$
$$\text{upper limit of } Y_i = \sqrt{\frac{3}{N}}$$

then

$$dY_i = \frac{dX_i}{\sqrt{N}}$$

then, in the interval $[-\sqrt{3/N}, \sqrt{3/N}]$

$$f_{Y_i}(y_i) = f_{X_i}(x_i) \left| \frac{dx_i}{dy_i} \right|$$

$$= \frac{1}{2}\sqrt{\frac{N}{3}}$$

and zero elsewhere. Therefore the characteristic function of $Z$ is,

$$E\left[\exp\left\{jv\sum_1^N Y_i\right\}\right] = \prod_1^N E\left[\exp\{jvY_i\}\right]$$

$$=$$

$$= \prod_1^N \sqrt{\frac{N}{12}} \int_{-\sqrt{3/N}}^{\sqrt{3/N}} e^{jvy_i} dy_i$$

$$\Phi_{cl}(v) = \left\{ \frac{\sin\left(\frac{\sqrt{3}v}{\sqrt{N}}\right)\sqrt{N}}{\sqrt{3}v} \right\}^N \qquad (3.159)$$

Similarly characteristic function of the $\mathcal{N}(0, 1)$ distribution is (from Eq. (3.73))

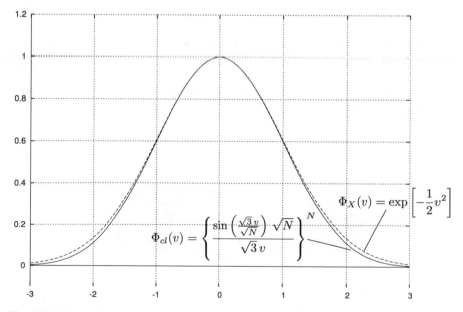

**Fig. 3.32** Plot of $\Phi_{cl}(v)$ for the uniform distributions (for $N = 5$) with $\Phi_X(v)$ for the $\mathcal{N}(0, 1)$ distribution

$$\Phi_X(v) = \exp\left[-\frac{1}{2}v^2\right] \tag{3.160}$$

We plot $\Phi_{cl}(v)$ for the uniform distributions (for $N = 5$) with $\Phi_X(v)$ for the $\mathcal{N}(0, 1)$ distribution in Fig. 3.32 which shows the close agreement between these two functions.

## 3.19   Narrowband Noise

If we pass white noise, $w(t)$, through a narrowband filter with frequency response $H(\omega)$ centred at $\omega_c$ and bandwidth $\omega_B$ as shown in Fig. 3.33, we generate narrowband noise, $n_{NB}(t)$. Though the filter shown in the figure is an ideal BP filter, in practice it may use any practical filter centred at $\omega = \omega_c$, and the theory which follows still applies.

It is clear that the noise generated by passing the white noise through the narrowband filter will have the characteristic given by

$$n_{\text{NB}}(t) \Leftrightarrow N_{\text{NB}}(\omega)$$

where $N_{\text{NB}}(\omega)$ is centred at $\omega = \omega_c$ and having a bandwidth of $\omega_B$, and since the PSDs of the input and output noise processes are related by

$$S_{\text{NB}}(\omega) = S_W(\omega) |H(\omega)|^2 \tag{3.161}$$

and hence the PSD of $n_{NB}(t)$ is centred at $\omega = \pm\omega_c$. Since $w(t)$ is a wide-sense stationary random process, $n_{\text{NB}}(t)$ is also a wide-sense stationary random process. A sample function is shown in Fig. 3.34.

To investigate the properties of the narrowband noise we define its pre-envelope in the usual fashion by (see Sect. 2.17),

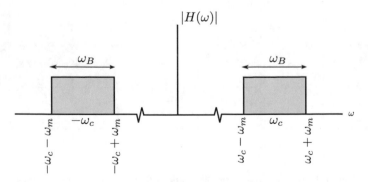

**Fig. 3.33** Bandpass filter used to generate narrowband noise

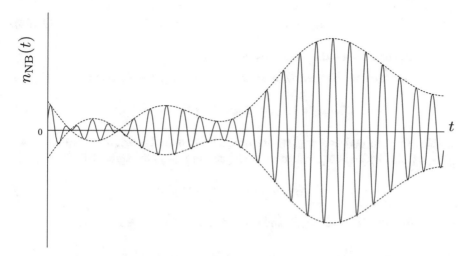

**Fig. 3.34** Sample function of narrowband noise

$$n_{\mathrm{NB}+}(t) = n_{\mathrm{NB}}(t) + jn_{\mathrm{NB}h}(t) \tag{3.162}$$

where $n_{\mathrm{NB}h}(t)$ is the Hilbert transform of $n_{\mathrm{NB}}(t)$. $N_{\mathrm{NB}+}(\omega)$ clearly consists of all the positive frequencies of $N_{\mathrm{NB}}(\omega)$ centred at $\omega = \omega_c$ and multiplied by two. Translating these frequencies to $\omega = 0$ from $\omega = \omega_c$,

$$
\begin{aligned}
\tilde{n}(t) &= n_{\mathrm{NB}+}(t)e^{-j\omega_c t} \\
&= [n_{\mathrm{NB}}(t) + jn_{\mathrm{NB}h}(t)][\cos(\omega_c t) - j\sin(\omega_c t)] \\
&= [n_{\mathrm{NB}}(t)\cos(\omega_c t) + n_{\mathrm{NB}h}(t)\sin(\omega_c t)] + j[n_{\mathrm{NB}h}(t)\cos(\omega_c t) - n_{\mathrm{NB}}(t)\sin(\omega_c t)] \\
&= n_I(t) + jn_Q(t)
\end{aligned}
\tag{3.163}
$$

In these equations, $\tilde{n}(t)$, $n_I(t)$ and $n_Q(t)$ are centred at $\omega = 0$. $n_I(t)$ is the in-phase component of the noise while $n_Q(t)$ is the quadrature phase component of the noise. Therefore

$$
\begin{aligned}
n_I(t) &= n_{\mathrm{NB}}(t)\cos(\omega_c t) + n_{\mathrm{NB}h}(t)\sin(\omega_c t) \\
n_Q(t) &= n_{\mathrm{NB}h}(t)\cos(\omega_c t) - n_{\mathrm{NB}}(t)\sin(\omega_c t)
\end{aligned}
\tag{3.164}
$$

We now obtain an expression for $n_{\mathrm{NB}}(t)$ by multiplying the first expression of Equation set (3.164) by $\cos(\omega_c t)$ and the second by $-\sin(\omega_c t)$ and adding the two resultant expressions,

$$n_{\mathrm{NB}}(t) = n_I(t)\cos(\omega_c t) - n_Q(t)\sin(\omega_c t) \tag{3.165}$$

Similarly we may obtain an expression for $n_{\mathrm{NB}h}(t)$ by multiplying the first expression of Equation set (3.164) by $\sin(\omega_c t)$ and the second by $\cos(\omega_c t)$ and adding the two resultant expressions to get

$$n_{\text{NB}h}(t) = n_I(t)\sin(\omega_c t) + n_Q(t)\cos(\omega_c t) \qquad (3.166)$$

We now concentrate on the properties of $n_{\text{NB}}(t)$.

1. Since we obtained the narrowband noise, $n_{\text{NB}}(t)$, from Expression 3.161, then

$$E[n_{\text{NB}}(t)n_{\text{NB}}(t+\tau)] = R_{\text{NB}}(\tau) \Leftrightarrow S_W(\omega)|H(\omega)|^2 \qquad (3.167)$$

2. Also since $n_{\text{NB}}(t)$ and $n_{\text{NB}h}(t)$ are linked though the Hilbert transform with the equation, (which is equivalent through passing $n_{\text{NB}}(t)$ through a LTI filter)

$$N_{\text{NB}h}(\omega) = -j\text{sgn}(\omega)N_{\text{NB}}(\omega) \qquad (3.168)$$

therefore the PSDs of $n_{\text{NB}}(t)$ and $n_{\text{NB}h}(t)$ are exactly the same,

$$E[n_{\text{NB}h}(t)n_{\text{NB}h}(t+\tau)] = R_{\text{NB}h}(\tau) \Leftrightarrow S_W(\omega)|H(\omega)|^2 \qquad (3.169)$$

And since the two PSDs are exactly the same, therefore both $n_{\text{NB}}(t)$ and $n_{\text{NB}h}(t)$ have the same average power.

$$E[n_{\text{NB}}(t)^2] = E[n_{\text{NB}h}(t)^2] = \mathbb{P}_{\text{NB}} \qquad (3.170)$$

We also assume, without proof that $n_{\text{NB}}(t)$ and $n_{\text{NB}h}(t)$ are uncorrelated. That is

$$E[n_{\text{NB}}(t)n_{\text{NB}h}(t)] = 0 \qquad (3.171)$$

We can, using these expressions to show the following:

1. The in-phase and quadrature phase random processes have zero mean. This we can show by using Equation set (3.164),

$$E[n_I(t)] = E[n_{\text{NB}}(t)]\cos(\omega_c t) + E[n_{\text{NB}h}(t)]\sin(\omega_c t)$$
$$= 0 \qquad (3.172)$$

since both $n_{\text{NB}}(t)$ and $n_{\text{NB}h}(t)$ have zero mean. Similarly we can show that $E[n_Q(t)] = 0$.

2. The in-phase and quadrature phase components have the same average power. To prove this result we use Equation set (3.164) (see Appendix I, for the proof),

$$E\left[n_I(t)^2\right] = \mathbb{P}_{\text{NB}}$$

where we have used Eq. (3.171). Similarly we can show that

$$E\left[n_Q(t)^2\right] = \mathbb{P}_{\text{NB}} \qquad (3.173)$$

We show, symbolically, the PSDs of the four random processes in Fig. 3.35.

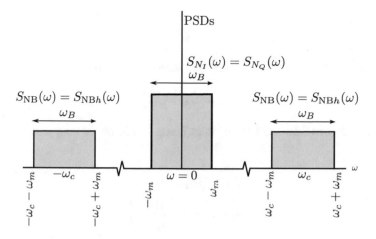

**Fig. 3.35** Power spectral densities of all the various noise phenomena

3. The two random processes, $n_I(t)$ and $n_Q(t)$ are also uncorrelated (see Appendix I for the proof).

$$E\left[n_I(t)n_Q(t)\right] = 0 \tag{3.174}$$

### 3.19.1 Rayleigh Distribution

The probability density function of either $n_I(t)$ or $n_Q(t)$ is given by

$$f_N(n) = \frac{1}{\sigma_N\sqrt{2\pi}}e^{-[n^2/(2\sigma_N{}^2)]} \tag{3.175}$$

where $\sigma_N$ is the variance of the noise and is equal to

$$\sigma_N^2 = E\left[N^2(t)\right] = \frac{N_0 \times 2B_{\text{eq}}}{4\pi} \tag{3.176}$$

where $B_{\text{eq}}$ (in rad/s) is the equivalent noise bandwidth of either the in-phase or quadrature phase noise.

The joint PDF of these two WSS processes is given by

$$f_{N_I}(n_I)f_{N_Q}(n_Q) = \left(\frac{1}{\sigma_{N_I}\sqrt{2\pi}}e^{-[n_I^2/(2\sigma_{N_I}{}^2)]}\right)\left(\frac{1}{\sigma_{N_Q}\sqrt{2\pi}}e^{-[n_Q^2/(2\sigma_{N_Q}{}^2)]}\right)$$

which simplifies to

$$f_{N_I, N_Q}(n_i, n_Q) = \frac{1}{\sigma_N^2 (2\pi)} e^{-\left[(n_I^2 + n_Q^2)/(2\sigma_N^2)\right]} \tag{3.177}$$

where these two processes are considered independent, with

$$\sigma_N = \sigma_{N_I} = \sigma_{N_Q}$$

If we want to consider the PDF of joint process in terms of amplitude and phase, where

$$a(t) = \sqrt{n_I^2(t) + n_Q^2(t)}$$

$$\phi(t) = \tan^{-1}\left[\frac{n_I(t)}{n_Q(t)}\right] \tag{3.178}$$

which is a polar coordinate representation. Then

$$n_I(t) = A(t)\cos\phi(t)$$
$$\text{and } n_Q(t) = A(t)\sin\phi(t) \tag{3.179}$$

and

$$dn_I dn_Q = a \, da \, d\phi$$

so Eq. (3.177) becomes

$$f_{A,\Phi}(a, \phi) a da d\phi = \frac{a}{\sigma_N^2 (2\pi)} e^{-[a^2/(2\sigma_N^2)]} da d\phi \quad 0 \leq a \leq \infty; \quad -\pi \leq \phi \leq \pi \tag{3.180}$$

where $\Phi$ is uniformly distributed in the interval $[0, 2\pi]$. After integration over $\phi$,

$$f_A(a) = \frac{a}{\sigma_N^2} e^{-[a^2/(2\sigma_N^2)]}$$

$A$ is Rayleigh distributed. Figure 3.36 shows the graph of $af_{A,\Phi}(a, \phi)$ for $\sigma_N = 1$. For different values of $\sigma_N$ we get scaled versions of the same figure.

## 3.20  Important Concepts and Formulae

- The *sample space*, $S$ is the set of all possible outcomes of an experiment
- The set, $S$, may have

  - A finite number of (elementary) outcomes
  - or an infinite number of (elementary) outcomes.

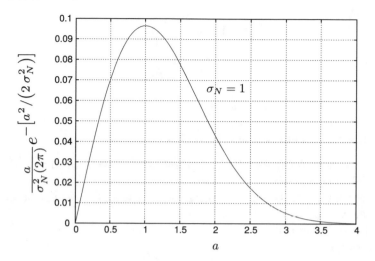

**Fig. 3.36** Graph of $af_{A,\Phi}(a, \phi)$. $A$ is Rayleigh distributed and $\Phi$ is uniformly distributed

- The *sample points* are all possible outcomes $x$, such that $x \in S$ are the *elementary outcomes*.
- For an event $E$

  - $1 \geq P(E) \geq 0$ and which is true for all events, $E \subset S$.
  - $P(S) = 1$. Here the example is the event "getting any number on die '1' and any number on die '2'" is the sample space itself: it will always happen, therefore probability of this event is $P(S) = 1$.
  - If two events are mutually exclusive,

$$E_1 \cap E_2 = \phi \qquad (3.181)$$

  then the probability of occurrence of $E_1$ *or* $E_2$ is given by

$$P(E_1 \cup E_2) = P(E_1) + P(E_2)$$

- If we consider the event $E$ and its complement, $E^c$,

$$P(E^c) = 1 - P(E)$$

- $P(E_1 \cup E_2) = P(E_1) + P(E_2) - P(E_1 \cap E_2)$
- $P(A \cap B) = P(A|B)P(B)$
- Baye's rule

$$P(A|B) = \frac{P(A \cap B)}{P(B)}$$
$$= \frac{P(B|A)P(A)}{P(B)}$$

- A random variable $x$ is the outcome of random experiment such that $x$ is defined on the real line, $-\infty < x < \infty$.
- The probability density function (PDF) has the properties,

  - The function, $f_X(x)$, (the PDF) takes real values and is defined on the real line, $x \in \mathbb{R}$, generally $-\infty < X < \infty$
  - On some parts of the line the function may be zero. For example in the screw example $f_L(l) = 0$ for $-\infty < L < 0$.
  - $f_X(x) \geq 0$ everywhere.
  - The integral

$$P(-\infty < X < \infty) = \int_{-\infty}^{\infty} f_X(\sigma)d\sigma = 1 \qquad (3.182)$$

- The cumulative distribution function (CDF) is defined as

$$F_X(x) = \int_{-\infty}^{x} f_X(\sigma)d\sigma$$

it is the probability

$$P(-\infty < X < x) = F_X(x) \qquad (3.183)$$

  - $F_X(x)$ is bounded by zero and one.

$$0 \leq F_X(x) \leq 1$$

  - The CDF is monotonically increasing

$$F_X(x_2) > F_X(x_1) \qquad \text{when } x_2 > x_1$$

- The mean of a random variable $x$ is given by

$$E[X] = m_X = \int_{-\infty}^{\infty} x f_X(x)dx$$

- The $n$th moment of a random variable, $X$ is given by $E[X^n]$
- The $n$th central moment of a random variable is $E[(X - m_X)^n]$
- The characteristic function of a random variable $x$ is given by $\Phi_X(v) = E[e^{jvX}]$
- For two random variables, the PDF is

$$f_{X,Y}(x, y)$$

- Also
$$f_{X,Y}(x, y) = f_{X|Y}(x|y) f_Y(y) = f_{Y|X}(y|x) f_X(x)$$

- And in the case of *independent random variables,*

$$f_{X|Y}(x|y) = f_X(x)$$

$$f_{X,Y}(x, y) = f_X(x) f_Y(y)$$

- A random process is of random function of time.
- A wide-sense stationary process has a mean which is constant for all values of time and an autocorrelation function which is dependant on the difference $\tau = t_1 - t_2$ where $t_1 \geq t_2$ or $t_2 \geq t_1$.
- A random process is ergodic if the ensemble mean of the process is equal to its mean over time. Also variance of the ergodic mean is equal to zero.
- For a process to be ergodic it must be WSS.
- The power spectral density of a random process is the FT of the autocorrelation function using the variable $\tau$ as the transform variable

$$S_X(\omega) \Leftrightarrow R_X(\tau)$$

- The autocorrelation function is an even function of $\tau$.
- The autocorrelation function at $\tau = 0$

$$R_N(0) > 0$$

- Also
$$R_N(0) \geq |R_N(\tau)|$$

- The autocorrelation function for a random binary wave is

$$R_N(\tau) = \begin{cases} 0 & \tau < -T \\ A^2(1 + \frac{\tau}{T}) & -T \leq \tau \leq 0 \\ A^2(1 - \frac{\tau}{T}) & 0 \leq \tau \leq T \\ 0 & \tau > T \end{cases}$$

- For transmission through a linear system,

$$S_Y(\omega) = |H(\omega)|^2 S_X(\omega)$$

- The noise equivalent bandwidth is given by

$$B = \frac{\int_0^\infty |H(\omega)|^2 d\omega}{|H_{max}|^2}$$

- For the WSS random signal

$$E[X^2(t)] = R(0) = \frac{1}{2\pi} \int_{-\infty}^{\infty} S_X(\omega)d\omega$$

- The PSD of a random process is non-negative,

$$S_X(\omega) \geq 0$$

- And

$$S_X(\omega) = S_X(-\omega)$$

- White noise has an autocorrelation function

$$R_W(\tau) = \frac{N_0}{2}\delta(\tau)$$

with

$$S_W(\omega) = \frac{N_0}{2}$$

- The PDF of a Gaussian process is given by

$$f_X(x) = \frac{1}{\sigma_X\sqrt{2\pi}}e^{-[(x-\mu_X)^2/(2\sigma_X{}^2)]}$$

- The Chebyshev inequality states that

$$P(|X - m| \geq \epsilon) \leq \left(\frac{\sigma}{\epsilon}\right)^2$$

for any continuous random variable. Another way of writing this is

$$P(|X - m| \leq \epsilon) \geq 1 - \left(\frac{\sigma}{\epsilon}\right)^2$$

- The characteristic function of a random variable is given by

$$\Phi_X(v) = E[e^{jvx}] = \int_{-\infty}^{\infty} f_X(x)e^{jvx}dx$$

- The two dimensional Rayleigh distribution is

$$f_{A,\Phi}(a, \phi)adad\phi = \frac{a}{\sigma_N^2(2\pi)}e^{-[a^2/(2\sigma_N{}^2)]}dad\phi \quad 0 \leq a \leq \infty; \quad -\pi \leq \phi \leq \pi$$

$$(3.184)$$

where $A$ is the amplitude and $\Phi$ is the phase random variables.

## 3.21 Self Assessment

### 3.21.1 Review Questions

1. Why must $P(S) = 1$ be true?
2. If $E \subseteq S$ then why is the following equation true $0 \leq P(E) \leq 1$?
3. When is $P(E_1 \cup E_2) = P(E_1) + P(E_2)$ true?
4. If the probability of an event $E$ is $p$, then what is the probability that the event $E$ *does not happen?*
5. If $A \subseteq B$ what is the probability of $A|B$?
6. State Baye's theorem.
7. Define what a PDF and CDF is.
8. Give an example of a uniformly distributed RV.
9. Why is the Gaussian distribution very important?

### 3.21.2 Numerical Problems

1. Define the sample space of two dice.

   (a) Are all the elementary events mutually exclusive and exhaustive?
   (b) Are the events $E_1 = \{(1,1),(2,2),(3,3)\}$ and $E_2 =$ getting a '2' in one throw mutually exclusive?

2. For two dice.

   (a) What is the probability of getting two '6's?
   (b) What is the probability of getting one '2' at least in three throws?
   (c) What is the probability of not getting a '1' in three throws?

3. You are given $P[A \cup B] = 0.7$; $P[A \cup B^c] = 0.9$. Then what is $P[A]$?.
4. In a class of 64 students 12 take Maths, 22 take History and 5 take both History and Maths. How many students do not take History or Maths. Find you answer making a Venn Diagram.
5. In a group of 12, everybody plays football or hockey. 5 members of the group play both sports and 2 members of the group play only hockey. How many play only football?
6. In a binary symmetric channel, if $x$ is the transmitted binary digit and $y$ is the received bit, $P_{y|x}(1|0) = P_{y|x}(0|1) = 0.1$ and $P_x(0) = 0.4$, determine $P_y(0)$ and $P_x(0)$ giving the proper explanation.
7. In a binary symmetric channel, if $x$ is the transmitted binary digit and $y$ is the received bit, $P_{y|x}(1|0) = P_{y|x}(0|1) = 0.1$, find $P_{x|y}(0|1)$ and $P_{x|y}(0|0)$.
8. The duration of a telephone conversation over a telephone is a random variable $T$. The PDF of $T$ is given by

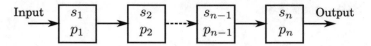

**Fig. 3.37** $n$ systems in cascade

$$f_T(t) = \begin{cases} Ate^{-t} & t \geq 0 \\ 0 & \text{elsewhere} \end{cases}$$

(a) What is the value of $A$?
(b) Calculate $P(t \leq 5)$
(c) Calculate the mean and standard deviation of $t$
(d) Is this a good PDF for $t$. Modify the PDF to get a more realistic one. Do not look at the next question to influence your answer.

9. The duration of a telephone conversation over a telephone is a random variable $T$ in minutes. The PDF of $T$ is given by

$$f_T(t) = \begin{cases} Ate^{-Bt} & t \geq 0 \\ 0 & \text{elsewhere} \end{cases}$$

(a) Find the relation between $A$ and $B$
(b) If the average telephone conversation duration is 1 min, calculate $A$ and $B$
(c) Calculate the CDF of the random variable

10. The exponential time-to-failure PDF in years is given by

$$f(t) = \frac{1}{\lambda} \exp(-t/\lambda), \qquad t > 0.$$

(a) If the average time to failure is 1 year, calculate $\lambda$
(b) Find the probability that the device will fail within two years
(c) Find CDF of this PDF.

11. If at the very start of using a device (burn in), the very instant it is connected to the power supply, out of every 100 devices, 1 device fails, write the PDF of this random variable based on Example 10.

12. A system consists of $n$ systems in cascade as shown in Fig. 3.37, and the

$$p_i = P(t \leq T) \qquad \text{for the } i\text{th system, } s_i$$

is the probability that the system will fail in time $T$. Find the probability that the full system works in the interval $0 \leq t \leq T$

13. system consists of $n$ systems in parallel as shown in Fig. 3.38, and the

**Fig. 3.38** $n$ systems in parallel

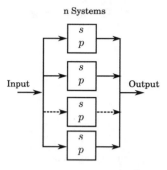

n Systems

Input

Output

$$p = P(t \leq T) \quad \text{for the } i\text{th system, } s$$

is the probability that the system will fail in time $T$. Find the probability that the full system works in the interval $0 \leq t \leq T$

14. Show that the mean of the Gaussian with nonzero mean with

$$f_X(x) = \sqrt{\frac{B}{\pi}} e^{-B(x-x_0)^2}$$

has an average

$$m_X = x_0$$

15. Show that the average time of arrival of a particle which follows the Poisson process, whose probability density function is

$$f_T(t) = \frac{1}{\tau} e^{-t/\tau} \quad 0 \leq T < \infty$$

is $\tau$. Find $P(0 < t < \tau)$, $P(\tau < t < 2\tau)$ and $P(\infty > t > 2\tau)$.

16. Find the second moment of the random variable (RV) $x$ using the PDF

$$f_X(x) = \sqrt{\frac{B}{\pi}} e^{-Bx^2}$$

**Hint:** Use integral as explained in Example 3.12. Find $P(-B < x < B)$.

17. Show why the PDF given by Eq. (3.81) in Example 3.24 is a valid PDF.

18. The CDF of the Weibull distribution is

$$F_X(x) = 1 - e^{-(x/\lambda)^k}$$

for $x \geq 0$ and zero for $x < 0$. Find the PDF for this distribution. Where is it used? How is it connected with the exponential and Rayleigh distributions?

**Fig. 3.39** PDF for Example
19

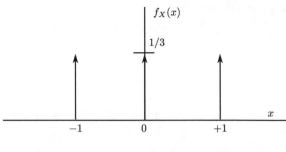

**Fig. 3.40** Detector circuit
for Problem 20

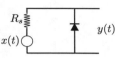

19. A stationary random process has a PDF given by Fig. 3.39. Draw a sample wave of the process

20. A Gaussian random process $x(t)$ has a PDF given by

$$f_X(x) = \frac{1}{\sigma\sqrt{2\pi}}e^{-x^2/2\sigma^2}$$

is passed through a detector circuit shown in Fig. 3.40. Find the PDF of the output random process.

21. The PDF of two jointly distributed RV is

$$f_{XY}(x, y) = Ae^{-x/3}e^{-y/4}u(x)u(y)$$

(a) Find the value of $A$.
(b) Find $f_X(x)$, $f_Y(y)$, $f_{X|Y}(x|y)$ and $f_{Y|X}(y|x)$
(c) Are $x$ and $y$ independent?

22. If two Gaussian RVs

$$f_{X_i}(x_i) = \frac{1}{\sigma_i\sqrt{2\pi}}e^{-(x-\mu_i)/2\sigma_i^2}$$

where $i = 1, 2$. Show that $y = x_1 + x_2$ is also Gaussian with mean $\mu = \mu_1 + \mu_2$ and $\sigma = \sigma_1 + \sigma_2$.

23. Consider patients coming to a dentist's office at random points in time. Let $X_n$ denote the time (in minutes) that the $n$th patient has to wait before being admitted to see the doctor. (a) Describe the random process $X_n; n \geq 1$. (b) Sketch a typical sample of $X_n$.

24. Sketch the ensemble of the waveform of Example 3.26.

$$N(t) = A\cos(\omega t + \Phi) \tag{3.185}$$

where $\Phi$ is randomly distributed in the interval, $0 \leq \Phi \leq 2\pi$. $A$ and $\omega$ are constants. The probability density function of $\Phi$ is

$$f_\Phi(\phi) = \begin{cases} \frac{1}{2\pi} & 0 \leq \phi \leq 2\pi \\ 0 & \text{otherwise} \end{cases}$$

25. For the random process of Eq. (3.185), where $A$ and $\phi$ are constants and $\omega$ is taken two values $\omega = 10$ and $\omega = 20$ uniformly with probability equal to 0.5, find out whether it is WSS.

26. Find the autocorrelation function of

$$N(t) = A \cos(\omega t + \Phi)$$

where $\Phi$ is randomly distributed in the interval, $0 \leq \Phi \leq 2\pi$. $A$ and $\omega$ are constants. The probability density function of $\Phi$ is

$$f_\Phi(\phi) = \begin{cases} \frac{1}{2\pi} & 0 \leq \phi \leq 2\pi \\ 0 & \text{otherwise} \end{cases}$$

and find its power spectral density.

27. Show that

$$\int_{-\infty}^{\infty} h(\tau) d\tau = H(0)$$

where $h(t)$ is the impulse response and $H(0)$ is the frequency response of the system at $\omega = 0$.

28. Show that the PSDs of $n_{\text{NB}}(t)$ and $n_{\text{NB}h}(t)$ are exactly the same no matter which BPF we employ.

# References

1. Johnson J (1928) Thermal agitation of electricity in conductors. Phys Rev 32:97
2. Nyquist H (1928) Thermal agitation of electric charge in conductors. Phys Rev 32:110

# Chapter 4
# Introduction to Select Communication Circuits

## 4.1 Phasors

In communication systems, there is a need to consider sinusoidal functions in a different form from that of the time domain (where we need to consider $t$) to a framework where we do not have to deal with the variable $t$. This is done elegantly using complex variables: consider the sinusoidal function

$$\tilde{\phi}(t) = \phi_0 \cos(\omega t + \theta) \tag{4.1}$$

where $\omega$ is the frequency in rad/s, $\theta$ is a constant phase in radians and $\phi_0$ is the amplitude of the signal. The tilde– $\cdot\tilde{\cdot}\cdot$ – signifies that the entity is an sinusoidal signal–notice that tilde looks like a sine wave. This equation can be written as

$$
\begin{aligned}
\tilde{\phi}(t) &= \Re\left[\phi_0 e^{j(\omega t + \theta)}\right] \\
&= \Re\left[\phi_0 e^{j\theta} e^{j\omega t}\right] \\
&= \Re\left[\Phi\, e^{j\omega t}\right]
\end{aligned}
\tag{4.2}
$$

where $\Re$ signifies taking the real part of the function. Referring to Fig. 4.1 and Eq. (4.2), $\phi_0 e^{j(\omega t + \theta)}$ represents a rotating complex variable with a frequency of rotation equal to $\omega$. In this equation, $\Phi$ is a complex phasor corresponding to $\phi(t)$

$$\Phi = \phi_0\, e^{j\theta} \tag{4.3}$$

A further requirement of sinusoidal circuits is that the amplitude should be converted to root mean square (rms) values. That is $\phi_0$ be replaced by their corresponding rms values.

$$\phi_0 \rightarrow \phi_0/\sqrt{2} \tag{4.4}$$

S. Bhooshan, *Fundamentals of Analogue and Digital Communication Systems*,
Lecture Notes in Electrical Engineering 785,
https://doi.org/10.1007/978-981-16-4277-7_4

**Fig. 4.1** Showing the
relation between
time-domain and frequency-
domain entities

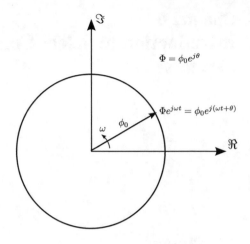

Some interesting results follow if we use this representation:

1. If we differentiate Eq. (4.1) with $t$,

$$\frac{d\tilde{\phi}(t)}{dt} = -\phi_0 \, \omega \sin(\omega t + \theta) \qquad (4.5)$$

and compare with the result of the differentiation of

$$\frac{d}{dt} \Re \left[ \phi_0 \, e^{j \, (\omega t + \theta)} \right] \overset{?}{=} \Re \left\{ \frac{d}{dt} \left[ \phi_0 \, e^{j \, (\omega t + \theta)} \right] \right\} \qquad \text{(interchanging d/dt with } \Re\text{)}$$

where the question mark is asking the valid question: can we interchange $\Re$ and $d/dt$? Continuing,

$$\frac{d}{dt} \Re \left[ \phi_0 \, e^{j \, (\omega t + \theta)} \right] = \Re \left\{ j\omega\phi_0 e^{j \, (\omega t + \theta)} \right\}$$
$$= \Re \left\{ j\omega\phi_0 \left[ \cos \, (\omega t + \theta) + j \sin \, (\omega t + \theta) \right] \right\}$$
$$= \Re \left\{ j\omega\phi_0 \cos \, (\omega t + \theta) - \omega\phi_0 \sin \, (\omega t + \theta) \right\}$$
$$= -\omega\phi_0 \sin \, (\omega t + \theta)$$

which is the same as differentiating Eq. (4.1). Therefore

$$\frac{d}{dt} \tilde{\phi} = \frac{d}{dt} \Re \left( \Phi e^{j\omega t} \right)$$
$$= \Re \left[ j\omega\Phi e^{j\omega t} \right]$$

2. In the same manner, we can show that

$$\int \tilde{\phi}(t)\, dt = \int \phi_0 \cos(\omega t + \theta)\, dt$$

$$= \int \Re\left[\phi_0 e^{j\,(\omega t + \theta)}\right] dt$$

$$= \Re\left[\int \phi_0 e^{j\,(\omega t + \theta)}\, dt\right]$$

$$= \Re\left[\frac{\Phi e^{j\omega t}}{j\omega}\right]$$

**Example 4.1** Find the phasors for (a) $A\cos(\omega_0 t)$ (b) $A\sin(\omega_0 t)$ (c) $A\cos(10^6 t + 2)$
(d) $25\sin(2\pi 10^3 t + 45)$

(a) The term

$$A\cos(\omega_0 t) = \Re\left[Ae^{(\omega_0 t + 0)}\right]$$

That is $\theta = 0$. So $Ae^{j0} \rightarrow (A/\sqrt{2})\angle 0$ which is the phasor.

(b) The term

$$A\sin(\omega_0 t) = A\cos(\pi/2 - \omega_0 t)$$
$$= A\cos(\omega_0 t - \pi/2)$$
$$= \Re\left[Ae^{j\,(\omega_0 t - \pi/2)}\right]$$

so $Ae^{j(-\pi/2)} \rightarrow (A/\sqrt{2})\angle(-\pi/2)$ which is the phasor.

(c) $A\cos(10^6 t + 2)$ is by inspection $= \Re\left[Ae^{j(10^6 t + 2)}\right]$, which gives the phasor $(A/\sqrt{2})e^{j2}$.

(d) The phasor is by inspection $(25/\sqrt{2})e^{j\,(45 - \pi/2)}$.

**Example 4.2** Convert the phasors (a) $A\angle 10$ (b) $10e^{j25}$ to their real-time counterparts.

(a) $A\angle 10 \rightarrow (\sqrt{2}A)e^{j10}$. The conversion to time domain is $\sqrt{2}Ae^{j(10+\omega_0 t)}$. The real-time function is therefore

$$\Re[\sqrt{2}Ae^{j(10+\omega_0 t)}] = \sqrt{2}A\cos(10 + \omega_0 t)$$

(b) $1e^{j15}$ converts to $\sqrt{2}\cos(15 + \omega_0 t)$.


## 4.2 Passive Elements and Filters

### 4.2.1 Resistors Capacitors and Inductors

Using the phasor notation, if

$$\tilde{i}(t) = I_0 \cos \omega t \tag{4.6}$$

is the current through a resistor $R$ then the voltage across it is

$$\tilde{v}(t) = \tilde{i}(t)R$$
$$= RI_0 \cos \omega t \tag{4.7}$$

in phasor notation this is

$$V = IR \tag{4.8}$$

where

$$I = I_0 \angle 0$$
$$\text{and } V = I_0 R \angle 0 \tag{4.9}$$

In general we can show that if $I = I_0 \angle \theta$ then $V = I_0 R \angle \theta$ and so using phasors, this is represented more conveniently:

$$\boxed{V = IR} \tag{4.10}$$

If now we look at the governing equation of a capacitor, then

$$\tilde{v}(t) = \frac{1}{C} \int \tilde{i}(t) dt$$
$$= \frac{1}{C\omega} I_0 \sin \omega t$$
$$= \frac{1}{C\omega} I_0 \cos(\omega t - 90°) \tag{4.11}$$

which using phasors is

$$V = \frac{I \angle -90}{\omega C}$$

$$\boxed{V = \frac{1}{j\omega C} I \quad \text{(remember that integration causes the } 1/j\omega \text{ factor)}} \tag{4.12}$$

similarly for an inductor we get

$$\boxed{V = j\omega L I} \tag{4.13}$$

since $\tilde{v}(t) = L d\tilde{i}(t)/dt$.

The complex quantities $R$, $1/j\omega C$ and $j\omega L$ are the impedances of a resistor, capacitor and inductor, respectively.

$$Z_R = R$$

$$Z_C = \frac{1}{j\omega C}$$

$$Z_L = j\omega L \tag{4.14}$$

## 4.2.2 Mutual Inductances

Considering inductors, if there are many inductors near each other, then the inductor magnetic fields may be linked to each other. Thus if there are inductances $L_i$, for $i = 1, \ldots, N$ is the self inductance of each inductor in a circuit; $M_{ij}$ is the mutual inductance between inductors $L_i$ and $L_j$; and $i_i$ is the current through each of the inductors, then the voltage $v_i$ across each of the inductors is given by

$$v_i = \sum_{j=1}^{N} M_{ij} \frac{di_j}{dt} \tag{4.15}$$

with $M_{ii} = L_i$ and $M_{ij} = M_{ji}$. If these quantities (the voltages and currents) are sinusoidal, then using phasors

$$V_i = \sum_{j=1}^{N} M_{ij} \left( j\omega I_j \right) \tag{4.16}$$

Thus, for instance, for only two inductors with mutual inductance $M$,

$$V_1 = j\omega L_1 I_1 + j\omega M I_2$$
$$V_2 = j\omega M I_1 + j\omega L_2 I_2 \tag{4.17}$$

inverting this equation

$$I_1 = \frac{j\omega L_2}{\Delta} V_1 - \frac{j\omega M}{\Delta} V_2$$
$$I_2 = -\frac{j\omega M}{\Delta} V_1 + \frac{j\omega L_1}{\Delta} V_2 \tag{4.18}$$

where

$$\Delta = \omega^2 M^2 - \omega^2 L_1 L_2 \tag{4.19}$$

Applying these equations to two inductors $L_1$ and $L_2$ with mutual inductances $M$, in the case of Fig. 4.2a

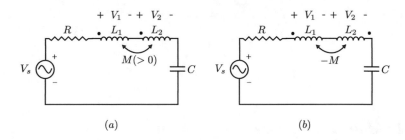

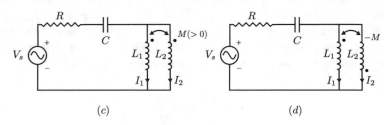

**Fig. 4.2** RLC circuits with mutual inductances

$$V_1 = j\omega L_1 I_1 + j\omega M I_2 = j\omega (L_1 + M) I$$
$$V_2 = j\omega(M + L_2)I \qquad\qquad (4.20)$$

where $I = I_1 = I_2$, the current through the inductors. In this case, the inductors are placed such that their windings have the same direction. One notices that the inductance of each inductor is effectively *increased* slightly. For inductors which are oppositely wound, the equations are

$$V_1 = j\omega L_1 I_1 + j\omega M I_2 = j\omega (L_1 - M) I$$
$$V_2 = j\omega(-M + L_2)I \qquad\qquad (4.21)$$

In a similar manner, the equations applicable to Fig. 4.2c and d are

$$I_1 = \left[ \frac{j\,\omega\,L_2}{\Delta} - \frac{j\,\omega\,(\pm)M}{\Delta} \right] V$$
$$I_2 = \left[ -\frac{j\,\omega\,(\pm)M}{\Delta} + \frac{j\,\omega\,L_1}{\Delta} \right] V \qquad\qquad (4.22)$$

where $\Delta = \omega^2 M^2 - \omega^2 L_1 L_2$; $V$ is the voltage across each of the two inductors and the plus sign is used for the case of (c) part of the figure, while the minus sign is used for the (d) part of the figure.

As far as the 'dot' convention is concerned, when the current enters the dotted terminal of an inductance, it induces a voltage of positive polarity at the dotted

**Fig. 4.3** Basic series RC circuit

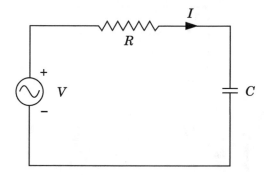

terminal of the second inductor. Mutual inductance between two inductors $L_1$ and $L_2$ is given by

$$M = k\sqrt{L_1 L_2} \qquad 0 \le k \le 1 \tag{4.23}$$

### 4.2.3 RC Circuits

See Fig. 4.3.

If we look at a basic series RC circuit shown in the figure, the phasor equation applicable to the circuit is

$$V = Z_R I + Z_C I$$
$$= I R + \frac{I}{j\omega C} \tag{4.24}$$

so

$$I = \frac{V}{R + \frac{I}{j\omega C}}$$
$$= \frac{j\omega C V}{j\omega R C + 1} \tag{4.25}$$

therefore the voltages across the capacitor, $V_C$, and across the resistor $V_R$ are

$$V_C = \frac{I}{j\omega C} = \frac{V}{j\omega R C + 1} \tag{4.26}$$
$$V_R = I R = \frac{j\omega C R V}{j\omega R C + 1} \tag{4.27}$$

from these equations it is obvious that the two voltages are functions of frequency, and whose magnitudes are shown in Fig. 4.4.

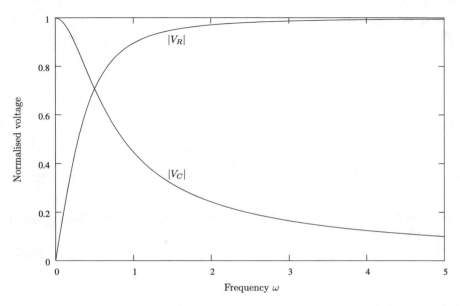

**Fig. 4.4** $|V_C|$ and $|V_R|$ as functions of frequency $\omega$

From the figure it is clear the if the output is taken across the resistor, then the low frequency components are attenuated, while if the output is taken across the capacitor, the high-frequency components are similarly decreased. This is called filtering action. The filter attenuates certain frequencies and passes other frequencies. Thus, a circuit which passes low frequencies is called a *low-pass filter* (LPF), while a circuit which passes high frequencies is called a *high-pass filter* or HPF.

### 4.2.4  Low-Pass and High-Pass Filters

We realise some of the concepts may be used to give us an understanding of low- and high-pass filters. We notice two points concerning the impedances of capacitors and inductors:

$$Z_C = \frac{1}{j\omega C}$$

is infinite at zero frequency and zero at high frequency; while

$$Z_L = j\omega L$$

shows the opposite behaviour.

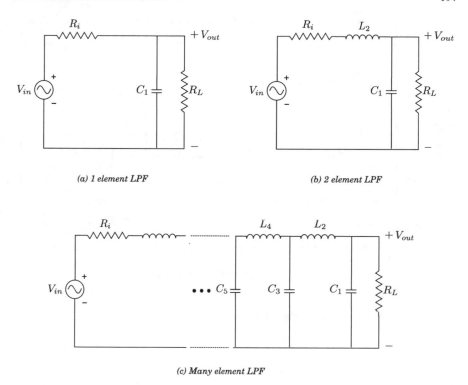

(a) 1 element LPF                                    (b) 2 element LPF

(c) Many element LPF

**Fig. 4.5**  LPF ladder network

Thus, a ladder network with series capacitors and shunt inductors will behave like a high-pass network while a ladder with series inductors and shunt capacitors will behave like a low-pass filter. Such a LPF ladder is illustrated in Fig. 4.5. At low frequencies, the series inductors have low impedances and so allow more current to flow to the load, and more power is dissipated there. The opposite reasoning applies to the shunt capacitors. In this case, at low frequencies *less* current is shunted off to the ground lead (since the capacitor has high impedance at low frequencies) and as the frequency is increased, each capacitor bleeds off more and more current and less current reaches the load.

Take the case of a single shunt capacitor with $C_1 = 1F$ and $R_i = R_L = 1\ \Omega$ as shown in Fig. 4.5a. The circuit shows a voltage $V_{in}$ with a source resistor $R_i$ connected to a shunt capacitor $C_1$. The load resistor, $R_L$, is connected in parallel to the capacitor. For this circuit, the frequency response $|V_{out}/V_{in}|$ is computed and plotted in Fig. 4.6 the case being considered is $n = 1$ which is saying only one element is present in the LPF.

The figure shows that as the frequency is increased, the magnitude of the output voltage falls as compared to the input voltage. When the output voltage falls to $|V_{in}|_{max}/\sqrt{2}$ the power consumed by the load resistor is half of its maximum value (at $\omega = 0$). That is

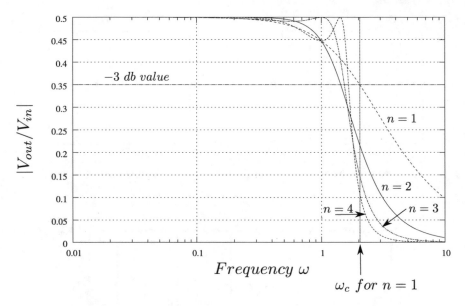

**Fig. 4.6**  The frequency response of many element LPFs

$$20 \log_{10} |\{V_{\text{out}}\}_{\text{max}} / V_{\text{in}}| = 20 \log_{10} \left(1/\sqrt{2}\right) = -3 \text{ db} \qquad (4.28)$$

the frequency, $\omega$, at which this happens is traditionally known as the cutoff frequency or $\omega_c$. For $n = 1$ the value of $\omega_c = 2$ r/s. For the higher values of $n$ the $\omega_c$ values lie between 1 and 2 r/s.

**Example 4.3**  As a first attempt an exceptionally simple approach is used to design the LPFs. We make each inductor $L_i = 1$ H, each capacitor $C_i = 1$ F, and both the resistors $R_i$ and $R_L$ equal to 1 $\Omega$. The value of $n$ is the number of elements used. That is if $n = 1$ then only a single capacitor is used; when $n = 2$ then one capacitor and one inductor is used, and so forth.

The simulation of the LPFs designed by this approach is shown in Fig. 4.6. We can see that all these circuits do indeed behave like LPFs. That is the pass the low frequencies, from $\omega = 0$ to about $\omega = 1$ r/s. After that we can see that $V_{\text{out}}$ falls off as the frequency is increased.

By close examination of this figure it is clear that as the number of elements is increased the important point to be noted is that the output falls more sharply after crossing $\omega = 1$. At $\omega = 1$ r/s the output for $n = 1$ to $n = 4$ the output ratio $|V_{\text{out}}/V_{\text{in}}|$ is 0.45, 0.45, 0.5 and 0.45, respectively. When we observe the response for $\omega = 2$ r/s the output ratios for $n = 1$ to $n = 4$ are 0.35, 0.275, 0.135, and 0.1 respectively. At a third frequency, $\omega = 3$ the output ratios for $n = 1$ to 4 are about 0.225, 0.075, 0.025 and 0, respectively.

Are we satisfied by these results? The price of not designing these filters properly is that the response around $\omega = 1$ r/s becomes badly behaved as the number of elements are increased. For example, for $n = 4$ the response shows an unwelcome ripple which will become worse as the number of elements are increased still further. The reader is referred to the Appendix, and Schaumann et al. [2] where a discussion of Butterworth filters is given.

## 4.2.5  LC Circuits and Bandpass Filters

An LC tuned circuit is an electrical circuit which consists of an inductor, and a capacitor, connected in series or in parallel. The circuit may be part of a harmonic oscillator due to its having resonance properties. Since some resistance is unavoidable in real circuits due to losses in wires (in the case of an inductor) or in the dielectric (in the case of a capacitor). A pure LC circuit has no existence due to these reasons.

There are many applications of LC circuits.

1. As the tuning element of an oscillator circuit.
2. As the tuning element, in the first stage of radio receivers or television sets, where they are used to select a narrow range of frequencies.
3. An LC circuit can be used as a bandpass filter or a band-stop filter.

One important point to be noted is the need to take into account resistance of the inductor. Inductors are constructed from coils of wire, the resistance of which is not desirable, has a significant effect on the circuit.

In Fig. 4.7, a voltage $V$ is impressed on the circuit and current $I$ is entering into the circuit in all cases. Looking at the figure if we analyse the impedance, $Z_{LparC}$, of the parallel LC combination,

$$
\begin{aligned}
Z_{LparC} &= \frac{Z_L Z_C}{Z_L + Z_C} \\
&= \frac{(j\omega L / j\omega C)}{j\omega L + \frac{1}{j\omega C}} \\
&= \frac{j\omega L}{1 - \omega^2 LC}
\end{aligned}
\tag{4.29}
$$

we can observe that at the frequencies $\omega = 0$ and $\infty$ the impedance $|Z_{LparC}| = 0$; and $|Z_{LparC}| = 0$. At the frequency $\omega = \omega_r = 1/\sqrt{LC}$, we find that the impedance is infinity.

$$
|Z_{LparC}|_{\omega=\omega_r} = \infty
\tag{4.30}
$$

where $\omega_r$ is called the resonance frequency.

Now observing the figure we see that the parallel combination is present in (a) and (d). In (a) at low and high frequencies, most of the current, $I$ flows to ground since the impedance is small, and so very little power is delivered to the load. On the other hand at frequencies around $\omega_r$ the impedance acts like an open circuit and

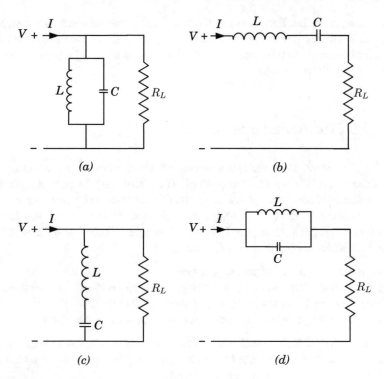

**Fig. 4.7** LC circuits in various configurations in two port networks

all the current flows to the load. The circuit acts like a bandpass filter. For the same reasons, (d) behaves like a band-stop filter. Here, it is the voltage which comes into play and the voltage divider theorem plays a role.

To be specific in (a) part of the figure, to simplify matters, we make $L = 1$ H, $C = 1$ F and $R_L = 1\ \Omega$ then by the current divider theorem the current delivered to the load is

$$I_L = I \frac{Z_{LparC}}{Z_{LparC} + R_L}$$

$$= I \frac{\frac{j\omega}{1-\omega^2}}{\frac{j\omega}{1-\omega^2} + 1}$$

$$= I \frac{j\omega}{j\omega + 1 - \omega^2} \tag{4.31}$$

and the power delivered to the load is $|I_L|^2 R_L$ , which is

$$\mathbb{P}_L = |I|^2 \frac{|j\omega|^2}{|j\omega + 1 - \omega^2|^2} = |I|^2 \frac{\omega^2}{\left[1 - \omega^2\right]^2 + \omega^2} \tag{4.32}$$

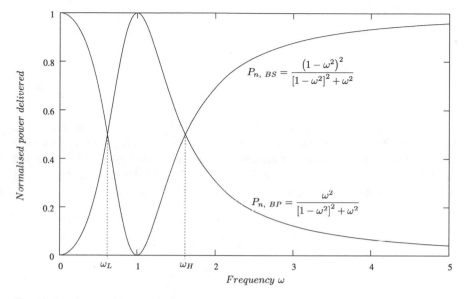

**Fig. 4.8** Bandpass and Band-stop frequency responses

observing this equation we can see that at $\omega = 0$, $\infty$ the power delivered to the load is zero, and at $\omega = \omega_r = 1$ the power delivered is maximum which is $|I|^2$, and the normalised power delivered is 1.

Proceeding along the same lines, analysing (d) part of the figure, the voltage across the load is given by the voltage divider theorem

$$V_L = V \frac{R_L}{Z_{LparC} + R_L}$$

$$= V \frac{1}{\frac{j\omega}{1-\omega^2} + 1}$$

$$= V \frac{1 - \omega^2}{j\omega + 1 - \omega^2} \tag{4.33}$$

and the power delivered to the load is $|V_L|^2 / R_L$, which is

$$\mathbb{P}_L = |V|^2 \frac{\left|1 - \omega^2\right|^2}{\left|j\omega + 1 - \omega^2\right|^2} = |V|^2 \frac{\left(1 - \omega^2\right)^2}{\left[1 - \omega^2\right]^2 + \omega^2} \tag{4.34}$$

observing this equation we can see that at $\omega = 0$, $\infty$ the power delivered to the load is maximum, which is $|V|^2$, and the normalised power delivered is 1; at $\omega = \omega_r = 1$ the power delivered is minimum which zero. These ideas are shown in the figure.

We now turn our attention to the series connection of LC circuits.

$$Z_{LserC} = Z_L + Z_C$$

$$= j\omega L + \frac{1}{j\omega C}$$

$$= \frac{1 - \omega^2 LC}{j\omega C} \tag{4.35}$$

we can, on observing this equation that for $\omega = 0$, $\infty$, the series impedance magnitude $|Z_{LserC}| = \infty$ and for $\omega = \omega_r = 1/\sqrt{LC}$ series impedance magnitude $|Z_{LserC}| = 0$. This fact when applied to Fig. 4.8b and c makes (b) a BP filter and (c) a BS filter and we can show that the normalised power delivered to the load are identical as earlier. That is for $L = 1$ H, $C = 1$ F and $R_L = 1\ \Omega$

$$\mathbb{P}_{n,\ BP}(\omega) = \frac{\omega^2}{\left[1 - \omega^2\right]^2 + \omega^2} \tag{4.36}$$

$$\mathbb{P}_{n,\ BS}(\omega) = \frac{\left(1 - \omega^2\right)^2}{\left[1 - \omega^2\right]^2 + \omega^2} \tag{4.37}$$

we may want to quantify the bandwidth of the filter. We quantify the bandwidth by defining those frequencies at which more than half of the normalised power is either passed or stopped. These frequencies are called the 3 dB frequencies since the power in decibels (dB) is given by

$$\mathbb{P}_{dB} = 10 \log_{10}\left(\frac{1}{2}\right) = -3.010\ \text{dB} \tag{4.38}$$

(See Appendix F). Or, continuing,

$$\mathbb{P}_{n,\ BP\ half} = \frac{\omega_{3db}^2}{\left[1 - \omega_{3db}^2\right]^2 + \omega_{3db}^2} = \frac{1}{2}$$

or

$$2\omega_{3db}^2 = \left[1 - \omega_{3db}^2\right]^2 + \omega_{3db}^2 \tag{4.39}$$

solving this equation

$$\omega_{3db}^2 = 2.6180\ \text{or}\ 0.3820$$
$$\omega_H = 1.6180 \tag{4.40}$$
$$\omega_L = 0.6181 \tag{4.41}$$

where $\omega_H$ and $\omega_L$ are the upper and lower 3dB frequencies, respectively. The $Q$ or quality of these filters is defined as

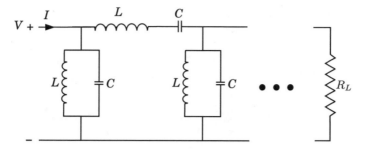

*(a) Multistage Band-pass filter*

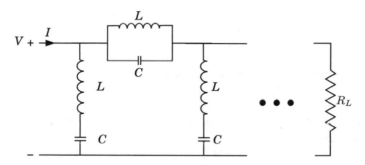

*(b) Multistage Band-stop filter*

**Fig. 4.9** Multistage BS and BP filters

$$Q - \frac{\omega_r}{\omega_H - \omega_L} - 1 \quad \text{(in these cases)} \tag{4.42}$$

how do we improve the performance of these filters? We improve the performance by adding more LC stages of the same type: for BP filters the stages are in series and parallel alternately of the BP type till we achieve the desired performance. These ideas are shown in Fig. 4.9. The same remarks apply to BS filters. There are of course more advanced methods to design filters which are beyond the scope of this book. Some further information about inductors is given in Appendix E.

## 4.3 Operational Amplifiers

Many times in our circuits we require operational amplifiers (Op Amps) which essentially may be used as linear circuits. Without going into the internal details of the Op Amp, the equivalent circuit of an Op Amp is shown in Fig. 4.10. In this circuit, $V_+$ is the voltage applied to the 'non-inverting' terminal, $V_-$ is the voltage applied to the 'inverting' terminal, $V_d = V_+ - V_-$ is the difference between these two voltages, $R_i$

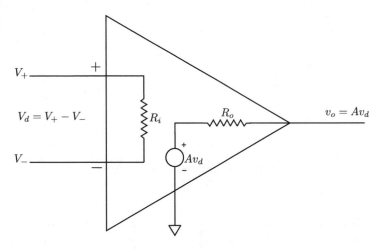

**Fig. 4.10**  Equivalent circuit of an operational amplifier

is the input impedance, $R_0$ is the output impedance and $A$ is the large signal gain. The output voltage,

$$V_0 = AV_d = A(V_+ - V_-) \tag{4.43}$$

In a general Op Amp, the large signal gain is very large, $A \approx 10^3 - 10^4$, the input impedance is also very large, $R_i \approx 10^6 - 10^9$ ($\Omega$), while the output impedance is small, $R_0 \approx 100$ ($\Omega$). A most important point in using Op Amps is that they do not 'load' the next Op Amp circuit following them, and therefore they can be used like block diagrams, each stage performing its own function without changing the function of the next stage or previous stage. A more comprehensive treatment of Op Amps is given in Gayakwad [1].

The most important aspect about an Op Amp is that when it is used in a (negative) feedback configuration, the input terminals may be treated as possessing identical voltages.

$$V_+ - V_- \approx 0 \tag{4.44}$$

since the open loop gain, $A$ is very large.

The Op Amp may be used in a very large number of configurations, such as adders, integrators, differentiators, and active filters to only mention some of the applications. Op Amps may also be used in oscillators and other nonlinear applications.

### 4.3.1   Low-Pass Filter

To do an analysis of the circuit shown in Fig. 4.11, we know that the circuit has feedback through resistors $R_1$ and $R_2$. Therefore, the voltages at the inverting and

**Fig. 4.11** A low-pass filter using an Op Amp

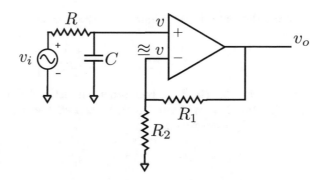

non-inverting terminals are about the same. This is true since the input impedance of the Op Amp, $R_i$, is very large as compared to these resistors. Using this fact, at the non-inverting terminal,

$$v(= V_+) = v_i \times \frac{1/j\omega C}{R + 1/j\omega C}$$

$$= v_i \frac{1}{1 + j\omega RC} \tag{4.45}$$

We may write this equation as if terminal $+$ is an open circuit, since the input impedance of the Op Amp is very large, and very little current flows into the terminal. At the inverting terminal, for the same reason,

$$v(= V_-) = v_o \times \frac{R_2}{R_1 + R_2} \tag{4.46}$$

Equating these two equations,

$$\frac{v_0}{v_i} = \underbrace{\left[\frac{R_2}{R_1 + R_2}\right]^{-1}}_{A} \underbrace{\left[\frac{1}{1 + j\omega RC}\right]}_{B} \tag{4.47}$$

Here, $A$ is the amplification factor and $B$ is the frequency response of the low-pass filter.

We may also design second-order filters by using Op Amps. For example, if we wish to design a Butterworth filter of higher orders we may cascade a number of stages, each stage realising only a single pole or a pair of complex conjugate poles and thereby we may get any filter response. For example, in the low-pass filter which we designed,

$$\left[\frac{1}{1 + j\omega RC}\right] \xrightarrow{\mathcal{L}} \left[\frac{1}{1 + sRC}\right]$$

where $\mathcal{L}$ stands for Laplace transform. The pole here is

$$s_p = -\frac{1}{RC}$$

Another fact which is of great importance is that in all Op Amp circuits, no inductors (and their problems with construction) are used. (See Gayakwad [1] for more details on various Op Amp circuits.)

## 4.4  Linear Time-Invariant Circuits

Linear time-invariant circuits are such that the impulse response, input and output are related by Eq. (2.148) which is

$$y(t) = \int_{-\infty}^{\infty} x(\tau)h(t - \tau)d\tau \tag{4.48}$$

where $h(t)$ is the impulse response of the system, $x(t)$ is the input and $y(t)$ is the output. Using the convolution theorem, (Eq. 2.151)

$$Y(\omega) = H(\omega)X(\omega) \tag{4.49}$$

where $H(\omega)$ is the frequency response of the system. If the system is distortionless, then

$$H(\omega) = Ae^{-j\omega t_0} \tag{4.50}$$

where $A$ and $t_0$ are constants (or in simple language, the amplitude is constant and the phase is linear). With these parameters,

$$Y(\omega) = Ae^{-j\omega t_0}X(\omega)$$
$$\text{which implies that } y(t) = Ax(t - t_0) \tag{4.51}$$

Or that the output is a time-shifted version of the input.

Similarly if the input to the system is $x(t) = A\cos(\omega_0 t)$, ($A$ and $\omega_0$ being constants) then

$$y(t) = \int_{-\infty}^{\infty} A \cos\left[\omega_0(t-\tau)\right] h(\tau) d\tau$$

$$= A \int_{-\infty}^{\infty} \left[\frac{e^{j\omega_0(t-\tau)} + e^{-j\omega_0(t-\tau)}}{2}\right] h(\tau) d\tau$$

$$= \frac{A}{2} \int_{-\infty}^{\infty} e^{j\omega_0(t-\tau)} h(\tau) d\tau + \frac{A}{2} \int_{-\infty}^{\infty} e^{-j\omega_0(t-\tau)} h(\tau) d\tau$$

$$= \frac{Ae^{j\omega_0 t}}{2} \underbrace{\int_{-\infty}^{\infty} e^{-j\omega_0\tau} h(\tau) d\tau}_{H(\omega_0)} + \frac{Ae^{-j\omega_0 t}}{2} \underbrace{\int_{-\infty}^{\infty} e^{-(-j\omega_0)\tau} h(\tau) d\tau}_{H(-\omega_0)}$$

but

$$H(\omega_0) = \int_{-\infty}^{\infty} e^{-j\omega_0\tau} h(\tau) d\tau$$

$$\text{and } H(-\omega_0) = \int_{-\infty}^{\infty} e^{j\omega_0\tau} h(\tau) d\tau \tag{4.52}$$

also since $h(\tau)$ is real, therefore $|H(\omega_0)|$ is a symmetric function while $\angle H_0(\omega_0)$ is an antisymmetric function. Therefore,

$$y(t) = A\frac{e^{j\omega_0 t}}{2} |H(\omega_0)| e^{j\angle H_0(\omega_0)} + A\frac{e^{-j\omega_0 t}}{2} |H(\omega_0)| e^{-j\angle H_0(\omega_0)}$$

$$= A|H(\omega_0)| \frac{e^{j[\omega_0 t + \angle H_0(\omega_0)]} + e^{-j[\omega_0 t + \angle H_0(\omega_0)]}}{2}$$

$$= A|H(\omega_0)| \cos\left[\omega_0 t + \angle H(\omega_0)\right] \tag{4.53}$$

This equation simply put says that if the input is sinusoidal, then the output is also sinusoidal but amplitude is multiplied by the magnitude of the frequency response, at $\omega_0$

$$|H(\omega)|_{\omega=\omega_0}$$

while a phase term is added which is equal to the phase of the frequency response at $\omega_0$:

$$\angle H(\omega)|_{\omega=\omega_0}$$

This result is a very important result, since in our discussions on filters, the frequency responses shown in Sect. 4.2.4 and beyond are the Fourier transforms of the impulse responses of the concerned filters.

**Fig. 4.12**  The filtering
operation

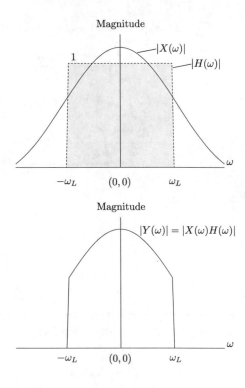

## 4.4.1  A Filter as Treated as a Linear System

Let us take another look at the convolution equation, (4.49) and let $H(\omega)$ be the
frequency response of a low-pass filter. Referring to Fig. 4.12, let $X(\omega)$ be the Fourier
transform of a real but arbitrary signal, and $H(\omega)$ be the Fourier transform of a low-
pass filter which for simplicity may be taken to be

$$H(\omega) = \begin{cases} 1 & \text{for } -\omega_L < \omega < \omega_L \\ 0 & \text{otherwise} \end{cases} \tag{4.54}$$

The Fourier transform of $H(\omega)$ is 1 in the range $-\omega_L < \omega < \omega_L$ and zero otherwise.
Since $|H(\omega)|$ is zero for all values of frequency in the ranges $-\infty < \omega < -\omega_L$
and $\omega_L < \omega < \infty$ and it multiplies $X(\omega)$ in these ranges, therefore in these ranges
the Fourier transform of the output is zero. In the light of the result of Eq. (4.53),
if the Fourier transform is zero in any range means that when we use an input
$A\cos(\omega_0 t)$ then there is no output when $\omega_0$ is in the concerned frequency ranges.
On the other hand in the frequency range $-\omega_L < \omega < \omega_L$ the output has the same
Fourier transform as the input signal and when we use a cosine signal as input, the
output is the same as the input.

## 4.5 Nonlinear Circuit Elements[1]

When we consider communication systems, we have often to use circuits which use nonlinear elements. The simplest of such elements is the diode. The governing equation of a diode is

$$i = i_0 \left[ \exp\left(\frac{q}{kT} v\right) - 1 \right] \qquad (4.55)$$

where $v$ is the voltage across the diode and $i$ is the current through it as shown in Fig. 4.13. Here, the current $i_0$ is the saturation current, which is of the order of $10^{-10}$ A, $q$ is the electron charge, ($=1.602 \times 10^{-19}$ C), $k$ is the Boltzmann constant ($=1.381 \times 10^{-23}$ MKS units) and $T$ is the absolute temperature in $^\circ K$. Substituting these values, for $T = 300^\circ$K,

$$kT/q \cong 0.026 \text{ V} \qquad (4.56)$$

and the diode equation becomes

$$i = i_0 \left[ \exp\left(v/0.026\right) - 1 \right] 2000 \qquad (4.57)$$

If we plot Eq. (4.55), we find that the essentially the diode equation is nonlinear in nature. These plots are shown in Fig. 4.14. In the (a) part of the figure, $i$ is obtained for values of $-0.1 \le v \le 0.65$ V, while in the (b) part $i$ is plotted for $-0.1 \le v \le 0.1$ V. We can see from these figures that the diode is a nonlinear element. If we take a Taylor series about the point $v = 0$ of Eq. (4.57), we get

$$i \cong i_0 \left[ 38.46v + 739.4v^2 + 9478v^3 + \cdots \right] \qquad (4.58)$$

The importance of this equation is that if we give an input voltage, *the current is in terms of powers of that voltage*. This result may be shown in the form of a diagram (Fig. 4.15)

In this figure, $x(t)$ is the input, while $y(t)$ is the output,

$$y(t) = ax(t) + b(x(t)^2 + \cdots \qquad (4.59)$$

**Example 4.4** Using the diode as a nonlinear element, and using only two terms find the output when the input is

$$x(t) = \cos\left(\omega_1 t\right) + \cos\left(\omega_2 t\right)$$

The nonlinear device may be characterised as

$$i = b v^2 + a v$$

---

[1] This section and the following sections must be read when they are required in the other chapters.

**Fig. 4.13**  A diode circuit

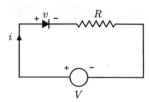

**Fig. 4.14**  Diode equation,
Eq. (4.55) plotted on two
scales: **a** $-0.1 \le v \le 0.65$ V,
and **b** $-0.1 \le v \le 0.1$ V

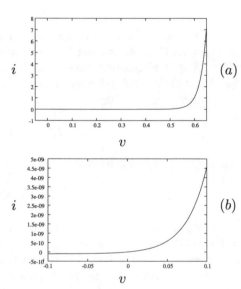

**Fig. 4.15**  Block diagram of
a nonlinear device

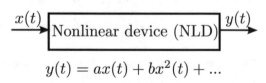

$$y(t) = ax(t) + bx^2(t) + \ldots$$

where $v$ is the voltage across the device and $i$ is the output. Then

$$i = b\left[\cos(\omega_2 t) + \cos(\omega_1 t)\right]^2 + a\left[\cos(\omega_2 t) + \cos(\omega_1 t)\right]$$

$$= b + a\left[\cos(\omega_2 t) + \cos(\omega_1 t)\right] + \frac{b}{2}\left[\cos(2\omega_2 t) + \cos(2\omega_1 t)\right]$$

$$+ b\left[\cos(\omega_2 + \omega_1)t + \cos(\omega_1 - \omega_2)t\right]$$

from this example we can see that for signals centred around two frequencies: $\omega_1$ and $\omega_2$ are served as input to the device, then in general the following signals are generated:

1. a (DC) low frequency term;
2. signals at $\omega_1$ and $\omega_2$;
3. signals at $2\omega_1$ and $2\omega_2$;

4. a signal at $\omega_1 + \omega_2$; and
5. a signal at $|\omega_1 - \omega_2|$.

By using a BPF, we can obtain the frequency range which we desire. For example, if $f_1 = 1$ MHz $f_2 = 5$ MHz, then the output frequencies will be at 0, 1, 2, 4, 5, 6 and 10 MHz, respectively. (the student should work this out)

## 4.6 Introduction to Mixers[2]

### 4.6.1 Frequency Shifting

Mixing is the translation of a frequency $\omega_{rf}$ to another frequency $\omega_{if}$ through a third frequency $\omega_{lo}$. The subscript 'rf' is a short form for 'radio frequency' and the subscripts 'if' and 'lo' are short for 'intermediate frequency' and 'local oscillator', respectively. The relation between these frequencies is

$$\omega_{if} = \left|\omega_{lo} - \omega_{rf}\right| \tag{4.60}$$

or some times

$$\omega_{if} = \omega_{lo} + \omega_{rf} \tag{4.61}$$

In Eq. (4.60), the absolute value is taken to mean the following two equations:

$$
\begin{aligned}
\omega_{if} &= \omega_{lo} - \omega_{rf} \quad &(\text{If } \omega_{lo} > \omega_{rf}) \\
\text{or} \quad &= \omega_{rf} - \omega_{lo} \quad &(\text{If } \omega_{rf} > \omega_{lo})
\end{aligned}
\tag{4.62}
$$

Generally $\omega_{rf} = \omega_c$, the carrier frequency.

The principal circuit which we will be considering here is shown in Fig. 4.16, which is technically called a double balanced mixer. The circuit consists of two centre tapped transformers connected with ring connected diodes, as shown. The RF (radio frequency) signal is generally weak and comes from an antenna, therefore

$$A \gg B$$

or $$A \gg B \,|m(t)|_{\max} \tag{4.63}$$

depending whether the signal is

$$B \cos\left(\omega_{rf} t\right)$$

or $$Bm(t) \cos\left(\omega_{rf} t\right)$$

---

[2] This section must be read when they are required in the other chapters.

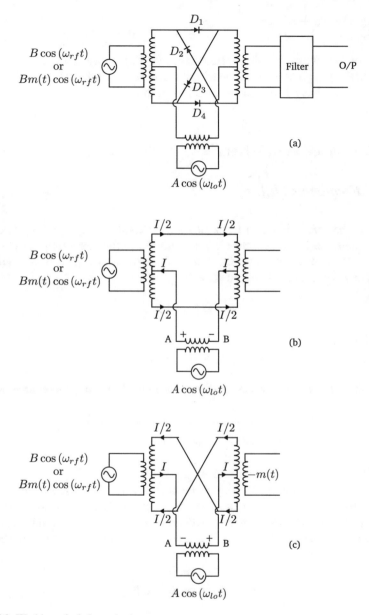

**Fig. 4.16**  Working of a balanced mixer

Assuming that the input waveform is $B \cos\left(\omega_{rf} t\right)$, and since Eq. (4.63) is valid, the switching properties of the diodes are controlled by $A$. Referring to Fig. 4.16b, which is the case of the positive half of the cycle of the local oscillator, that is when the point A is at a higher potential than point B, diodes $D_1$ and $D_4$ are forward biased (they are ON) while diodes $D_2$ and $D_3$ are reverse biased (they are OFF). Hence, the leads containing the ON diodes may be replaced by short circuits while the leads containing the OFF diodes may be open circuited. Since the current $I$ splits into two equal values, namely $I/2$ in each arm of the transformer winding, no voltage due to the local oscillator appears on the output. In this case, on the other hand, the input waveform appears on the output as it is, namely $B \cos\left(\omega_{rf} t\right)$.

Now consider the case when point B of the LO is at a higher potential than point A, we find that diodes $D_1$ and $D_4$ are reverse biased (they are OFF) while diodes $D_2$ and $D_3$ are forward biased (they are ON). This case is shown in Fig. 4.16c when $D_1$ and $D_4$ are replaced by open circuits and $D_2$ and $D_3$ are replaced by short circuits which results in the input appearing reversed on the output, namely $-B \cos\left(\omega_{rf} t\right)$. This repeats every cycle at frequency of $f_{lo} = \omega_{lo}/(2\pi)$ Hz.

These ideas are shown in Fig. 4.17. The pulse train of the upper figure represents how one set of diodes are switched ON and OFF, which in particular are diodes $D_1$ and $D_4$. If this signal is considered as

$$p(t, \omega_{lo}) \equiv p(t) \qquad (4.64)$$

(similar to the pulse train shown in Fig. 2.21 but where $2\tau = T/2$ and which is shifted down by 1/2) then when it is +1 the diodes $D_1$ and $D_4$ are ON, and when $p(t) = -1$, $D_1$ and $D_4$ are OFF. The RF cosine signal is also shown on the same diagram, while

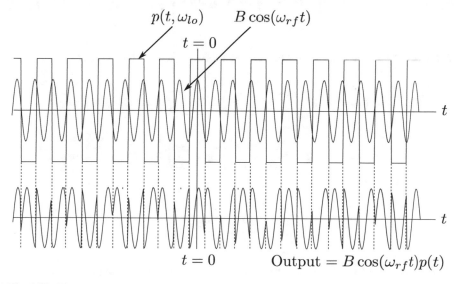

**Fig. 4.17** Waveforms for a balanced mixer

the output is shown below. After a little careful thought it should be clear that the output is a product of $p(t)$ and $B \cos(\omega_{rf} t)$

$$\text{Output} = Bp(t) \cos(\omega_{rf} t) \tag{4.65}$$

It seems like we can get little sense from the output. But let us conduct a little analysis on this configuration. Referring back to the Example 2.13, we find that we consider exactly the same pulse train as we are using here, except that in that example, the pulse train is shifted down. That is

$$p(t) = 2\ p_T(t)|_{\text{example}} - 1 \tag{4.66}$$

and

$$p_T(t)|_{\text{example}} = 0.5 + \sum_{n=1}^{\infty} \left(\frac{2}{n\pi}\right) \sin\left(\frac{n\pi}{2}\right) \cos(n\omega_0 t) \tag{4.67}$$

where $\omega_0$ is the fundamental frequency which we may identify as $\omega_{lo}$. Substituting this Eq. (4.67) in the previous Eq. (4.66), we get

$$p(t) = \sum_{n=1}^{\infty} \left(\frac{4}{n\pi}\right) \sin\left(\frac{n\pi}{2}\right) \cos(n\omega_{lo} t) \tag{4.68}$$

a little thought should inform us that all the terms when $n$ is even are all zero since the sine function is zero for integral values of $\pi$. Therefore,

$$p(t) = \sum_{n=1,3,5,\ldots}^{\infty} \left(\frac{4}{n\pi}\right) \sin\left(\frac{n\pi}{2}\right) \cos(n\omega_{lo} t) \tag{4.69}$$

Notice that the decay as $1/n$. Multiplying this equation by $B \cos(\omega_{rf} t)$

$$Bp(t) \cos(\omega_{rf} t) = B \sum_{n=1,3,5,\ldots}^{\infty} \left(\frac{4}{n\pi}\right) \sin\left(\frac{n\pi}{2}\right) \cos(n\omega_{lo} t) \cos(\omega_{rf} t)$$

$$= B \sum_{n=1,3,5,\ldots}^{\infty} \left(\frac{2}{n\pi}\right) \sin\left(\frac{n\pi}{2}\right) \left\{ \cos\left[ \underbrace{(n\omega_{lo} + \omega_{rf})t}_{\omega_{if}} \right] \right.$$

$$\left. + \cos\left[ \underbrace{(n\omega_{lo} - \omega_{rf})t}_{\omega_{if}} \right] \right\} \tag{4.70}$$

This equation tells us that the output consists of a series of frequencies:

$$n = 1, \omega_{if} = \omega_{lo} + \omega_{rf} \text{ or } |\omega_{lo} - \omega_{rf}|$$
$$n = 3, \omega_{if} = 3\omega_{lo} + \omega_{rf} \text{ or } |3\omega_{lo} - \omega_{rf}|$$
$$\cdots\cdots \tag{4.71}$$

Generally, one is interested in the $n = 1$ term:

$$\omega_{if} = \omega_{lo} + \omega_{rf} \text{ or } |\omega_{lo} - \omega_{rf}| \tag{4.72}$$

one may obtain the desired frequency by using an appropriate LO frequency and bandpass filter on the output. If the input is $Bm(t)\cos(\omega_{rf}t)$ then analysis is exactly the same except that we can extract terms such as

$$m(t)\cos\left[\left(n\omega_{lo} \pm \omega_{rf}\right)t\right] \tag{4.73}$$

where $m(t)$ typically has a bandwidth much less than $\omega_{if}$.

If we very closely examine the performance of the mixer, we come to realise after a little thought, that the mixer really behaves like a *multiplier*. Thus, the two input signals are

$$\cos(\omega_{lo}t) \text{ and } \cos(\omega_{lo}t)$$
$$\text{and the ouput is } \tfrac{1}{2}\left\{\cos\left[\left(\omega_{lo} + \omega_{rf}\right)t\right] + \cos\left[\left(\omega_{lo} - \omega_{rf}\right)t\right]\right\} \tag{4.74}$$

We will remember this point in later.

## 4.7  Important Concepts and Formulae

- For a resister, in phasor notation
$$V = IR$$

- For a capacitor,
$$V = \frac{1}{j\omega C}I$$

- and for an inductor,
$$V = j\omega LI$$

- The complex quantities $R$, $1/j\omega C$ and $j\omega L$ are the impedances of a resistor, capacitor and inductor, respectively.

$$Z_R = R$$
$$Z_C = \frac{1}{j\omega C}$$
$$Z_L = j\omega L$$

- For mutual inductances, with $M_{ii} = L_i$ and $M_{ij} = M_{ji}$. If these quantities (the voltages and currents) are sinusoidal, then using phasors

$$V_i = \sum_{j=1}^{N} M_{ij} \left( j\omega I_j \right)$$

- Linear time-invariant circuits are such that the impulse response, input and output are related by

$$y(t) = \int_{-\infty}^{\infty} x(\tau)h(t - \tau)d\tau$$

where $h(t)$ is the impulse response of the system, $x(t)$ is the input and $y(t)$ is the output.
- Using the convolution theorem,

$$Y(\omega) = H(\omega)X(\omega)$$

where $H(\omega)$ is the frequency response of the system.
- If the system is distortionless, then

$$H(\omega) = Ae^{-j\omega t_0} \tag{4.75}$$

where $A$ and $t_0$ are constants
- With these parameters,

$$Y(\omega) = Ae^{-j\omega t_0} X(\omega)$$

which implies that $y(t) = Ax(t - t_0)$

Or that the output is a time-shifted version of the input.
- For a nonlinear element, if $x(t)$ is the input, then

$$y(t) = ax(t) + b(x(t)^2 + \cdots \tag{4.76}$$

## 4.8  Self-Assessment

### 4.8.1  Review Questions

1. What is the difference between a passive element and an active element? Give the example of a passive element and an active element.
2. What is the relation between the voltage and current in an inductor, capacitor and resistor?

3. If the voltage across a inductor is $v(t)$ and the current through it is $i(t)$ what is the energy stored in the inductor? Of a capacitor?

4. Investigate the concept of a corner frequency and its importance in the frequency response of a filter.

5. After doing Problems 5, 6, 7 of Sect. 4.8.2, what is the importance of these results?

6. In communication system theory why is the frequency response more important than the impulse response?

7. In Eq. (4.52), '$h(\tau)$ is real, therefore $|H(\omega_0)|$ is a symmetric function while $\angle H_0(\omega_0)$ is an antisymmetric function' why has this statement been made?

### 4.8.2 Numerical Problems

1. Write in phasor notation:

   (a) $m(t) = 10\sin(10^6 t + 50°)(V)$ write in terms of rms
   (b) $x(t) = 100\cos(10^6 t + \pi/10)(mV)$ write in terms of amplitude

2. Write the following phasors in time domain

   (a) $100\angle 10°$ (rms V), . Write in terms of amplitude.
   (b) $1000\angle \pi/2$ (mV)

3. Write in phasor form. If $m(t) = 1000\cos(10^6 t + \pi/6)$ (mV)

   (a) Write $d/dt\,[m(t)]$ in phasor form
   (b) Write $\int [m(t)]\,dt$ in phasor form

4. A source of $V(t) = 10\cos(10^6 t + \pi/4)$ with internal resistance 50 $\Omega$ is placed across an impedance combination of a capacitor with $C = 10^{-6}$ F in series with an inductance $L = 10^{-3}$ H, what is the current through the combination in terms of time $(I(t) = \cdots)$? As a phasor?

5. Find the impedance of

   (a) a series combination of a resistor $R$ and capacitor $C$ as a function of frequency. Comment on the value of the impedance as $\omega \to 0$ and $\omega \to \infty$. Plot the impedance versus frequency. What is the result of placing this combination in a series arm of a ladder network (Fig. 4.18) for the frequency $\omega = 0$, and $\omega = \infty$?

   (b) a parallel combination of a resistor $R$ and capacitor $C$ as a function of frequency. Comment on the value of the impedance as $\omega \to 0$ and $\omega \to \infty$. Plot the impedance versus frequency. What is the result of placing this combination in a shunt arm of a ladder network (Fig. 4.18) for the frequency $\omega = \infty$, and $\omega = 0$?

**Fig. 4.18**  Ladder network

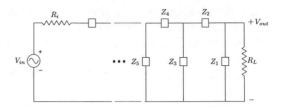

6. Find the impedance of

   (a) a series combination of a resistor $R$ and inductor $L$ as a function of frequency. Comment on the value of the impedance as $\omega \to 0$ and $\omega \to \infty$. Plot the impedance versus frequency. If a source of $V(t) = 10\cos(2\pi\,10^6 t + \pi/4)$ with internal resistance $50\,\Omega$ is placed across an impedance of $5 \times 10^{-6}$ F in series with an inductance of $6 \times 10^{-3}$ H, what is the voltage across the capacitor in terms of time($V_C(t) = \cdots$)?
   (b) Find the impedance of a parallel combination of a resistor $R$ and inductor $L$ as a function of frequency. Comment on the value of the impedance as $\omega \to 0$ and $\omega \to \infty$. Plot the impedance versus frequency.

7. Find the impedance of

   (a) a series combination of a capacitor $C$ and inductor $L$ as a function of frequency. Comment on the value of the impedance as $\omega \to 0$ and $\omega \to \infty$ and $\omega_0 = 1/\sqrt{LC}$. Plot the impedance versus frequency. What is the result of placing this combination in a shunt arm of a ladder network (Fig. 4.18) at $\omega = \omega_0$?
   (b) Find the impedance of a parallel combination of a capacitor $C$ and inductor $L$ as a function of frequency. Comment on the value of the impedance as $\omega \to 0$ and $\omega \to \infty$ and $\omega_0 = 1/\sqrt{LC}$. Plot the impedance versus frequency. What is the result of placing this combination in a series arm of a ladder network (Fig. 4.18) at $\omega = \omega_0$?

8. For the circuit of Fig. 4.19, find the voltage drop across the resistor $R_L$ and the value of $I_1$ and $I_2$.
9. A two port circuit is to be used to pass all the frequencies from 0 1000 Hz (approximately) and reject the frequencies above 1000 Hz. What is the circuit to be used to accomplish this?
10. Define the quality factor of a resonant circuit. A tuned circuit is to be used to (a) reject the 1000 Hz and its neighbour frequencies, (b) pass the 1000 Hz and its neighbour frequencies. What are the circuits we will use to accomplish these results and in what configuration in terms of series and parallel elements?
11. Find $V_L(s)/V(s)$ as a function of $s$ for the RC filter shown in Fig. 4.20. What is the frequency response and impulse response of the circuit? What is the value of the $-3$ dB frequency?

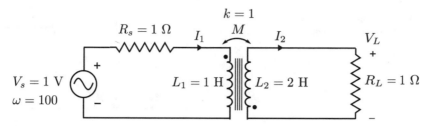

**Fig. 4.19**  Circuit for Problem 8

**Fig. 4.20**  RC filter for
Problem 11

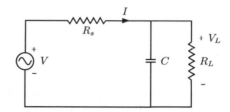

12. Find the impulse response of the ideal LP filter

$$H(\omega) = \begin{cases} 1 & \text{for } -\omega_L < \omega < \omega_L \\ 0 & \text{otherwise} \end{cases}$$

then show why such a filter is not realisable. From this result also show that an ideal BP filter is not realisable.

13. A filter is characterised by the magnitude frequency response

$$|H(\omega)|^2 = \frac{0.25}{1 + \left(\frac{\omega}{\omega_0}\right)^{10}}$$

find the magnitude of the output voltage with an input voltage of 1 V at (a) $\omega = 0$, (b) $\omega \to \infty$ (c) $\omega = 0.95\omega_0$ and (d) $\omega = 1.01\omega_0$. What kind of filter is it? What is the $-3$ dB frequency?

14. It is required that a filter be designed to meet the following requirements: the output voltage, $V_0$, in the frequency band [0, 1000] Hz be

$$0.7071 \, |V_i(\omega)| \le |V_0(\omega)| \le |V_i(\omega)|$$

and in the frequency band [1100,∞) Hz be

$$|V_0(\omega)| \le 0.01 \, |V_i(\omega)|$$

**Fig. 4.21** Circuit with
nonlinear element to go with
Problem 20

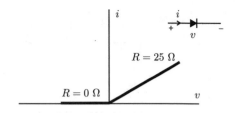

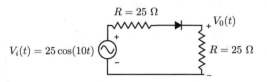

where $V_i(\omega)$ is the input voltage. What is the magnitude squared characteristic
of such a filter? (**Hint:** see Appendix D.1)

15. Draw the circuit to accomplish the design of Problem 14.

16. It is required that a filter be designed to meet the following requirements: the
output voltage, $V_0$, in the frequency band [900, 1100] Hz be

$$0.7071\,|V_i(\omega)| \le |V_0(\omega)| \le |V_i(\omega)|$$

and in the frequency bands outside this band, [0 850] Hz, and [1200,$\infty$) Hz be

$$|V_0(\omega)| \le 0.01\,|V_i(\omega)|$$

where $V_i(\omega)$ is the input voltage. Write the values of the -3 dB frequencies.
How do they map to the low-pass frequencies? What is the magnitude squared
characteristic of such a filter? (**Hint:** see Appendix D.4)

17. Draw the circuit to accomplish the design of Problem 16.

18. Design a first-order low-pass op amp filter to have a cutoff frequency 1000 Hz.
Write the magnitude squared characteristic of such a filter. If we place two such op
amp circuits in cascade, what is the magnitude characteristic of the combination?

19. A LTI filter having the frequency response

$$H(s) = \frac{s}{(s^2 + 5s + 30)(s^2 + 10s + 25)}$$

has an input

$$v_i(t) = 10\cos(7t)$$

find the output, $v_0(t)$.

20. A nonlinear device with the characteristic shown in Fig. 4.21 is used to generate an output voltage as shown in the figure. Find the sinusoidal components in the output, $V_0(t)$, at the frequencies of $\omega = 20$ and $\omega = 30$ rad/sec.
21. In a mixer, if the value $\omega_{lo} = 955$ kHz and $\omega_{if} = 455$ kHz, then what are the possible values of $\omega_{rf}$?

## References

1. Gayakwad RA (2000) Op-amps and linear integrated circuits. Prentice Hall
2. Schaumann R, Xiao H, Mac VV (2009) Design of analog filters, 2nd edn. Oxford University Press, Inc., 2nd edition

# Part II
# Analogue or Continuous Wave Modulation

# Chapter 5
# Amplitude Modulation

## 5.1 Introduction

Communication implies that we have a *source of information*, which may be voice, music or moving pictures, which we want to send to a distant place, called the *receiver*. Since we cannot send the raw information, the first step in the communication process is to convert this information to electrical signals. This signal is called the *message signal*, and it is written in mathematical terms as $m(t)$.

To communicate using these signals, waves are transmitted from a source, and which travel through a convenient medium which is available to the receiver. The medium is technically named as *a channel*. For example, in AM or FM radio broadcast, we use the Earth's atmosphere or the ground as a channel. In satellite television, the channel is also the Earth's atmosphere, but here the link is a satellite in outer space and a receiving dish on the ground. Another example is the transmission of waves using transmission lines which are used in some parts of the Internet as in the case of internet radio. In this case, the channel is a transmission line or optical cables, and, in some cases, even a satellite relay station.

A very important consideration in the transmission of information by use of sinusoidal waves is that such a signal be used which is *not attenuated* and *not distorted* when it is used to communicate from the transmitter to the receiver. For example if we resort to satellite communication, we have to take into account the absorption of waves by the atmosphere of the earth. The absorption is mainly due to the presence of molecules of oxygen, nitrogen, ozone, water vapour and carbon dioxide among others. On observation of Fig. 5.1 we find that for this type of communication we may only use those wavelengths which suffer low attenuation. This is the case, for example, of the window of wavelengths near 1 m, frequencies around 300 MHz. As a counter example if we use long waves ($\lambda \approx 100$ m) for satellite communication we would find that the waves never reach the surface of the Earth as a result of heavy attenuation.

© The Author(s), under exclusive license to Springer Nature Singapore Pte Ltd. 2022    225
S. Bhooshan, *Fundamentals of Analogue and Digital Communication Systems*,
Lecture Notes in Electrical Engineering 785,
https://doi.org/10.1007/978-981-16-4277-7_5

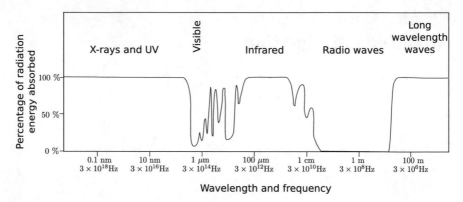

**Fig. 5.1** Absorption characteristics of the electromagnetic spectrum due to the atmosphere. [Credit: NASA, (Public Domain)]

## 5.2   Baseband Communication

Baseband communication consists of transmitting a signal without any *modulation* or *modification*. Baseband communication runs into problems because if there are two or more transmitters and one or more receivers, using the same channel, the signals interfere with each other when they arrive at the receiver. This picture is shown in Fig. 5.2a. In this figure there are $N$ transmitters, $Tx_1$ to $Tx_N$, (and where $N \geq 2$) and $M$ receivers, $Rx_1, \ldots, Rx_M$, ($M \geq 1$) using the same channel. Since the channel is an *additive* one, all the signals get superimposed on each other. That is, if transmitter $i$ transmits the signal $s_i(t)$ then the received signal at the $j$th receiver is

**Fig. 5.2** Baseband communication

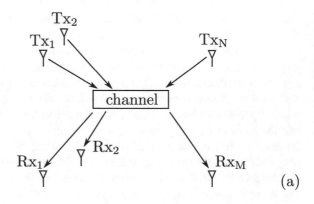

$$r_j(t) = \sum_{i=1}^{N} a_{ji} s_i(t) \tag{5.1}$$

where the $a_{ji}$ are the channel attenuation constants which ensures that all the receivers do not receive the same signal level. From this equation, there is very little possibility of recovering the signal from any particular transmitter.

Referring to Fig. 5.2b where there is a single transmitter and single receiver, we may recover the signal quite easily and baseband communication may be used. An example such a communication is transmission of a signal using transmission lines where there is a single transmitter and single receiver.

## 5.3 Amplitude Modulation: DSB/SC

Amplitude modulation (AM) is a method to transmit and receive messages, using a single high-frequency sinusoidal carrier frequency ($\omega_c$) whose amplitude is varied according to the message signal, $m(t)$. Communication using amplitude modulation was invented in the first quarter of the twentieth century, namely 1900–1925. Many inventors were involved in its conception and execution (see [3] for more details).

A sinusoidal wave as a function of time is given by

$$c(t) = A\cos(\omega_c t) \tag{5.2}$$

where $\omega_c$ is the frequency of the wave in rad/s, $A$, which is a constant, is the amplitude of the wave and $t$ is the time. $\omega_c$ is also called the *carrier* frequency, while $c(t)$ is the carrier signal.

This signal may be written (by Euler's identity) as

$$A\cos(\omega_c t) = \frac{A}{2}\left(e^{j\omega_c t} + e^{-j\omega_c t}\right) \tag{5.3}$$

and its Fourier transform is

$$A\cos(\omega_c t) \Leftrightarrow A\pi\left[\delta(\omega - \omega_c) + \delta(\omega + \omega_c)\right] \tag{5.4}$$

where we have used

$$e^{j\omega_c t} \Leftrightarrow 2\pi\delta(\omega - \omega_c) \tag{5.5}$$

The message signal, which is the signal containing the information, is $m(t)$, and has a Fourier transform, $M(\omega)$, written mathematically as

$$m(t) \Leftrightarrow M(\omega) \tag{5.6}$$

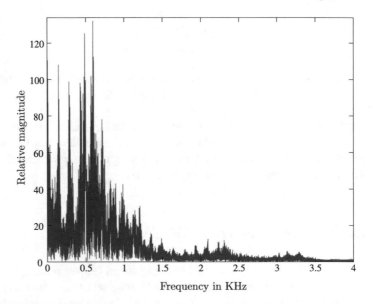

**Fig. 5.3** Magnitude of the Fourier transform of a case of the human voice versus frequency in KHz. The voice is that of the author

The bandwidth of the message is assumed to be limited to some value. For example, if the human voice is considered to be the message signal, $m(t)$, then we find that it has an *approximate* bandwidth of about 4 KHz. An example of this is shown in Fig. 5.3, which depicts the magnitude of the Fourier transform of human speech. We can see from the figure that the magnitude consists of a series of peaks at in the frequency range 0–1 kHz and then falls to low values as the frequency is increased. By 4 kHz the magnitude falls to a negligible value.

Just like in the case of human speech, each information bearing signal has a limited bandwidth, and in technical terms these are called *band limited* signals. For television, the bandwidth of the signal is 4.2–6 MHz, for music it is about 15–20 KHz and so on. Each of these signals are *baseband* signals and signifies the occupation of the bandwidth from zero to the upper limit of the frequency, $\omega = \omega_m$. In the light of the above remarks, the message signal and the magnitude of its Fourier transform are shown, symbolically, in Fig. 5.4.

The figure that shows the message signal is on the left of the figure, and the magnitude of its Fourier transform is shown on the right. The message signal is a time varying signal starting at $t = 0$ and continuing till the message is completed, which is not shown in the figure. In general, the signal is *time limited*. The Fourier transform magnitude $|M(\omega)|$ is clustered around $\omega = 0$, and it is symmetric since the signal is a real-time signal. Also the signal is band limited, lying between $\omega = \mp\omega_m$, which are the minimum and maximum frequencies of the spectrum of the message signal. Note that here we are contradicting one of the properties of the Fourier transform that a time limited signal has an infinite duration FT, but for engineering purposes the

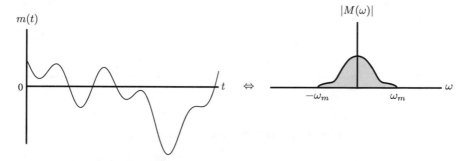

**Fig. 5.4** The message signal and its Fourier transform

extension of the FT beyond $\omega = \mp\omega_m$ is neglected since the tail is too small (many tens of dB less than the maximum value). If $B$ is the bandwidth of the signal in Hz, then $B = \omega_m/2\pi$ where $\omega_m$ is the frequency in rad/s.

In *baseband communication*, the information is transmitted without any modulation over a channel which supports (does *not* attenuate or distort) communication starting from zero hertz up to the upper limit of the signal, $\omega_m$. Typically, baseband communication would occur over coaxial cables or two wire lines since these two media support such a type of communication. An example of baseband communication is communication between a server and a computer over Ethernet LAN using pulses of energy.

As opposed to baseband communication there exists carrier communication, where the message signal is shifted to another (higher) band. The basis of this type of communication is the simple expression

$$m(t)e^{j\omega_c t} \Leftrightarrow \frac{1}{2\pi}[M(\omega) * 2\pi\delta(\omega - \omega_c)] = M(\omega - \omega_c) \tag{5.7}$$

where the $*$ represents convolution, remembering that multiplication of two signals in the time domain is equivalent to convolution in the frequency domain.

$$\boxed{f(t)g(t) \Leftrightarrow \frac{1}{2\pi}F(\omega) * G(\omega)} \tag{5.8}$$

We can see from Eq. (5.7) that the baseband Fourier transform has been shifted to $\omega_c$. This is shown in Fig. 5.5. The figure shows that when we multiply the message signal, $m(t)$, by the complex exponential signal, $e^{j\omega_c t}$, then its spectrum is shifted intact to the centre frequency $\omega_c$. The zero frequency of $M(\omega)$ has been shifted to the frequency $\omega_c$ of $M(\omega - \omega_c)$; the frequency $-\omega_m$ has been shifted to $\omega_c - \omega_m$; and the frequency $+\omega_m$ has been shifted to $\omega_c + \omega_m$. In short every frequency in the spectrum has been shifted to the right by $\omega_c$ without distortion. It must be noted that $\omega_c \gg \omega_m$.

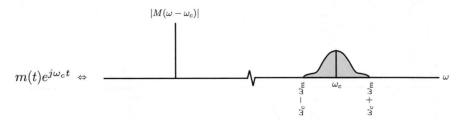

**Fig. 5.5** Shift of $M(\omega)$ to $M(\omega - \omega_c)$ in the frequency domain

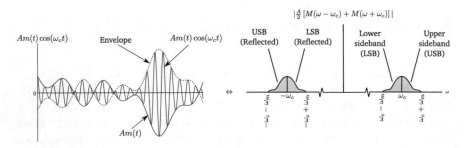

**Fig. 5.6** The function $Am(t)\cos(\omega_c t)$ and its Fourier transform.

It is clear from the observation of the above (Eq. (5.7)) expression that it cannot be put into practice since the function $e^{j\omega_c t}$ is not a real world function:

$$e^{j\omega_c t} = \cos \omega_c t + j \sin \omega_c t \tag{5.9}$$

where the complex number $j$ implies that we cannot realise Eq. (5.7) in reality. But there is a way around this: if we use Eq. (5.3), then

$$Am(t)\cos(\omega_c t) = \frac{A}{2}m(t)\left(e^{j\omega_c t} + e^{-j\omega_c t}\right) \Leftrightarrow \frac{A}{2}\left[M(\omega - \omega_c) + M(\omega + \omega_c)\right] \tag{5.10}$$

Referring to Fig. 5.6, the function,

$$s(t) = Am(t)\cos(\omega_c t) \tag{5.11}$$

which is the transmitted signal, and the magnitude of its spectrum,

$$S(\omega) = (A/2)\left[M(\omega - \omega_c) + M(\omega + \omega_c)\right] \tag{5.12}$$

are both shown. The figure requires a little explanation. The drawing on the left shows $Am(t)$ and $Am(t)\cos(\omega_c t)$, where $A$ is a constant. We notice that

$$-1 \le \cos \omega_c t \le +1 \tag{5.13}$$

**Fig. 5.7** Amplitude
modulation, DSB/SC

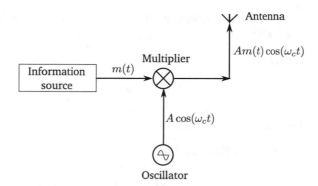

when $\cos \omega_c t = 1$ then $A m(t) \cos(\omega_c t)$ is equal to $A m(t)$, and when $\cos \omega_c t = -1$
then $A m(t) \cos(\omega_c t)$ is equal to $-A m(t)$. Therefore $A m(t) \cos(\omega_c t)$ oscillates within
an envelope of $A m(t)$ and $-A m(t)$.

In the same figure, the drawing on the right shows the FT of the signal on the left.
There are two spectra, each of bandwidth $2\omega_m$ (from $-\omega_m$ to $+\omega_m$) centred at the
two carrier frequencies $\pm\omega_c$. As usual $\omega_c \gg \omega_m$.

This signal is ready for transmission. The signal is called an *amplitude modulated*
*signal* since the carrier amplitude has been changed from $A$ to $A m(t)$. The carrier
has to be of course chosen with care so that, as far as possible, the channel should
not either attenuate or distort the transmitted signal.

Some terms need to be considered here: $m(t)$ is the *modulating signal,* while

$$s(t) = A m(t) \cos(\omega_c t) \tag{5.14}$$

is the *modulated signal* and $A \cos \omega_c t$ is the carrier. That part of the spectrum of
$M(\omega - \omega_c) + M(\omega + \omega_c)$ (centred at $\pm\omega_c$), which lies between $\omega_c$ and $\omega_c + \omega_m$
and that of its mirror image, lying between $-\omega_c - \omega_m$ and $-\omega_c$ are both called the
*upper sidebands* (USBs) as indicated in the figure. These are called upper sidebands,
since they represent the sideband with higher frequencies.

Similarly, the spectra of $M(\omega - \omega_c) + M(\omega + \omega_c)$ centred at $\pm\omega_c$ lying between
the frequencies $\omega_c - \omega_m$ and $\omega_c$, and $-\omega_c$ and $-\omega_c + \omega_m$, respectively, are both
called the *lower sidebands* (LSBs), and which are also shown in the figure. These are
called the lower sidebands since they represent the sideband of the lower frequencies.

Since both sidebands are transmitted, this is a *double sideband signal.* Furthermore, the carrier is *not* sent along with the modulated signal, and the signal is a
'*suppressed carrier*' signal. The whole scheme is therefore called *amplitude modu-*
*lation with double sidebands and suppressed carrier* (AM/DSB/SC).

The block diagram to implement this simple communication system is shown in
Fig. 5.7. The block diagram shows two sources: an information source producing the
message signal $m(t)$ and an oscillator which produces a carrier frequency $A \cos \omega_c t$.
The output of these two sources are fed into a nonlinear element which behaves as

a multiplier and which produces $Am(t)\cos\omega_c t$ on the output. This signal is then transmitted and shown here via an antenna.

**Example 5.1** For the signal,

$$f(t) = u(t)e^{-t}e^{j50t}$$

find the Fourier transform.

The FT of this signal is

$$F(\omega) = \int_{-\infty}^{\infty} u(t)e^{-t+j50t}e^{-j\omega t}\,dt$$

$$= \int_{0}^{\infty} e^{[-1+j(50-\omega)]t}\,dt$$

$$= \frac{e^{[-1+j(50-\omega)]t}}{-1+j(50-\omega)}\Big|_{t=0}^{t=\infty}$$

$$= \frac{1}{1+j(\omega-50)} \tag{5.15}$$

We may recall that

$$u(t)e^{-t} \Leftrightarrow \frac{1}{1+j\omega} \tag{5.16}$$

and we observe that Eq. (5.15) is a right shifted version of (5.16) to $\omega = 50$.

**Example 5.2** Find the Fourier transform of $f(t)\cos(50t)$ where $f(t)$ is the same as Example 5.1 (see Fig. 5.8).

The function

$$f(t)\cos(50t) = u(t)e^{-t}\cos(50t)$$

$$= \frac{u(t)}{2}\left[e^{-t}e^{j50t} + e^{-t}e^{-j50t}\right]$$

$$\Leftrightarrow \frac{1}{2}\left[\frac{1}{1+j(\omega-50)} + \frac{1}{1+j(\omega+50)}\right]$$

One can see the two shifted functions at the two frequencies, $\omega = \pm 50$. The magnitude of the Fourier transform is shown in Fig. 5.9.

**Example 5.3** Investigate the radio frequency pulse as shown in Fig. 5.10.

The RF pulse is derived from the pulse,

$$p_s(t) = \begin{cases} 1 & -\tau \le t \le \tau \\ 0 & \text{otherwise} \end{cases}$$

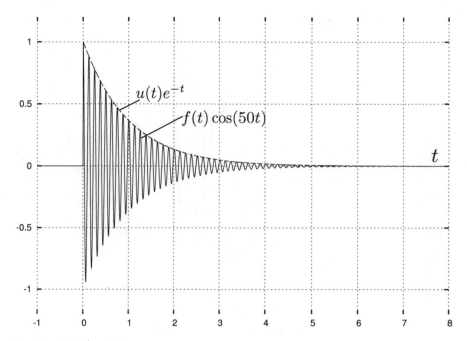

**Fig. 5.8** $u(t)e^{-t}\cos(50t)$

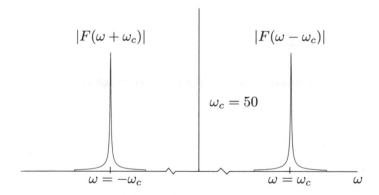

**Fig. 5.9** Fourier transform magnitude of Example 5.2

and its Fourier transform is given by

$$\mathcal{F}\{p_s(t)\} = 2\tau \frac{\sin(\omega\tau)}{\omega\tau}$$

Now to obtain the RF pulse,

$$p_{\text{RF}}(t) = p_s(t)\cos\omega_c t \tag{5.17}$$

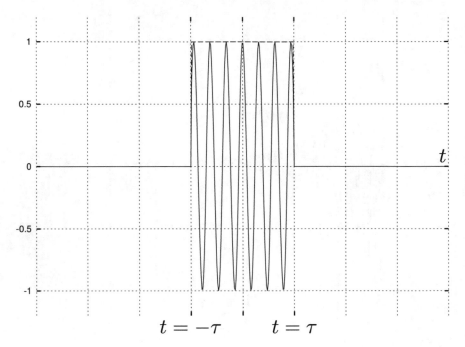

**Fig. 5.10**  RF Pulse

To obtain its Fourier transform,

$$\mathcal{F}\{p_{\text{RF}}(t)\} = \left(\frac{1}{2\pi}\right)\mathcal{F}\{p_s(t)\} * \mathcal{F}\{\cos \omega_c t\}$$

and

$$\mathcal{F}\{\cos \omega_c t\} = \pi \{\delta(\omega - \omega_c) + \delta(\omega + \omega_c)\}$$

and therefore,

$$\mathcal{F}\{p_{\text{RF}}(t)\} = \tau \frac{\sin\left[(\omega - \omega_c)\,\tau\right]}{(\omega - \omega_c)\,\tau} + \tau \frac{\sin\left[(\omega + \omega_c)\,\tau\right]}{(\omega + \omega_c)\,\tau} \tag{5.18}$$

and the Fourier transform of the RF pulse is shown diagrammatically in Fig. 5.11.

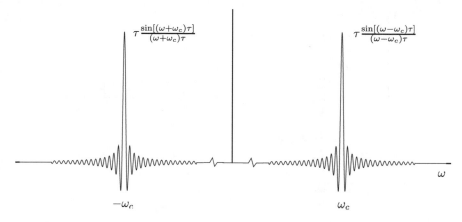

**Fig. 5.11** Fourier transform of the RF pulse

## 5.4 Demodulation of AM/DSB/SC

### 5.4.1 Introduction

Demodulation is the process by which we extract the original message signal, $m(t)$, from the modulated carrier. A demodulator generally consists of an electronic circuit that recovers the message signal from the modulated signal. In the case of AM/DSB/SC, there are various methods and circuits employed to recover the original message signal. Demodulation is also called detection in layman terms.

### 5.4.2 Coherent Detection

To recover or *demodulate* the received signal, we employ the block diagram of Fig. 5.12. Essentially, (referring to the figure), at point 1 (enclosed in a circle) the signal is the modulated signal, $m(t)\cos(\omega_c t)$ which is obtained from the receiving antenna. This signal is then multiplied with $A\cos\omega_c t$, and the result is shown at point 2 which is $A\cos^2\omega_c t$. This signal is passed through a low-pass filter and the original message signal recovered the details of which are discussed next.

Figure 5.13 shows the time and frequency- domain pictures of the demodulation process. Referring to the (a) part of the figure, the message signal is shown by the solid line (which constitutes one part of the envelope). The received signal is then multiplied by the carrier signal $A\cos\omega_c t$, as discussed earlier, and the details of the result are shown in Fig. 5.13b in time and frequency domain. The signal is

$$Am(t)\cos^2\omega_c t = \frac{A}{2}m(t)\left[1 + \cos\left(2\omega_c t\right)\right] \tag{5.19}$$

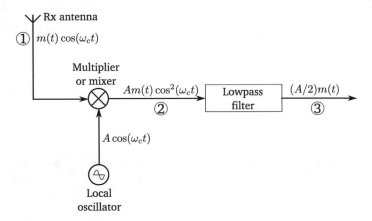

**Fig. 5.12** Demodulation of the received signal AM/DSB/SC

Notice that the envelope of $-Am(t)$ has disappeared and the x-axis is the other part of the envelope, since $Am(t) \cos^2 \omega_c t$ oscillates between zero and $m(t)$. Now taking Fourier transform of Eq. (5.19),

$$\mathcal{F}\left\{\frac{A}{2}m(t)\left[1 + \cos\left(2\omega_c t\right)\right]\right\} = \mathcal{F}\left[\frac{A}{2}m(t)\right] + \mathcal{F}\left[\frac{A}{2}m(t)\cos\left(2\omega_c t\right)\right]$$

$$= \frac{A}{2}M(\omega) + \frac{A}{4}\left[M(\omega - 2\omega_c) + M(\omega + 2\omega_c)\right]$$

$$(5.20)$$

where we have used Eq. (5.10). In the spectrum, there are three components: the spectrum at $\omega = -2\omega_c$, the spectrum at $\omega = 0$ and the spectrum at $\omega = 2\omega_c$. All these three spectra are shown in Fig. 5.13c. Notice that only one spectrum is of use to us: the one clustered at $\omega = 0$. To obtain this spectrum, we employ a low-pass filter to filter out the spectra at $\omega = \pm 2\omega_c$, and the output of the low-pass filter is the message signal, since

$$\mathcal{F}\left\{\frac{A}{2}m(t)\left[1 + \cos\left(2\omega_c t\right)\right]\right\} \times \mathcal{F}\left\{\text{Lowpass filter}\right\} =$$

$$\left\{\frac{A}{2}M(\omega) + \frac{A}{4}\left[M(\omega - 2\omega_c) + M(\omega + 2\omega_c)\right]\right\} \times \mathcal{F}\left\{\text{Lowpass filter}\right\} = \frac{A}{2}M(\omega) + 0 \quad (5.21)$$

The spectrum of the low-pass filter is zero around the frequencies $\omega = \pm 2\omega_c$, and nonzero and constant in the frequency range $-\omega_m \leq \omega \leq \omega_m$. Therefore, the frequencies around $\omega = \pm 2\omega_c$ are eliminated but those in the interval $[-\omega_m, \ \omega_m]$ are preserved and just multiplied by a constant.

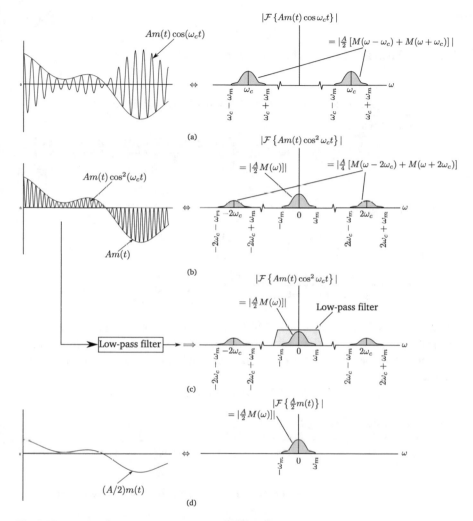

**Fig. 5.13**  Details of the demodulation of the AM/DSB/SC

## 5.4.3   Coherent Detection—Discussion

Two points must be noticed in the demodulation procedure just discussed:

1. The frequency of the local oscillator (LO) must exactly match the frequency of the carrier
2. The phase of the LO also should be the same as the carrier phase.

Let us look at these two problems separately. Suppose the local oscillator (LO) frequency is not equal to the carrier frequency,

$$\omega_{LO} = \omega_c + \Delta\omega_c \tag{5.22}$$

then after the multiplier at circled point 2 in Fig. 5.12,

$$\underbrace{A\,m(t)\cos(\omega_c t)}_{\text{Received signal}}\cos(\omega_{LO} t) = Am(t)\cos(\omega_c t)\cos(\omega_c t + \Delta\omega_c t)$$

$$= Am(t)\left[\frac{\cos(2\omega_c t + \Delta\omega_c\,t)}{2} + \frac{\cos(\Delta\omega_c\,t)}{2}\right] \tag{5.23}$$

the FT of this signal is

$$\mathcal{F}\left\{Am(t)\left[\frac{\cos(2\omega_c t + \Delta\omega_c\,t)}{2} + \frac{\cos(\Delta\omega_c\,t)}{2}\right]\right\}$$

$$= \frac{A}{4}\underbrace{[M(\omega - 2\omega_c - \Delta\omega_c) + M(\omega + 2\omega_c + \Delta\omega_c)]}_{\text{Signal centred around } \pm 2\omega_c} + \frac{A}{4}\underbrace{[M(\omega - \Delta\omega_c) + M(\omega + \Delta\omega_c)]}_{\text{Distorted message signal}} \tag{5.24}$$

We notice that when the frequency of the local oscillator does not match the frequency of the carrier the recovered signal is distorted. Let us now concentrate on the other problem: when the phase of the LO is not equal to the incoming received signal. Let the LO signal be

$$f_{LO}(t) = A\cos(\omega_c t + \theta) \tag{5.25}$$

then at circled point 2 in Fig. 5.12,

$$\underbrace{A\,m(t)\cos(\omega_c t)}_{\text{Received signal}}\underbrace{\cos(\omega_c t + \theta)}_{\text{LO signal}} = Am(t)\cos(\omega_c t)\left[\cos(\theta)\cos(\omega_c t) - \sin(\theta)\sin(\omega_c t)\right]$$

$$= Am(t)\left[\cos(\theta)\cos^2(\omega_c t) - \sin(\theta)\cos(\omega_c t)\sin(\omega_c t)\right]$$

$$= Am(t)\left[\cos(\theta)\frac{\cos(2\omega_c t) + 1}{2} - \sin(\theta)\frac{\sin(2\omega_c t)}{2}\right]$$

$$= \frac{A}{2}m(t)\left[\cos(2\omega_c t + \theta) + \cos(\theta)\right] \tag{5.26}$$

the first term in this expression has a spectrum centred at the two frequencies $\omega = \pm\omega_c$, while the second term is $(A/2)m(t)\cos\theta$. The first term is filtered out by the low-pass filter, while the second one appears at the output of the receiver at point 3 in Fig. 5.12. Obviously, the maximum power on the output is when $\theta = 0$, and there is no output when $\theta = \pi/2$.

From this analysis, it is clear that to demodulate the signal very expensive circuitry is required at the receiver in the form of a local oscillator whose output must be strictly controlled both in terms of its frequency and phase. The multiplier is also called a *mixer* which we considered in Sect. 4.6.

## 5.5  Circuits to Modulate and Demodulate DSB/SC Signals

### 5.5.1  Modulation Circuits

To modulate the signal in the DSB/SC scheme various circuits may be employed. Most of these circuits employ nonlinear elements such as diodes and transistors since multipliers *are* nonlinear elements. In these circuits that part of the *v-i* characteristic is used which is nonlinear.

#### 5.5.1.1  Nonlinear Modulation

If we use a nonlinear element with input $x(t)$ and output $y(t)$, then the relation between the two signals is

$$y(t) = ax(t) + bx(t)^2 + \cdots \tag{5.27}$$

where $a \gg b$ etc. Now if we let

$$x_1(t) = A \cos(\omega_c t) + m(t) \quad \text{and}$$
$$x_2(t) = A \cos(\omega_c t) - m(t)$$

then if we input $x_1(t)$ and $x_2(t)$ into exactly two identical nonlinear devices, then

$$y_1(t) = b\,[\cos(t\,\omega_c)\,A + m(t)]^2 + a\,[\cos(t\,\omega_c)\,A + m(t)] + \ldots$$
$$= \frac{b\cos(2t\,\omega_c)\,A^2}{2} + \frac{b\,A^2}{2} + 2\,b\,m(t)\cos(t\,\omega_c)\,A +$$
$$+ a\cos(t\,\omega_c)\,A + b\,m(t)^2 + a\,m(t) + \ldots$$

and

$$y_2(t) = b\,[\cos(t\,\omega_c)\,A - m(t)]^2 + a\,[\cos(t\,\omega_c)\,A - m(t)] + \ldots$$
$$= \frac{b\cos(2t\,\omega_c)\,A^2}{2} + \frac{b\,A^2}{2} - 2\,b\,m(t)\cos(t\,\omega_c)\,A$$
$$+ a\cos(t\,\omega_c)\,A + b\,m(t)^2 - a\,m(t) + \ldots$$

forming the expression

$$y(t) = y_1(t) - y_2(t)$$
$$= 4Abm(t)\cos(t\,\omega_c) + 2am(t) + \ldots$$

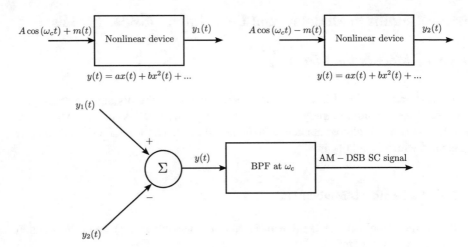

**Fig. 5.14** AM/DSB/SC modulation using nonlinear elements

which we may obtain just by inspection. The second term may be eliminated using a BPF centred at $\omega_c$. The block diagram of this modulator is shown in Fig. 5.14.

### 5.5.1.2  Bridge Modulator

The bridge modulator produces a modulated output in a similar manner as the ring modulator does. The modulator circuit diagram is shown in Fig. 5.15a. In this diagram the amplitude of the oscillator $A$ in the oscillator output, $A \cos (\omega_c t)$, is much greater than the maximum value of the modulus of $m(t)$

$$A \gg |m(t)|_{\max} \tag{5.28}$$

The bridge operates in the following way. When the terminal B is at a higher potential than A, then due to condition given in (5.28), all the diodes are *reverse biased* and $m(t)$ appears at CD (Fig. 5.15b). When the reverse is true, then all diodes are forward biased as shown in Fig. 5.15c and then $m(t)$ is effectively shorted, and the output between C and D is zero. These waveforms are shown in Fig. 5.16.

The maths of the whole process is similar to what we have considered earlier (see Eqs. (4.67) and (4.68)). The output at CD is given by

$$m(t) \left[ 0.5 + \sum_{n=1}^{\infty} \left( \frac{2}{n\pi} \right) \sin \left( \frac{n\pi}{2} \right) \cos (n\omega_c t) \right]$$
$$= m(t) \left[ 0.5 + \sum_{n=1}^{\infty} \frac{\sin \left( \frac{n\pi}{2} \right)}{\left( \frac{n\pi}{2} \right)} \cos (n\omega_c t) \right] \tag{5.29}$$

**Fig. 5.15** Bridge modulator

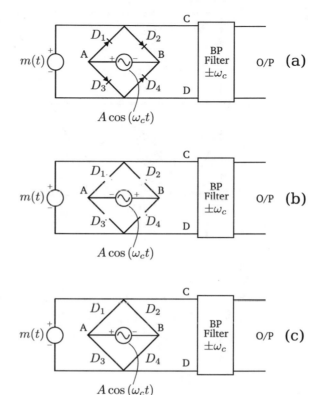

where Eq. (4.67) has been used. The BPF gives us the $n = 1$ term which is the cosine term centred at $\omega = +\omega_c$ in the frequency domain.

$$s(t) = m(t) \left(\frac{2}{\pi}\right) \cos(\omega_c t) \tag{5.30}$$

shown in Fig. 5.17.

### 5.5.1.3 Double Balanced Modulator

To modulate and demodulate, the signal various circuits may be employed. Such a circuit is the double balanced mixer shown in Fig. 5.18a. Four diodes are placed in the configuration shown with two centre tapped transformers and a bandpass filter with its centre frequency at $\omega_c$. The working of this mixer is as follows.

The major point of importance is that amplitude of the local oscillator, $A$, should be about 20 dB higher than the maximum absolute value of $m(t)$. Or

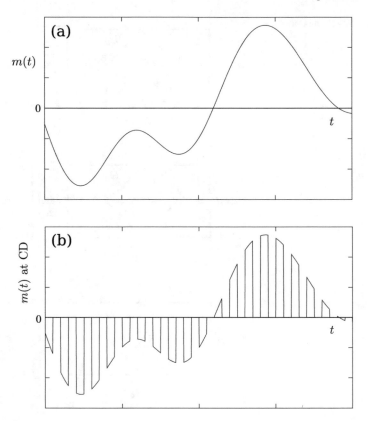

**Fig. 5.16** Bridge modulator waveforms

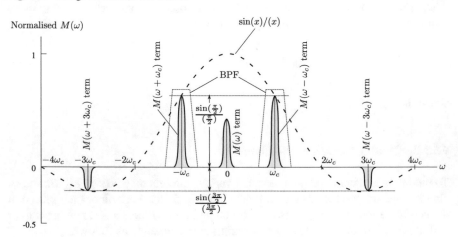

**Fig. 5.17** Figure showing spectral output of a bridge modulator

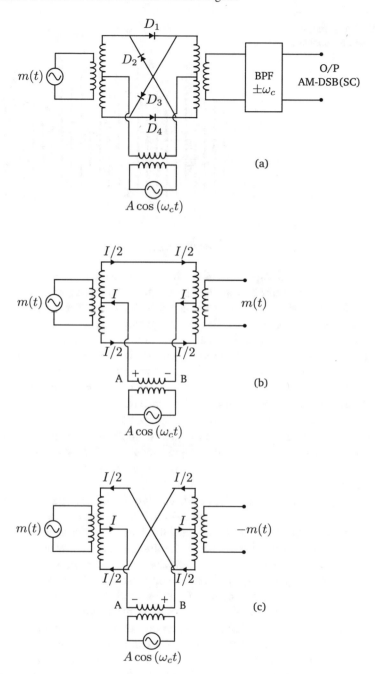

**Fig. 5.18**  Circuit to produce AM/DSB/SC signals

**Fig. 5.19** Output of the double balanced modulator

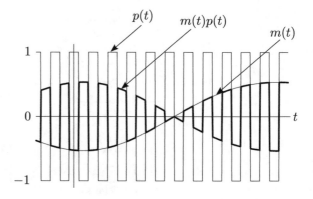

$$A \gg |m(t)|_{\max} \tag{5.31}$$

This is a point of vital importance since in the discussion that follows, the amplitude of the local oscillator controls the behaviour of the diodes, and not $m(t)$. The performance of the mixer hinges on the fact that the diodes behave like switching elements, and are switched ON and OFF at a frequency of $f_c = \omega_c/2\pi$ Hz. Referring to Fig. 5.18b, where point A of the local oscillator is at a higher potential than point B. Due to this, diodes $D_1$ and $D_4$ conduct (that is they are ON) and diodes $D_2$ and $D_3$ do not conduct (or they are OFF). This situation is shown in the figure where the conductors containing $D_1$ and $D_4$ are short circuited while the conductors containing $D_2$ and $D_3$ are effectively open circuited. The current $I$ entering the left transformer winding from point A splits into two equal values, each of $I/2$ ampere, due to the symmetry of the circuit. These two currents now combine in the right transformer winding but produce no voltage on the output because the two voltages are opposite each other and therefore cancel each other on the output. The message signal on the other hand appears unchanged on the output, $m(t)$.

We now examine the circuit in the negative half of the cycle of the local oscillator. Now the point B is at a higher potential as compared to A as shown in Fig. 5.18c. Now $D_1$ and $D_4$ are OFF and $D_2$ and $D_3$ are ON. Therefore, the arms containing $D_2$ and $D_3$ may be short circuited and those containing $D_1$ and $D_4$ may be removed. The current now flows from the point B into the winding of the right transformer. Since the current, $I$, splits equally ($I/2$) into two in opposite directions, the total potential due to the local oscillator across the winding is zero and therefore no voltage appears on the output due to the LO. However, this time the message signal appears on opposite sign, namely $-m(t)$, due the fact that the current from the upper end of the transformer winding on the left enters the lower end of the winding on the right.

Based on this discussion, it is clear that every half cycle output oscillates between $m(t)$ and $-m(t)$ as shown in Fig. 5.19a. The figure shows pulses of frequency $f_c = \omega_c/2\pi$ Hz whose amplitude is modulated by $m(t)$. If we examine the figure it is clear that $m(t)$ appears in one half cycle while $-m(t)$ appears in the other half cycle. If $p(t)$ is periodic train of pulses, shown in Fig. 5.19b, then the output can be modelled as $m(t)p(t)$. Notice that when $m(t)p(t)$ are both positive or both negative the output is positive, and when the two signals are of opposite signs, the output is negative.

Referring to Example 2.13, we know, by inspection, that

$$p(t) = \sum_{n=1,3,5}^{\infty} \frac{4}{n\pi} \sin\left(\frac{n\pi}{2}\right) \cos(n\omega_c t) \tag{5.32}$$

which when multiplied by $m(t)$ becomes

$$m(t)p(t) = \sum_{n=1,3,5}^{\infty} \frac{4m(t)}{n\pi} \sin\left(\frac{n\pi}{2}\right) \cos(n\omega_c t)$$

$$= \sum_{n=1,3,5}^{\infty} \frac{2}{n\pi} \sin\left(\frac{n\pi}{2}\right) [2m(t)\cos(n\omega_c t)]$$

$$= \sum_{n=1,3,5}^{\infty} \frac{\sin\left(\frac{n\pi}{2}\right)}{\left(\frac{n\pi}{2}\right)} [2m(t)\cos(n\omega_c t)] \tag{5.33}$$

Now if we use the modulation theorem,

$$p(t)m(t) \Leftrightarrow \sum_{n=1,3,5}^{\infty} \underbrace{\frac{\sin\left(\frac{n\pi}{2}\right)}{\left(\frac{n\pi}{2}\right)}}_{\sin x/x \text{ term}} \underbrace{[M(\omega - n\omega_c) + M(\omega + n\omega_c)]}_{\text{Sidebands}} \tag{5.34}$$

since

$$2m(t)\cos(n\omega_c t) \Leftrightarrow M(\omega - n\omega_c) + M(\omega + n\omega_c)$$

Referring to Fig. 5.20, Expression 5.34 is shown graphically. The $\sin x/x$ term is drawn as the dotted line, while the spectrum of the message signal is shown translated to $\omega = \pm\omega_c$ and $\omega = \pm 3\omega_c$ are shown. The bandpass filter on the output of the circuit shown in Fig. 5.18a filters out the higher sidebands, where $n \geq 3$, and allows only the bands at $n = 1$ to pass.

## 5.5.2 Demodulator Circuits

### 5.5.2.1 Double Balanced Demodulator

Demodulation with a double balanced diode circuit has the same reasoning which we used for modulation. Using the reasoning applied to Figs. 5.18, and 5.19, the input signal in Fig. 5.21 is

$$m(t)\cos(\omega_c t)$$

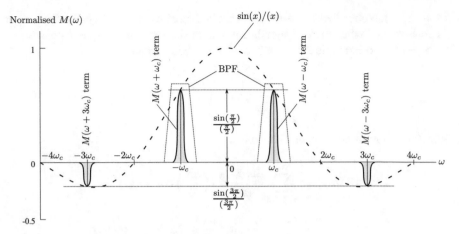

**Fig. 5.20** Details of the amplitude modulation of a signal

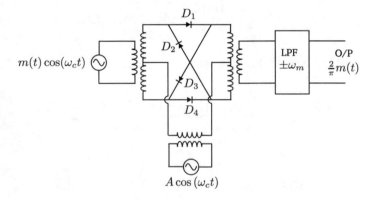

**Fig. 5.21** Demodulation circuit for AM/DSB/SC wave

which is when multiplied by $p(t)$, then can be simplified using Eq. (5.32):

$$m(t)\cos(\omega_c t)\,p(t) = m(t)\cos(\omega_c t)\left[\sum_{n=1,3,5}^{\infty} \frac{\sin\left(\frac{n\pi}{2}\right)}{\left(\frac{n\pi}{2}\right)}\left[2\cos(n\omega_c t)\right]\right]$$

$$= \sum_{n=1,3,5}^{\infty} \frac{\sin\left(\frac{n\pi}{2}\right)}{\left(\frac{n\pi}{2}\right)}\left[2m(t)\cos(\omega_c t)\cos(n\omega_c t)\right]$$

$$= \sum_{n=1,3,5}^{\infty} \frac{\sin\left(\frac{n\pi}{2}\right)}{\left(\frac{n\pi}{2}\right)}\left[m(t)\cos((n+1)\,\omega_c t) + m(t)\cos((n-1)\,\omega_c t)\right] \quad (5.35)$$

if we set $n = 1$ in this equation, we get the first term and its Fourier transform,

$$\frac{2}{\pi}[m(t) + m(t)\cos(2\omega_c t)] \Leftrightarrow \underbrace{\frac{2}{\pi}M(\omega)}_{\text{Near } \omega=0} + \underbrace{\frac{1}{\pi}[M(\omega - 2\omega_c) + M(\omega + 2\omega_c)]}_{\text{Shifted to } \omega=\pm 2\omega_c}$$

(5.36)

similarly setting $n = 3, 5, \ldots$ we get frequency-domain terms at $\pm 2\omega_c$, $\pm 4\omega_c$, $\pm 6\omega_c, \ldots$ and so on. If now use a low-pass filter on the output, we only get one term, namely

$$\frac{2}{\pi}m(t) \Leftrightarrow \frac{2}{\pi}M(\omega) \tag{5.37}$$

which is the message signal.

### 5.5.2.2 Costas Receiver

The message signal, $m(t)$, of the DSB/SC signal may be recovered by a circuit known as the Costas receiver. The block diagram of the Costas receiver is shown in upper figure of Fig. 5.22. The incoming signal, $Am(t)\cos(\omega_c t)$, is split into two parts. The upper part is multiplied by the output, $\cos(\omega_c t)$ of the voltage controlled oscillator (VCO). In the lower part, the VCO signal is phase shifted by $-\pi/2$ which multiplies the incoming signal. The output of the upper multiplier is

$$Am(t)\cos(\omega_c t)\cos(\omega_{c1}t + \theta) = \frac{Am(t)}{2}\left\{\cos\left[(\omega_{c1} - \omega_c)t + \theta\right] + \cos\left[(\omega_{c1} + \omega_c)t + \theta\right]\right\} \tag{5.38}$$

where we have used

$$\cos(a)\cos(b) = \frac{\cos(b+a)}{2} + \frac{\cos(b-a)}{2}$$

where $b = \omega_{c1}t + \theta$ and $a = \omega_c t$, while that of the lower multiplier is

$$Am(t)\cos(\omega_c t)\sin(\omega_{c1}t + \theta) = \frac{Am(t)}{2}\left\{\sin\left[(\omega_{c1} - \omega_c)t + \theta\right] + \sin\left[(\omega_{c1} + \omega_c)t + \theta\right]\right\} \tag{5.39}$$

where we have used the identity

$$\sin(a)\cos(b) = \frac{\sin(b+a)}{2} - \frac{\sin(b-a)}{2}$$

where $a = \omega_{c1}t + \theta$ and $b = \omega_c t$. Both these signals are then passed through low-pass filters, so that in the upper figure, in the upper branch $[Am(t)/2]\cos[(\omega_{c1} - \omega_c)t + \theta]$ is passed, while in the lower branch $[Am(t)/2]\sin[(\omega_{c1} - \omega_c)t + \theta]$ is passed. Both these signals are then fed into a phase detector, which is another multiplier and low-pass filter as shown in the lower diagram. At the multiplier output, (in the lower diagram) is

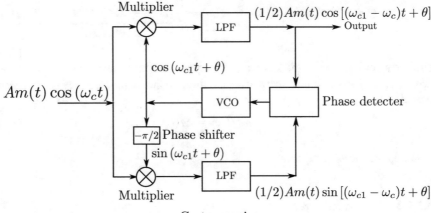

Costas receiver

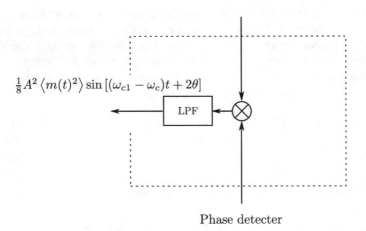

Phase detecter

**Fig. 5.22** The Costas receiver

$$\frac{[Am(t)]^2 \left[\cos\left(\theta + \omega_{c1} t - \omega_c t\right) \sin\left(\theta + \omega_{c1} t - \omega_c t\right)\right]}{4}$$
$$= \frac{[Am(t)]^2 \sin\left(2\theta + 2\omega_{c1} t - 2\omega_c t\right)}{8}$$

The output of this LPF is

$$\frac{1}{8} A^2 \left\langle m(t)^2 \right\rangle \sin\left[(\omega_{c1} - \omega_c)t + 2\theta\right]$$

where $\left\langle m(t)^2 \right\rangle$ is the average value of the signal, a DC term. The VCO is designed so that when $\omega_{c1} > \omega_c$ the frequency is *decreased* and when $\omega_{c1} < \omega_c$ the frequency is *increased*. At the nominal frequency $\omega_{c1} \approx \omega_c$, the VCO stays locked.

Similar remarks apply to the phase, when $\theta < 0$ the frequency is increased for a short period time $t_p$ so that $\omega t_p = \theta$. In other words, the phase of the VCO is increased, and similarly when $\theta > 0$ the VCO phase decreases through a very small frequency change which is applied for a shot period of time. The signal locks completely when both $\omega_{c1} = \omega_c$ and also $\theta = 0$ and the input to the VCO is zero. After locking no further changes take place.

## 5.6 Amplitude Modulation–DSB with Carrier

AM/DSB with carrier, or just AM is one of the most important methods of commercial broadcasting during most of the twentieth century and still remains widely used today, in the twenty-first century. AM broadcast radio began commercial operations in the early 1920s due to experimental verification of this form of communication in the preceding years (1900–1920). In USA and Canada, various AM broadcast radio stations operated such as KCBS, XWA and KDKA. In Europe, BBC (noncommercial) was formed in early 1920s and with other commercial radio stations such as Radio Luxembourg and Radio Normandie came into existence. Interestingly, one of the oldest radio stations in the world operated from Sri Lanka, known as Radio Ceylon.

In contrast with AM/DSB communication, amplitude modulation (DSB) with carrier (or just *amplitude modulation*) is easier to implement and demodulate. The AM signal is

$$s(t) = A\left[1 + k_a m(t)\right] \cos\left(\omega_c t\right) \tag{5.40}$$

where $k_a$ is the amplitude sensitivity of the modulator and $s(t)$ is the modulated signal. In amplitude modulation, we would like to have $k_a m(t) < 1$ for ease of demodulation. Figure 5.23 shows the amplitude modulated signal with $A = 1$. The (a) part of the figure shows $m(t)$, while (b) shows the signal when $1 + k_a m(t) > 0$. (c) shows the signal when $1 + k_a m(t) \not> 0$ for all values of time, and where the zero crossings are clearly shown.

If we take a look at the spectrum of this signal,

$$\mathcal{F}\left\{A\left[1 + k_a m(t)\right]\cos\left(\omega_c t\right)\right\} = \mathcal{F}\left\{A\cos\omega_c t + A k_a m(t)\cos\left(\omega_c t\right)\right\}$$

$$= A\pi\left[\delta\left(\omega - \omega_0\right) + \delta\left(\omega + \omega_0\right)\right] + \frac{A k_a}{2}\left[M(\omega - \omega_c) + M(\omega + \omega_c)\right] \tag{5.41}$$

this spectrum is shown in Fig. 5.24. The figure shows the term

$$\left[(A k_a)/2\right]\left[M(\omega - \omega_c) + M(\omega + \omega_c)\right]$$

centred at $\pm\omega_c$, and the spectrum of the cosine term which consists of two impulse functions also centred at $\pm\omega_c$.

**Example 5.4** Single tone modulation

**Fig. 5.23** Figure showing amplitude modulation

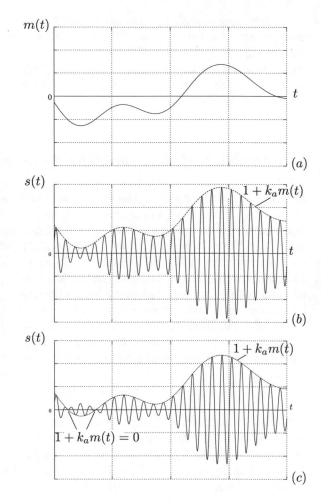

Single tone modulation is the modulated signal when

$$m(t) = A_m \cos(\omega_m t) \tag{5.42}$$

and so

$$s(t) = A(1 + k_a A_m \cos \omega_m t) \cos \omega_c t \tag{5.43}$$

Referring to Fig. 5.25, the message and the carrier are shown in ($a$) and ($b$) parts of the figure both of which are sinusoidal, and $\omega_c \gg \omega_m$. The general modulated signal is shown in ($c$) part of the figure. In the figure the two extremas, $A_{\max}$ and $A_{\min}$ are shown. The modulation index $\mu$ for single tone modulation is

$$0 < \mu = \frac{A_{\max} - A_{\min}}{A_{\max} + A_{\min}} < 1 \tag{5.44}$$

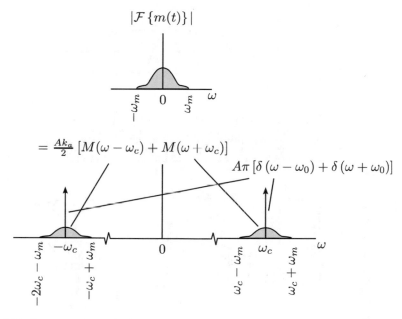

**Fig. 5.24** Spectrum of am amplitude modulated signal

since the envelope of the carrier is

$$1 + k_a A_m \cos \omega_m t \qquad (5.45)$$

the maximum and minimum values are

$$A_{\max} = A (1 + k_a A_m)$$
$$\text{and } A_{\min} = A (1 - k_a A_m)$$
$$\text{therefore } A_{\max} + A_{\min} = 2A$$
$$\text{and } A_{\max} - A_{\min} = 2A k_a A_m$$
$$\text{or } \mu = k_a A_m \qquad (5.46)$$

if we simplify Eq. (5.43), we get

$$s(t) = \underbrace{\frac{\mu A \cos (\omega_m t + \omega_c t)}{2} + \frac{\mu A \cos (\omega_m t - \omega_c t)}{2}}_{\text{sidebands}} + \underbrace{A \cos (\omega_c t)}_{\text{carrier}} \qquad (5.47)$$

we can see from the expression that there are two sidebands, at $\pm (\omega_c \pm \omega_m)$. We also know that the sidebands carry the information. The FT of Eq. (5.47) is

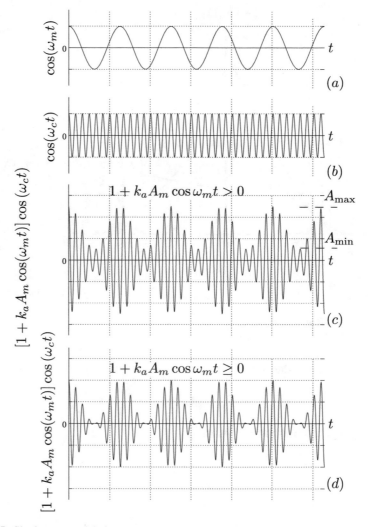

**Fig. 5.25** Single tone modulation

$$
S(\omega) = \frac{\mu A}{2} \left\{ \frac{1}{2} \left[ \delta(\omega - (\omega_m + \omega_c)) + \delta(\omega + (\omega_m + \omega_c)) \right] \right\}
$$

$$
+ \frac{\mu A}{2} \left\{ \frac{1}{2} \left[ \delta(\omega - (\omega_c - \omega_m)) + \delta(\omega + (\omega_c - \omega_m)) \right] \right\}
$$

$$
A \left\{ \frac{1}{2} \left[ \delta(\omega - \omega_c)) + \delta(\omega + \omega_c)) \right] \right\}
$$

$$
(5.48)
$$

We recall that the average power in any cosine waveform, $B \cos (\omega_0 t)$ is $B^2/2$ therefore the power in the sidebands is

$$\mathbb{P}_{av,sb} = \frac{\mu^2 A^2}{8} + \frac{\mu^2 A^2}{8} = \frac{\mu^2 A^2}{4} \tag{5.49}$$

while the power in the carrier is

$$\mathbb{P}_{av,c} = \frac{A^2}{2} \tag{5.50}$$

so the ratio of the average power in the sidebands to the power in the carrier is

$$\frac{\mathbb{P}_{av,sb}}{\mathbb{P}_{av,c}} = \frac{\mu^2}{2} \tag{5.51}$$

which is a maximum when $\mu = 1$. This case is shown in Fig. 5.25d. This exercise gives us an idea about the ratio of the useful power, the sidebands to the carrier power (the power which is unused).

**Exercise 5.1** Plot the spectrum of the single tone modulated signal.

## 5.7 Circuits to Modulate and Demodulate DSB Signals

### 5.7.1 Modulation

#### 5.7.1.1 Nonlinear Modulator

We now look to circuits which will serve as modulators of DSB signals (that is signals *with* carrier). A DSB signal with carrier has the equation

$$f_c(t) = [A + Bm(t)] \cos (\omega_c t) \tag{5.52}$$
$$= A \left[ 1 + \frac{B}{A} m(t) \right] \cos (\omega_c t)$$
$$= A [1 + k_a m(t)] \cos (\omega_c t)$$

where we can see that the carrier signal $A \cos(\omega_c t)$ is sent along with the modulated signal $Bm(t) \cos(\omega_c t)$, where generally

$$|Bm(t)|_{\max} < A \tag{5.53}$$

The simplest of such modulators are the nonlinear modulators as discussed in Sect. 4.5. The block diagram is shown in Fig. 5.26.

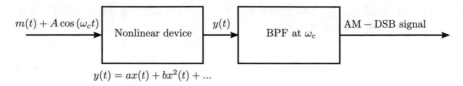

$$y(t) = ax(t) + bx^2(t) + \ldots$$

**Fig. 5.26** Nonlinear modulator using a single element

The analysis of this figure is as follows: the input to the modulator is $x(t)$ which is the sum of two terms

$$x(t) = m(t) + A\cos(\omega_c t) \tag{5.54}$$

where $m(t)$ is the message signal and $A\cos(\omega_c t)$ is the carrier term. This signal is fed into the nonlinear modulator

$$y(t) = ax(t) + bx(t)^2 + \cdots \tag{5.55}$$

hence

$$y(t) = \underbrace{\frac{b\cos(2t\,\omega_c)\,A^2}{2}}_{\text{centred at } 2\omega_c} + \frac{b\,A^2}{2} + \underbrace{2\,b\,m(t)\cos(t\,\omega_c)\,A + a\cos(t\,\omega_c)\,A}_{\text{centred at } \omega_c} + b\,m(t)^2 + a\,m(t)$$

$$\tag{5.56}$$

as we can see that the signal,

$$\underbrace{2A\,b\,m(t)\cos(t\,\omega_c)}_{\text{modulated carrier}} + \underbrace{a\,A\cos(t\,\omega_c)}_{\text{carrier}} \tag{5.57}$$

is a DSB signal (with carrier) which passes through the BPF.

## 5.7.2  Demodulation

### 5.7.2.1  Envelope Detector

The demodulation of the amplitude modulation is a diode and $RC$ circuit combination as shown in Fig. 5.27. In this figure, $R_s$ is the source resistance, $r_D$ is the equivalent forward biased diode resistance, $R$ is the resistance on the output end of the circuit and $s(t)$ is the modulated carrier.

Referring to the figure in conjunction with Fig. 5.28, we notice that the carrier waveform oscillates rapidly between $A[k_a m(t) + 1]$ and $-A[k_a m(t) + 1]$. From observation of the circuit, it is apparent that the diode is forward biased and effectively short circuit in the positive half cycles of the carrier. During the positive half cycles,

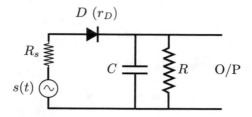

**Fig. 5.27** Envelope detector

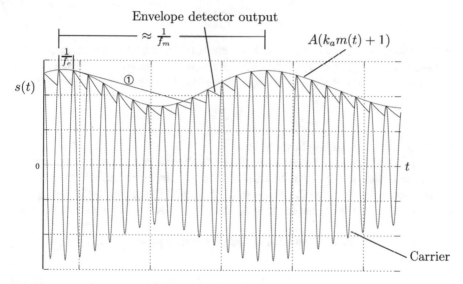

**Fig. 5.28** Envelope detector waveforms

the capacitor 'sees' a resistance ($R_{eq}$) of

$$R_{eq} = (R_s + r_D)||R \tag{5.58}$$

Or

$$\frac{1}{R_{eq}} = \frac{1}{R_s + r_D} + \frac{1}{R} \tag{5.59}$$

if we impose the condition that the sum of the source and diode resistances are much smaller than $R$,

$$R_s + r_D \ll R \tag{5.60}$$

then

$$\frac{1}{R_s + r_D} \gg \frac{1}{R} \tag{5.61}$$

and therefore,

$$R_{eq} \cong R_s + r_D \qquad (5.62)$$

and the charging time constant, $\tau_c$ is given by

$$\tau_c = (R_s + r_D)C \qquad (5.63)$$

The carrier period is

$$T_c = \frac{1}{f_c}$$

and we design $\tau_c \ll 1/f_c$ so that the capacitor easily charges to its maximum value.

During each negative half cycle, the diode is reverse biased and effectively open circuited, hence the capacitor discharges through the resistor, $R$, whose time constant is

$$\tau_d = RC \qquad (5.64)$$

where $\tau_d$ is the discharging time constant.

As the capacitor discharges, it continues to discharge as long as the potential difference across it is greater than the modulated carrier, effectively keeping the diode reverse biased during the negative half cycles. The capacitor is then charged to the maximum value when the carrier amplitude is greater than the capacitor potential difference.

If we compare the time constant $\tau_d$ with the carrier period, $T_c(= 1/f_c)$, then

$$\tau_d \gg \frac{2\pi}{\omega_c} = T_c \qquad (5.65)$$

otherwise the discharge rate would be too large and the envelope detector waveforms would fall too low as shown in Fig. 5.29 effective causing oscillations on the output of the envelope detector which are much too large and therefore unacceptable.

Another case which is of importance is when $\tau_d$ is too large. If we compare $\omega_m$ which is the maximum frequency in the message signal, with $\tau_d$ then we should satisfy

$$\frac{2\pi}{\omega_m} \gg \tau_d \ (\gg T_c) \qquad (5.66)$$

where $2\pi/\omega_m$ is the shortest period of the wave. If this condition is not satisfied then the capacitor discharges too slowly and may miss the next positive cycle as shown in Fig. 5.28 where the discharge is shown circled: ①. The envelope output is then passed through a low-pass filter which smooths the ripple and gives us back the analogue signal.

To understand this point, if the envelope detector output waveform is $e_d(t)$, then its spectrum will have two components, one of the message signals centred near $\omega = 0$, and the other centred at $\omega_c$ (due to the ripple) as shown in Fig. 5.30. The LPF allows the message signal to pass and blocks the unwanted signal centred at $\pm\omega_c$.

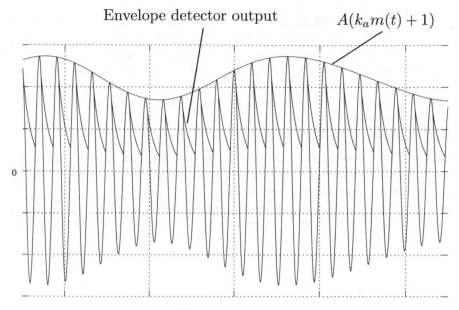

**Fig. 5.29** Envelope detection with too small an $RC$ time constant

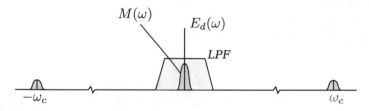

**Fig. 5.30** Envelope detector spectrum

## 5.8 Amplitude Modulation—Discussion

Amplitude modulation (AM) is the oldest modulation which is still used in broadcast radio. Its principal desirable features are ease of generation and demodulation. The generator requires a nonlinear element as explained in Sect. 5.5.1.1, and the demodulation requires an envelope detector. The bands used for AM broadcasting for commercial purposes are

1. Medium wave: 530–1600 kHz and
2. Short wave: 2.3–26 MHz

Medium wave communication is principally guided by ground wave, while short wave uses ionospheric propagation (see [1] or [2]). Medium wave transmitters use broadcasting power as low as 250 W or may be as high as 500 kW. The reach of medium waves is about a few hundred kilometres from the transmitting station and the wave is guided by diffraction off the surface of the earth. At night, medium wave radio stations may be tuned into as far as 800–1000 km away, and restrictions may be placed on the transmitting station concerning the transmitting power at night.

Short wave communication, on the other hand, uses ionospheric propagation and was first investigated by amateur radio enthusiasts. Distances in the range of 6000 km are accessible by shortwave radio. Again AM in shortwave is very easy to implement in terms of transmission and reception.

The disadvantages of AM are

1. It is difficult to extract it from noise, and the only remedy is higher signal power.
2. It is wasteful in terms of power, since a great deal of power is consumed by the carrier (which is sent along with the signal).
3. It is wasteful in terms of bandwidth, since though the useful bandwidth is $\omega_m$ (where $\omega_m$ is the maximum frequency present in the baseband signal), the AM bandwidth is $2\omega_m$ (which is $\omega_c \pm \omega_m$).

## 5.9  Single Side-Band Modulation[1]

In single side-band modulation, we use only one side-band of the message signal. This implies that we may use the side-band of $M(\omega)$ for $\omega \geq 0$ (or $\omega \leq 0$) only for the purposes of communication. We may then write

$$M_p(\omega) = M(\omega)u(\omega)$$
$$= \frac{1}{2}M(\omega)\left[1 + \text{sgn}(\omega)\right]$$
and $M_n(\omega) = M(\omega)u(-\omega)$
$$= \frac{1}{2}M(\omega)\left[1 - \text{sgn}(\omega)\right] \tag{5.67}$$

where these two spectra are shown in Fig. 5.31a. The sidebands here are related to the pre-envelope discussed in Sect. 2.17 by a factor of $1/2$. Now let us concentrate on

$$M(\omega)\text{sgn}(\omega) = \begin{cases} M(\omega) & \omega \geq 0 \\ -M(\omega) & \omega \leq 0 \end{cases} \tag{5.68}$$

and therefore, using Property (2.127),

$$\mathcal{F}\{x(t) * y(t)\} = X(\omega)Y(\omega)$$

we have

---

[1] Read Sect. 2.17 and onward till the end of the chapter.

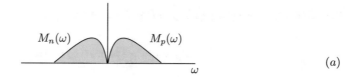

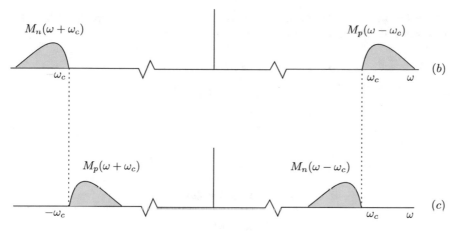

**Fig. 5.31** Figure concerning single side-band modulation

$$M(\omega)\mathrm{sgn}(\omega) \Leftrightarrow m(t) * \frac{j}{\pi t}$$

$$= \frac{j}{\pi} \int_{-\infty}^{\infty} \frac{m(\tau)}{t - \tau} d\tau \tag{5.69}$$

(since

$$\frac{j}{\pi t} \Leftrightarrow \mathrm{sgn}(\omega)$$

for the Fourier inverse of the $\mathrm{sgn}(\omega)$ function). The integral in Eq. (5.69) is known as the Hilbert transform of $m(t)$ (read Sect. 2.17):

$$\mathcal{H}\{m(t)\} = m_h(t) = \frac{1}{\pi} \int_{-\infty}^{\infty} \frac{m(\tau)}{t - \tau} d\tau \tag{5.70}$$

which was introduced in that section. From the previous equation, (5.69),

$$M(\omega)\mathrm{sgn}(\omega) = j M_h(\omega) \tag{5.71}$$

where $M_h(\omega)$ is the Fourier transform of the Hilbert transform, $m_h(t)$

$$m_h(t) \Leftrightarrow M_h(\omega) \tag{5.72}$$

therefore substituting Eq. (5.71) in Equation set (5.67),

$$M_p(\omega) = \frac{1}{2}[M(\omega) + jM_h(\omega)]$$

$$\text{and } M_n(\omega) = \frac{1}{2}[M(\omega) - jM_h(\omega)] \tag{5.73}$$

now we want to transmit only $M_p(\omega - \omega_c) + M_n(\omega + \omega_c)$ (ask yourself: why this spectrum?) therefore using Eq. (5.5),

$$M_p(\omega - \omega_c) = M_p(\omega) * \delta(\omega - \omega_c)$$

$$\Leftrightarrow \frac{1}{2\pi}m_p(t)e^{j\omega_c t}$$

$$= \frac{1}{4\pi}[m(t) + jm_h(t)]e^{j\omega_c t} \tag{5.74}$$

and

$$M_n(\omega + \omega_c) = M_n(\omega) * \delta(\omega + \omega_c)$$

$$\Leftrightarrow \frac{1}{2\pi}m_n(t)e^{-j\omega_c t}$$

$$= \frac{1}{4\pi}[m(t) - jm_h(t)]e^{-j\omega_c t} \tag{5.75}$$

adding the previous two equations, ((5.74) and (5.75))

$$s_{\text{USB}}(t) = \frac{1}{2\pi}m(t)\cos(\omega_c t) + \frac{m_h(t)}{4\pi}\{j[\cos\omega_c t + j\sin\omega_c t] - j[\cos\omega_c t - j\sin\omega_c t]\}$$

$$= \frac{1}{2\pi}m(t)\cos(\omega_c t) - \frac{1}{2\pi}m_h(t)\sin\omega_c t \tag{5.76}$$

where $s_{\text{USB}}(t)$ is the single side-band transmission using the upper side-band as shown in Fig. 5.31b. Similarly, we can show that when we transmit the signal whose FT is shown in Fig. 5.31c,

$$s_{\text{LSB}}(t) = \frac{1}{2\pi}m(t)\cos(\omega_c t) + \frac{1}{2\pi}m_h(t)\sin\omega_c t \tag{5.77}$$

to recover the message signal we multiply with $\cos\omega_c t$ and then pass the resultant signal through a LPF. This part is explained in the section on coherent detection. (Sect. 5.4.2)

The single side-band modulation scheme is used in amateur (HAM) radio communications. The bands used are according to Table 5.1.

**Table 5.1**  HAM radio frequencies using SSB

| Frequency band (MHz) | Wavelength (m) | Type |
|---|---|---|
| 1.87–4 | 160–75 | LSB |
| 5 | 60 | USB |
| 7.5 | 40 | LSB |
| >7.5 | <40 | USB |

### 5.9.1  The Hilbert Transformer

The Hilbert transform of a function, from Eq. (5.71),

$$M_h(\omega) = -jM(\omega)\text{sgn}(\omega) \tag{5.78}$$

so the filter to be used has a frequency response,

$$H_h(\omega) = -j\text{sgn}(\omega)$$
$$= \begin{cases} -j & \omega > 0 \\ j & \omega < 0 \end{cases} \tag{5.79}$$

now

$$-j = -1 \times e^{j\pi/2} = e^{-j\pi/2} \tag{5.80}$$

and

$$j = e^{j\pi/2} \tag{5.81}$$

so the frequency-domain representation of the Hilbert transformer is

$$H_h(\omega) = \begin{cases} e^{-j\pi/2} & \omega > 0 \\ e^{j\pi/2} & \omega < 0 \end{cases} \tag{5.82}$$

which leaves the amplitude unchanged but only changes the phase of the signal. The frequency characteristics of this filter are shown in Fig. 5.32.

**Example 5.5**  Thus if the input is a cosine function,

$$\cos \omega_0 t = \frac{1}{2}\left\{e^{j\omega_0 t} + e^{-j\omega_0 t}\right\}$$
$$\Leftrightarrow \frac{2\pi}{2}\left\{\delta(\omega - \omega_0) + \delta(\omega + \omega_0)\right\}$$

then passing it through a Hilbert transformer, $\mathcal{H}\{\cos \omega_0 t\}$, then in the frequency domain,

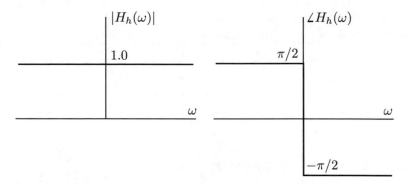

**Fig. 5.32** Frequency response of a Hilbert transformer

$$\mathcal{F}\left[\mathcal{H}\left\{\cos\omega_0 t\right\}\right] = -j\,\text{sgn}\left[\mathcal{F}\left\{\cos\omega_0 t\right\}\right]$$

which is

$$\mathcal{F}\left[\mathcal{H}\left\{\cos\omega_0 t\right\}\right] = \frac{2\pi}{2}\left\{\delta(\omega-\omega_0)(-j) + \delta(\omega+\omega_0)(+j)\right\}$$

$$= \frac{2\pi}{2j}\left\{\delta(\omega-\omega_0) - \delta(\omega+\omega_0)\right\}$$

and taking the inverse FT,

$$\mathcal{F}^{-1}\left[\frac{2\pi}{2j}\left\{\delta(\omega-\omega_0) - \delta(\omega+\omega_0)\right\}\right] = \frac{2\pi}{2j}\left\{\frac{e^{-j\omega_0 t}}{2\pi} - \frac{e^{j\omega_0 t}}{2\pi}\right\}$$

$$= \sin\omega_0 t$$

$$= \cos\left(\omega_0 t - \frac{\pi}{2}\right)$$

It basically means that this is equivalent to passing any signal $x(t)$ through a filter which affects the positive frequencies by a phase delay of $-\pi/2$, and therefore, *for a realisable filter*, the negative frequencies are advanced in phase by $\pi/2$. In short, *if we pass any signal through a* $-90°$ *phase shifter, we get its Hilbert transform on the output.*

### 5.9.2  Circuits to Modulate SSB Signals

#### 5.9.2.1  Filter Method

There are two methods of generation of SSB signals, one is the filter method and the other is the phase shift method. The filter method is outlined as per Fig. 5.33, (*a*)

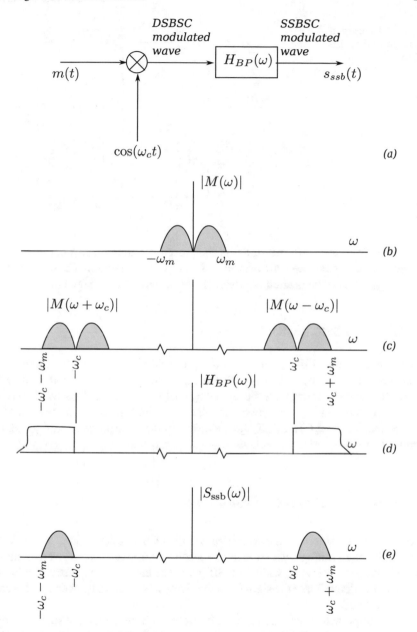

**Fig. 5.33** SSB generation with a filter

through $(e)$. In the $(a)$ part of the figure, the block diagram of the system is shown. The figure referred to shows the message signal, $m(t)$, is multiplied by $\cos(\omega_c t)$ to produce DSB amplitude modulated signal. The spectrum of DSB/SC is shown;

**Fig. 5.34** Generation of
SSB signal using phase
shifters

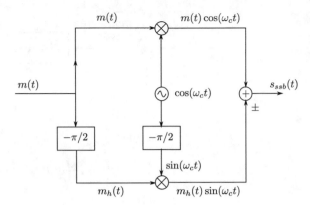

$M(\omega)$ is shifted to $\pm\omega_c$. The signal is the passed through a sharp cutoff BP filter whose characteristics are shown in the $(d)$ part of the diagram. The output of the filter is $s_{ssb}(t)$, and the desired SSB signal whose spectrum is shown in $(e)$.

#### 5.9.2.2   Phase Shift Method

In the generation of SSB signals using the phase shift method, we proceed as given in Sect. 5.9, Eqs. (5.76) and (5.77). Referring to Fig. 5.34, in the upper half of the figure, message signal, $m(t)$, is multiplied by $\cos(\omega_c t)$, while in the lower half of the figure, $m_h(t)$ and $\sin(\omega_c t)$ are generated. These two signals are also multiplied. Then, $m(t)\cos(\omega_c t)$ and $m_h(t)\sin(\omega_c t)$ are added or subtracted to produce the desired SSB signal.

### 5.10   Vestigial Sideband Modulation

In SSB, only one sideband is transmitted, while in DSB both the sideband are transmitted. The disadvantage of SSB is that expensive equipment is required to modulate and demodulate the signal, while in DSB (with carrier) very inexpensive equipment is required to demodulate the signal, but the sidebands consume power and therefore power is wasted.

A third approach which utilises the advantages of both is vestigial sideband modulation. In this type of modulation, (refer to Fig. 5.35) the spectrum of the original signal is shown in $(a)$ where the spectrum has a bandwidth of $\omega_m$. The signal is multiplied by $\cos(\omega_c t)$ and so is translated to the band extending from $\omega_c - \omega_m$ to $\omega_c + \omega_m$ which is shown in $(b)$.

$$m(t)\cos(\omega_c t) \Leftrightarrow \frac{1}{2}\left[M(\omega + \omega_c) + M(\omega - \omega_c)\right]$$

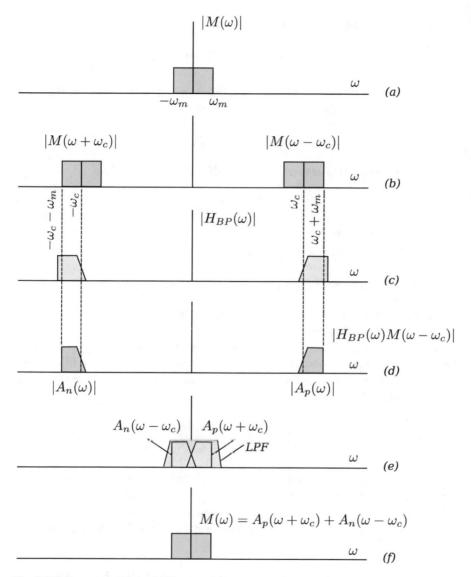

**Fig. 5.35** Diagram explaining VSB

The signal is next passed through a filter, $H_{BP}(\omega)$, shown in (c), so that

$$s_{vsb}(t) = \frac{1}{2}\left[M(\omega + \omega_c) + M(\omega - \omega_c)\right] H_{BP}(\omega)$$

to demodulate this signal a further multiplication by $\cos(\omega_c t)$ is performed and passed through a LPF, shown in (e). So

$$x(t) = s_{vsb}(t) \cos(\omega_c t)|_{LPF}$$

$$= \frac{1}{4} M(\omega) [H_{BP}(\omega - \omega_c) + H(\omega + \omega_c)]$$

Normally the spectrum of the BPF, $H_{BP}(\omega)$ is centred at $\pm\omega_c$. Hence, $H_{BP}(\omega + \omega_c)$ is shifted to the left so that

$$H_{BP}(\omega + \omega_c)|_{\omega_c = \omega + \omega_c}$$

is shifted to $\omega = 0$. Also $H_{BP}(\omega - \omega_c)$ is shifted to the right so that

$$H_{BP}(\omega - \omega_c)|_{-\omega_c = \omega - \omega_c}$$

is also shifted to $\omega = 0$.

The BPF is designed with the following criteria in mind

$$H_{BP}(\omega - \omega_c) + H_{BP}(\omega + \omega_c) = k \quad (\text{for } -\omega_m \leq \omega \leq \omega_m) \quad (5.83)$$

where $k$ is a constant. This characteristic (called antisymmetric roll off) is kept in mind so that $x(t)$ is proportional to $m(t)$.

The characteristic of the bandpass filter is shown in $(c)$. When we pass the waveform of $(b)$ through the filter of $(c)$ the resulting spectrum is shown in $(d)$. We can notice that one sideband is included, and part of the other is also included. We call these sidebands $A_p(\omega)$ and $A_n(\omega)$, where

$$A_p(\omega) = M(\omega - \omega_c) H_{BP}(\omega)$$
$$\text{and } A_n(\omega) = M(\omega + \omega_c) H_{BP}(\omega)$$

When the signal is received, we perform synchronous detection on the signal, multiplying it with $\cos(\omega_c t)$. One part of the signal is then translated to near $\omega = 0$, while the other part is translated to $2\omega_c$. The spectrum near $\omega = 0$ then superimposes to give us the original signal, where the condition imposed in Eq. (5.83) ensures that this happens. That is, near $\omega = 0$,

$$A_p(\omega + \omega_c) + A_n(\omega - \omega_c) = M(\omega) H_{BPF}(\omega + \omega_c) + M(\omega) H_{BPF}(\omega - \omega_c)$$
$$= kM(\omega) \quad \text{Using Eqn (5.83)}$$

This is shown in part $(e)$ of the figure. An LPF is then used to get the signal which is shown in the figure.

The transmitter and receiver of a VSB system are shown in Fig. 5.36. The $(a)$ part of the figure shows the transmitter, while the $(b)$ part of the figure shows the receiver. The transmitter and receiver are standard parts of a synchronous detection transmitter and receiver.

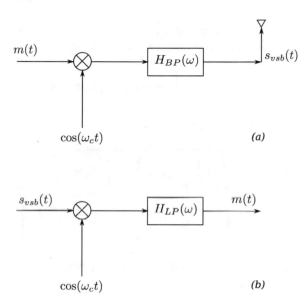

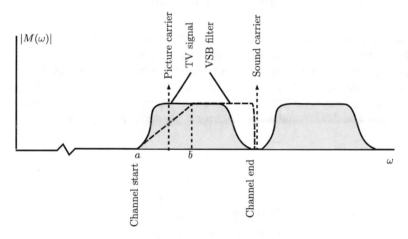

**Fig. 5.36** VSB transmitter and reciever

**Fig. 5.37** Details of a transmitted television signal

## 5.10.1 Television Signals

Baseband television signals occupy a band of frequencies from DC (0 Hz) to about
4.2–6 MHz depending on the type of TV format used. The type of modulation is
VSB, and the details are shown in Fig. 5.37. Basically, the picture carrier and sound
carrier are separated. The picture carrier is set at the point midway between $a$ and $b$,
while the sound carrier is set beyond the point where the picture signal ends along

the frequency band. The VSB filter characteristics are shown by the dotted line. If $v(t)$ is the transmitted television signal, then (see Sect. 5.6)

$$v(t) = \left\{1 + \frac{1}{2}k_a m(t)\right\} \cos(\omega_c t) - \frac{1}{2}k_a m'(t) \sin(\omega_c t)$$

then the envelope detector output is

$$En[v(t)] = \sqrt{\left\{1 + \frac{1}{2}k_a m(t)\right\}^2 + \left\{\frac{1}{2}k_a m'(t)\right\}^2}$$

$$= \left\{1 + \frac{1}{2}k_a m(t)\right\} \sqrt{1 + \left\{\frac{\frac{1}{2}k_a m'(t)}{1 + \frac{1}{2}k_a m(t)}\right\}^2}$$

which causes distortion. To decrease the distortion, $k_a$ must be made small.

## 5.11   Quadrature Amplitude Modulation

In quadrature amplitude modulation, two message signals $m_1(t)$ and $m_2(t)$ are used with carriers $\cos(\omega_c t)$ and $\sin(\omega_c t)$ which are then sent in the same region of the frequency domain, and therefore occupy the same bandwidth. Generally, the two message signals have similar bandwidths. When we demodulate the signals, demodulation is successful due to the fact that the two carriers are $\pi/2$ out of phase. The two signals and their modulated counterparts are shown in Fig. 5.38.

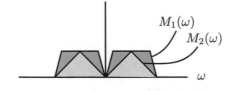

**Fig. 5.38**  Quadrature modulation message signals

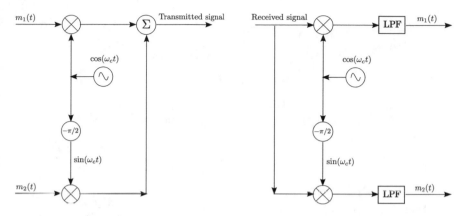

**Fig. 5.39**  Transmitter and receiver of a quadrature modulation system

The demodulation procedure requires that we use a LO which is in phase with the transmitter LO. Both the transmitter and receiver are shown in Fig. 5.39. The advantage of a quadrature modulated signal is a saving in bandwidth equivalent to an SSB signal.

The mathematics uses the two expressions

$$m_2(t)\sin^2(\omega_c t) = m_2(t)\frac{1 - \cos(2\omega_c t)}{2}$$

$$\text{and } m_1(t)\cos^2(\omega_c t) = m_1(t)\frac{1 + \cos(2\omega_c t)}{2} \tag{5.84}$$

$$\text{while } m_1(t) \text{ or } m_2(t) \times \sin(\omega_c t)\cos(\omega_c t) = m_1(t) \text{ or } m_2(t) \times \frac{\sin(2\omega_c t)}{2} \tag{5.85}$$

where the LPF allows only the low frequency components to pass through.

## 5.12   Frequency Division Multiplexing

After modulation of the message signal, it occupies the band $\omega_c \pm \omega_m$, where $\omega_c$ is the carrier frequency and $\omega_m$ is the maximum bandwidth occupied by the message signal. We use these ideas to introduce communication of a number of message signals *simultaneously* and receive them simultaneously. The spectra of the message signals are shown in Fig. 5.40d, and the spectra of the modulated signals are shown in Fig. 5.40e.

Consider $N$ message signals

$$m_i(t) \; i = 1, \ldots, N$$

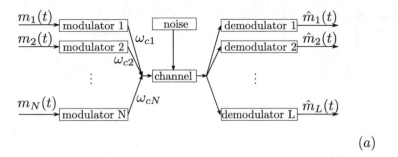

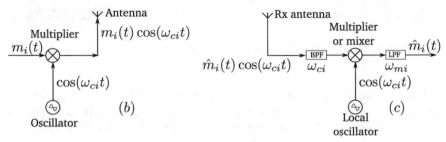

**Transmitter reciever combination**

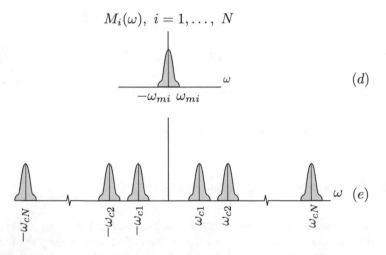

**Fig. 5.40**  Figure showing the concepts behind frequency division multiplexing

Let $i$th message signal has a bandwidth of $\omega_{mi}$ and is used to modulate carriers with frequencies $\omega_{ci}$. The $\omega_{mi}$s are approximately all equal though this has not a necessary condition for the validity of the present discussion. For example, if all the message signals are voice signals, then $f_{mi} \cong 4$ kHz, and if the signals consist of music, then the various bandwidths are approximately 22 kHz, and so on. Details of the spectrum of the $i$th message are shown in Fig. 5.40d.

These modulated signals are then transmitted simultaneously over the channel. Figure 5.40a shows the block diagram of the system, where the particulars of the modulation system are shown in the figure, in part (b). The spectrum of the composite transmitted signal is shown in the (e) part of the figure. Notice that the spectra of each individual modulated carrier have a bandwidth of $2\omega_{mi}$ and are separated from its neighbours by what is referred to as a *guard-band,* $\omega_{gb}$, so that the spectra do not overlap.

On reception, the signal is passed through a bandpass filter, or tuner, which is centred at $\omega_{ci}$ with a bandwidth, $B_i$, such that

$$2\omega_{mi} + \omega_{gb} \geq B_i \geq 2\omega_{mi} \tag{5.86}$$

The receiver is shown in part (c) of the figure. In our diagram for each transmitter there is a single receiver, though this is not a necessity; there may be a number of receivers for a single transmitter. The bandpass filter is generally a capacitor inductor combination so the signal may be 'tuned into' using a variable capacitor. This requirement (on $B_i$) ensures that the bandwidth of the BPF (shown in the figure) does not encroach upon the neighbouring spectra. After the mixing operation, the signal is translated to $\omega = 0$ with a bandwidth of $\omega_{mi}$. This signal is recovered with an LPF of the appropriate bandwidth.

Since each received signal is corrupted by noise therefore the demodulated message signal is referred to as $\hat{m}_i(t)$ (note that $\hat{m}_i(t) \approx m_i(t)$). This process, outlined above, of transmitting modulated signals with different carriers is called *frequency division multiplexing*. The receiver which is used with an Frequency division multiplexed (FDM) signal is shown in Fig. 5.41.

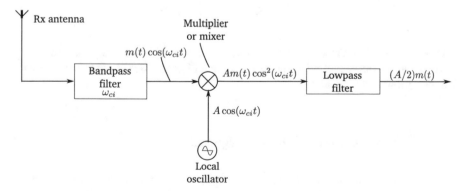

**Fig. 5.41** Receiver for a FDM signal

## 5.13   Performance of AM Systems in the Presence of Noise[2]

### 5.13.1   Introduction

Going back to the introductory chapter, in Sect. 1.3 we discussed the block diagram of a typical communication system. Once the signal is transmitted, noise is added to the modulated message while in transit through the channel. Typically, the noise which is added to the signal is referred to as *additive white Gaussian noise* (AWGN). The noise is termed 'additive' since it adds to the signal; the noise is 'white', since the PSD of the noise contains all frequencies; it is assumed to be Gaussian to denote the fact that its amplitude pdf is Gaussian with zero mean. The noise is assumed to have the property of being wide sense stationary.

Typically we analyse the system, assuming ideal white noise whose autocorrelation function is given by

$$R_W(\tau) = \frac{N_0}{2}\delta(\tau) \tag{5.87}$$

with the PSD

$$S_W(\omega) = \frac{N_0}{2} \tag{5.88}$$

since

$$R_W(\tau) \Leftrightarrow S_W(\omega) \tag{5.89}$$

So we next ask the relevant question: how do we evaluate the noise performance of the system? The answer to this lies in looking at the ratio of the signal power to noise power. The average signal power may be calculated from the equation,

$$\langle m^2(t) \rangle = \frac{1}{T} \int_{-T/2}^{T/2} m^2(t)dt \tag{5.90}$$

where $T$ is a fairly large interval of time. Therefore by observing the parameters involved in the output signal-to-noise ratio, we will get an idea about how to improve the performance of the system.

### 5.13.2   Performance of DSB/SC Systems

The block diagram of the DSB/SC communication system is shown in Fig. 5.42, which shows the basic elements of such a system. At the extreme left, the figure shows the message signal, $m(t)$ and its processing to give the modulated signal marked ① which is transmitted using the channel shown in the figure. The transmitter

---

[2] Read Sect. 3.11 onward till the end of the chapter.

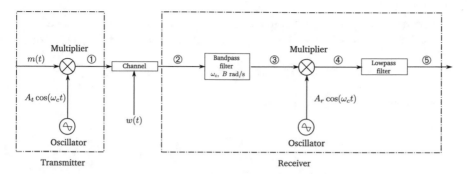

**Fig. 5.42** Block diagram of a transmitter and receiver of a AM/DSB/SC system

is assumed to be ideal (the resistors within the transmitter are assumed to produce negligible thermal noise). At the extreme right, at the output, ⑤, the demodulated message signal is shown though corrupted by noise. At circled 1, ①, the signal is

$$s_1(t) = A_t m(t) \cos(\omega_c t) \qquad (5.91)$$

which is transmitted through the channel. When the signal exits the channel, at ②, at the receiving end, the signal is now corrupted by additive while noise,

$$s_2(t) = A'_t m(t) \cos(\omega_c t) + w(t) \qquad (5.92)$$

where $w(t)$ is the noise whose characteristics given by Eqs. (5.87) and (5.88), and $A'_t$ is a constant (which, in the general case, is not necessarily equal to $A_t$.) The constant $A'_t$ takes into account the attenuation due to the channel. For example, when the communication is a radio broadcast, the attenuation in the received voltage is $\propto 1/d$, $d$ being the distance between the transmitting antenna and receiving antennas.

We also know that the signal is now passed through a BPF centred at $\omega = \omega_c$ having a bandwidth[3]

$$B = 2\omega_m$$

At circled 3, ③ therefore, the signal is

$$s_3(t) = A'_t m(t) \cos(\omega_c t) + n_{NB}(t) \qquad (5.93)$$

where $n_{NB}(t)$ is narrowband noise of bandwidth $B$ rad/s. The characteristics of the narrowband noise are given by Eq. (3.165) in Sect. 3.19,

$$n_{\text{NB}}(t) = n_I(t) \cos(\omega_c t) - n_Q(t) \sin(\omega_c t)$$

---

[3] $\omega_m$ is the maximum frequency of $M(\omega)$, the FT of $m(t)$.

where $n_I(t)$ and $n_Q(t)$ are the in-phase and quadrature phase components of noise (which have a PSD at low frequency of bandwidth $B/2(= \omega_m)$ rad/s.) Therefore,

$$s_3(t) = \left[A_t' m(t) + n_I(t)\right] \cos(\omega_c t) - n_Q(t) \sin(\omega_c t)$$

The signal power at ③ is given by

$$
\begin{aligned}
S_i &= \left(A_t'\right)^2 \left[\frac{1}{T} \int_{-T/2}^{T/2} m^2(t) \cos^2(\omega_c t) dt\right] \\
&= \frac{\left(A_t'\right)^2}{2} \left[\frac{1}{T} \int_{-T/2}^{T/2} m^2(t)[1 + \cos(2\omega_c t)\, dt\right] \\
&= \frac{\left(A_t'\right)^2}{2} \left[\frac{1}{T} \int_{-T/2}^{T/2} m^2(t) dt\right] \\
&= \frac{\left(A_t'\right)^2}{2} \langle m^2(t)\rangle
\end{aligned}
\tag{5.94}
$$

where the term in the square brackets is the average noise power in the interval $T$. $S_i$ is considered as the input signal power to the receiver, and

$$\langle m^2(t)\rangle = \frac{1}{T} \int_{-T/2}^{T/2} m^2(t) dt \tag{5.95}$$

is the average power of the message signal. The noise power is given by

$$N_i = 2N_0 \omega_{m(Hz)} \tag{5.96}$$

where

$$\omega_{m(Hz)} = \omega_m/2\pi$$

is the maximum frequency of the message signal *in hertz*. Therefore,

$$\frac{S_i}{N_i} = \frac{\frac{\left(A_t'\right)^2}{2} \langle m^2(t)\rangle}{2N_0 \omega_{m(Hz)}} \tag{5.97}$$

The signal is now passed through a mixer where it is multiplied by $A_r \cos(\omega_c t)$; so therefore at ④ the signal is

$$
\begin{aligned}
s_4(t) &= A_r \left[A_t' m(t) + n_I(t)\right] \cos^2(\omega_c t) - n_Q(t) \cos(\omega_c t) \sin(\omega_c t) \\
&= \frac{A_r}{2} \left[A_t' m(t) + n_I(t)\right] [\cos(2\omega_c t) + 1] - \frac{A_r}{2} n_Q(t) \sin(2\omega_c t)
\end{aligned}
\tag{5.98}
$$

At this stage, the signal has a high-frequency signal and a low frequency one. We then pass $s_4(t)$ through a low-pass filter of bandwidth $B/2$ to get

$$s_5(t) = \frac{A_r}{2} \left[ A'_t m(t) + n_I(t) \right] \tag{5.99}$$

where the high-frequency components have been filtered out. The output signal power is therefore,

$$
\begin{aligned}
S_o &= \left[ \frac{A_r A'_t}{2} \right]^2 \left[ \frac{1}{T} \int_{-T/2}^{T/2} m^2(t) dt \right] \\
&= \frac{A_r^2}{2} \times \frac{(A'_t)^2}{2} \left[ \frac{1}{T} \int_{-T/2}^{T/2} m^2(t) dt \right] \\
&= \frac{A_r^2}{2} S_i
\end{aligned} \tag{5.100}
$$

and the noise power is

$$N_0 = (A_r/2)^2 2 N_0 \omega_{m(Hz)} \qquad (\omega_m \text{ in Hz})$$

Hz) (see Sect. 3.16.1). The signal-to-noise ratio at the output end is

$$
\begin{aligned}
\frac{S_o}{N_o} &= \frac{\frac{A_r^2}{2} S_i}{\left[ \frac{A_r}{2} \right]^2 N_0 (2\omega_{m(Hz)})} \\
&= \frac{2 S_i}{N_0 B}
\end{aligned}
$$

where $B$ is in hertz. Therefore,

$$\left( \frac{S_o}{N_o} \right)_{DSB-SC} = \frac{2 S_i}{N_0 B} = \frac{S_i}{N_0 \omega_{m(Hz)}} \tag{5.101}$$

where $B$ and $\omega_{m(Hz)}$ are in Hertz and $S_i$ is the input signal power.

### 5.13.3   Performance of SSB-SC Systems

We now take a look at the noise performance of the SSB-SC systems which was discussed in Sect. 5.9. The block diagram of the transmitter, channel and receiver is shown in Fig. 5.43 where the details of the transmitter are not shown. The output of the SSB transmitter is the signal at ①, and which is given by

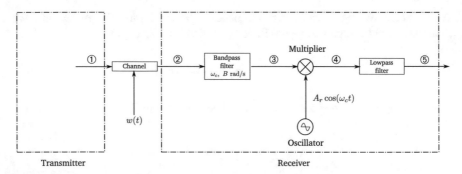

**Fig. 5.43**  Block diagram of a transmitter and receiver of a AM/SSB/SC system

$$s_1(t) = A'_t [m(t) \cos(\omega_c t) \pm m_h(t) \sin \omega_c t] \qquad (5.102)$$

where the positive sign is used when we transmit the lower sideband, and the negative sign is used when we transmit the upper sideband. It does not matter whether we produce this signal by the filter method or the phase shift method (see Sects. 5.9.2.1 and 5.9.2.2) the signal has this general form.

Working with the upper sideband SSB signal,

$$s_1(t) = A'_t [m(t) \cos(\omega_c t) - m_h(t) \sin \omega_c t]$$

which we use to illustrate the evaluation of the noise performance of the system, and, when we use the lower sideband signal, the results are substantially the same.

After we pass the signal through the channel, we get at ②

$$s_2(t) = A'_t [m(t) \cos(\omega_c t) - m_h(t) \sin \omega_c t] + w(t) \qquad (5.103)$$

where $w(t)$ is additive white noise.

We then pass the signal through the bandpass filter centred at $\omega = \omega_c$ with bandwidth, $B = \omega_m$ since only one sideband has been transmitted. The filter must be such that it must cover the full sideband. The narrowband noise added to the signal has the analytic expression given by (3.165) and therefore the signal at ③ is

$$s_3(t) = A'_t [m(t) \cos(\omega_c t) - m_h(t) \sin \omega_c t] + \left[ n_I(t) \cos(\omega_c t) - n_Q(t) \sin(\omega_c t) \right]$$

or

$$s_3(t) = \left[ A'_t m(t) + n_I(t) \right] \cos(\omega_c t) - \left[ A'_t m_h(t) + n_Q(t) \right] \sin \omega_c t \qquad (5.104)$$

The input signal power is

$$S_i = \left(A_t'\right)^2 \frac{1}{T} \int_{-T/2}^{T/2} [m(t)\cos(\omega_c t) - m_h(t)\sin\omega_c t]^2\, dt$$

$$= \left(A_t'\right)^2 \frac{1}{2T} \int_{-T/2}^{T/2} \left[m^2(t) + m_h^2(t)\right] dt$$

$$= \left(A_t'\right)^2 \left\langle m^2(t)\right\rangle$$

where we have used

$$\left\langle m^2(t)\right\rangle = \left\langle m_h^2(t)\right\rangle$$

We now multiply this signal with $\cos(\omega_c t)$ and pass it through a low-pass filter. The signal at ④ is

$$s_4(t) = A_r \left[A_t' m(t) + n_I(t)\right]\cos^2(\omega_c t) - A_r \left[A_t' m_h(t) + n_Q(t)\right]\sin\omega_c t\cos\omega_c t$$

$$= \frac{A_r}{2}\left\{\left[A_t' m(t) + n_I(t)\right][1 + \cos(2\omega_c t)] - \left[A_t' m_h(t) + n_Q(t)\right]\sin 2\omega_c t\right\}$$

$$\tag{5.105}$$

When we pass the signal through a low-pass filter, the high-frequency terms are blocked, and only the low frequency terms are passed through. So at ⑤ we have

$$s_5(t) = \frac{A_r}{2}\left[A_t' m(t) + n_I(t)\right] \tag{5.106}$$

The output signal power is

$$S_o = \frac{A_r^2 \left(A_t'\right)^2}{4}\langle m^2(t)\rangle = \frac{A_r^2}{4}S_i \tag{5.107}$$

and the output noise power is

$$N_0 = \frac{A_r^2}{4}N_0 B$$

where $B$ is in hertz. The output signal-to-noise ratio can then be calculated as

$$\left(\frac{S_o}{N_o}\right)_{\text{SSB−SC}} = \frac{S_i}{N_0 B} = \frac{S_i}{N_0\omega_{m(Hz)}} \tag{5.108}$$

where both $B$ and $\omega_{m(Hz)}$ are in hertz. If we work with the lower sideband, then we get identical results. The results show that whether we DSB/SC modulation or SSB-SC modulation, the noise performance of the two are identical (see Eq. (5.101)).

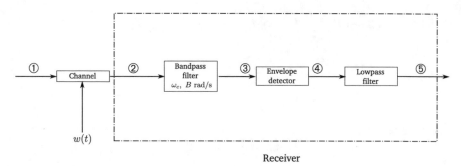

Receiver

**Fig. 5.44** Performance of AM signal in noise

### 5.13.4 Amplitude Modulation with Carrier

Referring to Fig. 5.44, received AM signal at ② is given using Eq. (5.40),

$$s(t) = A'_t [1 + k_a m(t)] \cos(\omega_c t) + w(t) \tag{5.109}$$

where $w(t)$ is white noise. This signal is passed through a BPF, and at ③ the signal is

$$
\begin{aligned}
s(t) &= A'_t [1 + k_a m(t)] \cos(\omega_c t) + n_{NB}(t) \\
&= A'_t [1 + k_a m(t)] \cos(\omega_c t) + n_I(t) \cos(\omega_c t) - n_Q(t) \sin(\omega_c t) \\
&= \left\{ A'_t [1 + k_a m(t)] + n_I(t) \right\} \cos(\omega_c t) - n_Q(t) \sin(\omega_c t) \tag{5.110}
\end{aligned}
$$

the average signal power is a sum of the carrier power plus the signal power. Therefore,

$$S_i = \frac{\left(A'_t\right)^2}{2} \left[1 + k_a^2 \langle m^2(t) \rangle \right] \tag{5.111}$$

and the average noise power is

$$N_i = N_0 B \tag{5.112}$$

where $B$ is in hertz. So the input signal-to-noise ratio, SNR, is

$$\left(\frac{S_\rangle}{N_\rangle}\right) = \frac{\frac{(A'_t)^2}{2} \left[1 + k_a^2 \langle m^2(t) \rangle \right]}{N_0 B} \tag{5.113}$$

Now at ④ the output of the envelope detector is

$$s_4(t) = \left\{ \left\{ A'_t [1 + k_a m(t)] + n_I(t) \right\}^2 + n_Q^2(t) \right\}^{1/2}$$

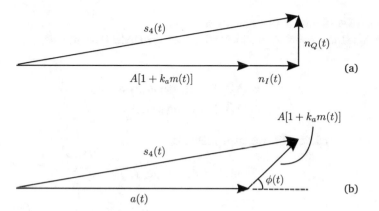

**Fig. 5.45** Noise considerations in AM. **a** Low noise case **b** High noise case

### 5.13.4.1 Low Noise Case

We now assume that the noise power is much smaller than the signal power. In this case, from Fig. 5.45a,

$$s_4(t) = \left\{ \left\{ A_t' \left[ 1 + k_a m(t) \right] + n_I(t) \right\}^2 + n_Q^2(t) \right\}^{1/2}$$

$$\approx \left\{ \left\{ A_t' \left[ 1 + k_a m(t) \right] + n_I(t) \right\}^2 \right\}^{1/2}$$

$$\approx s_4(t) = A_t' \left[ 1 + k_a m(t) \right]$$

and

$$S_o \approx \frac{\left( A_t' \right)^2}{2} k_a^2 \left\langle m^2(t) \right\rangle$$

where the DC term is missing, since it has been blocked by a capacitor. The noise power is, as earlier,

$$N_o = B N_0$$

so

$$\left( \frac{S_o}{N_o} \right) = \frac{\frac{\left( A_t' \right)^2}{2} k_a^2 \left\langle m^2(t) \right\rangle}{B N_0}$$

$$= \left( \frac{S_i}{N_o \omega_{m(Hz)}} \right) \frac{k_a^2 \left\langle m^2(t) \right\rangle}{\left[ 1 + k_a^2 \left\langle m^2(t) \right\rangle \right]} \tag{5.114}$$

hence there is a definite fall in the performance as compared to AM/DSB/SC and AM/SSB/SC. (Compare Eq. (5.101))

**Example 5.6** Single tone modulation

In single tone modulation with carrier, the signal to be transmitted is (see Example 5.4)

$$s(t) = A(1 + k_a A_m \cos \omega_m t) \cos \omega_c t$$
$$= A(1 + \mu \cos \omega_m t) \cos \omega_c t$$

where $0 < \mu \leq 1$, and the message signal is

$$m(t) = A_m \cos (\omega_m t)$$

so

$$\langle m^2(t) \rangle = \frac{A_m}{2}$$

from Eq. (5.114),

$$\left(\frac{S_o}{N_o}\right) = \left(\frac{S_i}{N_o \omega_m}\right) \frac{k_a^2 \langle m^2(t) \rangle}{\left[1 + k_a^2 \langle m^2(t) \rangle\right]}$$
$$= \left(\frac{S_i}{N_o \omega_m}\right) \frac{\mu^2/2}{\left[1 + \mu^2/2\right]}$$
$$= \left(\frac{S_i}{N_o \omega_m}\right) \frac{\mu^2}{\left[2 + \mu^2\right]} \tag{5.115}$$

the maximum value of the right-hand side of this expression is

$$\left(\frac{S_i}{N_o \omega_m}\right) \frac{1}{3}$$

for $\mu = 1$. Hence to obtain the same performance as the other modulations, the signal level has to be at least three times higher. On the other hand to get improved performance, $\mu$ should be small, the minimum value being zero. Therefore,

$$\left(\frac{S_i}{N_o \omega_m}\right) \frac{1}{3} \geq \left(\frac{S_o}{N_o}\right) > 0$$

### 5.13.4.2   Large Noise Case

We now look at the case when the noise is large: when the carrier to noise ratio is small compared to unity. The noise term can be written as (see Sect. 3.19.1)

$$n_{\text{NB}}(t) = a(t) \cos [\omega_c t + \phi(t)]$$

then from Fig. 5.45b, applying the triangle law of cosines,

$$s_4(t) = \sqrt{a^2(t) + \{A[1 + k_a m(t)]\}^2 + 2a(t)\{A[1 + k_a m(t)]\}\cos\phi(t)}$$

$$= a(t)\sqrt{1 + \left\{\frac{A[1 + k_a m(t)]}{a(t)}\right\}^2 + 2\left\{\frac{A[1 + k_a m(t)]}{a(t)}\right\}\cos\phi(t)}$$

$$\approx a(t)\left\{1 + \frac{1}{a(t)}\{A[1 + k_a m(t)]\}\cos\phi(t)\right\}$$

$$= a(t) + A[1 + k_a m(t)]\cos\phi(t)$$

from this analysis we conclude that in the large noise case, there is no component which is proportional to the message signal, and therefore we get no sensible output. If we observe this expression we notice that the $\cos(\phi(t))$ takes on random values between $+1$ and $-1$. However, as the signal level increases and approximates the low noise case, an output proportional to the message signal appears. This is called the threshold effect.

## 5.14 Important Concepts and Formulae

- To communicate using message signals, waves are transmitted from a source, and which travel through a convenient medium which is available to the receiver
- The medium is technically named as *a channel.*
- A very important consideration in the transmission of information by use of sinusoidal waves is that such a signal be used which is *not attenuated* and *not distorted* when it is used to communicate from the transmitter to the receiver.
- Baseband communication consists of transmitting a signal without any *modulation* or *modification.*
- Effective baseband communication is the transmission of a signal using a channel where there is a single transmitter and single receiver.
- In DSB/SC

$$s(t) = Am(t)\cos(\omega_c t)$$

and

$$S(\omega) = (A/2)[M(\omega - \omega_c) + M(\omega + \omega_c)]$$

- That part of the spectrum of $M(\omega - \omega_c) + M(\omega + \omega_c)$ (centred at $\pm\omega_c$), which lies between $\omega_c$ and $\omega_c + \omega_m$ and that of its mirror image, lying between $-\omega_c - \omega_m$ and $-\omega_c$ are both called the *upper sidebands* (USBs)
- Similarly, the spectra of $M(\omega - \omega_c) + M(\omega + \omega_c)$ centred at $\pm\omega_c$ lying between the frequencies $\omega_c - \omega_m$ and $\omega_c$, and $-\omega_c$ and $-\omega_c + \omega_m$, respectively, are both called the *lower sidebands* (LSBs) (see Fig. 5.6)

- Coherent detection is the case where the LOs of the transmitter and receiver are synchronised.
- In the demodulation

$$Am(t)\cos^2\omega_c t = \frac{A}{2}m(t)\left[1 + \cos(2\omega_c t)\right]$$

- The demodulation process is shown in Fig. 5.13.
- In coherent detection, the frequency and phase must be matched. Discussion in Sect. 5.4.3.
- Modulation of DSB/SC is achieved through

  ▶ Nonlinear modulation
  ▶ Bridge modulator
  ▶ Double balanced modulator

- Demodulation is achieved through

  ▶ Double balanced demodulator
  ▶ Costas receiver

- Many transmitters can transmit using AM through FDM
- The DSB AM signal (with carrier) is

$$s(t) = A\left[1 + k_a m(t)\right]\cos(\omega_c t)$$

where $k_a$ is the amplitude sensitivity of the modulator and $s(t)$ is the modulated signal. In amplitude modulation, we would like to have $k_a m(t) < 1$ for ease of demodulation.

- The FT of this signal is

$$A\left[1 + k_a m(t)\right]\cos(\omega_c t) \Leftrightarrow A\pi\left[\delta(\omega - \omega_0) + \delta(\omega + \omega_0)\right] + \frac{Ak_a}{2}\left[M(\omega - \omega_c) + M(\omega + \omega_c)\right]$$

(see Fig. 5.24)

- The modulation index, $\mu$ for single tone modulation is

$$0 < \mu = \frac{A_{\max} - A_{\min}}{A_{\max} + A_{\min}} < 1$$

- The power in the sidebands is

$$\mathbb{P}_{av,sb} = \frac{\mu^2 A^2}{8} + \frac{\mu^2 A^2}{8} = \frac{\mu^2 A^2}{4} \tag{5.116}$$

while the power in the carrier is

$$\mathbb{P}_{av,c} = \frac{A^2}{2} \tag{5.117}$$

- Modulation of DSB signals with carrier

  ▶ Nonlinear modulator

- Demodulation is done with an envelope detector
- AM frequencies are Medium wave: 530–1600 kHz and Short wave: 2.3–26 MHz
- The signals for SSB are

$$s_{\text{USB}}(t) = \frac{1}{2\pi} m(t) \cos(\omega_c t) - \frac{1}{2\pi} m_h(t) \sin \omega_c t$$

and

$$s_{\text{LSB}}(t) = \frac{1}{2\pi} m(t) \cos(\omega_c t) + \frac{1}{2\pi} m_h(t) \sin \omega_c t$$

- The Hilbert transformer filter is

$$H_h(\omega) = -j\,\text{sgn}(\omega)$$

- SSB can be generated by using

  ▶ Sharp cutoff filters
  ▶ Phase shift method

- The output signal-to-noise ratio for DSB-SC is given by

$$\left(\frac{S_o}{N_o}\right)_{\text{DSB-SC}} = \frac{2S_i}{N_0 B}$$

- The output signal-to-noise ratio for SSB-SC is given by

$$\left(\frac{S_o}{N_o}\right)_{\text{SSB-SC}} = \frac{S_i}{N_0 B}$$

- For low noise, AM signal-to-noise ratio is given by

$$\left(\frac{S_o}{N_o}\right) = \left(\frac{S_i}{N_o \omega_{m(Hz)}}\right) \frac{k_a^2 \langle m^2(t) \rangle}{[1 + k_a^2 \langle m^2(t) \rangle]}$$

## 5.15  Self Assessment

### 5.15.1  Review Questions

1. Why do we require modulation of a signal to transmit. Why not transmit it directly?

2. What is the meaning of the bandwidth of a baseband signal? What is the bandwidth of voice?
3. What is the difference between amplitude modulation DSB/SC and amplitude modulation, DSB/wC?
4. Plot the upper and lower sidebands in a DSB spectrum.
5. What is the bandwidth of music? Television?
6. Why is it easier to demodulate an amplitude modulated signal (with carrier) rather than an amplitude modulated signal with supressed carrier?
7. What is the advantage of amplitude modulation (with carrier) over amplitude modulation (suppressed carrier) and where is this advantage put into practice?
8. Why do we want to transmit $M_+(\omega - \omega_c) + M_-(\omega + \omega_c)$ in SSB modulation? Why not transmit $M_+(\omega - \omega_c) + M_+(\omega + \omega_c)$?
9. Explain why to recover the message signal of SSB modulation, we need to multiply it with $\cos \omega_c t$ and pass the resultant signal through an LPF.

## 5.15.2  Numerical Problems

1. If $m(t)$ is given by

$$m(t) = \begin{cases} A & -1 \le t \le 1\mu s \\ 0 & \text{elsewhere} \end{cases}$$

find the effective bandwidth of this signal, using the bandwidth between first nulls. Use a carrier of appropriate value so that the modulation is that of DSB/SC so that demodulation can take place. What is the minimum value of the carrier frequency? Plot the waveform in time domain. Also show the spectrum in the frequency domain.
2. Find the FT of the square wave shown in Fig. 5.46

3. The spectrum of a signal lies in the range $300 \le f \le 5000$ Hz. The maximum amplitude of the signal is 1 V. What should be the minimum amplitude of the carrier to have AM/DSB signal. For this case with carrier frequency equal to 25 KHz plot the AM/DSB signal in time and frequency domains.
4. If an amplitude modulated signal is

$$s(t) = 10 \cos(2\pi \times 9000t) + 50 \cos(2\pi \times 10,000t) + 10 \cos(2\pi \times 11,000t)$$

(a) Plot $s(t)$.
(b) Plot the approximate spectrum of this waveform.
(c) How draw the block diagram of the demodulator.

5. If
$$s(t) = 10 \cos(2\pi \times 80,000t) + 50 \cos(2\pi \times 100,000t) + 10 \cos(2\pi \times 120,000t)$$

**Fig. 5.46** Square wave

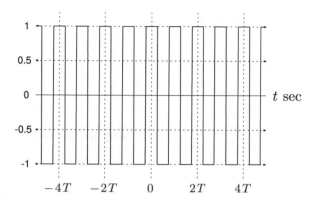

(a) Find the modulation index.
(b) Find the carrier power.
(c) Find the power in the sidebands.
(d) What should be the carrier amplitude for a modulation index of 0.2?

6. If the message signal is given in Fig. 5.47, plot the AM modulated signal with carrier and with supressed carrier.

7. If the message signal is given in Fig. 5.48, plot the AM modulated signal with carrier and with supressed carrier.

8. For the modulated waveform of Fig. 5.49a, find

   (a) the expression for $m(t)$ in the time duration 0–4 s,
   (b) the energy in $m(t)$ in the time duration 0–4 s,
   (c) the average power in $m(t)$ in the time duration 0–4 s.

9. A modulated message waveform is shown in Fig. 5.49a.

   (a) If the signal is passed through an envelope detector what would be the output?
   (b) What type of modulation is it?
   (c) What is the frequency in Hertz of the carrier?
   (d) Give a mathematical expression for the modulated waveform.
   (e) Find the energy and average power in the modulated waveform in the duration 0–4 s.

10. For the case of Fig. 5.49b, give an expression for the modulated waveform. How much average power is in the message signal, and how much is in the carrier?

11. For the case of Fig. 5.49a, b, give an expression for the magnitude of the spectrum of modulated waveforms for the interval 0–2 s.

12. What is the type of modulation in Fig. 5.49b? The modulated waveform of Fig. 5.49b is passed through a coherent detector. What is the output?

**Fig. 5.47** Message signal

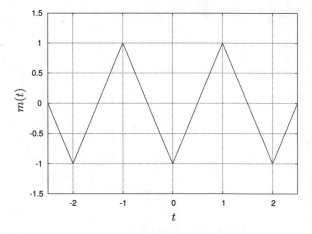

**Fig. 5.48** Message signal

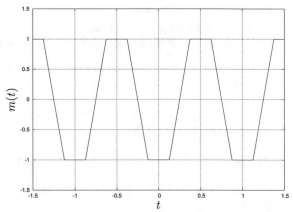

13. Show that when we transmit the lower sideband, the equation which we obtain is Eq. (5.77):

$$s_{\text{LSB}}(t) = \frac{1}{2\pi}m(t)\cos(\omega_c t) + \frac{1}{2\pi}m_h(t)\sin\omega_c t$$

14. If the signal

$$m(t) = 0.1\cos(2\pi t)$$

is passed through a Hilbert transformer, find the output.

15. Create a AM/SSB upper sideband signal using

$$m(t) = 10\cos(2\pi t) + 2\cos(3\pi t)$$

and carrier

$$c(t) = 10\cos(20\pi t)$$

**Fig. 5.49** Modulated
waveforms

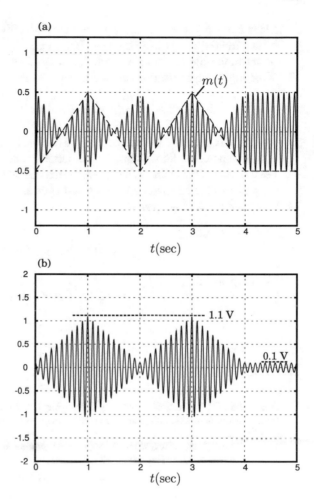

Plot $|M(\omega)|$, $|C(\omega)|$. Using

(a) Hilbert transformers.

(b) A sharp cutoff Filters.

Plot the magnitude of the AM/SSB waveform. Using a coherent detector show
you would get back the original waveform.

16. For the real-time waveform $m(t)$ having the magnitude spectrum for positive
values of $\omega$

$$|M(\omega)| = \begin{cases} 0 & \omega \leq 100 \text{ r/s} \\ 10 & 100 < \omega \leq 4000 \text{ r/s} \\ 0 & \text{elsewhere} \end{cases}$$

Mathematically describe $|M_h(\omega)|$. Plot the AM/SSB signal using the lower side-band. Find the energy in SSB signal. If the signal lasts for 10 sec find the average power in the signal. Compare with the energy in the message signal.

17. If the input/output relation of a nonlinear element is given by

$$y(t) = ax(t) + bx^2(t)$$

we use this element to 'demodulate' a DSB/SC signal without using a synchronous detector. Show what are the output products.

18. For the case of Problem 17, if the input signal is AM/DSB with carrier, is demodulation possible? Explain with a block diagram.

19. Calculate the average power

(a) in the signal
$$s(t) = m(t)\cos(\omega_c t)$$

(b) in the signal
$$s(t) = [A + m(t)]\cos(\omega_c t)$$

# References

1. Bhooshan S (2013) Fundamentals of engineering electromagnetics. Oxford University Press
2. Jordan EC, Balmain KG (1968) Electromagnetic waves and radiating systems, 2nd edn. Prentice Hall of India
3. Tucker DG (1971) The early history of amplitude modulation, sidebands and frequency-division-multiplex. Radio Electron Eng 41:43–47

# Chapter 6
# Angle Modulation

## 6.1 Introduction

We have studied amplitude modulation, where the amplitude of the sinusoidal signal is varied in accordance with the message signal,

$$s(t) = A(t)\cos(\omega_c t) \tag{6.1}$$

where $A(t)$ is the time-varying amplitude which contains information about the message signal. In another kind of modulation, we may concentrate on the phase of the signal. For example, in Eq. (6.1), the phase angle is

$$\theta(t) = \omega_c t$$

If we want to modulate the signal in a way different from that in the case of amplitude modulation, then we may consider including the message information in the phase or frequency, and which would give us two kinds of new modulations, namely, phase and frequency modulations. We now write the sinusoidal signal in two ways:

1. The first way would be to concentrate on the frequency. To do this we consider the general signal whose phase is a function of time.

$$s(t) = A\cos[\theta(t)] \tag{6.2}$$

where $A$ is a constant, and $\theta(t)$ is the instantaneous phase at time $t$. Then we may then use $\theta(t)$ to define what would be the instantaneous frequency,

$$\omega(t) = \frac{d}{dt}\theta(t) \tag{6.3}$$

In frequency modulation, we make the $\omega(t)$ to be proportional to the message signal,

© The Author(s), under exclusive license to Springer Nature Singapore Pte Ltd. 2022
S. Bhooshan, *Fundamentals of Analogue and Digital Communication Systems*,
Lecture Notes in Electrical Engineering 785,
https://doi.org/10.1007/978-981-16-4277-7_6

**Fig. 6.1** Frequency and
phase in angle modulation

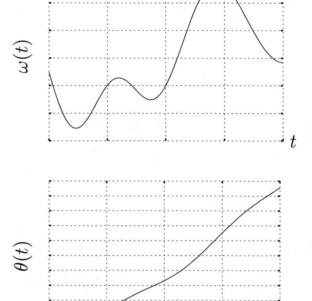

$$\omega(t) = \omega_c + k_f m(t) \tag{6.4}$$

where $\omega_c$ would be the carrier, $m(t)$ the message signal, and $k_f$ is a constant called
the *frequency sensitivity*, and which implies that the *instantaneous* frequency,
$\omega(t)$, is a function of the message signal. The phase in this case is

$$\theta(t) = \int \left[ \omega_c + k_f m(t) \right] dt$$

$$= \omega_c t + k_f \int m(t) dt \tag{6.5}$$

Figure 6.1 shows a representative sample of the frequency and phase of such a
signal. The transmitted signal is

$$s(t) = A_c \cos \left[ \omega_c t + k_f \int m(t) dt \right]$$

where

$$\omega_c \gg k_f m(t)$$

The instantaneous frequency ranges between

$$\omega_c - k_f \, |m(t)|_{\max} \le \omega \le \omega_c + k_f \, |m(t)|_{\max}$$

and

$$\Delta\omega = k_f \, |m(t)|_{\max} \tag{6.6}$$

is the frequency deviation. A frequency modulator is shown in Fig. 6.2 using a phase modulator. Essentially, if we refer to the diagram, the phase modulator adds the two phases and gives the cosine of the added phases: $\cos\left[\omega_c t + k_f \int m(t)dt\right]$, and the instantaneous frequency is $\omega(t) = \omega_c + k_f m(t)$.

2. In the second case, which is phase modulation,

$$s(t) = A\cos[\theta(t)] \tag{6.7}$$

where

$$\theta(t) = \omega_c t + k_p m(t) \tag{6.8}$$

where $k_p$ is a constant called the phase sensitivity. In phase modulation, the phase of the transmitted signal is a function of the message signal, and $\omega_c$ is the carrier frequency. The instantaneous phase at time $t$ is given by Eq. (6.8), while the instantaneous frequency is given by

$$\frac{d\theta(t)}{dt} = \omega_c + k_p \frac{dm(t)}{dt} \tag{6.9}$$

and the instantaneous frequency has the range

$$\omega_c - k_p \left| \frac{dm(t)}{dt} \right|_{\max} \le \omega \le \omega_c + k_p \left| \frac{dm(t)}{dt} \right|_{\max} \tag{6.10}$$

A diagram of a phase modulator using just phase modulator, and also a frequency modulator is shown in Fig. 6.3. If we use a frequency modulator, the output frequency is the sum of the two input frequencies,

$$\omega(t) = \omega_c + k_p \frac{dm(t)}{dt} \tag{6.11}$$

Examples of the same message signal being modulated in two different ways is shown in Fig. 6.4. The interesting point about this figure is that the amplitude of the sinusoidal signal is constant though the frequency/phase is a time-varying function of time.

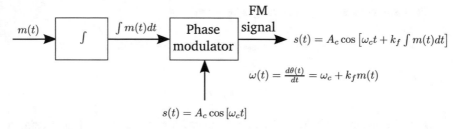

**Fig. 6.2** FM generation block diagram

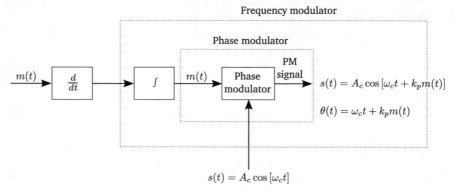

**Fig. 6.3** PM generation using an FM modulator

**Fig. 6.4** Examples of angle modulations of the same message signal. **a** Frequency modulation. **b** Phase modulation

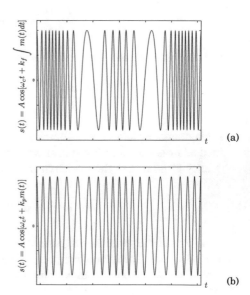

Edwin Armstrong published a paper in 1936 which introduced FM to the world.[1] In the paper he showed that in FM the interference through noise is greatly reduced, though the bandwidth is increased.

## 6.2 Sinusoidal Message Signal

Let us take, as an example which will illustrate many of the properties of FM, a message signal which is sinusoidal in nature. Let the message signal be

$$m(t) = A_m \cos(\omega_m t) \tag{6.12}$$

then the frequency-modulated signal is

$$s(t) = A \cos \left[ \omega_c t + A_m k_f \int \cos(\omega_m t) \right]$$

Further in accordance with Eq. (6.4),

$$\omega(t) = \omega_c + k_f A_m \cos(\omega_m t) \tag{6.13}$$

and then since $\cos(\omega_m t)$ has a maximum value of $+1$ and a minimum value of $-1$,

$$\omega_c - k_f A_m \leq \omega \leq \omega_c + k_f A_m \tag{6.14}$$

where we can set $k_f A_m = \Delta\omega$, the frequency deviation, by Eq. (6.6). So

$$\omega(t) = \omega_c + \Delta\omega \cos(\omega_m t) \tag{6.15}$$

and

$$\theta(t) = \int \omega(t) dt$$
$$= \omega_c t + \frac{\Delta\omega}{\omega_m} \sin(\omega_m t) \tag{6.16}$$

where $\Delta\omega/\omega_m$ is called the *modulation index*. We set

$$\beta = \frac{\Delta\omega}{\omega_m} \tag{6.17}$$

so that

---

[1] Armstrong, E. H., "A Method of Reducing Disturbances in Radio Signalling by a System of Frequency Modulation". Proceedings of the IRE, May 1936, 689–740.

$$\theta(t) = \omega_c t + \beta \sin(\omega_m t) \tag{6.18}$$

which is the phase of the wave. Therefore the transmitted signal is

$$s(t) = A_c \cos[\omega_c t + \beta \sin(\omega_m t)] \tag{6.19}$$

The units of $\beta$ are in radians.

### 6.2.1   Narrowband FM

In narrowband frequency modulation, $\beta$ is small, $(\beta < 1)$, and

$$
\begin{aligned}
s(t) &= A_c \cos[\omega_c t + \beta \sin(\omega_m t)] \\
&= A_c \cos(\omega_c t) \cos[\beta \sin(\omega_m t)] - A_c \sin(\omega_c t) \sin[\beta \sin(\omega_m t)] \tag{6.20} \\
&\approx A_c \cos(\omega_c t) - A_c \sin(\omega_c t)\beta \sin(\omega_m t) \tag{6.21}
\end{aligned}
$$

where we have used the approximations,

$$\cos[\beta \sin(\omega_m t)] \approx 1 \tag{6.22}$$

and

$$\sin[\beta \sin(\omega_m t)] \approx \beta \sin(\omega_m t) \tag{6.23}$$

setting

$$\sin(\omega_c t)\sin(\omega_m t) = -\frac{\cos(\omega_m t + \omega_c t) - \cos(\omega_m t - \omega_c t)}{2} \tag{6.24}$$

then

$$s(t) = A_c \cos(\omega_c t) + \frac{A_c\beta}{2}[\cos(\omega_m t + \omega_c t) - \cos(\omega_c t - \omega_m t)] \tag{6.25}$$

the block diagram of the circuits to produce narrowband FM is shown in Fig. 6.5. If we apply the diagram to single tone modulation, $\omega = \omega_m$, the message signal, $m(t)$ is

$$m(t) = A_m \cos(\omega_m t) \tag{6.26}$$

integrated in accordance with Eq. (6.16), to give

$$\frac{A_m}{\omega_m}\sin\omega_m t$$

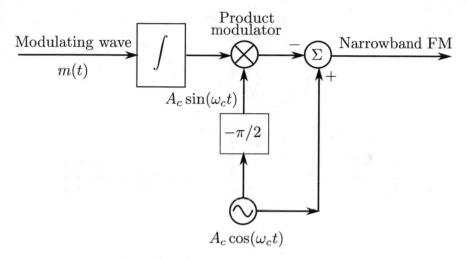

**Fig. 6.5** Block diagram to produce narrowband FM

This signal is then multiplied with $A_c \sin \omega_c t$ in the product modulator. The output of the product modulator is

$$\frac{A_m k_f}{\omega_m} A_c \sin \omega_m t \sin \omega_c t$$

which is then subtracted from the oscillator signal, $A_c \cos(\omega_c t)$.

$$A_c \cos(\omega_c t) - \frac{A_m k_f}{\omega_m} A_c \sin \omega_m t \sin \omega_c t = A_c \cos(\omega_c t) - \beta A_c \sin \omega_m t \sin \omega_c t$$

(6.27)

The spectrum of this signal is given in Fig. 6.6.

### 6.2.2  Wideband FM

Let us proceed to understand wideband FM, without making the assumption that $\beta$ is small. We start with Eq. (6.19).

$$s(t) = A \cos [\omega_c t + \beta \sin(\omega_m t)]$$
$$= \Re \{A \exp [j\omega_c t + j\beta \sin(\omega_m t)]\}$$

(6.28)

Let us concentrate on

$$s_{ce}(t) = e^{j\beta \sin(\omega_m t)}$$

(6.29)

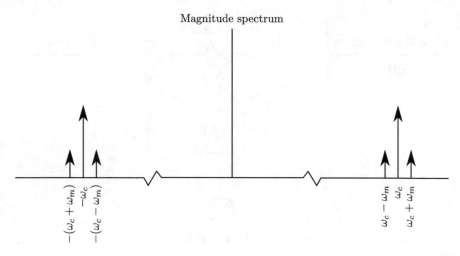

**Fig. 6.6** Narrowband FM spectrum

which is the complex envelope of the earlier function. We write this *periodic* function as a Fourier series, with

$$T = \frac{1}{f_m} = \frac{2\pi}{\omega_m} \tag{6.30}$$

and

$$e^{j\beta \sin(\omega_m t)} = \sum_{k=-\infty}^{k=\infty} F_k e^{jk\omega_m t} \tag{6.31}$$

with

$$F_k = \frac{1}{T} \int_{t_0}^{t_0+T} e^{j\beta \sin(\omega_m t)} e^{-jk\omega_m t} dt$$

$$= \left[ \frac{2\pi}{\omega_m} \right]^{-1} \int_{-\frac{\pi}{\omega_m}}^{\frac{\pi}{\omega_m}} e^{j[\beta \sin(\omega_m t) - k\omega_m t]} dt \tag{6.32}$$

defining a new variable,

$$x = \omega_m t$$

$$dt = \frac{dx}{\omega_m}$$

$$\text{when } t = \pm \frac{\pi}{\omega_m}; \quad x = \pm \pi \tag{6.33}$$

then

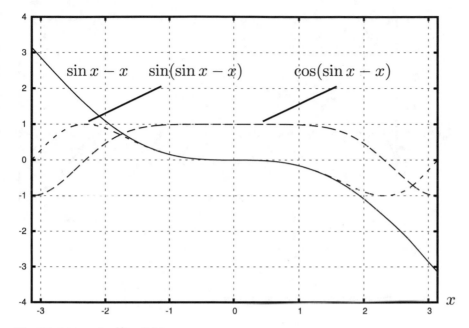

**Fig. 6.7** Integrands of Eq. (6.34)

$$F_k = \frac{1}{2\pi} \int_{-\pi}^{\pi} \exp\{j [\beta \sin x - kx]\} \, dx$$

$$= \frac{1}{2\pi} \int_{-\pi}^{\pi} \cos [\beta \sin x - kx] + j \sin [\beta \sin x - kx] \, dx \qquad (6.34)$$

$$= J_k(\beta) \qquad (6.35)$$

where $J_k(.)$ is the Bessel function of the first kind of order $k$ (see [1]). To understand the integration of Eq. (6.34), we observe the integrands given in Fig. 6.7, which shows that $\sin [\beta \sin x - kx]$ is an odd function, while $\cos [\beta \sin x - kx]$ is an even function in the range $[-\pi, \pi]$ and therefore

$$J_k(\beta) = \frac{1}{2\pi} \int_{-\pi}^{\pi} \cos [\beta \sin x - kx] \, dx$$

The first few Bessel functions of integer order are shown in Fig. 6.8. Bessel functions as defined by a power series are

$$J_m(x) = \sum_{l=0}^{\infty} \frac{(-1)^l x^{2l+m}}{2^{2l+m} l! (l + m)!} \qquad (6.36)$$

where in our case, $m$ is an integer. Returning to our earlier equation, (6.31)

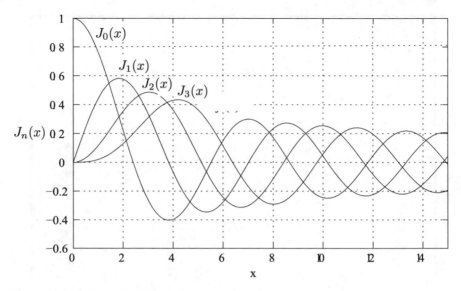

**Fig. 6.8** First few Bessel functions of integer order

$$e^{j\beta \sin(\omega_m t)} = \sum_{k=-\infty}^{k=\infty} J_k(\beta) e^{jk\omega_m t} \tag{6.37}$$

and then, from Eq. (6.28)

$$s(t) = \Re\{A \exp[j\omega_c t + j\beta \sin(\omega_m t)]\}$$

$$= \Re\left\{A e^{j\omega_c t} \times \sum_{k=-\infty}^{k=\infty} J_k(\beta) e^{jk\omega_m t}\right\}$$

$$= A \sum_{k=-\infty}^{k=\infty} J_k(\beta) \Re\left\{e^{j(\omega_c + k\omega_m)t}\right\}$$

$$= A \sum_{k=-\infty}^{k=\infty} J_k(\beta) \cos[(\omega_c + k\omega_m)t]$$

with the Bessel functions of integer order,

$$J_{-m}(x) = (-1)^m J_m(x) \tag{6.38}$$

and using these properties,

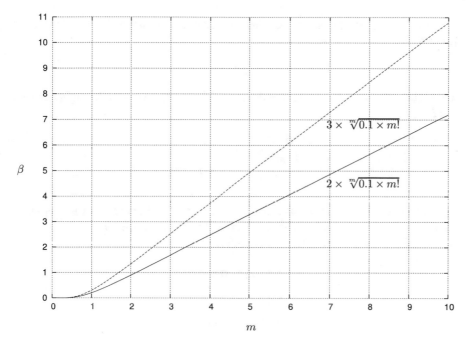

**Fig. 6.9** $\beta$ versus $m$. $m!$ here was replaced by $\Gamma(m+1)$

$$s(t) = AJ_0(\beta)\cos(\omega_c t) + AJ_1(\beta)\cos[(\omega_c+\omega_m)t] - AJ_1(\beta)\cos[(\omega_c-\omega_m)t] + \cdots$$

$$= AJ_0(\beta)\cos(\omega_c t) + A\sum_{k=1}^{\infty}J_k(\beta)\left\{\cos[(\omega_c+k\omega_m)t] + (-1)^k\cos[(\omega_c-k\omega_m)t]\right\}$$

(6.39)

The first point which is of great importance is that the wideband FM signal has infinite bandwidth. However, the Bessel functions which are the coefficients of the cosine terms become smaller and smaller, with increasing value of $k$, as the following table shows. The table shows the values of Bessel functions, with orders 0 to 8 for $\beta = 0.2, 0.4, 0.8, 1.6, 3.2$ and $6.4$.

|            | $\beta = 0.2$ | 0.4       | 0.8       | 1.6       | 3.2       | 6.4       |
|------------|---------------|-----------|-----------|-----------|-----------|-----------|
| $J_0(\beta)$ | 0.9900      | 0.9604    | 0.8463    | 0.4554    | −0.3202   | 0.2433    |
| $J_1(\beta)$ | 0.0995      | 0.1960    | 0.3688    | 0.5699    | 0.2613    | −0.1816   |
| $J_2(\beta)$ | 0.0050      | 0.0197    | 0.0758    | 0.2570    | 0.4835    | −0.3000   |
| $J_3(\beta)$ | 0.0002      | 0.0013    | 0.0102    | 0.0725    | 0.3431    | −0.0059   |
| $J_4(\beta)$ | $\approx 0$ | $\approx 0$ | 0.0010    | 0.0150    | 0.1597    | 0.2945    |
| $J_5(\beta)$ | $\approx 0$ | $\approx 0$ | $\approx 0$ | 0.0025    | 0.0562    | 0.3741    |
| $J_6(\beta)$ | $\approx 0$ | $\approx 0$ | $\approx 0$ | 0.0003    | 0.0160    | 0.2899    |
| $J_7(\beta)$ | $\approx 0$ | $\approx 0$ | $\approx 0$ | $\approx 0$ | 0.0038    | 0.1696    |
| $J_8(\beta)$ | $\approx 0$ | $\approx 0$ | $\approx 0$ | $\approx 0$ | 0.0008    | 0.0810    |

From Eq. (6.36), we can see that the first term ($l = 0$) in the higher order Bessel functions is

$$\frac{x^m}{2^m m!}$$

and if allow that this term to be less than 0.1, then we may neglect the higher orders. That is

$$\frac{\beta^m}{2^m m!} < 0.1$$

$$\beta < 2 \times \sqrt[m]{0.1 \times m!}$$

A plot of $2 \times \sqrt[m]{0.1 \times m!}$ versus $\beta$ is shown in Fig. 6.9, where $m!$ has been replaced by $\Gamma(m + 1)$. To use this inequality, suppose $\beta = 5.2$ then, from Fig. 6.9, $m = 7$ or 8. The values of these functions at $\beta = 5.2$ is

$$J_8(5.2) = 0.02369$$

and

$$J_7(5.2) = 0.06545$$

but both these are too small. The power contained in these amplitudes are very small.

$$[J_8(5.2)]^2 = 7.236 \times 10^{-4}$$
$$[J_7(5.2)]^2 = 4.284 \times 10^{-3}$$

From the curve

$$\beta = 2 \times \sqrt[m]{0.1 \times m!}$$
$$\text{we get } B = 2.53\omega_m (\beta + 0.69) \tag{6.40}$$

where $B$ is the required bandwidth about $\omega_c$ in rad/s. A better, or perhaps different, approximation is the (empirical) curve

$$\beta = 3 \times \sqrt[m]{0.1 \times m!} \tag{6.41}$$

which is shown in the same figure. Using this curve, $m = 5$ or 6 and

$$J_5(5.2) = 0.2865$$
$$J_6(5.2) = 0.1525 \tag{6.42}$$

which has higher overall power in the last terms which are not neglected. Once $m$ is obtained, then the bandwidth, $B$, is $2m\omega_m$ about $\omega_c$ (which, for example, for $\beta = 5.2$ is about $12\omega_m$).

**Fig. 6.10** Magnitude spectrum of the FM signal for $\beta = 3.2$

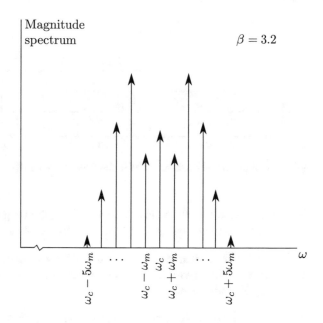

An empirical rule due to Carson known as *Carson's rule* [2] is that the bandwidth $B$ is approximated by

$$B = 2\Delta\omega + 2\omega_m$$
$$B = 2\omega_m(\beta + 1) \tag{6.43}$$

which agrees with our second result ($B = 12.4\omega_m$). The spectrum for $\beta = 3.2$ is shown in Fig. 6.10 for the positive frequencies.

Now the pertinent question which arises is that how do we use results which we have obtained for $m(t) = A\cos(\omega_m t)$ apply to the general case where $m(t)$ is a general band-limited signal? We know that

$$\omega_c - k_f \,|m(t)|_{\max} \leq \omega \leq \omega_c + k_f \,|m(t)|_{\max} \tag{6.44}$$

(Compare with Relation (6.14)) so

$$\Delta\omega = k_f \,|m(t)|_{\max} \tag{6.45}$$

and the maximum frequency in the signal $m(t)$ is $\omega_m$ so

$$\beta \approx \frac{k_f \,|m(t)|_{\max}}{\omega_m} \tag{6.46}$$

then we may use the results which we obtained to get an approximate idea of the bandwidth through Carson's rule or Eq. (6.40).

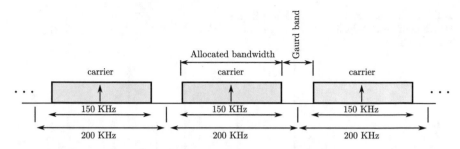

**Fig. 6.11**  Banwidth allocation for FM in the frequency range 87.5–108 MHz

Throughout the world, with a few exceptions, the frequency band, 87.5–108 MHz is used for FM broadcasting. The slots for FM are shown in Fig. 6.11.

**Example 6.1**  FM channels are placed 200 KHz apart and occupy a bandwidth of 150 KHz. For audio (with music) with bandwidth of 20 KHz, what is the maximum frequency deviation $\Delta\omega$?

Since $\omega_m = 2\pi \times 20$ Krad/s, and $B = 2m\omega_m = 2\pi \times 150$ Krad/s, (see Fig. 6.11) therefore

$$
\begin{aligned}
m &= \frac{B}{2\omega_m} \\
&= \frac{2\pi \times 150}{2 \times 2\pi \times 20} \\
&= 3.75
\end{aligned}
$$

so from the lower curve given in Fig. 6.9, $\beta = 2.4$. Therefore

$$\Delta f = 2.4 f_m = 48 \text{ KHz}$$

this result is on the conservative side. If we use the upper curve, $\beta = 3.6$ then

$$\Delta f = 72 \text{ KHz}$$

However, if we use Carson's rule,

$$
\begin{aligned}
B &= 2\Delta f + 2 f_m \\
150 &= 2\Delta f + 40 \\
\Delta f &= 55 \text{ KHz}
\end{aligned}
\tag{6.47}
$$

It is best to use the most conservative result of 48 KHz.

## 6.3  Generation of Wideband FM

There are two methods of generation of FM signals:

1. The so-called *indirect* method and
2. The direct method.

In the case of the indirect method, a narrowband FM signal is first produced, which then is converted into a wideband FM signal by frequency multiplication. On the other hand, in the direct method, a voltage controlled oscillator is used to directly produce the FM signal.

### 6.3.1  Indirect Method of FM Generation

In the indirect methods of FM generation, first a narrowband FM signal is generated, (see Fig. 6.5) which is then followed by a frequency multiplier (see Sect. 5.5.1.1).

In a frequency multiplier circuit, we use a nonlinear element, where the output $y(t)$ is related to the input $x(t)$ by the equation,

$$y(t) = a_1 x(t) + a_2 x(t)^2 + \cdots$$

$$= \sum_{i=1}^{\infty} a_i \, [x(t)]^i \quad i = 1, 2, \dots \tag{6.48}$$

From this equation it is clear that when $x(t) = \cos[\omega_c t + \theta(t)]$ then $x^i(t)$ will contain a term proportional to $\cos[i\omega_c t + i\theta(t)]$ (which will have a frequency $i\omega_c$). For example

$$\cos^4[\omega_c t + \theta(t)] = \frac{\cos\{4[\omega_c t + \theta(t)]\} + 4\cos\{2[\omega_c t + \theta(t)]\} + 3}{8} \tag{6.49}$$

where

$$\theta(t) = k_f \int m(t)dt$$

then, when we use a BPF to get $\cos\{4[\omega_c t + \theta(t)]\}/8$ term on the output.

From the above discussion, it is clear that if the input signal has a bandwidth of $\omega_c \pm \omega_m$ then after passing the signal through the frequency multiplier the bandwidth will be

$$n(\omega_c \pm \omega_m) \; n = 1, 2, \dots$$

rad/s, where $n$ is an integer, and which we extract by using a BPF on the output. The band is obtained without distortion of the message signal. This is shown in Fig. 6.12.

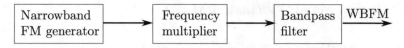

**Fig. 6.12**   Generation of wideband FM through the indirect method

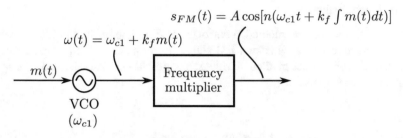

**Fig. 6.13**   Generation of WBFM with a VCO

### 6.3.2   Direct Methods of FM Generation

The easiest method of generating an FM signal is by using a voltage controlled oscillator (VCO). A block diagram is shown in Fig. 6.13. The method consists of using a VCO whose input is the message signal, $m(t)$. The output signal has an instantaneous frequency,

$$\omega(t) = \omega_{c1} + k_f m(t) \tag{6.50}$$

and therefore an instantaneous phase given by

$$\theta(t) = \int [\omega_{c1} + k_f m(t)] dt$$

$$= \omega_{c1} t + k_f \int m(t) dt \tag{6.51}$$

the signal is then passed through a frequency multiplier, so that the final frequency,

$$\omega_{\text{final}}(t) = n[\omega_{c1} + k_f m(t)] \tag{6.52}$$

so that

$$\Delta\omega = n k_f \, |m(t)|_{\text{max}} \tag{6.53}$$

$$\omega_c = n \omega_{c1} \tag{6.54}$$

and

$$\beta = \frac{n k_f \, |m(t)|_{\text{max}}}{\omega_m} \tag{6.55}$$

**Fig. 6.14** Hartley oscillator

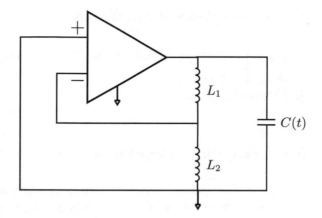

where $\omega_m$ is the maximum frequency in $M(\omega)$.

An example of a fairly simple VCO is the capacitor controlled Hartley oscillator shown in Fig. 6.14. The capacitor in turn is controlled by a voltage. The instantaneous frequency is given by

$$\omega(t) = \frac{1}{\sqrt{(L_1 + L_2)C(t)}} \tag{6.56}$$

where $C(t)$ is the variable capacitor. In this case, the capacitor variation is controlled by $m(t)$. That is

$$C(t) = C_0 + \Delta C m(t) \tag{6.57}$$

where $C_0$ is a fixed capacitance and $\Delta C m(t)$ is the variable part of the capacitance. In general

$$\Delta C \ll C_0 \tag{6.58}$$

setting $L = L_1 + L_2$, we have

$$\begin{aligned}
\omega(t) &= \frac{1}{\sqrt{L(C_0 + \Delta C m(t)}} \\
&= \frac{1}{\sqrt{LC_0}\sqrt{1 + \frac{\Delta C m(t)}{C_0}}} \\
&\approx \omega_c \times \left(1 - \frac{\Delta C m(t)}{2C_0}\right)
\end{aligned} \tag{6.59}$$

where $\omega_c = 1/\sqrt{LC_0}$. From this equation,

$$k_f = -\frac{\omega_c \Delta C}{2C_0} \tag{6.60}$$

## 6.4   Demodulation of FM Signals

In general there are two methods of demodulating FM signals.[2] These are

1. Frequency discriminator
2. Phase-locked loop.

### 6.4.1   Frequency Discriminator

The frequency discriminator method is in effect a differentiator circuit followed by an amplitude demodulator or envelope detector. The block diagram of a frequency discriminator is shown in Fig. 6.15. If the incoming signal is

$$s(t) = A \cos \left[ \omega_c t + k_f \int m(t)dt \right]$$

then the output of the differentiator is

$$\frac{ds}{dt} = -A \left[ \omega_c + k_f m(t) \right] \sin \left[ \omega_c t + k_f \int m(t)dt \right]$$

If we use an envelope detector on this signal, it gives us the message signal. The waveforms of the discriminator are shown in Fig. 6.16.

In practice, to produce a differentiator circuit (which is multiplication by $j\omega$ in the frequency domain) we need a filter which has a linear slope near $\omega = \omega_c$ as shown in Fig. 6.17. The filter is such that locally near $\omega_c$, $\omega \propto \omega_c$,

$$H(\omega)|_{\omega \approx \omega_c} = ja\,(\omega - \omega_c) \tag{6.61}$$

therefore if $F(\omega)$ is the Fourier transform of the input signal,

$$F(\omega) \Leftrightarrow A \cos \left[ \omega_c t + k_f \int m(t)dt \right] \tag{6.62}$$

then

$$F(\omega)H(\omega) = \underbrace{ja\omega F(\omega)}_{\text{Differentiated term}} - \underbrace{ja\omega_c F(\omega)}_{\text{Term multiplied by a constant}}$$

or

---

[2] To understand the meaning of demodulation, see Sect. 5.4.1.

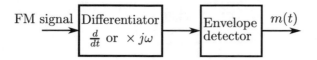

**Fig. 6.15** Block diagram of an FM discriminator

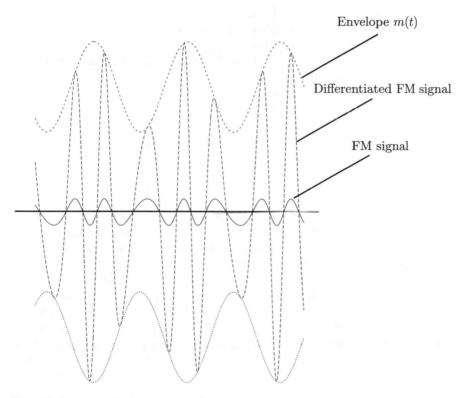

**Fig. 6.16** Frequency discriminator wave forms

$$F(\omega)H(\omega)|_{\omega \approx \omega_c} \Leftrightarrow a\frac{d}{dt}A\cos\left[\omega_c t + k_f \int m(t)dt\right]$$

$$- a\omega_c A\cos\left[\omega_c t + k_f \int m(t)dt + \frac{\pi}{2}\right]$$

$$\Leftrightarrow -aA\left[\omega_c + k_f m(t)\right]\sin\left[\omega_c t + k_f \int m(t)dt\right]$$

$$+ aA\omega_c \sin\left[\omega_c t + k_f \int m(t)dt\right]$$

$$\Leftrightarrow -aAk_f m(t)\sin\left[\omega_c t + k_f \int m(t)dt\right] \qquad (6.63)$$

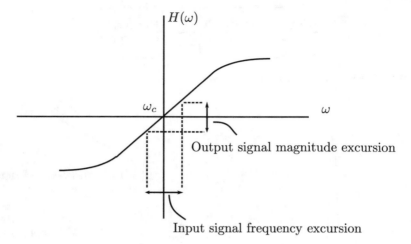

**Fig. 6.17** A linear discriminator

where, in the first of these equations, a multiplication by $j\omega$ results in differentiation, and a multiplication by $j$ in the frequency-domain results in an addition of $\pi/2$ in the phase in the time domain. Equation set (6.63) tells us that the amplitude of the output signal is proportional to $m(t)$ which we may obtain through an envelope detector.

The Foster–Seeley discriminator [4] or an FM detector circuit was proposed in 1936 by D. E. Foster and S. W. Seeley. The circuit found application in the demodulation of an FM signal. It uses a tuned RF transformer to convert frequency changes into amplitude changes. The Foster Seeley discriminator, whose circuit is shown in Fig. 6.18, is made using two linear slope filters whose outputs are subtracted. If the input has a frequency equal to the carrier frequency, the two halves of the tuned transformer circuit produce the same rectified voltage and the output is zero.

To understand its operation,

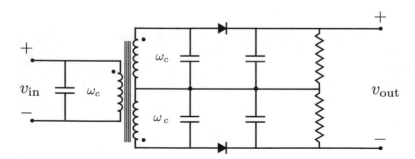

**Fig. 6.18** Foster–Seeley discriminator

$$V_1(\omega) = H_1(\omega)V_{in}$$
$$V_2(\omega) = H_2(\omega)V_{in}$$
$$V_{out}(\omega) = V_1(\omega) - V_2(\omega)$$
$$= V_{in}(\omega)\left[H_1(\omega) - H_2(\omega)\right] \tag{6.64}$$

where $V_{in}(\omega)$ is the input FM signal, and $V_{out}(\omega)$ is the output 'differentiated' signal. The output signal, $V_{out}(\omega)$ is identical to Eq. (6.63). As the frequency of the carrier changes (as it is in the case of FM), the balance between the two halves of the transformer secondary changes, and the result is a voltage proportional to the frequency deviation of the carrier.

To illustrate the working of the circuit, in the frequency range $0 < \omega < 2$, consider a band pass filter (see Sect. 4.2.5) with

$$H_1(\omega) = \frac{j\omega}{-\omega^2 + j\omega + 1} \qquad 0 \le \omega \le 2 \tag{6.65}$$

which has a resonant frequency at $\omega = 1$. If we now use the second transfer function,

$$H_2(\omega) = H_1(-\omega) = -\frac{j\omega}{-\omega^2 - j\omega + 1} \qquad 0 \le \omega \le 2 \tag{6.66}$$

which also has a resonant frequency, $\omega = 1$. Then near $\omega = 1$ the Taylor series expansion of $H_1(\omega) - H_2(\omega)$ is

$$H_1(\omega) - H_2(\omega) \approx -4j(\omega - 1) + 2j(\omega - 1)^2 + \cdots \tag{6.67}$$

where the first term is a linear term, (which behaves like a differentiator) while the second term is a nonlinear (quadratic) term. Near $\omega = 1$, the first term predominates, while the second term is very small. A plot of $H_1(\omega) - H_2(\omega)$ is shown in Fig. 6.19.

**Fig. 6.19** Foster Seeley response

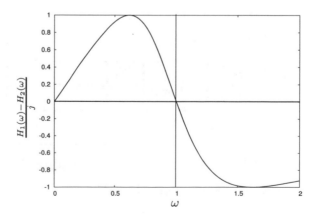

The Foster Seeley circuit shows two slope detectors which are back to back with resonant frequencies of $\omega = \omega_c$ with the Taylor series expansion of

$$H_1(\omega) \cong \alpha(\omega - \omega_c) + \text{Higher order terms}$$
$$\text{and } H_2(\omega) \cong H_1(-\omega) \cong -\alpha(\omega - \omega_c) + \text{Higher order terms}$$

where $\alpha$ is a constant. So

$$H_1(\omega) - H_2(\omega) = \alpha\left[\omega - \omega_c + (\omega - \omega_c)\right] + \text{Higher order terms}$$
$$= 2\alpha\omega - 2\alpha\omega_c + \text{Higher order terms} \qquad (6.68)$$

and from Eq. (6.64),

$$V_{\text{out}} \cong 2\alpha(\omega - \omega_c) + \text{Higher order terms}$$

### 6.4.2  *Phase-Locked Loop*

A phase-locked loop (PLL) [3] is a control system that locks the phase of an input sinusoidal signal to an output sinusoidal signal. While there are several differing types of PLLs, the Costas loop being one of them, the bare bones PLL consists of a variable frequency oscillator and a phase detector (see Sect. 5.5.2.2). The phases here means generalised phases, $\theta(t)$.

In this book, an extremely simplified explanation of the phase locked loop is given. A block diagram of the phase-locked loop is shown in Fig. 6.20. The oscillator generates a sinusoidal signal, while the phase detector compares the phase of oscillator signal with the phase of the input FM signal and adjusts the voltage controlled oscillator (VCO) to keep the phases matched. The default frequency of the VCO with no

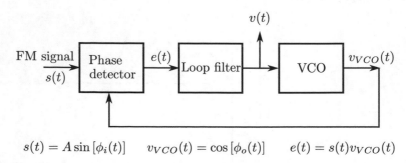

$$s(t) = A\sin\left[\phi_i(t)\right] \qquad v_{VCO}(t) = \cos\left[\phi_o(t)\right] \qquad e(t) = s(t)v_{VCO}(t)$$

**Fig. 6.20**  Block diagram of a phase-locked loop

input, $v(t) = 0$ is $\omega_c$. Feeding the output signal back towards the phase detector for comparison with the input FM signal forms a loop, thus giving the PLL its name.

The input to the phase detector, $s(t)$, is given by

$$s(t) = A \sin [\phi_i(t)]$$

$$= A \sin \left[ \omega_c t + k_f \int m(t)dt \right] \tag{6.69}$$

where we designate

$$\phi_2(t) = k_f \int m(t)dt$$

$$\text{and } \phi_i(t) = \omega_c t + \phi_2(t) \tag{6.70}$$

The second input to the phase detector is

$$v_{VCO}(t) = \cos [\phi_o(t)]$$

$$= \cos [\omega_c t + \phi_1(t)] \tag{6.71}$$

Notice that *the two signals, $s(t)$ and $v_{VCO}(t)$, under steady state conditions, are out of phase by $\pi/2$.* The phase detector is essentially a multiplier, and the output of the phase detector is

$$e(t) = s(t)v_{VCO}(t)$$

$$= A \sin [\phi_i(t)] \cos [\phi_o(t)]$$

$$= A \left\{ \frac{\sin (2\omega_c t + \phi_2 + \phi_1)}{2} - \frac{\sin (\phi_2 - \phi_1)}{2} \right\} \tag{6.72}$$

which is the input to the lowpass loop filter. The first term is centred at $\omega = 2\omega_c$ (which is filtered out by the loop filter) while the second term, which is centred near $\omega = 0$ is available at the output of the filter. Also we know that when $\phi_2 - \phi_1$ is small,

$$\sin (\phi_2 - \phi_1) \approx \phi_2 - \phi_1 \tag{6.73}$$

$$\text{when } \phi_e(t) = \phi_2(t) - \phi_1(t) \ll 1 \tag{6.74}$$

and

$$\frac{d\phi_e(t)}{dt} = \frac{d\phi_2(t)}{dt} - \frac{d\phi_1(t)}{dt}$$

$$= k_f m(t) - kv(t) \tag{6.75}$$

where we have used the fact that the input to the VCO is $v(t)$ controls the output frequency.

$$\phi_1(t) = k \int v(t)dt \tag{6.76}$$

$$\frac{d\phi_1}{dt} = kv(t) \tag{6.77}$$

Under locked conditions,

$$\frac{d\phi_e(t)}{dt} \approx \phi_e(t) \approx 0$$

so

$$v(t) = \frac{k_f}{k}m(t)$$

The essential point to understand, is that in the steady state condition or locked condition, the output of the properly designed loop filter $v(t)$ is the demodulated FM signal

$$v(t) \propto k_f m(t) \tag{6.78}$$

Having got an idea about how the phase-locked loop works, let us develop a model. We start with $\phi_e(t)$, which is

$$\phi_e(t) = \phi_2(t) - \phi_1(t)$$
$$\text{or } \frac{d\phi_e(t)}{dt} = \frac{d\phi_2(t)}{dt} - \frac{d\phi_1(t)}{dt}$$
$$= k_f m(t) - kv(t) \tag{6.79}$$

Now the input voltage to the VCO, $v(t)$ is related to the error signal,

$$v(t) = K_o \int_{-\infty}^{\infty} \sin[\phi_e(\tau)] h(t - \tau)d\tau \tag{6.80}$$

where $h(t)$ is the impulse response of the loop filter. So Eq. (6.79) becomes

$$\frac{d\phi_e(t)}{dt} = \frac{d\phi_2(t)}{dt} - kK_o \int_{-\infty}^{\infty} \sin[\phi_e(\tau)] h(t - \tau)d\tau \tag{6.81}$$

which is the nonlinear model of the PLL.

We next linearise the equation by

$$\sin[\phi_e(\tau)] \approx \phi_e(\tau)$$

or

$$\frac{d\phi_e(t)}{dt} \approx \frac{d\phi_2(t)}{dt} - kK_o \int_{-\infty}^{\infty} \phi_e(\tau)h(t-\tau)d\tau$$

then, working with LT of this equation, we get

$$s\Phi_e(s) = s\Phi_2(s) - kK_0\Phi_e(s)H(s)$$
$$= k_f M(s) - kK_0\Phi_e(s)H(s)$$

or

$$\Phi_e(s)[s + kK_0 H(s)] = k_f M(s)$$
$$\text{or } \Phi_e(s) = \frac{k_f M(s)}{s + kK_0 H(s)} \tag{6.82}$$

We understand the operation of the PLL as follows

1. The condition to make $\Phi_e$ small is to make the constant $kK_0$ large, and $H(s)$, is designed to be a lowpass filter, so that

$$\Phi_e(s) \approx 0$$
$$\Rightarrow \phi_e(t) \approx 0$$

   or, Eq. (6.79) becomes true.
2. We set

$$H(s) = \frac{1}{s+a}$$

which is a lowpass filter, in Eq. (6.82). Therefore,

$$\Phi_e(s) = \frac{(s+a)k_f M(s)}{kK_0 + s^2 + as}$$

which by the final value theorem

$$\lim_{t \to \infty} \phi_e(t) = \lim_{s \to 0} s\Phi_e(s)$$

is zero. Though this is a simplified analysis, we get some insight into the PLL.

## 6.5 Superheterodyne Receiver

The superheterodyne receiver was invented by Edwin Armstrong in 1918, and who later invented FM for radio transmission. The principle of the superheterodyne receiver is to down-convert the received RF (radio frequency) signal to what is known

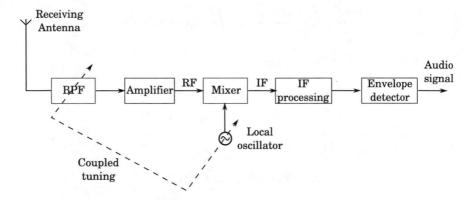

**Fig. 6.21** Superheterodyne receiver

as IF (intermediate frequency) which is a fixed frequency so the demodulation takes place at that frequency. This is a more convenient way to handle the signal, since the IF, which appears after the mixer, is constant (see Sect. 4.6). The block diagram of the superheterodyne receiver is shown in Fig. 6.21.

This receiver which is now used in almost AM and FM receivers uses a local oscillator frequency depending on the frequency range of modulation. The local oscillator is coupled to the BPF through two coupled ganged capacitors generally connected through the same tuning knob in the older designs.

In AM the carrier lies in the frequency range 0.535–1.605 MHz, and the IF is at 0.455 MHz, while in FM the carrier range is 88–108 MHz and the IF is at 10.7 MHz. The equation of relevance is

$$\omega_{IF} = |\omega_{LO} - \omega_{RF}| \tag{6.83}$$

where $\omega_{IF}$ is the intermediate frequency, $\omega_{RF}$ is the input RF and $\omega_{LO}$ is the local oscillator frequency. The problem with Eq. (6.83), is that corresponding to the (desirable) RF there is an (undesirable) image frequency which would give the same IF. To be specific, if the desired RF is 500 KHz, and the LO frequency is 955 KHz then,

$$\omega_{IF} = |\omega_{LO} - \omega_{RF}| = |955 - 500| = 455 \text{ KHz}$$

similarly, if the RF is 1410 KHz, then

$$\omega_{IF} = |\omega_{LO} - \omega_{RF}| = |955 - 1410| = 455 \text{ KHz}$$

To make sure that we only target the desirable RF, we use a BPF which extracts only the desirable RF signal, of the relevant bandwidth, which is amplified and is used as one of the inputs to the mixer. In the case of AM the filter bandwidth is about 10 KHz, while for FM it is around 200 KHz. These frequencies are given in the form of a Table 6.1

**Table 6.1** Details of the frequencies in the superheterodyne receiver

|  | AM (KHz) | FM (MHz) |
|---|---|---|
| RF carrier range | 535–1605 | 88–108 |
| LO frequency range | 990–2060 | 98.7–118.7 |
| IF | 455 | 10.7 |
| IF Bandwidth | 10 | 0.2 |

The IF is then processed to further convert the signal down to baseband using an additional mixer and LO combination, which is centred at the appropriate frequency. Generally the IF is passed through an additional BPF, of the desired bandwidth, whose cutoff is much sharper than the input BPF. For example in the case of AM, if the IF is 455 KHz, then LO of the IF section is also 455 KHz with a BPF of 10 KHz. The IF section LO is of higher quality than the tunable LO if the input section.

## 6.6 Nonlinear Effects in FM Systems

Consider a channel or a system which causes nonlinearities to be introduced in an FM signal. For example, if the input output relationship of a system or channel is given by

$$y(x) = a_1 x + a_2 x^2 + \cdots \tag{6.84}$$

then if we introduce a signal,

$$x(t) = A \cos\left[\omega_c t + \theta(t)\right] \tag{6.85}$$

where

$$\theta(t) = k_f \int m(t) dt \tag{6.86}$$

then we know that

$$y(t) = A a_1 \cos\left[\omega_c t + \theta(t)\right] + A^2 a_2 \cos\left[\omega_c t + \theta(t)\right]^2 + A^3 a_3 \cos\left[\omega_c t + \theta(t)\right]^3 + \cdots \tag{6.87}$$

recalling that

$$\cos^m(\phi) = A_m \cos(m\phi) + A_{m-2} \cos\left[(m-2)\phi\right] + \cdots \tag{6.88}$$

where the $A_i$s are constants, then in the general case,

$$y(t) = B_0 + B_1 \cos\left[\omega_c t + \theta(t)\right] + B_2 \cos\left[2\omega_c t + 2\theta(t)\right] + \cdots$$

To obtain the desired FM signal, $B_1 \cos[\omega_c t + \theta(t)]$, we pass the signal represented by (6.87) through an appropriate BPF and recover the signal. This discussion tells us that angle modulated signals are insensitive to system and channel nonlinearities.

When we apply the results of this section to the specific case of Eq. (6.87), we get

$$y(t) = Aa_1 \cos[\omega_c t + \theta(t)] + \frac{A^2 a_2 \cos[2\omega_c t + 2\theta(t)]}{2} + \frac{A^2 a_2}{2} + \dots$$

where $Aa_1 \cos[\omega_c t + \theta(t)]$ is the original FM signal.

## 6.7  Interference in Angle Modulated Signals

If two signals, $s(t)$ the angle modulated signal and $s_i(t)$ an interference signal, are present in the same channel whose carrier frequencies are close together,

$$s(t) = A_c \cos[\omega_c t + \phi(t)]$$
$$\text{and } s_i(t) = A_i \cos[\omega_c t + \Delta\omega t]$$

where

$$\phi(t) = k_p m(t) \qquad \text{for PM}$$
$$\text{and} = k_f \int m(t) dt \text{ for FM}$$

then

$$s(t) + s_i(t) = [-A_i \sin(\Delta\omega t) - A_c \sin(\phi)] \sin(\omega_c t)$$
$$+ [A_i \cos(\Delta\omega t) + A_c \cos(\phi)] \cos(\omega_c t)$$

then setting

$$C(t) \sin\theta(t) = A_i \sin(\Delta\omega t) + A_c \sin(\phi)$$
$$C(t) \cos\theta(t) = A_i \cos(\Delta\omega t) + A_c \cos(\phi)$$

so

$$s(t) + s_i(t) = -C(t) \sin\theta(t) \sin(\omega_c t) + C(t) \cos\theta(t) \cos(\omega_c t) \qquad (6.89)$$

then

$$s(t) + s_i(t) = C(t) \cos[\omega_c t + \theta(t)] \qquad (6.90)$$

where

$$\theta(t) = \tan^{-1} \left\{ \frac{[A_i \sin(\Delta\omega t) + A_c \sin(\phi)]}{[A_i \cos(\Delta\omega t) + A_c \cos(\phi)]} \right\} \tag{6.91}$$

In the detection process we are not concerned about changes in amplitude of the signal but only with the changes in phase. If we expand $\theta(t)$ in powers of $A_i$ where $A_i$ is small, then

$$\theta(t) \cong \phi + \frac{A_i \sin(\Delta\omega t - \phi)}{A_c} + \text{Higher powers of } A_i/A_c \tag{6.92}$$

where the second term is the interference signal. From this equation, we conclude that for PM,

$$\theta(t) \cong k_p m(t) + \frac{A_i \sin\left[\Delta\omega t - k_p m(t)\right]}{A_c} \tag{6.93}$$

the interference depends on the value of the ratio $A_i/A_c$: the smaller this ratio, the less is its effect.

When we consider FM, the output of the FM demodulator is

$$\frac{d}{dt}\theta(t) \cong k_f m(t) + \frac{A_i}{A_c}\left[\Delta\omega + k_f m(t)\right]\cos\left[\Delta\omega t - k_p \int m(t)dt\right] \tag{6.94}$$

where the second term in this equation is the interference term.

## 6.7.1 Capture Effect

If two FM signals are present in the channel, with practically the same carrier frequency,

$$s_1(t) = A_{c1} \cos[\omega_c t + \phi(t)]$$
$$\text{and } s_2(t) = A_{c2} \cos[\omega_c t + \psi(t)] \tag{6.95}$$

then these two signals interfere with each other.

In these equations,

$$\phi(t) = k_{f1} \int m_1(t)dt$$

$$\text{and } \psi(t) = k_{f2} \int m_2(t)dt + \Delta\omega t \tag{6.96}$$

How do we evaluate this interference? Proceeding as earlier,

$$s_1(t) + s_2(t) = [-A_{c2} \sin(\psi) - A_{c1} \sin(\phi)] \sin(t\omega_c)$$
$$+ [A_{c2} \cos(\psi) + A_{c1} \cos(\phi)] \cos(t\omega_c)$$

setting

$$C(t) \sin\theta(t) = A_{c2} \sin(\psi) + A_{c1} \sin(\phi)$$
$$C(t) \cos\theta(t) = A_{c2} \cos(\psi) + A_{c1} \cos(\phi) \tag{6.97}$$

so

$$\theta(t) = \tan^{-1} \left\{ \frac{A_{c2} \sin(\psi) + A_{c1} \sin(\phi)}{A_{c2} \cos(\psi) + A_{c1} \cos(\phi)} \right\} \tag{6.98}$$

Now if we calculate $d\theta/dt$, then, (after some trigonometric manipulation)

$$\frac{d\theta}{dt} = \frac{\left\{ A_{c1}A_{c2}\left(\frac{d}{dt}\psi\right) + A_{c1}A_{c2}\left(\frac{d}{dt}\phi\right)\right\} \cos[\psi(t) - \phi(t)] + A_{c2}^2\left(\frac{d}{dt}\psi\right) + A_{c1}^2\left(\frac{d}{dt}\phi\right)}{2A_{c1}A_{c2}\cos[\psi(t) - \phi(t)] + A_{c2}^2 + A_{c1}^2} \tag{6.99}$$

On examining this equation we see that it is symmetric in $\{A_{c1}, \phi\}$ and $\{A_{c2}, \psi\}$. For small values of $A_{c2}$,

$$\frac{d\theta}{dt} \approx \frac{d}{dt}\phi + \frac{\left(\frac{d}{dt}\psi - \frac{d}{dt}\phi\right)\cos(\psi - \phi)A_{c2}}{A_{c1}} + \cdots$$

That is, if $A_{c1} \gg A_{c2}$ then

$$\dot{\theta} \approx \frac{d}{dt}\phi = k_{f1}m(t)$$

and if $A_{c2} \gg A_{c1}$ then

$$\dot{\theta} \approx \frac{d}{dt}\psi = k_{f2}m(t)$$

this is called the capture effect. The FM signal with higher power 'captures' the receiver.

## 6.8   Performance of FM Systems in the Presence of Noise

We first need to read Sect. 5.13.1, before considering FM systems. Referring to Fig. 6.22, the incoming FM signal at ① is given by

$$s_1(t) = A_c \cos\left[\omega_c t + k_f \int m_1(t)dt\right]$$
$$= A_c \cos[\omega_c t + \phi(t)] \tag{6.100}$$

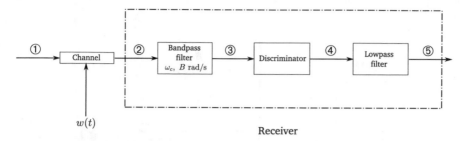

**Fig. 6.22** FM receiver considering noise

which is a typical FM signal. After the addition of white noise, the signal at ②, is

$$s_2(t) = A_c \cos [\omega_c t + \phi(t)] + w(t) \tag{6.101}$$

where $w(t)$ is white noise with an autocorrelation function

$$R_W(\tau) = \frac{N_0}{2} \delta(\tau)$$

After this, the signal is passed through a BPF of bandwidth $B$ (Hz or rad/s, whichever units are used), the signal at ③ is

$$s_4(t) = A_c \cos [\omega_c t + \phi(t)] + n_{NB}(t) \tag{6.102}$$

which has a PSD

$$S_{NB}(\omega) = \begin{cases} N_0/2 & \text{for } \omega_c - \frac{B}{2} \leq |\omega| \leq \omega_c + \frac{B}{2} \\ 0 & \text{otherwise} \end{cases}$$

where the bandwidth, $B$, is the bandwidth of the FM signal.

At this point we calculate the input signal-to-noise ratio, which is

$$\frac{S_i}{N_i} = \frac{A_c^2/2}{N_0 B} \tag{6.103}$$

where $B$ is in hertz.

The PSD for narrowband noise resulting from the filtering action may be written in terms of amplitude and phase and is shown in Fig. 6.23a. (Further, see Sect. 3.19.1)

$$n_{NB}(t) = a(t) \cos [\omega_c t + \psi(t)] \tag{6.104}$$

where $a(t)$ and $\psi(t)$ are both stationary random processes.

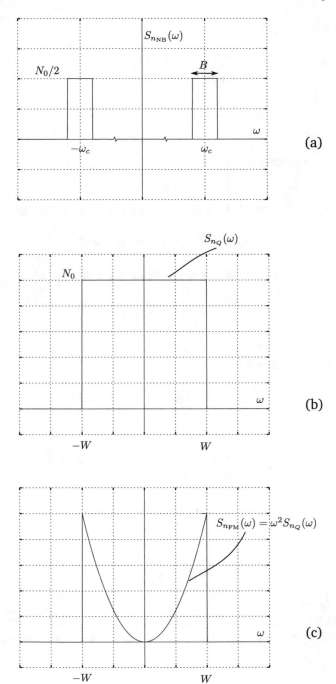

**Fig. 6.23** PSDs for the calculation of noise power

$$a(t) = \sqrt{n_I^2(t) + n_Q^2(t)}$$

$$\text{and } \psi(t) = \tan^{-1}\left[\frac{n_I(t)}{n_Q(t)}\right] \tag{6.105}$$

Therefore

$$
\begin{aligned}
s_4(t) &= A_c \cos[\omega_c t + \phi(t)] + a(t)\cos[\omega_c t + \psi(t)] \\
&= [-a(t)\sin\psi(t) - A_c \sin\phi(t)]\sin(\omega_c t) \\
&\quad + [a(t)\cos\psi(t) + A_c \cos\phi(t)]\cos(\omega_c t)
\end{aligned} \tag{6.106}
$$

as earlier we make the following substitutions,

$$
\begin{aligned}
C(t)\sin\theta(t) &= a(t)\sin\psi(t) + A_c \sin\phi(t) \\
C(t)\cos\theta(t) &= a(t)\cos\psi(t) + A_c \cos\phi(t)
\end{aligned} \tag{6.107}
$$

so that

$$s_4(t) = C(t)\cos[\omega_c t + \theta(t)] \tag{6.108}$$

where

$$\theta(t) = \tan^{-1}\left[\frac{a(t)\sin\psi(t) + A_c \sin\phi(t)}{a(t)\cos\psi(t) + A_c \cos\phi(t)}\right] \tag{6.109}$$

Since we are considering angle modulation, we are mostly concerned with the phase, $\theta(t)$, of the output signal and the amplitude, $C(t)$, is of little concern to us. When the phase is expanded in a Taylor series expansion for small values of $a(t)$, (we are assuming here that the carrier power, $A_c/2$, is much greater than $a^2(t)$, the noise amplitude) we get

$$\theta \approx \phi(t) + \frac{\sin[\psi(t) - \phi(t)]a(t)}{A_c}$$

$$\text{and therefore } \frac{d\theta}{dt} \approx \underbrace{k_f m(t)}_{\text{message}} + \underbrace{\frac{d}{dt}\frac{\sin[\psi(t) - \phi(t)]a(t)}{A_c}}_{\text{noise}}$$

which is the message signal plus noise.

If we take a look at the noise, we notice the term

$$\sin[\psi(t) - \phi(t)]$$

now the argument, which is the phase,

$$\psi(t) - \phi(t) = \psi(t) - k_f \int m(t)dt$$

behaves very much like a uniformly distributed random process, in the interval $[0, 2\pi]$, which is exactly the same as the behaviour of $\psi(t)$, so we may replace $\psi(t) - \phi(t)$ by $\psi(t)$ and therefore the noise term may be written without any doubt as

$$
\begin{aligned}
n_{\text{FM}}(t) &= \frac{d}{dt} \frac{\sin[\psi(t)] a(t)}{A_c} \\
&= \frac{1}{A_c} \frac{d}{dt} n_Q(t)
\end{aligned} \tag{6.110}
$$

since

$$
n_Q(t) = a(t) \sin \psi(t)
$$

Now the PSD of the quadrature component is

$$
S_{n_Q}(\omega) = \begin{cases} N_0 & -W \leq \omega \leq W \\ 0 & \text{otherwise} \end{cases} \tag{6.111}
$$

as shown in Fig. 6.23b, and when we differentiate $n_Q(t)$ in the time domain, we have to multiply the PSD of the quadrature component by $\omega^2$ in the frequency domain, since differentiation may be treated as a filter with frequency response $j\omega$. The bandwidth $W$ is the bandwidth of the message signal, that is, $W = \omega_m$. Therefore, as shown in Fig. 6.23c,

$$
S_{n_{\text{FM}}}(\omega) = \omega^2 S_{n_Q}(\omega) = \begin{cases} \omega^2 N_0 & -W \leq \omega \leq W \\ 0 & \text{otherwise} \end{cases} \tag{6.112}
$$

The noise power is

$$
\begin{aligned}
\mathcal{N}_o &= \frac{1}{A_c^2} \int_{-W}^{W} f^2 N_0 df \\
&= \frac{N_0 2 W^3}{3 \times A_c^2} \quad (W \text{ in Hz})
\end{aligned}
$$

the signal power on the other hand is

$$
S_o = k_f^2 \langle m^2(t) \rangle
$$

so the output signal-to-noise ratio is

$$\frac{S_o}{N_o} = \frac{k_f^2 \langle m^2(t) \rangle \times 3A_c^2}{2N_0 W^3} \quad (W \text{ in Hz})$$

$$= \frac{k_f^2 \langle m^2(t) \rangle \times 3B}{W^3} \left( \frac{A_c^2/2}{N_0 B} \right)$$

$$= \frac{3k_f^2 \langle m^2(t) \rangle B}{W^3} \left( \frac{S_i}{N_i} \right) \tag{6.113}$$

$$= \frac{3k_f^2 \langle m^2(t) \rangle B}{\omega_m^3} \left( \frac{S_i}{N_i} \right) \tag{6.114}$$

we notice here that the signal-to-noise ratio improves drastically as we increase the FM bandwidth, $B$, of the system.

## 6.8.1 Pre-emphasis and De-emphasis in FM

Due to the fact that the noise adding to FM systems is given by Eq. (6.112) shown in Fig. 6.23b, while the message signal $m(t)$ has the bulk of its power concentrated near $\omega = 0$ and very little near $\omega = W$, the higher frequencies of the message signal suffer due to this phenomenon. Therefore it is necessary that the higher frequencies of the message signal be boosted by a 'pre-emphasis' filter, $H_p(\omega)$. After reception, we employ a de-emphasis filter, $H_d(\omega)$, at the final stage after demodulation, but with the condition that

$$H_p(\omega) H_d(\omega) \approx \text{a constant}, \quad \omega \leq W \tag{6.115}$$

An example of these two filters are shown in Fig. 6.24a, b and the frequency responses are shown for $0 \leq \omega \leq 1$ in Fig. 6.24c. An analysis of Fig. 6.24a yields

$$H_p(\omega) = \frac{(j\omega C_1 R_1 + 1) R_2}{(j\omega C_1 R_1 + 1) R_2 + R_1} \tag{6.116}$$

and

$$|H_p(\omega)| = \frac{\sqrt{\omega^2 C_1^2 R_1^2 + 1} R_2}{\sqrt{(R_2 + R_1)^2 + \omega^2 C_1^2 R_1^2 R_2^2}} \tag{6.117}$$

the de-emphasis filter is shown, on the other hand, in Fig. 6.24b whose analysis gives us

$$H_d(\omega) = \frac{1}{jcr\omega + 1} \tag{6.118}$$

and

$$|H_d(\omega)| = \frac{1}{\sqrt{c^2 r^2 \omega^2 + 1}} \tag{6.119}$$

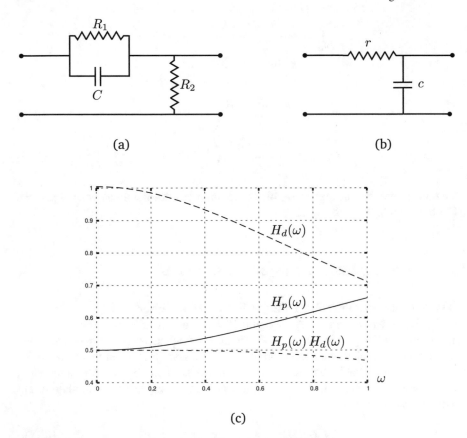

(a)                                                    (b)

(c)

**Fig. 6.24** Pre-emphasis and De-emphasis filters with their frequency responses

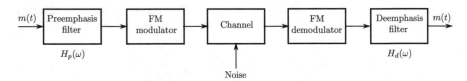

**Fig. 6.25** Block diagram showing the pre-emphasis and de-emphasis filters

after adjusting parameters, we get the result of Fig. 6.24c where the product

$$H_p(\omega) H_d(\omega) \approx 0.5 \tag{6.120}$$

The block diagram of the system is shown in Fig. 6.25. We notice from the figure that the pre-emphasis filter is placed before the FM modulator, while the de-emphasis filter is placed after demodulation takes place.

## 6.9 Performance of PM Systems in the Presence of Noise

Referring to Fig. 6.26, the incoming PM signal at ① is given by

$$
\begin{aligned}
s_1(t) &= A_c \cos\left[\omega_c t + k_p m_1(t)\right] \\
&= A_c \cos\left[\omega_c t + \phi(t)\right]
\end{aligned}
\tag{6.121}
$$

After the addition of white noise, the signal at ②, is

$$
s_2(t) = A_c \cos\left[\omega_c t + \phi(t)\right] + w(t)
\tag{6.122}
$$

where $w(t)$ is white noise with an autocorrelation function

$$
R_W(\tau) = \frac{N_0}{2}\delta(\tau)
$$

After this, the signal is passed through a BPF of bandwidth $B$ (Hz or rad/s, whichever units are used), the signal at ③ is

$$
s_4(t) = A_c \cos\lfloor\omega_c t + \phi(t)\rfloor + n_{\text{NB}}(t)
\tag{6.123}
$$

which has a PSD

$$
S_{\text{NB}}(\omega) =
\begin{cases}
N_0/2 & \text{for } \omega_c - \frac{B}{2} \le |\omega| \le \omega_c + \frac{B}{2} \\
0 & \text{otherwise}
\end{cases}
$$

where the bandwidth, $B$, is the bandwidth of the PM signal.

At this point we calculate the input signal-to-noise ratio, which is

$$
\frac{S_i}{N_i} = \frac{A_c^2/2}{N_0 B}
\tag{6.124}
$$

where $B$ is in hertz.

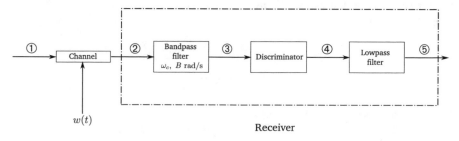

**Fig. 6.26** Analysis of PM systems in the presence of noise

The PSD for narrowband noise resulting from the filtering action may be written in terms of amplitude and phase and is shown in Fig. 6.23a.

We can write the narrowband noise, as earlier,

$$n_{NB}(t) = a(t) \cos\left[\omega_c t + \psi(t)\right] \tag{6.125}$$

where $a(t)$ and $\psi(t)$ are both stationary random processes, where

$$a(t) = \sqrt{n_I^2(t) + n_Q^2(t)}$$

$$\text{and } \psi(t) = \tan^{-1}\left[\frac{n_I(t)}{n_Q(t)}\right] \tag{6.126}$$

Therefore

$$\begin{aligned} s_4(t) &= A_c \cos\left[\omega_c t + \phi(t)\right] + a(t) \cos\left[\omega_c t + \psi(t)\right] \\ &= \left[-a(t) \sin \psi(t) - A_c \sin \phi(t)\right] \sin(\omega_c t) \\ &\quad + \left[a(t) \cos \psi(t) + A_c \cos \phi(t)\right] \cos(\omega_c t) \end{aligned} \tag{6.127}$$

as earlier we make the following substitutions,

$$\begin{aligned} C(t) \sin \theta(t) &= a(t) \sin \psi(t) + A_c \sin \phi(t) \\ C(t) \cos \theta(t) &= a(t) \cos \psi(t) + A_c \cos \phi(t) \end{aligned} \tag{6.128}$$

so that

$$s_4(t) = C(t) \cos\left[\omega_c t + \theta(t)\right] \tag{6.129}$$

where

$$\theta(t) = \tan^{-1}\left[\frac{a(t) \sin \psi(t) + A_c \sin \phi(t)}{a(t) \cos \psi(t) + A_c \cos \phi(t)}\right] \tag{6.130}$$

Since we are considering angle modulation, we are mostly concerned with the phase, $\theta(t)$, of the output signal and the amplitude, $C(t)$, is of little concern to us. When the phase is expanded in a Taylor series expansion for small values of $a(t)$, (we are assuming here that the carrier amplitude, $A_c$, is much greater than $a(t)$, the noise amplitude) we get

$$\theta(t) \approx \underbrace{\phi(t)}_{\text{Signal}} + \underbrace{\frac{\sin\left[\psi(t) - \phi(t)\right] a(t)}{A_c}}_{\text{Noise}} \tag{6.131}$$

which is the message signal plus noise.

As earlier, if we take a look at the noise, we notice the term

$$\sin\left[\psi(t) - \phi(t)\right]$$

now the argument, which is the phase,

$$\psi(t) - \phi(t) = \psi(t) - k_p m(t)$$

behaves very much like a uniformly distributed random process, in the interval $[0, 2\pi]$, which is exactly the same as the behaviour of $\psi(t)$, so we may replace $\psi(t) - \phi(t)$ by $\psi(t)$ and therefore the noise term may be written as

$$n_{PM} = \frac{a(t) \sin[\psi(t)]}{A_c}$$

$$= \frac{n_Q(t)}{A_c} \tag{6.132}$$

So

$$\theta(t) \approx \underbrace{\phi(t)}_{\text{Signal}} + \underbrace{\frac{n_Q(t)}{A_c}}_{\text{Noise}} \tag{6.133}$$

the signal power is therefore

$$S_o = k_p^2 \langle m^2(t) \rangle$$

and the noise power is

$$N_o = \frac{N_0 B}{A_c^2} \tag{6.134}$$

and

$$\left(\frac{S_o}{N_o}\right) = \frac{k_p^2 \langle m^2(t) \rangle A_c^2}{N_0 B}$$

$$= \left(\frac{S_i}{N_i}\right) \frac{k_p^2 \langle m^2(t) \rangle}{2} \tag{6.135}$$

## 6.10  Important Concepts and Formulae

- For an angle modulated signal,

$$s(t) = A \cos[\theta(t)]$$

- The instantaneous frequency is

$$\omega(t) = \frac{d}{dt}\theta(t)$$

- In frequency modulation, we make the $\omega(t)$ to be proportional to the message signal,

$$\omega(t) = \omega_c + k_f m(t)$$

where $\omega_c$ would be the carrier, $m(t)$ the message signal, and $k_f$ is a constant called the *frequency sensitivity*, and which implies that the *instantaneous* frequency, $\omega(t)$, is a function of the message signal. The phase in this case is

$$\theta(t) = \int \left[\omega_c + k_f m(t)\right] dt$$

$$= \omega_c t + k_f \int m(t)dt$$

- The frequency excursion is

$$\omega_c - k_f \, |m(t)|_{\max} \le \omega \le \omega_c + k_f \, |m(t)|_{\max}$$

and

$$\Delta\omega = k_f \, |m(t)|_{\max}$$

- In the second case, which is phase modulation,

$$s(t) = A\cos[\theta(t)]$$

where

$$\theta(t) = \omega_c t + k_p m(t)$$

where $k_p$ is a constant called the phase sensitivity. In phase modulation, the phase of the transmitted signal is a function of the message signal, and $\omega_c$ is the carrier frequency.
- The instantaneous frequency is given by

$$\frac{d\theta(t)}{dt} = \omega_c + k_p \frac{dm(t)}{dt}$$

- The instantaneous frequency has the range

$$\omega_c - k_p \left|\frac{dm(t)}{dt}\right|_{\max} \le \omega \le \omega_c + k_p \left|\frac{dm(t)}{dt}\right|_{\max}$$

## 6.11   Self Assessment

### 6.11.1   Review Questions

1. What is designated by the time-varying phase of a cosine signal?
2. How is the time-varying frequency related to the time-varying phase?
3. In angle modulation does the carrier frequency change with time?
4. In an angle modulated wave does the amplitude of the wave change?
5. In an FM wave the is phase proportional to $m(t)$? Explain how is the phase related to $m(t)$.
6. Explain the working of a Foster Seeley discriminator.
7. Explain the working of a superheterodyne receiver.
8. Why are pre-emphasis and de-emphasis used in an FM signal?

### 6.11.2   Numerical Problems

1. (a) Plot $\theta(t)$ when
$$\theta(t) = 2\pi \times t + 2\sin(2\pi t)$$

   (b) Plot $\omega(t)$ for the same $\theta(t)$, and
   (c) Plot
$$2\cos[\theta(t)]$$

2. For the phase $\theta(t)$ as shown in Fig. 6.27, find the frequency $f(t)$ as a function of time
3. For the frequency $\omega(t)$, shown in Fig. 6.28, which is a function of time, obtain the analytic expression for $\theta(t)$ and then plot $\cos[\theta(t)]$. Does this wave have a carrier?
4. Obtain an analytic expression for $m(t)$ for the wave shown in Fig. 6.29.
5. For the wave shown in Fig. 6.29, obtain the analytic expression for the PM signal and plot it for one cycle choosing a convenient value of $\omega_c$ and $k_p$. Plot the frequency as a function of time for one cycle.
6. For the wave shown in Fig. 6.29, obtain the analytic expression for the FM signal and plot it for one cycle choosing a convenient value of $\omega_c$; also choose a convenient value of $k_f$. Plot the frequency as a function of time for one cycle.
7. Obtain the analytic expression, $m(t)$, of the for the wave shown in Fig. 6.30.
8. For the wave shown in Fig. 6.30, obtain the analytic expression for the PM signal and plot it for one cycle choosing a convenient value of $\omega_c$; also choose a convenient value of $k_p$. Plot the frequency as a function of time for one cycle.
9. For the wave shown in Fig. 6.30, obtain the analytic expression for the FM signal and plot it for one cycle choosing a convenient value of $\omega_c$; also choose a convenient value of $k_f$. Plot the frequency as a function of time for one cycle.

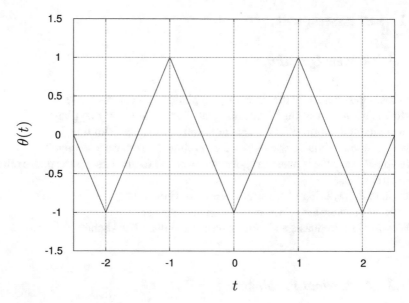

**Fig. 6.27** Phase as a function of time

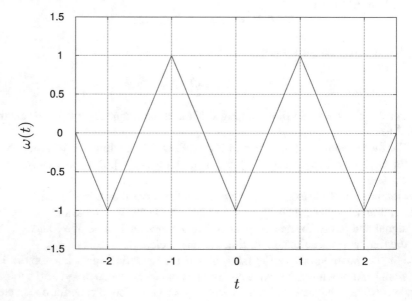

**Fig. 6.28** Frequency as a function of time

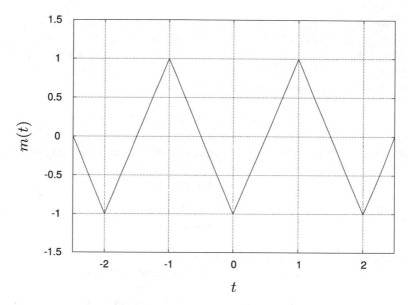

**Fig. 6.29** Triangular periodic wave

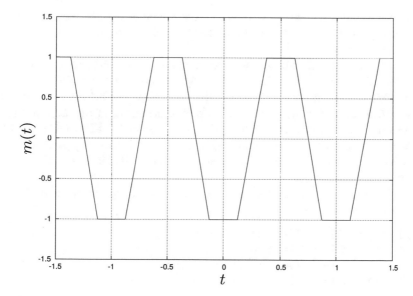

**Fig. 6.30** Blunt triangular periodic wave

10. Using the exponential Fourier series estimate the maximum frequency of the message signal of Fig. 6.29 so that the energy in the highest frequency is less than 50% of the harmonic of greatest amplitude (3 dB bandwidth). Having determined the highest frequency estimate the bandwidth needed for FM.

11. For a VCO with characteristic given by

$$f = 10^4 + 2 \times 10^3 v_{in} \quad \text{(Hz)}$$

where $v_{in}$ is in volts. Find the expression for the instantaneous frequency in the range $-5 \leq t \leq 5$ for

$$m(t) = \begin{cases} 0 & t < 0 \\ 1 & 0 \leq t \leq 1 \\ 0 & t > 1 \end{cases}$$

where $m(t)$ is in volts and $t$ is in seconds.

12. A nonlinear element is described by the input–output relation

$$y(t) = 0.9x(t) + 0.1x^2(t) + 0.05x^3(t)$$

Starting with a suitable low-frequency FM signal of $\omega_{c1}$, and using three such nonlinear elements in series obtain the value of $\omega_{c1}$ to give an output at $\omega_c = 2\pi \times 100$ Mr/s.

13. In a Hartley oscillator, the carrier frequency is to be 100 MHz. Design the circuit giving the values of $L_1$, $L_2$, and $C_0$. If $\Delta C = 0.1C_0$ find the value of $k_f$.

14. In a Hartley oscillator we would like the value of $|k_f| = 10^7$ Hz/volt. The value of $\Delta C = 0.1C_0$ and $\omega_c = 2\pi \times 10^8$ r/s. Find the values of $L_1$, $L_2$, and $C_0$.

15. In a superheterodyne receiver the RF signal is $f_{RF} = 10^6$ KHz. What should be the frequency of the LO signal to make the IF signal frequency to be 455 kHz? What is the image frequency? What should be the bandwidth of the BPF to reject the image frequency?

16. With regard to Eq. (6.34), show that if $f_{o|e}(x)$ and $g_{o|e}(x)$ are odd and even functions of $x$, then

   (a) $h(x) = f_o[g_o(x)]$ is an odd function,
   (b) $h(x) = f_o[g_e(x)]$ is an even function,
   (c) $h(x) = f_e[g_o(x)]$ is an even function, and
   (d) $h(x) = f_e[g_e(x)]$ is an even function.

17. Analyse the circuits of Fig. 6.24, and design the two filters for a normalised bandwidth of $\omega_{max} = 1$ r/s. Hint: Let $R_1 = R_2 = R = 1\ \Omega$ and $C_1 R = 1/\omega_{01}$. Also let $cr = 1/\omega_{02}$. Obtain a relation between $\omega_{01}$ and $\omega_{02}$. Show that $\omega_{01} = 3/4 \times \omega_{02}$.

# References

1. Bowman F (1968) Introduction to Bessel functions. Dover Publications
2. Carson JR (1922) Notes on the theory of modulation. Proc Inst Radio Eng 10:57–64
3. de Bellescize H (1932) La reception synchrone. Revue de l'Electricite et de l'Electronique 11:230–240
4. Forster DW, Seeley SW (1937) Automatic tuning, simplified circuits, and design practice. Proc Inst Radio Eng 25(3):289–313

# Part III
# Digital Communication

# Chapter 7
# Digital Communication Basics

## 7.1 Introduction

In 1832, Samuel Morse a painter and later an inventor conceived of the telegraph, which was the first system to transmit digital messages. The system consisted of connecting different parts of the land with wires mounted on telegraph poles, and the messages consisted of a series of pulses of short duration known as 'dots' (one unit of time) and those of longer duration known as 'dashes', (three units of time). Every letter of the English alphabet had a unique code invented after the advent of telegraphy, known internationally as the Morse code shown in Fig. 7.1. Three units of time were placed between letters, and when a new word was to be sent, seven units of time were interspersed between words. By this means, telegrams were sent between cities, as long as telegraph wires connected the cities.

Taking a concrete example if we desired to send the word 'as' whose Morse code representation would be $\bullet - \bullet \bullet$, the pulses would be arranged as follows (note that $p \equiv \bullet$ (dot) and $ppp \equiv -$(dash)):

$$\underbrace{p\phi ppp}_{a} \; \underbrace{\phi\phi\phi}_{\text{new letter}} \; \underbrace{p\phi p\phi p}_{s} \tag{7.1}$$

In Expression 7.1, $p$ would be a pulse, while $\phi$ would be a space (no signal).

Telegraphy revolutionised communication, just as digital communication is revolutionising our time. Even after the invention of the radio, the Morse code was used in 'wireless' telegraphy, which was the goal of Marconi, the inventor of the radio, (early 1900s) when the first transatlantic radio message was the letter 'S': $\bullet \bullet \bullet$.

The main difference between analogue communication and digital communication is that in analogue communication the signal to be transmitted could be practically any signal which is continuous in time, while in digital communication the signal is (possibly) a series of pulses, signals which are discontinuous in time. For example, in binary communication, the signal consists of only two possibilities: we may either

S. Bhooshan, *Fundamentals of Analogue and Digital Communication Systems*,
Lecture Notes in Electrical Engineering 785,
https://doi.org/10.1007/978-981-16-4277-7_7

# International Morse Code

1. The length of a dot is one unit.
2. A dash is three units.
3. The space between parts of the same letter is one unit.
4. The space between letters is three units.
5. The space between words is seven units.

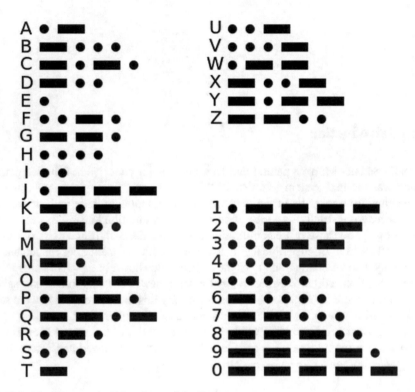

**Fig. 7.1** The Morse code. Taken from wikipedia.org

transmit a zero or a one. As an example, a zero could be a pulse of magnitude $-A$ and a one could be a pulse of $+A$ ($A$ having a convenient value). If we want to transmit a letter or a pixel, we transmit a particular combination of zeros and ones to give meaning to our message.

## 7.2   Sampling

Signals in nature are essentially analogue. Examples of such signals are voice, music, pictures (on film), ECG, EEG and so on. Some of these signals are electrical signals and others are converted to electrical signals by transducers. The question which arises is that how do we convert these signals which are essentially analogue to digital signals? An intuitively simple way of doing this would be to *periodically* find the value of the analogue signal and store it, then later convert those values into electrical signals. In particular, if $m(t)$ is the analogue signal, and if we 'sample' the signal at time $t = 10T$, then that particular value of that sample is

$$m_s[10] = m(t)|_{t=10T}$$

This method of conversion is done periodically at times

$$t = kT, \text{ where } k = \ldots - -1, 0, 1, \ldots \text{ and so on}$$

where $k$ is an integer and $T$ is the sampling period. Hence, under conditions of *ideal* sampling, as explained above, is

$$m_s[k] = m(t)|_{t=kT}, \qquad k = \ldots, -1, 0, 1, \ldots$$

This process, called 'sampling of the signal', is fundamental to digital communication.

An example of practical sampling is explained next. The concept of sampling applied through electrical pulses is to convert the value of the analogue signal, at times $t = kT$, into pulses of narrow width whose value is proportional to the analogue signal. An example of this type of sampling is shown in Fig. 7.2. The upper part of the figure shows the message signal, $m(t)$, which is to be communicated, and in the lower part of the graph, conversion of the data to pulses. The lower graph is actually a depiction of $m(t)$ multiplied by $p_T(t)$ (see Fig. 2.21) which is

$$m(t)p_T(t) = \sum_{k=-\infty}^{k=\infty} m(t)p_s(t - kT) \tag{7.2}$$

$p_s(t)$ being a single pulse of width $2\tau$ and amplitude 1, as shown in the upper figure of Fig. 8.1, and $T$ is the period of the pulse train. The relationship between rect$(t)$ and $p_s(t)$ is

$$p_s(t) = \text{rect}\left(\frac{t}{2\tau}\right) \tag{7.3}$$

But before we analyse this figure, we must understand sampling through the classical analysis, which is sampling the data with *impulses*. It is obvious that impulses are not practical and cannot be produced in the laboratory, but the insight which we obtain by this approach is well worth the analysis.

**Fig. 7.2** Message signal
converted to pulses

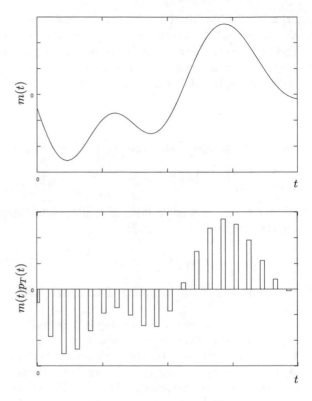

## 7.2.1  Fourier Transform of a Train of Impulses

Before we proceed any further, let us find the Fourier transform of a train of impulses
since this will be of importance in understanding the whole process of sampling with
impulses and eventually pulses. The impulse train is shown in Fig. 7.3. The impulse
train is

$$\delta_T(t) = \sum_{k=-\infty}^{k=\infty} \delta(t - kT) \tag{7.4}$$

since a single impulse at $t = kT$ is $\delta(t - kT)$; therefore, an in infinite series of such
impulses are represented by the summation.

As we can guess, this function is a periodic function, which we can represent by
means of its Fourier series,

$$\delta_T(t) = \sum_{k=-\infty}^{k=\infty} \Delta_k e^{jk\omega_s t}$$

$$\delta_T(t) = \sum_{k=-\infty}^{k=\infty} \delta(t - kT)$$

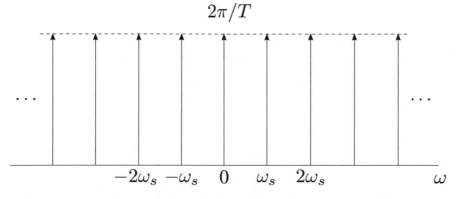

$$\Delta_T(\omega) = \frac{2\pi}{T} \sum_{k} \delta(\omega - k\omega_s)$$

**Fig. 7.3** Impulse train and its Fourier transform

where $\omega_s = 2\pi f_s = 2\pi/T$. From the chapter on the Fourier series, the $\Delta_k$ is given by

$$\Delta_k = \frac{1}{T} \int_{-T/2}^{T/2} \delta_T(t)dt = \frac{1}{T} \int_{-T/2}^{T/2} \delta(t)dt = \frac{1}{T} \quad \forall k \tag{7.5}$$

from Eq. (2.78); since we integrate over a single impulse, we get 1. Hence,

$$\delta_T(t) = \frac{1}{T} \sum_{k=-\infty}^{k=\infty} e^{jk\omega_s t} \tag{7.6}$$

taking the Fourier transform on both sides, and using Eq. (2.113),

$$\Delta_T(\omega) = \frac{2\pi}{T} \sum_{k=-\infty}^{k=\infty} \delta(\omega - k\omega_s) \tag{7.7}$$

This result is shown in Fig. 7.3. The upper figure shows the pulse train in time, while the lower figure shows the Fourier transform.

## 7.3   Fourier Transform of a Sampled Signal

We now apply the sampling process to an arbitrary message signal. To do this, let us observe Fig. 7.4. We can see here that there is a message signal shown in the top part of the figure, which is converted to impulses as shown in the lower part of the figure. Essentially, if we take two signals

$$m(t)$$

and

$$\delta_T(t) = \sum_{k=-\infty}^{k=\infty} \delta(t - kT)$$

then the lower figure is the multiplication of these two waveforms. That is,

$$m_\delta(t) = m(t) \times \delta_T(t)$$

$$= m(t) \times \left\{ \sum_{k=-\infty}^{k=\infty} \delta(t - kT) \right\}$$

$$= \sum_{k=-\infty}^{k=\infty} m(kT)\delta(t - kT) \tag{7.8}$$

where, at each value $t = kT$, the message signal multiplies the corresponding delta function, $\delta(t - kT)$. Taking the Fourier transform of the second of these equations

$$\mathcal{F}\{m_\delta(t)\} = \mathcal{F} \left\{ m(t) \times \left\{ \sum_{k=-\infty}^{k=\infty} \delta(t - kT) \right\} \right\}$$

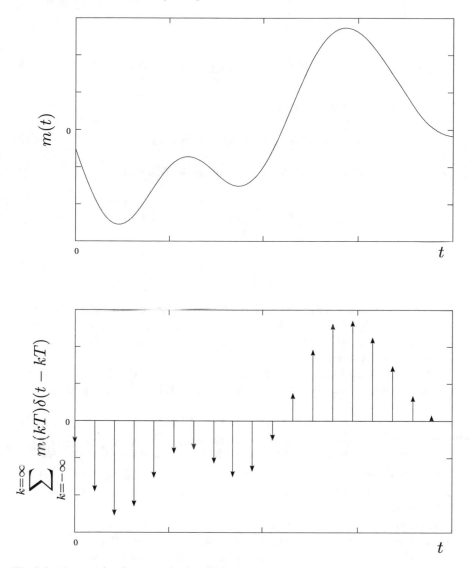

**Fig. 7.4** Message signal converted to impulses

which we simplify using the convolution theorem (Eq. (2.127)) which says

$$\mathcal{F}\{f_1(t) \times f_2(t)\} = F_1(\omega) * F_2(\omega)$$

Therefore,

$$M_\delta(\omega) = M(\omega) * \mathcal{F}\{\delta_T(t)\}$$

$$= M(\omega) * \left\{ \omega_s \sum_{k=-\infty}^{k=\infty} \delta(\omega - k\omega_s) \right\} \tag{7.9}$$

We know, however, that for any convolution

$$X(\omega) * \delta(\omega - k\omega_s) = X(\omega - k\omega_s) \tag{7.10}$$

(see Problem 25, Chap. 2) which is a standard result. Therefore,

$$M_\delta(\omega) = M(\omega) * \left\{ \omega_s \sum_{k=-\infty}^{k=\infty} \delta(\omega - k\omega_s) \right\}$$

$$= \omega_s \sum_{k=-\infty}^{k=\infty} M(\omega - k\omega_s) \tag{7.11}$$

If we take a look at the right-hand side of Eq. (7.11), find that the Fourier transform of the sampled signal is a periodic function with period $T$. On reflection, we realise this since it is a superposition of the same function, $M(\omega)$ shifted to the right by $k\omega_s$, infinite number of times.

This result is not surprising since this is proved in Example 2.31. These results are shown in Fig. 7.5. The upper figure shows the individual Fourier transforms

$$|M(\omega - k\omega_s)| \qquad k = 0, \pm 1 \text{ and } \pm 2 \tag{7.12}$$

where $\omega_s (= 2\pi/T)$ is the sampling frequency. We notice that the functions $|M(\omega)|$, $|M(\omega - \omega_s)|$ and $|M(\omega + \omega_s)|$ cross at $\pm\omega_s/2$. The sum $\left|\sum_{k=-\infty}^{k=\infty} M(\omega - k\omega_s)\right|$ is shown in the lower figure, which shows that the individual details of $M(\omega)$ are lost in the function $M_\delta(\omega)$.

Since $M_\delta(\omega)$ is the sum of all the shifted functions, this implies that we cannot extract (if we could think up a method of extraction) the analogue form, $m(t)$, of $M(\omega)$ from $M_\delta(\omega)$. Information is lost in addition of all the frequency-shifted functions, because at $\pm\omega_s/2$, the spectra are not zero. This phenomenon is called *aliasing*.

What is the solution? To find the solution, let us look at a slightly different situation shown in Fig. 7.6. Here, the upper diagram shows $|M(\omega)|$, the Fourier transform of a signal which is *band-limited* to $\omega_m$. The Fourier transform is symmetric and lies between $\pm\omega_m$ since $m(t)$ is real. Since in the earlier case, the spectra of the frequency-shifted functions crossed at $\omega_s/2$, we use this fact to sample the signal at a much higher rate as compared to $\omega_m$. A little reflection tells us that if we now sample $m(t)$ at a sampling frequency given by

$$\frac{\omega_s}{2} \geq \omega_m \tag{7.13}$$

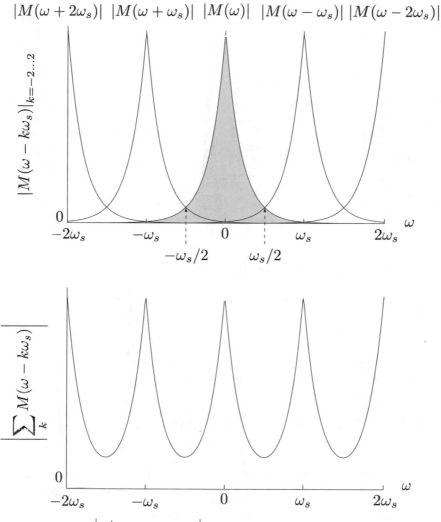

**Fig. 7.5**   $|M(\omega)|$ and $\left|\sum_{k=-\infty}^{k=\infty} M(\omega - k\omega_s)\right|$ with aliasing

then we find that there is no aliasing and we may recover the analogue form of $m(t)$ by passing $m_\delta(t)$ through a low-pass filter. The bandwidth of the filter is such that it satisfies the inequality

$$\omega_s - \omega_m \geq \mathrm{BW}_{\mathrm{LPF}} > \omega_m \tag{7.14}$$

as shown in Fig. 7.6. For the case

$$\boxed{\omega_s = 2\omega_m} \tag{7.15}$$

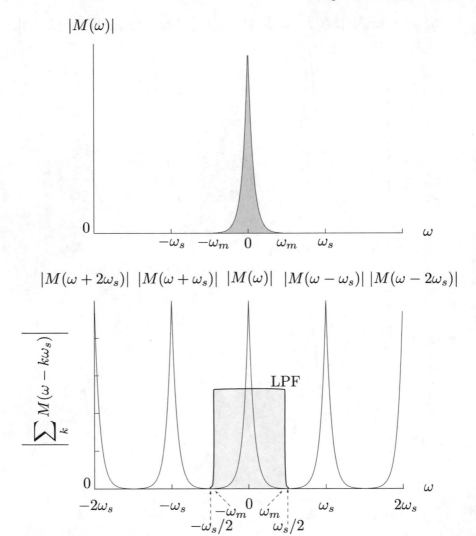

**Fig. 7.6**  $|M(\omega)|$ and $\left|\sum_{k=-\infty}^{k=\infty} M(\omega - k\omega_s)\right|$ without aliasing

the signals just touch at their tails. This frequency is called the *Nyquist* frequency, and the Nyquist sampling rate is

$$f_{\text{Nyquist}} = 2f_m \quad (\text{Hz})$$

after the Swedish–American, Nyquist.

Note that practically all signals occurring in nature are band-limited. A good example is the voice signal whose Fourier transform is shown in Fig. 5.3, which is band-limited to 4 KHz. Similarly, music is band-limited to about 22.5 KHz and so on.

## 7.3.1  Anti-aliasing Filter

To understand the problem created by aliasing, we take voice as an example signal. Though we have said that most signals in nature are band-limited, there is no signal which is *strictly* band-limited. Generally, a small tail of the spectrum is present as we go to higher frequencies. We have to make a choice: how much of the spectrum should we include? To include more spectrum, we have to sample the signal at a higher frequency, and in most cases, there exists a reasonable trade-off. From experimental evidence for voice, the bandwidth of 4 kHz appears as reasonable.

The Fourier transform of voice is shown in Fig. 5.3, which is symbolically shown in Fig. 7.7. Since the bulk of the FT lies in the region $-4$ (kHz) $\leq f \ (= \omega/2\pi) \leq$ 4 (kHz), the voice waveform is sampled at the Nyquist rate which is 8 kilosamples per second.

Examining Fig. 7.7 in some detail, the original FT of the voice signal, $M(\omega)$, is drawn as the shaded portion of the upper part of the figure, which is labelled ①. We then sample the waveform, and its Fourier transform is shown in the upper part labelled ②, ($M_\delta(\omega)$), where the aliasing is clearly shown (marked ③). If we now try to recover the original signal by passing it through a LPF of bandwidth 4 kHz, we find that the recovered signal will be fairly distorted. The distortion will vary depending on the sampling rate: the higher the sampling rate above the Nyquist rate, the lesser the distortion due to aliasing. Another point which is important is that aliasing affects both low as well as high frequencies in the signal (the reader should ask: why is this so?).

Can we improve the performance of our system? The answer to this question is yes. The solution to this problem is to pass the original signal, $m(t)$, through a sharp cutoff low-pass filter of bandwidth 4 kHz, *before* sampling takes place. The filter characteristic is shown as ④ in the upper part of the figure and is referred to as an anti-aliasing filter. The output of this filter is $m_{aa}(t)$.

The result of this operation (passing the signal through the anti-aliasing filter) is shown in the lower part of the figure, which is the shaded portion marked with a ⑤. Notice here that the signal spectrum is sharply cutoff at ⑦. When this signal (whose higher frequencies have been removed) is sampled at the Nyquist rate (which is eight kS/s), the resulting spectrum is shown marked ⑥, $M_{\delta aa}(\omega)$. On recovery of the signal, we get back our original signal minus the treble, $m_{\delta aa}(t)$. The block diagram of the system is shown in Fig. 7.8.

**Example 7.1** Since an anti-aliasing filter may introduce its own distortion, this example examines how an anti-aliasing filter distorts the signal. We show the effect of an anti-aliasing filter and recovery on the following spectrum:

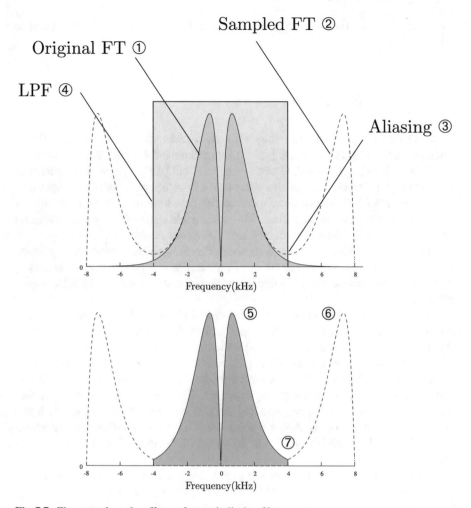

**Fig. 7.7** Figure to show the effects of an anti-aliasing filter

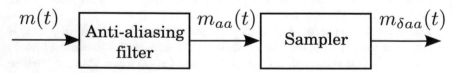

**Fig. 7.8** Block diagram in a system which includes an anti-aliasing filter

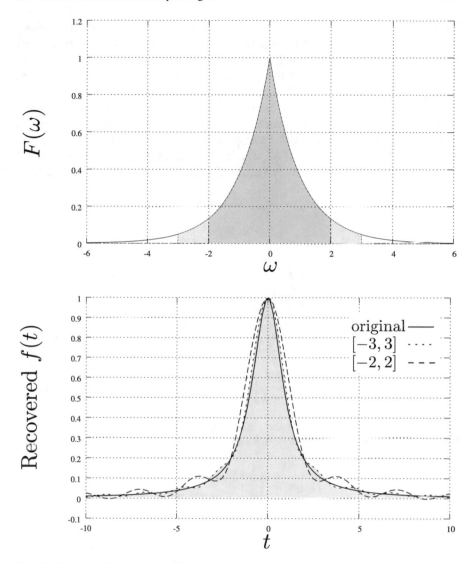

**Fig. 7.9** Example discussing anti-aliasing

$$F(\omega) = Ae^{-|\omega|}$$

Referring to Fig. 7.9, $F(\omega)$ in the frequency domain is shown in the upper figure, peaking at $\omega = 0$ and exponentially decaying on the positive and negative axes. The original waveform in the time domain which is the inverse FT of $F(\omega)$

$$f(t) \Leftrightarrow F(\omega)$$

normalised to unity is shown in the lower graph, as a solid curve marked 'original' in that figure. We notice that the signal is smooth throughout. We now pass $F(\omega)$ through an anti-aliasing filter with bandwidth of $\omega = 3$ rad/s. The spectrum is the lightly shaded curve in the upper graph. We then take its inverse FT (again normalised to one) which is marked '[−3, 3]' in the lower graph and is shown shaded. We notice that the original curve and the [−3, 3] curve are only slightly different from each other. In the third case, we use an anti-aliasing filter which has a bandwidth of $\omega = 2$ which is shown as heavily shaded in the top graph and whose time-domain representation is shown in the lower graph as '[−2, 2]'. The [−2, 2] signal and the original signal are significantly different from each other. The analysis of this problem is carried out in the following manner.

$$F(\omega) = Ae^{-|\omega|}$$

So

$$
\begin{aligned}
f(t) &= \frac{A}{2\pi} \left\{ \int_{-\infty}^{0} e^{\omega} e^{j\omega t} d\omega + \int_{0}^{\infty} e^{-\omega} e^{j\omega t} d\omega \right\} \\
&= \frac{A}{2\pi} \left\{ \int_{-\infty}^{0} e^{(jt+1)\omega} d\omega + \int_{0}^{\infty} e^{(jt-1)\omega} d\omega \right\} \\
&= \frac{A}{2\pi} \left\{ \frac{1 - jt}{t^2 + 1} + \frac{1 + jt}{t^2 + 1} \right\} \\
&= \frac{A}{2\pi} \left\{ \frac{2}{t^2 + 1} \right\}
\end{aligned}
\tag{7.16}
$$

Equation (7.16) is plotted with the solid line in the lower graph and marked as 'original' in the key.

We now represent

$$F_{aa}(\omega) = \begin{cases} e^{-|\omega|} & -a \leq \omega \leq a \\ 0 & \text{elsewhere} \end{cases}$$

and

$$
\begin{aligned}
f_{aa}(t) &= \frac{A}{2\pi} \left\{ \int_{-a}^{0} e^{\omega} e^{j\omega t} d\omega + \int_{0}^{a} e^{-\omega} e^{j\omega t} d\omega \right\} \\
&= \frac{A}{2\pi} \left\{ \int_{-a}^{0} e^{(jt+1)\omega} d\omega + \int_{0}^{a} e^{(jt-1)\omega} d\omega \right\} \\
&= \frac{A}{2\pi} \left\{ \frac{1}{jt + 1} - \frac{e^{-(jt+1)a}}{jt + 1} + \frac{e^{(jt-1)a}}{jt - 1} - \frac{1}{jt - 1} \right\} \\
&= \frac{A}{2\pi} \left\{ \frac{e^{-a} \left[ 2t \sin(ta) - 2\cos(ta) \right] + 2}{t^2 + 1} \right.
\end{aligned}
$$

These signals are then plotted with $a = 3$ (the $[-3, 3]$ curve) and $a = 2$ (the $[-2, 2]$ curve). Notice that the distortion is introduced by the factor $e^{-a}$.

**Exercise 7.1** Working with the function

$$F(\omega) = \begin{cases} 1 + \omega & -1 \le \omega \le 0 \\ 1 - \omega & 0 \le \omega \le 1 \\ 0 & \text{elsewhere} \end{cases}$$

carry out an exercise as in the previous example (Example 7.1) with $a = 0.9$ and $a = 0.8$. Show that when $a = 1$,

$$f(t) = \frac{2[1 - \cos(t)]}{t^2}$$

while for other values of $a$,

$$f_{aa}(t) = \frac{2 - 2t(a - 1)\sin(at) - 2\cos(at)}{t^2}$$

Plot these functions in their normalised form and compare. Here, the distortion is introduced by the factor $(a - 1)$.

## 7.4  Important Concepts and Formulae

- An impulse train is

$$\delta_T(t) = \sum_{k=-\infty}^{k=\infty} \delta(t - kT) \tag{7.17}$$

  since a single impulse at $t = kT$ is $\delta(t - kT)$; therefore, an infinite series of such impulses are represented by the summation.
- The FT of the impulse train given by Eq. (7.17) is

$$\Delta_T(\omega) = \frac{2\pi}{T} \sum_{k=-\infty}^{k=\infty} \delta(\omega - k\omega_s)$$

- For any convolution,

$$X(\omega) * \delta(\omega - k\omega_s) = X(\omega - k\omega_s) \tag{7.18}$$

- The Nyquist sampling rate is given by

$$\omega_s = 2\omega_m$$

where $\omega_m$ is the maximum frequency of the sampled signal.

## 7.5  Self Assessment

### 7.5.1  Review Questions

1. Explain why communication using the Morse code is part of digital communication?
2. If the signal

$$f(t) = e^{-at}u(t)$$

   is sampled at any rate,will there be aliasing?
3. What causes aliasing in sampling?
4. What is the function of an anti-aliasing filter?
5. After using an anti-aliasing filter is the signal distorted? Explain

### 7.5.2  Numerical Problems

1. Find the Morse code of the word SOS.
2. For

$$f(t) \Leftrightarrow F(\omega)$$

   shown in Fig. 7.10, find the minimum sampling frequency to avoid aliasing.
3. A sine wave

$$f(t) = 10\cos(2\pi \times 100t) \tag{7.19}$$

   is sampled at the rate of 10 samples/s with impulses. Plot the sampled signal using a convenient time axis.
4. Obtain a mathematical expression for the sampled signal for Problem 3.
5. Find the FT of the sampled signal for Problem 3 and plot its magnitude, using a convenient frequency axis.
6. If we sample the signal given by Eq. (7.19) at the rate of 100 samples/s, plot the output sampled waveform using a convenient time axis.
7. Obtain a mathematical expression for the sampled signal for Problem 6.
8. Find the FT of the sampled signal for Problem 6 and plot its magnitude using a convenient frequency axis.
9. If we pass the sampled signal of Problem 8 through a LPF of bandwidth (a) 10 Hz, (b) 1000 Hz and (c) 10,000 Hz, what will we get on the output?

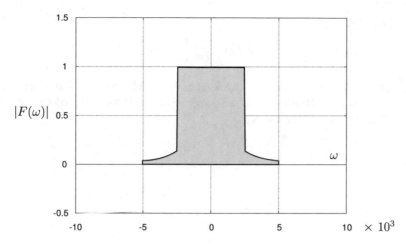

**Fig. 7.10** $|F(\omega)|$ for Problem 2

10. If we sample the signal given by Eq. (7.19) at the rate of 1000 samples/s, find a mathematical expression for the sampled waveform.
11. A signal

$$f(t) = A\frac{\sin(at)}{at}$$

is to be sampled at the Nyquist rate. What is the sampling rate for (a) $a = 10^3$ rad/s and (b) $a = 2\pi \times 10^3$ rad/s?
12. A signal

$$f(t) = \left[A\frac{\sin(at)}{at}\right]^2$$

is to be sampled at the Nyquist rate. What is the sampling rate for (a) $a = 10^3$ rad/s and (b) $a = 2\pi \times 10^3$ rad/s?
13. A signal

$$f(t) = A\frac{\sin(at)}{at} + \left[A\frac{\sin(at)}{at}\right]^2$$

is to be sampled at the Nyquist rate. What is the sampling rate for (a) $a = 10^3$ rad/s and (b) $a = 2\pi \times 10^3$ rad/s?
14. If two low-frequency signals $f_1(t)$ and $f_2(t)$ have bandwidths $\omega_{m1}$ rad/s and $\omega_{m2}$ rad/s, respectively, find the Nyquist sampling rates for the signal $f_1(t)f_2(t)$.
15. A signal of infinite bandwidth is passed through an anti-aliasing filter of bandwidth 22 kHz. The signal is then sampled at 40 ksamples/s. Show that aliasing is present in the sampled signal. What is the minimum sampling rate so that aliasing is not present and explain why.

16. The signal

$$f(t) = \frac{A}{2\pi} \left\{ \frac{2}{t^2 + 1} \right\}$$

is passed through an anti-aliasing filter of bandwidth $\omega = 2.5$ and then sampled at the rate of 5 samples/s. Will aliasing be present in the sampled signal? What will $f_{aa}(t)$ be after recovery?

# Chapter 8
# Pulse Modulation

We move from analogue modulations to digital modulations which are discrete in some sense or the other. In many cases, the message is such that it is possible to store the message communicated on a digital computer. In other cases, this may not be so, but these modulations use pulses which bring them into the area of discrete or digital modulations. With this in mind, we may subdivide the pulse modulations into those of the analogue type and those which are purely digital.

## 8.1  Pulse Amplitude Modulation (PAM)

We now take a look at pulse amplitude modulation (PAM), a sort of hybrid between analogue and digital modulations. If we observe Fig. 8.1, we see how a PAM signal is produced. At time $t = kT$ we hold the signal $m(t)$ at that value, $t = kT$ for a period of time equal to $2\tau$ (the width of the pulse; this convention is adopted as per Fig. 2.30). The circuit used here is called a sample and hold circuit, where the analogue waveform is sampled and then held at its sampled value for a period of time equal to $2\tau$ which is the width of the pulse.

Based on this statement we can then write the equation of the PAM signal as

$$\sum_{k=-\infty}^{k=\infty} m(kT)p_s(t - kT - \tau) \tag{8.1}$$

where $t$ is time, $k$ is the number of the pulse, $T$ is the period of the pulse and $\tau$ is the shift in the pulse shown in Fig. 8.1. We have seen a similar expression before in Eq. (7.8), where there are delta functions instead of pulse functions. We now look at another expression

© The Author(s), under exclusive license to Springer Nature Singapore Pte Ltd. 2022    355
S. Bhooshan, *Fundamentals of Analogue and Digital Communication Systems*,
Lecture Notes in Electrical Engineering 785,
https://doi.org/10.1007/978-981-16-4277-7_8

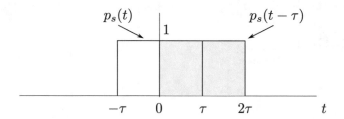

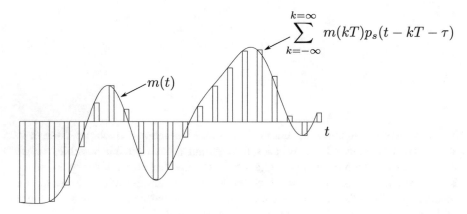

**Fig. 8.1** Pulse amplitude modulation

$$p_s(t - kT - \tau) = p_s(t - \tau) * \delta(t - kT) \tag{8.2}$$

which is true due to the shifting property of the convolution with delta functions. Multiplying both sides by $m(kT)$

$$m(kT)p_s(t - kT - \tau) = p_s(t - \tau) * m(kT)\delta(t - kT)$$

and the summing with respect to $k$,

$$\sum_{k=-\infty}^{k=\infty} m(kT)p_s(t - kT - \tau) = p_s(t - \tau) * \sum_{k=-\infty}^{k=\infty} m(kT)\delta(t - kT) \tag{8.3}$$

We recognise the right-hand side of this equation in Eq. (7.8),

$$\sum_{k=-\infty}^{k=\infty} m(kT)p_s(t - kT - \tau) = p_s(t - \tau) * m_\delta(t) \tag{8.4}$$

We have already studied the FT of $m_\delta(t)$. The FT is a periodic function given by

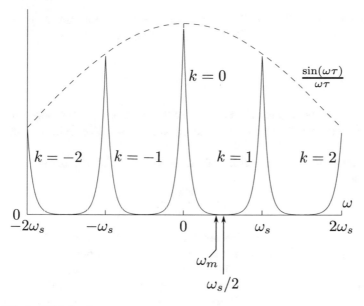

**Fig. 8.2** FT of a PAM signal

$$m_\delta(t) \Leftrightarrow \omega_s \sum_{k=-\infty}^{k=\infty} M(\omega - k\omega_s)$$

we now have to convolve $m_\delta(t)$ with $p_s(t - \tau)$ in the time domain, which becomes multiplication of the two Fourier transforms in the frequency domain. The steps in the computation of the FT are

$$\mathcal{F}\left\{\sum_{k=-\infty}^{k=\infty} m(kT)p_s(t - kT - \tau)\right\} = \mathcal{F}\{p_s(t - \tau)\}\mathcal{F}\{m_\delta(t)\}$$

$$= 2\tau \frac{\sin(\omega\tau)}{\omega\tau} e^{-j\omega\tau} \left\{\omega_s \sum_{k=-\infty}^{k=\infty} M(\omega - k\omega_s)\right\}$$

$$= 2\omega_s \tau e^{-j\omega\tau} \frac{\sin(\omega\tau)}{\omega\tau} \left[\sum_{k=-\infty}^{k=\infty} M(\omega - k\omega_s)\right]$$

$$(8.5)$$

where we have used Eqs. (2.101), and (7.11), and the time shifting property of the Fourier transform.

We have plotted the Fourier transform as per Eq. (8.5) shown in Fig. 8.2. There are some interesting observations:

1. Comparing Figs. 7.6 and 8.2 we find that the two graphs are very similar. The frequency of sampling in PAM must satisfy Inequality (7.13),

$$\omega_s \left( = \frac{2\pi}{T} \right) \geq 2\omega_m \tag{8.6}$$

where $\omega_m$ is the maximum frequency of the signal $m(t)$. This inequality must be met if are to reconstruct the analogue signal from the pulses using a reconstruction LPF.

2. The envelope of the spectrum of the PAM signal follows the sinc(.) function, namely

$$\frac{\sin(\omega\tau)}{\omega\tau}$$

which on we know that it becomes flatter with decreasing $\tau$ (see Eq. (2.30)). Therefore as small a value of $\tau$ may be used as is possible.

3. On observation of Fig. 8.2, it is clear that all frequency amplitudes of $M(\omega)$ are distorted by the sinc(.) envelope, so for proper reconstruction, this important point must be considered. To do this, we place an equaliser after the reconstruction filter. The frequency response of the equaliser is

$$H(\omega) = \begin{cases} \left[ \frac{\sin(\omega\tau)}{\omega\tau} \right]^{-1} & -\omega_m \leq \omega \leq \omega_m \\ \text{anything} & \omega < -\omega_m \text{ and } \omega > \omega_m \end{cases} \tag{8.7}$$

4. It is not necessary that only rectangular pulses be used in sampling the message signal, $m(t)$. For example, if we use exponential pulses, $e^{-at}u(t)$ (with large $a$) to sample $m(t)$, then the envelope will be of $M_\delta(\omega)$ will be

$$\frac{1}{\sqrt{(\omega^2 + a^2)}}$$

and further phase distortion

$$-\tan^{-1}\left(\frac{\omega}{a}\right)$$

will be introduced, both of which will have to be taken care of by a properly designed equaliser.

The block diagram of the PAM signal generation and recovery is shown in Fig. 8.3. If we observe the diagram, in (a), sampling is done after the anti-aliasing filter of the appropriate bandwidth, while the reconstruction filter has similar characteristics as the anti-aliasing filter for proper reconstruction.

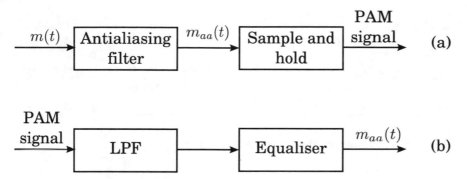

Fig. 8.3 The PAM block diagram

## 8.2 Time Division Multiplexing (TDM)

If we observe Fig. 8.1 carefully, we come to the conclusion that in the communication process, the width of the pulses ($2\tau$) may be made very small and the time between pulses ($T$) may be utilised by other message signals as shown in Fig. 8.4. In this figure are shown, as an example, three interleaved signals, $m_i(t), i = 1, \ldots, 3$. Each of these signals have comparable bandwidth (for example, voice having a bandwidth of 4 kHz). In this case $m_1(t)$ (unbroken line) is sampled with the unfilled pulses, $m_2(t)$ (short dashes) is sampled with the black pulses and $m_3(t)$ (long dashes) is sampled with the grey pulses. The three sets of pulses are interleaved at time intervals of $T/4$ seconds of each other. This whole process is call time division multiplexing (TDM)

The system which we may use to multiplex and demultiplex the messages is shown in Fig. 8.5. Here, $N$ messages are fed to the multiplexer which selects the signal to be

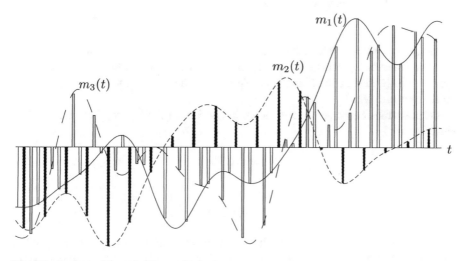

Fig. 8.4 Concept of time division multiplexing

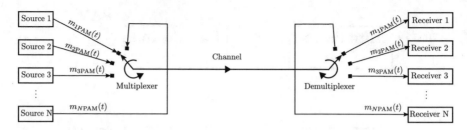

**Fig. 8.5** Various signals interleaved in the period $T$

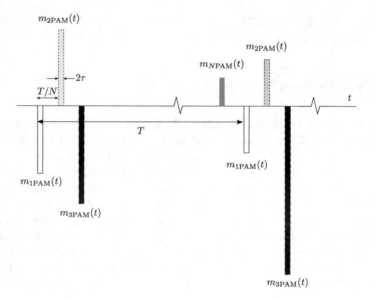

**Fig. 8.6** Details of the multiplexed PAM signal

transmitted in a serial order. The pulse to be transmitted is then sent over the channel. After an interval $T/N$ later, the next pulse is transmitted, and so on cyclically. On the receiving end the correct signal is selected which is sent to the correct destination by the demultiplexer.

Details of the interleaving are shown in Fig. 8.6. In this figure, there are $N$ PAM signals:

$$m_{1\text{PAM}}(t)$$

$$\vdots$$

$$m_{N\text{PAM}}(t)$$

The first pulse is of $m_{1\text{PAM}}(t)$ of width $2\tau$, the second pulse sent $T/N$ later from the start of the first pulse is $m_{2\text{PAM}}(t)$ and so on. The $N$th pulse of the signal $m_{N\text{PAM}}(t)$

is sent at $(N - 1)T/N$ seconds after the start of the first pulse. Note that $2\tau \ll T$ and $2\tau < T/N$ for time division multiplexing to be successful.

## 8.3  Pulse Width Modulation (PWM)

Pulse width modulation (PWM) or pulse duration modulation (PDM) is a kind of modulation where the duration of the pulse is proportional to the amplitude of the pulse. Figure 8.7 shows how PWM looks like for a message signal $m(t)$. If we study the figure, we can see that under the figure the sampling time instants are shown, which are $T$ seconds apart. When the message is sampled, the duration of the pulse (shown as $w$ in the figure) is made proportional to value of $m(t)$. If the value of $m(t) > 0$ then the duration is greater than $T/2$ and proportional to its value, otherwise it is less than $T/2$. For example, at time instant $a$ the $m(t) \approx 0$ so the duration of the pulse is $T/2$. At the time instant $b$, the sampled value of $m(t)$ is negative, and at its minimum value, hence the time duration is almost zero. On the other hand at time instant $c$ we have a global maximum, hence the time duration of the pulse is approximately equal to $T$. The drawback of PWM is that the channel may be used by a single transmitter and receiver, hence time is not used efficiently. The use of PWM is currently made use of in rheostats of electrical devices; these waveforms are also generated in many micro controllers. The bandwidth of PWM may be obtained from [1].

**Exercise 8.1**  If the message signal $m(t)$ is bounded by

$$-m_{\max} \leq m(t) \leq m_{\max}$$

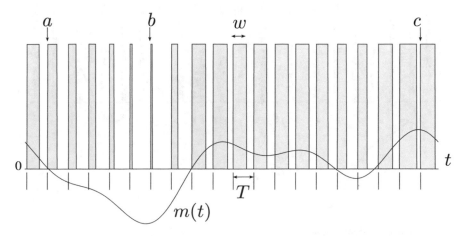

**Fig. 8.7**  Pulse width modulation generation

where $m_{\max}$ is a positive number, find the width of the pulse $w$ such that

$$w = \begin{cases} 0 & m(t) = -m_{\max} \\ T & m(t) = m_{\max} \end{cases}$$

## 8.4  Pulse Position Modulation (PPM)

Another modulation which is currently being used in communication is pulse position modulation (PPM). The generation of PPM is shown Fig. 8.8. The message signal is shown in the figure, which is sampled at instants of time $T$ seconds apart. The sampling instants are shown under the time axis. For example, at the very first instant, the message signal has a positive value so a pulse is generated at time instant $p_1(t)$ which is greater than $t = T/2$. At the point marked $d$ the message signal is at its global minimum, hence the pulse is generated very close to the sampling instant. The generation of a single pulse is shown in the $a - b$ combination of the figure: the sampling instant is at $a$ while the pulse is generated at $b$. Various possibilities exist in PPM systems which we will discuss later. In PPM, all the pulses are of uniform magnitude.

**Exercise 8.2**  If the message signal $m(t)$ is bounded by

$$-m_{\max} \leq m(t) \leq m_{\max}$$

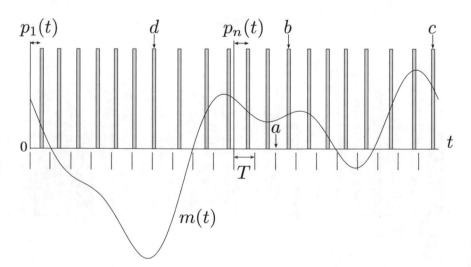

**Fig. 8.8**  Pulse position modulation

where $m_{max}$ is a positive number, find the position of the pulse with respect to the sampling moment, $p_n$ such that

$$p_n(nT + t) = \begin{cases} nT & m(t) = -m_{max} \\ (n + 1)T & m(t) = m_{max} \end{cases}$$

The spectrum analysis of pulse position modulation has been treated by various authors

## 8.5  Quantisation

As discussed earlier analogue signals are represented by $x(t)$ where $t$ is a continuous variable, and the same signal represented in digital form is $x(nT)$, $n = 0, \pm 1, \ldots$ where $T$ is the sampling period. We have also studied that to make sure that data is not lost, we must satisfy the Nyquist criterion:

$$\frac{2\pi}{T}(= \omega_s) \geq 2\omega_m \tag{8.8}$$

where $\omega_m$ is the bandwidth of the analogue signal in r/s. However, we have an additional complication: to store the signal digitally, or to perform operations on the signal using a microprocessor, the *value* of $x(nT)$ must be discrete. For example, if the value of the signal is $1.285|_{10}$ then in binary form it is $1.01001001$. However, the decimal number $1.285$ is itself truncated, but if we have to represent the number $\pi = 3.14159265359\ldots$ then there is no way by which we can do this. The nearest we can do this is to write

$$\pi \cong 11.001001000$$

to nine places of decimal, requiring eleven bits. Hence, we have a problem here. The whole problem of converting an analogue *value* of the signal to binary value of a fixed number of bits of binary represents a problem which we have to solve.

To do this consider a signal whose amplitude lies in the region

$$|m(t)|_{max} \geq |m(t)|$$

or

$$-|m(t)|_{max} \leq m(t) \leq +|m(t)|_{max} \tag{8.9}$$

To do the conversion process, we divide $2|m(t)|_{max}$ into $L$ levels each level being a binary number. For example, if we have four levels, then can use two bits to adequately describe these four levels.

We now define

**Fig. 8.9** The characteristic
of input and output for a
floor quantiser

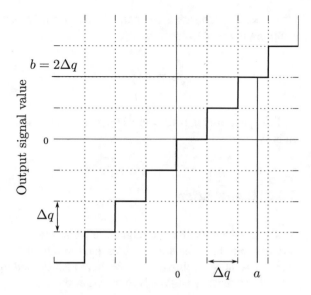

Input signal value

$$\Delta = \frac{2\,|m(t)|_{\max}}{L} \tag{8.10}$$

Referring to Fig. 8.9 and writing $m_m = |m(t)|_{\max}$, for five levels,

$$\underbrace{-2\Delta}_{-m_m}, \underbrace{-\Delta}_{m_m+\Delta}, 0, \underbrace{\Delta}_{m_m-\Delta}, \underbrace{2\Delta}_{m_m}$$

and

$$\underbrace{-2\Delta \ldots +2\Delta}_{4\Delta} \tag{8.11}$$

We may then do a coding in binary

$$-2\Delta \rightarrow 000$$
$$-\Delta \rightarrow 001$$
$$0 \rightarrow 010$$
$$\Delta \rightarrow 011$$
$$2\Delta \rightarrow 100 \tag{8.12}$$

which wasteful in terms of bits used. The bits 101, 110 and 111 are not being used. Hence for maximum utilisation of bits, we should use $L = 2^N$ where $N$ is a positive integer.

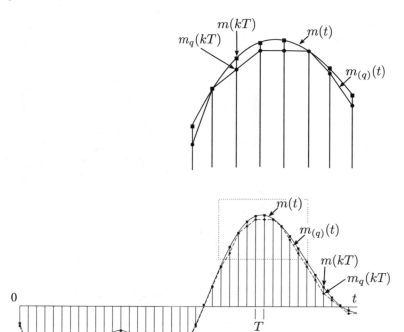

**Fig. 8.10**  The quantised signal

$$N = \log_2 L \tag{8.13}$$

or in general

$$N = \lceil \log_2 L \rceil \tag{8.14}$$

where $\lceil \cdots \rceil$ is the ceiling function. The quantiser is a device which converts a continuous analogue signal levels to discrete levels. For example, in Fig. 8.9, if the input level is $2\Delta \leq a \leq 3\Delta$, the output is

$$b = \Delta \left\lfloor \frac{a}{\Delta} \right\rfloor = 2\Delta \tag{8.15}$$

where $\lfloor \cdots \rfloor$ is the floor function.

Let us now go over to the conversion of the signal to digital format, to do this refer to Fig. 8.10. The figure shows the analogue signal $m(t)$ which is shown as an unbroken line, sampled with the appropriate sampling frequency, $\omega_s$ at intervals of time, $t = kT$, $k = 0, 1, \ldots$. The sampled signal is

**Fig. 8.11** Mid-tread and
mid-rise quantisers

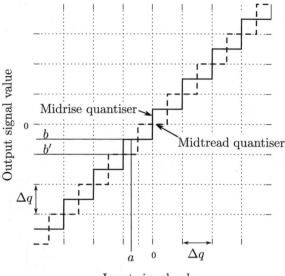

Input signal value

$$\sum m(kT) \tag{8.16}$$

we now quantise the signal with an appropriate quantiser,

$$m_q(kT) = \Delta \left\lfloor \frac{m(kT)}{\Delta} \right\rfloor \tag{8.17}$$

here the floor function is used, we may use other types of quantisers. In the figure the
signal $m_{(q)}(t)$ is not really a signal at all, but the linear interpolation of $\sum m_q(kT)$
but is shown only to make the explanation clearer. The sampled values of $m(t)$ are
shown with square heads, while those of $m_q(kT)$ are shown with round heads.

In our example we have used the floor quantiser (for ease of explanation), but
we may use quantisers which are more commonly used. For example, we may use
the mid-tread quantiser which is shown in Fig. 8.11. Another quantiser which is
frequently being used is the mid-rise quantiser which is also shown in the figure.
Examining the figure in more detail, if the input is $a$ then the output is

$$b = \lceil \frac{a}{\Delta} \rceil \Delta - \Delta/2 \tag{8.18}$$

for a mid-rise quantiser and it is

$$b' = \lceil \frac{a}{\Delta} \rceil \Delta - \Delta \tag{8.19}$$

for a mid-tread quantiser.

## 8.5.1 *Quantisation Noise*

Is obvious that the process of quantisation introduces an error into the message signal, since the message signal has values which are analogue and continuous, while the value of the quantised signal has discrete values. If we consider the error, $q$, as a random variable, we realise that this variable has a real value (for a mid-rise quantiser)

$$-\frac{\Delta}{2} \leq q \leq \frac{\Delta}{2}$$
$$q = 0 \quad \text{elsewhere}$$

we also realise that $q$ should be uniformly distributed,

$$f_Q(q) = \begin{cases} = \frac{1}{\Delta} & -\frac{\Delta}{2} \leq q \leq \frac{\Delta}{2} \\ = 0 & \text{elsewhere} \end{cases} \tag{8.20}$$

the mean of this distribution is

$$\begin{aligned} m_Q &= \int_{-\Delta/2}^{\Delta/2} q f_Q(q) dq \\ &= \int_{-\Delta/2}^{\Delta/2} \frac{q}{\Delta} dq \\ &= 0 \end{aligned} \tag{8.21}$$

while the variance,

$$\begin{aligned} \sigma_Q^2 &= \int_{-\Delta/2}^{\Delta/2} \underbrace{q^2}_{\text{sq of error}} \underbrace{f_Q(q) dq}_{\text{probability}} \\ &= \int_{-\Delta/2}^{\Delta/2} \frac{q^2}{\Delta} dq \\ &= \frac{\Delta^2}{12} \end{aligned} \tag{8.22}$$

If we look at the units of $\sigma_Q^2$, we realise that it is the *average* power of the noise signal. Going back to defining $\Delta$,

$$\Delta = \frac{2m_{\max}}{L} \tag{8.23}$$

where $m_{\max}$ is defined by

$$-m_{\max} \leq m(t) \leq m_{\max}$$

therefore, Signal-to-Noise Ratio (SNR) defined by the ratio of the (average) signal power ($S$) to the average noise power ($\sigma_Q^2$). The higher the SNR, the greater the fidelity to the original signal.

$$\text{SNR} = \frac{S}{\sigma_Q^2} = \frac{S}{\left(\frac{\Delta^2}{12}\right)} = \frac{12S}{\left(\frac{2m_{max}}{L}\right)^2} = \frac{3L^2S}{m_{max}^2} \tag{8.24}$$

which tells us that the SNR is proportional to the square of the number of quantisation levels. Also, $m_{max}^2$ is the maximum signal power, so

$$0 \leq \frac{S}{m_{max}^2} = (S_n) \leq 1$$

where $P_n$ is the normalised power of the signal. Therefore,

$$\text{SNR} = 3L^2 S_n \tag{8.25}$$

generally, $L$, the number of quantisation levels are connected to the number of bits to represent these levels, therefore we make the value of $L$ such that $L$ is one of $2^N$ $N$ being a positive integer. That is $L = \ldots 64, 128, 256, \ldots$

$$N = \log_2 L$$
$$\text{or } L = 2^N$$

therefore

$$\boxed{\text{SNR} = 3S_n 2^{2N} \qquad N = \ldots 4, 5, 6, \ldots} \tag{8.26}$$

if we look at this value in dB,

$$\text{SNR}_{dB} = 4.77 + 10\log_{10} S_n + 6.020N$$

## 8.6  Pulse Code Modulation (PCM)

Pulse code modulation in its essentials is the conversion of an analogue signal to binary pulses. In a previous section, namely Sect. 8.5 we discussed how when a signal is quantised the signal values will take on discrete levels. In this section, we will consider how the quantised signal can be converted into a binary format.

A block diagram of a pulse code modulator is given in Fig. 8.12. The signal is first passed through a sharp cutoff anti-aliasing low-pass filter and then sampled. After quantisation the signal has $L$ levels which generally is a power of two:

**Fig. 8.12** Block diagram of pulse code modulation

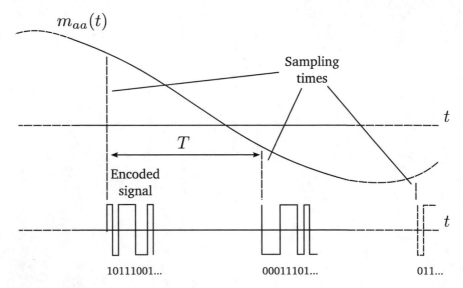

**Fig. 8.13** PAM to pulse code modulation

$$L = 2^N \qquad (8.27)$$

where $N$ is an integer. Recall Eq. (8.14),

$$N = \lceil \log_2 L \rceil$$

which connects the number of levels to the number of bits required per sample.

We now encode each level as a binary number starting from zero to $L - 1$, which then we encode into $N$ bits. This process is shown in Fig. 8.13. The rate of transmission is

$$R = SN \quad \text{bits/s} \qquad (8.28)$$

where $S$ is the sampling rate in samples/s and $N$ is the number of bits per sample.

**Example 8.1** Find the transmission rate in bits/s for voice and music using 256 levels of quantisation for voice and 32 bits for music.

1. Assuming a bandwidth of 4 kHz for voice, the sampling rate for voice would be 8 ks/s. If the number of levels are 256, then

$$2^8 = 256$$

so 8 bits are used per sample. Therefore, the transmission rate would be

$$8000 \times 8 = 64 \,\text{kbps}$$

2. If we assume a bandwidth of 22 kHz for music, and encoding at 32 bits per sample the transmission rate would be

$$44,000 \times 32 = 1408\,\text{kbps}$$

As we can see from this example we can have a very high transmission rate as we increase the number of quantisation levels. One major issue which arises is how to quantise the signal for large and small values of the signal.

### 8.6.1 Companding

Another problem which comes up in practice is that the magnitude of the signal is generally low, but sometimes it is large. How do we handle theses excursions? One solution is to compress the signal in some sort of logarithmic fashion, quantise it and then decompress it. One such method is the '$\mu$ law' method. The equation for the $\mu$ law is

$$v = \text{sgn}(u)\frac{\ln(1 + \mu|u|)}{\ln(1 + \mu)} \tag{8.29}$$

where $u$ is the *normalised* input to the compander and $v$ is the *normalised* output. After companding, the signal may be quantised.

The inverse transformation for the $\mu$ law is

$$u = \text{sgn}(v)\frac{e^{\ln(\mu+1)|v|} - 1}{\mu}$$

$$= \text{sgn}(v)\frac{(1 + \mu)^{|v|} - 1}{\mu} \tag{8.30}$$

which is the expander function. The compander function is shown in Fig. 8.14.

Another law which may be used for companding is the so-called 'A-law'. The compander function for the A-law is

$$v = \begin{cases} \text{sgn(u)}\frac{A|u|}{1+\ln A} & 0 \le |u| \le \frac{1}{A} \\ \text{sgn(u)}\frac{1+A|u|}{1+\ln A} & \frac{1}{A} \le |u| \le 1 \end{cases}$$

where $A$ is the parameter. The A-law compander function is depicted in Fig. 8.15. For the inverse function see Problem 16 Numerical Problems, at the end of the chapter. The block diagram showing the companding is shown in Fig. 8.16.

**Fig. 8.14** Figure showing the relationship between $|u|$ and $|v|$ for the $\mu$ law

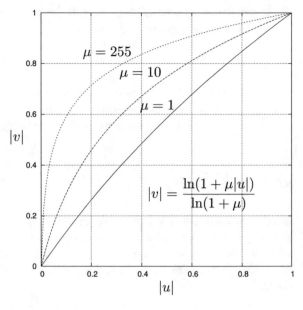

**Fig. 8.15** A-law compander

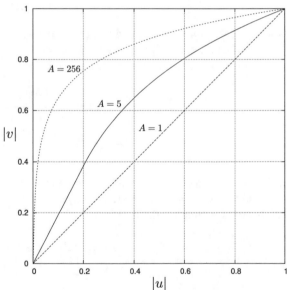

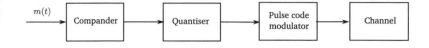

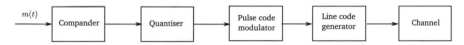

**Fig. 8.16** Block diagram showing the compander

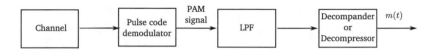

**Fig. 8.17** Block diagram showing the line code generator

## 8.6.2 Line Codes

After companding, quantisation and encoding, the encoded bits (for each sample) are then converted into an electrical signal, shown in Fig. 8.17. These signals are called line codes which are synchronised with a clock of appropriate frequency. Some of these line codes are shown in Fig. 8.18. In part (a) the binary bits are shown; in (b) *ON-OFF* signalling is used. When a '1' is to be transmitted, a positive pulse is generated, and when a '0' is to be transmitted, no signal is transmitted. In part (c) *Manchester* coding is used where when '1' is to be transmitted, then a positive pulse is followed by a negative pulse is generated, while when a '0' is to be transmitted, a negative pulse followed by a positive pulse is transmitted. In part (d) *Differential coding* is used, that is whenever a '1' is transmitted there is no change in the signal level, while when a '0' is transmitted, the signal level changes from high to low or low to high, and in (e) Non-return to zero is shown where a positive pulse denotes a '1' and a negative pulse denotes a '0'.

## 8.6.3 Power Spectral Density of Line Codes

Line codes, as we know, are used for digital baseband transmission in data communication applications. The digital data stream is encoded into a travelling wave voltage pulse sequences for transmission through baseband analogue channels. Here, the spectral properties of the line codes is of interest to us. Therefore, a procedure is required for finding the PSD of each line code. To investigate this procedure, we focus on Fig. 8.19.

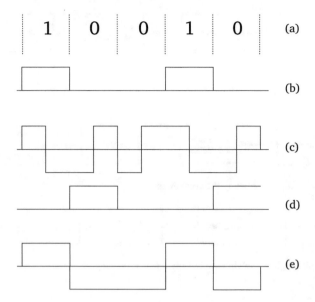

**Fig. 8.18** Line codes **a** Binary bits. **b** ON-OFF signalling. **c** Manchester coding. **d** Differential signalling and **e** Non-return to zero

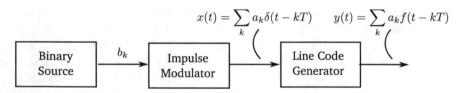

**Fig. 8.19** Line code block diagram

The input to the line code generator is $x(t)$ which is

$$x(t) = \sum_k a_k \delta(t - kT) \tag{8.31}$$

and the autocorrelation function is

$$R_x(\tau) = \frac{1}{T} \sum_{n=-\infty}^{n=\infty} R_n \delta(\tau - nT) \tag{8.32}$$

where

$$R_n = \lim_{N \to \infty} \frac{1}{N} \sum_n^N a_k a_{k+n} \tag{8.33}$$

with

$$R_n = R_{-n} \tag{8.34}$$

Based on these equations, the autocorrelation function,

$$S_x(\omega) = \frac{1}{T}\left(R_0 + 2\sum_n R_n \cos n\omega T\right) \tag{8.35}$$

Since the pulse generator has the impulse response $p(t)$, and

$$p(t) \Leftrightarrow P(\omega)$$

therefore,

$$S_y(\omega) = |P(\omega)|^2 S_x(\omega) \tag{8.36}$$

where $S_x(\omega)$ and $S_y(\omega)$ are the power spectral densities of the output and input signals, respectively, and $P(\omega)$ is the spectrum of the basic pulse used in the line code. An example of a line code is considered in Example 3.29.

In ON-OFF signalling, a 1 is transmitted as a pulse and a zero is transmitted as no pulse. Therefore, using Eq. (8.33), $N/2$ pulses have $a_k = 1$, and $N/2$ pulses have $a_k = 0$, therefore

$$R_0 = \lim_{N\to\infty} \frac{1}{N} \sum_n^N a_k a_k$$

or

$$R_0 = \lim_{N\to\infty} \frac{1}{N}\left[\frac{N}{2}(1) + \frac{N}{2}(0)\right] = \frac{1}{2}$$

To calculate $R_1$, $a_k = 0$ or 1, and $a_{k+1} = 0$ or 1. So $N/4$ cases are 00, $N/4$ cases are 01, $N/4$ cases are 10 and $N/4$ cases are 11. So

$$R_1 = \lim_{N\to\infty} \frac{1}{N}\left[\frac{N}{4}(1) + \frac{3N}{4}(0)\right]$$

$$= \frac{1}{4}$$

In the same manner,

$$R_n = \frac{1}{4} \quad n \geq 2$$

So,

$$S_x(\omega) = \frac{1}{T}\left(R_0 + 2\sum_n R_n \cos n\omega T\right)$$

$$= \frac{1}{T}\left(\frac{1}{2} + 2\sum_n R_n \cos n\omega T\right)$$

$$= \frac{1}{T}\left(\frac{1}{2} + \frac{1}{2}\sum_n \cos n\omega T\right)$$

$$= \frac{1}{T}\left(\frac{1}{4} + \frac{1}{4}\sum_n e^{jn\omega T}\right)$$

Now the series

$$C = \sum_{n=-\infty}^{n=\infty} e^{jn\omega T}$$

$$= \sum_{n=-\infty}^{n=\infty} \delta\left(\omega - \frac{2\pi n}{T}\right)$$

and

$$p(t) = \begin{cases} 1 & -T/2 \le t \le T/2 \\ 0 & \text{otherwise} \end{cases}$$

After proceeding in this way, for ON-OFF signalling, the normalised PSDs are

$$S_y(f) \approx \frac{T}{4}\text{sinc}^2\left(\frac{fT}{2}\right)\left(1 + \frac{\delta(f)}{T}\right) \tag{8.37}$$

For Manchester signalling,

$$R_0 = 1$$
$$R_n = 0 \quad n \ge 1$$

and the pulse shape is

$$p(t) = \begin{cases} 1 & -T/2 \le t \le 0 \\ -1 & 0 \le t \le T/2 \\ 0 & \text{otherwise} \end{cases}$$

so

$$S_y(f) \approx T\text{sinc}^2\left(\frac{fT}{2}\right)\sin^2\left(\frac{\pi fT}{2}\right) \tag{8.38}$$

Similarly, for non-return to zero signalling,

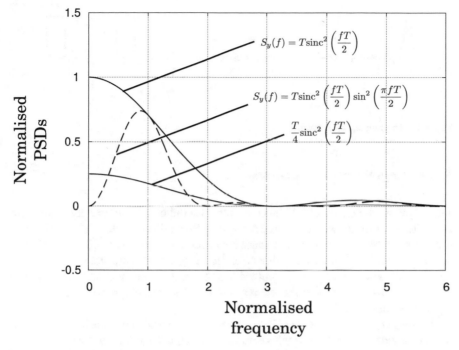

**Fig. 8.20** PSDs of line codes

$$R_0 = 1$$
$$R_n = 0 \quad n \geq 1$$

and

$$p(t) = \begin{cases} 1 & -T/2 \leq t \leq T/2 \\ 0 & \text{otherwise} \end{cases}$$

so

$$S_y(f) \approx T\text{sinc}^2\left(\frac{fT}{2}\right) \tag{8.39}$$

and so on. The plots of these spectra are shown in Fig. 8.20. NRZ encoding is used in RS232-based protocols, while Manchester encoding is used in Ethernet networks. Some authors have taken an alternative route in calculations of these PSDs, however the results are materially the same [2, 3].

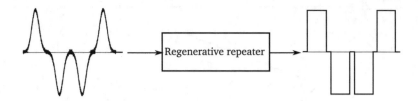

**Fig. 8.21**  The regenerative repeater

### 8.6.4  Regenerative Repeaters

The major advantage that PCM has over other modulations is immunity from noise and other distortions. After the signal has travelled a distance through the channel, it suffers distortions due to the channel characteristics, and the addition of AWGN noise. To remedy this situation, a device called a regenerative repeater is placed along the channel which examines the pulsed signal which is received and then issues a fresh set of pulses identical to the received pulses. The block diagram of a regenerative repeater is shown in Fig. 8.21.

The regenerative repeater consists of an equaliser, a timing device, and a decision-making device, which decides whether the incoming pulse is a zero or a one. A pulse generator then issues the pulses. The equaliser is to correct the distortions introduced by the channel, the timing device is present to demarcate the start of each pulse, and the decision-making device decides whether the pulse was a zero or a one. The decision-making device may make an error in the whether a zero or a one was received, in such a case an error develops. The probability of such an error depends upon the ratio, $E_b/N_0$ where $E_b$ is the energy per bit.

**Example 8.2**  Compute the ratio $E_b/N_0$ for a bit rate of 1 Mbits/s (Mbps), a noise temperature of $T_e \cong 300\,^\circ\text{K}$, and pulses of magnitude 1 V.

To solve this example,

$$N_0 = kT_e = 4.143 \times 10^{-21}\,\text{J}$$

and

$$E_b = 10^{-6}\,\text{J}$$

so

$$\frac{E_b}{N_0} = 2.414 \times 10^{15} = 153.8\,\text{dB}$$

if, on the other hand the pulses are of magnitude 1 mV, then

$$E_b = 10^{-12}\,\text{J}$$

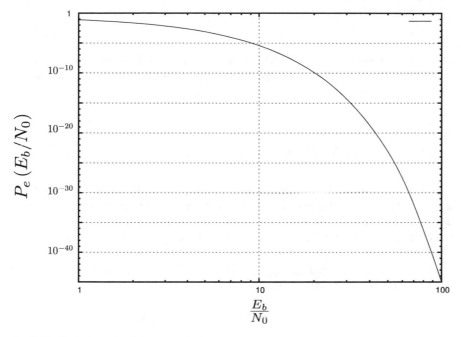

**Fig. 8.22** Typical graph of $P_e$ versus $E_b/N_0$

and

$$\frac{E_b}{N_0} = 93.8\,\text{dB}$$

In practical cases $N_0$ lies in the range $10^{-9}$ to $10^{-16}$ J.

If $x = E_b/N_0$, the probability of error is generally given by formulae of the type,

$$P_e(x) \cong \frac{1}{x\sqrt{2\pi}}e^{-x^2/2} \quad \text{for } x > 3 \tag{8.40}$$

or

$$P_e \cong \frac{1}{2}\text{erfc}(\sqrt{E_b/N_0}) \tag{8.41}$$

where

$$\text{erfc(x)} = \frac{2}{\sqrt{\pi}} \int_x^\infty e^{-t^2}\, dt \tag{8.42}$$

is called the complementary error function. A typical plot of the probability of error, $P_e$ versus $E_b/N_0$ is given in Fig. 8.22.

The meaning of the probability of error, $P_e$ is understood from the following example. If $E_b/N_0 = 10$ then the probability of error, $P_e \cong 10^{-5}$, which implies that about 1 bit in $10^5$ bits will be in error. For the case of 1 Mbps,

$$\frac{10^5}{10^6} = 0.1\,\text{s}$$

is the time interval when an error would be committed. However for the case of $E_b/N_0 = 100$, $P_e \approx 10^{-45}$, then there would be an error about every $10^{31}$ years, or practically never.

### 8.6.5   T-Carrier Systems

The T-carrier systems is a class of carrier systems originally developed by AT and T Bell Laboratories for digital transmission of time division multiplexed telephone calls. The T1 version was introduced in 1962 in the Bell telephone system in USA, which transmitted 24 telephone calls simultaneously over a single transmission medium.

The T1 system sampled the signal at 8000 samples/s, using 8 bits of quantisation giving a total bit rate of approximately

$$8000 \times 8 \times 24 = 1.536 \quad (\text{Mbits/s})$$

The specifications of the carrier systems are T1 (1.544 Mbit/s) with 24 channels, T2 (6.312 Mb/s $\approx$ 4 $\times$ T1) with 96 channels, and T3 (44.736 Mb/s $\approx$ 28 $\times$ T1) with 672 channels and so on.

## 8.7   Delta Modulation (DM)

Delta modulation (DM) is best understood by observing Fig. 8.23a. The waveform $m(t)$ is sampled at instants of time $t = nT$ and which at about four times the Nyquist rate which is the input to the DM modulator. The value of $m_q[(n-1)T]$, the quantised value of $m(t)$ at the time instant $t = (n-1)T$, is compared to $m(t)$, and the difference is the error, $e(nT)$,

$$e(nT) = m(t) - m_q[(n-1)T] \tag{8.43}$$

If the error is positive we add $\Delta$, a convenient value, to the signal, $m_q[(n-1)T]$ and if the error is negative we subtract $\Delta$ from the signal, $m_q[(n-1)T]$ to give us the new value of $m_q(nT)$

$$m_q(nT) = \begin{cases} m_q[(n-1)T] + \Delta & \text{for } e(nT) > 0 \\ m_q[(n-1)T] - \Delta & \text{for } e(nT) < 0 \end{cases} \tag{8.44}$$

or we may write

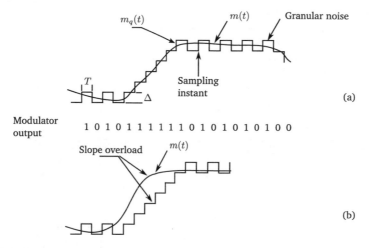

**Fig. 8.23** Delta modulation

$$e_q(nT) = \Delta \text{sgn}[e(nT)]$$
$$m_q(nT) - m_q[(n-1)T] + e_q(nT) \tag{8.45}$$

Over sampling implies that there is a high correlation between adjacent samples and $m_q(t)$ (shown in the figure) keeps in step with the signal $m(t)$. We can say that the DM signal is approximately a staircase approximation of the input signal. The block diagram of the DM output is shown in Fig. 8.24a. The advantage of DM is that we use only one bit per sample, as opposed to other schemes where we use a large number of bits. That is that if there are $L$ levels, we use approximately $\log_2(L)$ bits per sample, while in DM $L = 2$.

One of the problems encountered with using DM is that if the slope of the signal $m(t)$ increases very sharply, such that

$$\left| \frac{dm}{dt} \right| > \frac{\Delta}{T} \tag{8.46}$$

and the slope consistently satisfies Inequality (8.46) for a long period of time (compared to $T$), then we have the condition known as slope overload shown in the (b) part of the figure. To overcome this problem, we must have

$$\left| \frac{dm}{dt} \right|_{\text{max}} \leq \frac{\Delta}{T}$$

which can indicate either a large step size or a higher sampling rate. If we increase the step size, then there is a potential of creating another problem known as granular noise, where the signal oscillates between $m(t) \pm \Delta/2$ in regions where $|dm/dt| \approx$

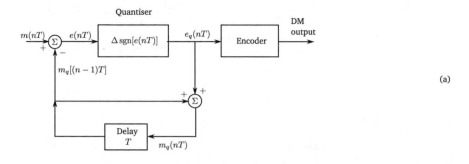

(a)

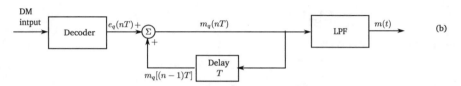

(b)

**Fig. 8.24**  Modulator and demodulator of a DM wave

0. This is shown in Fig. 8.23a. To demodulate the signal, we use the setup shown in Fig. 8.24b, where we implement Eq. (8.45).

Delta modulation is used in many voice communication applications like digital telephone and radio, and is very useful where delivery of the voice signal takes precedence over the quality of the signal.

## 8.8  Sigma-Delta Modulation ($\Sigma - \Delta$ Modulation)

The block diagram shown in Fig. 8.25 is called a Sigma-Delta ($\Sigma - \Delta$) Modulator. This modulator can be considered as being a 'smoothed out version' of a 1-bit delta modulator.

The name Sigma-Delta modulator comes from putting the integrator just before the delta modulator. The main advantage of the Sigma-Delta modulator is that the quantisation noise frequency characteristic is shifted to much higher frequency range, so that when the LPF is used to reconstruct the $\Sigma - \Delta$ signal, the output analogue signal is reconstructed to much higher fidelity. It is this characteristic that gives Sigma-Delta modulation a distinct edge over simple delta modulation.

Referring to Fig. 8.26, which shows the s-domain analysis of the $\Sigma - \Delta$ modulator, the input is $M(s)$. $E(s) = M(s) - \Delta(s)$, or

$$\frac{M(s) - \Delta(s)}{A^{-1}s} + N(s) = \Delta(s) \tag{8.47}$$

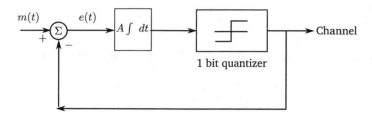

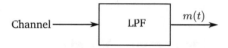

**Fig. 8.25**  Sigma-Delta modulator

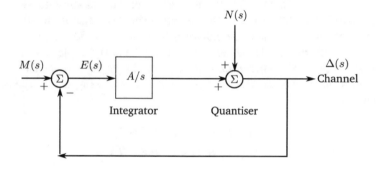

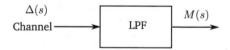

**Fig. 8.26**  S-Domain analysis of a $\Sigma - \Delta$ modulator

where $N(s)$ is the quantisation noise and $A$ is the integrator gain. Assuming that for the moment that $N(s) = 0$, then

$$\Delta(s) = \frac{M(s)}{A^{-1}s + 1} \qquad (8.48)$$

where $A/(s + A)$ is a low-pass filter. So $\Delta(s)$ is $M(s)$ passed through a low-pass filter.

On the other hand when $M(s) = 0$, the noise transfer function is

$$\Delta(s) = \frac{s}{s+A} N(s) \tag{8.49}$$

is noise passed through a high-pass filter. Therefore using superposition,

$$\Delta(s) = \left(\frac{1}{A^{-1}s+1}\right) M(s) + \left(\frac{A^{-1}s}{A^{-1}s+1}\right) N(s) \tag{8.50}$$

when we reconstruct the signal we pass it through a LPF and find that most of the quantisation noise is eliminated.


## 8.9   Differential Pulse Code Modulation (DPCM)

When we see the results of delta modulation we realise that since the signal varies slowly, the difference between one sample and the next is generally small, and therefore a properly designed delta modulator is a success. The idea then occurs to us that since the difference between one sample and the next sample is small, *why not transmit the differences between samples rather than the full sample itself?*

The block diagram of the DPCM modulator is shown in Fig. 8.27a. The pertinent equations are

$$e(kT) = m(kT) - \hat{m}(kT) \tag{8.51}$$

$$m_q(kT) = \hat{m}(kT) + e_q(kT) \tag{8.52}$$

In these equations, $\hat{m}(kT)$ is an *estimate* of $m(kT)$ which is predicted by the predictor, and the second equation is a quantised version of the first equation.

In general the predictor uses circuitry to implement, (Fig. 8.52)

$$\hat{m}(kT) = a_1 m_q [(k-1)T] + a_2 m_q [(k-2)T] + \cdots + a_N m_q [(k-N)T]$$

The simplest case is that of

$$\hat{m}(kT) = m_q [(k-1)T] \quad (a_1 = 1, \ a_2, a_3, \ldots a_N = 0)$$

which leads to

$$e_q(t) = m_q(kT) - m_q [(k-1)T]$$

which is direct differences in the two samples and this is similar to delta modulation (let the reader check this!).

Another more refined method would be to use a Taylor series expansion of two terms:

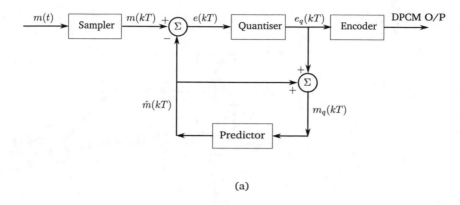

(a)

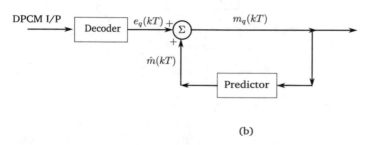

(b)

**Fig. 8.27** The DPCM **a** Modulator and **b** Demodulator

$$\hat{m}(kT) = m_q\left[(k-1)T\right] + \frac{m_q\left[(k-1)T\right] - m_q\left[(k-2)T\right]}{T} \times T$$
$$= 2m_q\left[(k-1)T\right] - m_q\left[(k-2)T\right] \quad (a_1 = 2, \ a_2 = -1, a_3, \ldots a_N = 0)$$

We may by this method more terms in the predictor to give a more accurate estimate of $m(kT)$ which leads to what known as a *linear* predictor. The predictor circuit is shown in Fig. 8.28.

The demodulator is similar to the DPCM modulator. In the demodulator, we use Eq. (8.52) to give us the original signal $m_q(kT)$ which is the quantised version of $m(kT)$.

## 8.9.1 Linear Prediction

We now concentrate on generalised linear prediction used in the DPCM. The general block diagram is given in Fig. 8.28. The equation of interest is

**Fig. 8.28**  Predictor circuit

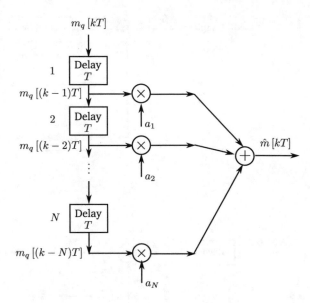

$$e(kT) = m(kT) - \hat{m}(kT)$$

$$= m(kT) - \sum_{n=1}^{N} a_k m[(k-n)T] \qquad (8.53)$$

we would like minimise the squared error function

$$J = E[e^2(kT)]$$

$$= E[m(kT) - \hat{m}(kT)]^2$$

$$= E\left[\left\{m(kT) - \sum_{n=1}^{N} a_n m[(k-n)T]\right\}^2\right] \qquad (8.54)$$

where the $a_n$ are the coefficients which minimise the function $J$, and $E(\cdot)$ is the expected value of the function in the brackets. Or

$$J = E[m^2(kT)] - E\left\{2 \sum_{n=1}^{N} a_n m(kT) m[(k-n)T]\right\}$$

$$+ E\left[\left\{\sum_{n=1}^{N} a_n m[(k-n)T]\right\}^2\right]$$

$$= E[m^2(kT)] - E\left\{2 \sum_{n=1}^{N} a_n m(kT) m[(k-n)T]\right\}$$

$$+ E\left\{\sum_{p=1}^{N}\sum_{n=1}^{N} a_n a_p m[(k-n)T]m[(k-p)T]\right\}$$

$$= E[m^2(kT)] - 2\sum_{n=1}^{N} a_n E\{m(kT)m[(k-n)T]\}$$

$$+ \sum_{p=1}^{N}\sum_{n=1}^{N} a_n a_p E\{m[(k-n)T]m[(k-p)T]\} \qquad (8.55)$$

In these equations,

$$E[m^2(kT)] = \sigma_M^2 \qquad (8.56)$$

where $\sigma_M^2$ is the variance of the message signal. This is so since the signal is assumed to have zero mean:

$$E[m(kT)] = 0 \quad \text{(zero mean)}$$

So

$$J = \sigma_M^2 - 2\sum_{n=1}^{N} a_n E\{m(kT)m[(k-n)T]\}$$

$$+ \sum_{p=1}^{N}\sum_{n=1}^{N} a_n a_p E\{m[(k-n)T]m[(k-p)T]\} \qquad (8.57)$$

To get the minimum of this function, we first define the autocorrelation functions,

$$r_M(n) = E\{m(kT)m[(k-n)T]\}$$
$$R_M(n-p) = E\{m[(k-n)T]m[(k-p)T]\} \qquad (8.58)$$

which are such that

$$r_M(n) = r_M(-n)$$
$$R_M(n-p) = R(p-n)$$
$$R_M(0) = \sigma_M^2$$
$$\text{and } r_M(0) = \sigma_M^2 \qquad (8.59)$$

differentiating (8.57) with respect to $a_1, a_2, \ldots, a_N$,

$$\frac{\partial J}{\partial a_1} = -2r_M(1) + 2\sum_{p=1}^{N} a_p R_M(1 - p) = 0$$

$$\frac{\partial J}{\partial a_1} = -2r_M(2) + 2\sum_{p=1}^{N} a_p R_M(2 - p) = 0$$

$$\vdots \quad = \quad \vdots$$

$$\frac{\partial J}{\partial a_N} = -2r_M(N) + 2\sum_{p=1}^{N} a_p R_M(N - p) = 0 \qquad (8.60)$$

which are $N$ equations in $N$ unknowns, and are known as the Wiener–Hopf equations for linear prediction. We may write these equations as

$$\mathbf{R}_M \mathbf{a} = \mathbf{r}_M$$

$$\text{or } \mathbf{a} = \mathbf{R}_M^{-1} \mathbf{r}_M \qquad (8.61)$$

*provided we know* $\mathbf{R}_M$ *and* $\mathbf{r}_M$.

## 8.9.2   Linear Adaptive Prediction

We now proceed to the case where $\mathbf{R}_M$ and $\mathbf{r}_M$ are *not known*. We start with

$$\frac{\partial J}{\partial a_1} = -2r_M(1) + 2\sum_{p=1}^{N} a_p R_M(1 - p) = 0$$

$$\frac{\partial J}{\partial a_1} = -2r_M(2) + 2\sum_{p=1}^{N} a_p R_M(2 - p) = 0$$

$$\vdots \quad = \quad \vdots$$

$$\frac{\partial J}{\partial a_N} = -2r_M(N) + 2\sum_{p=1}^{N} a_p R_M(N - p) = 0 \qquad (8.62)$$

We now set $g_l = \partial J/\partial a_l$, $l = 1 \ldots N$, and therefore,

$$g_1 = -2r_M(1) + 2\sum_{p=1}^{N} a_p R_M(1-p)$$

$$g_2 = -2r_M(2) + 2\sum_{p=1}^{N} a_p R_M(2-p)$$

$$\vdots = \vdots$$

$$g_N = -2r_M(N) + 2\sum_{p=1}^{N} a_p R_M(N-p) \qquad (8.63)$$

since we do not know $r_M(l)$ and $R_M(l-p)$ we instead work with the estimates of $g_l$:

$$\hat{g}_1[i] = -2m(kT)m[(k-1)T] + 2\sum_{p=1}^{N} a_p[i]m[(k-1)T]m[(k-p)T]$$

$$\hat{g}_2[i] = -2m(kT)m[(k-2)T] + 2\sum_{p=1}^{N} a_p[i]m[(k-?)T]m[(k-p)T]$$

$$\vdots = \vdots$$

$$\hat{g}_N[i] = -2m(kT)m[(k-N)T] + 2\sum_{p=1}^{N} a_p[i]m[(k-N)T]m[(k-p)T] \quad (8.64)$$

where $a_p[i]$ is the current value of the coefficients, and $\hat{g}_p[i]$ are the current value of the estimate of $g_p$. The value of these coefficients at the next iteration is

$$a_p[i+1] = a_p[i] - \frac{\mu}{2}\hat{g}_p[i]$$

$$= a_p[i] - \frac{\mu}{2}\left[-2m(kT)m[(k-p)T] + 2\sum_{q=1}^{N} a_q[i]m[(k-p)T]m[(k-q)T]\right]$$

$$= a_p[i] - \frac{\mu}{2}\left[-2m(kT)m[(k-p)T] + 2m[(k-p)T]\sum_{q=1}^{N} a_q[i]m[(k-q)T]\right]$$

$$= a_p[i] - \mu m[(k-p)T]\left[-m(kT) + \sum_{q=1}^{N} a_q[i]m[(k-q)T]\right]$$

$$= a_p[i] + \mu m[(k-p)T]e(kT) \qquad (8.65)$$

where

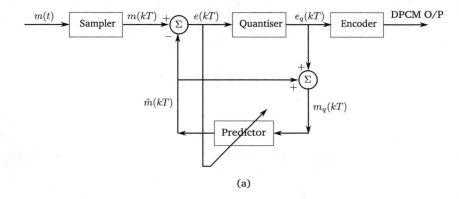

(a)

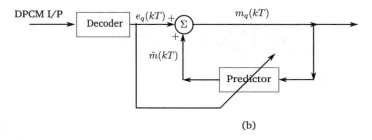

(b)

**Fig. 8.29** DPCM with adaptive prediction

$$e(kT) = m(kT) - \sum_{q=1}^{N} a_q[i]m[(k-q)T]$$
$$= m(kT) - \hat{m}(kT)$$

is the error signal and $\mu$ is the step size. The method resembles the method of steepest descent. DPCM with adaptive filtering is shown in Fig. 8.29.

## 8.10  Comparison Between PCM and ADPCM and Others

For audio computer applications, PCM works in intervals which are at sampling rates of 11, 22 and 44 kHz. In general, the data is quantised to either 16-bit or 8-bit PCM, and 8-bit PCM causes audible hiss in the waveform. It also requires half the disk space.

Adaptive differential PCM (ADPCM) uses only 4 bits per sample, taking up a quarter of the disk space of 16-bit PCM. However, the sound quality is inferior. In ADPCM, the computer first decompresses the signal, which takes up more CPU

| | Parameter | PCM | DM | DPCM |
|---|---|---|---|---|
| 1 | No. of bits | 8 or 16 bits / sample | 1 bit / sample | No of bits about half of PCM |
| 2 | Quantisation Problems | Depends on number of levels | Slope overload and granular noise | Depends on number of levels. Problems because of feedback |
| 3 | Bandwidth | Highest | Lowest for low quality applications. Same as PCM for high quality applications | < that for PCM |
| 4 | SNR due to processing | Good | Poor | Fair |
| 5 | Application | Audio and Video | Audio | Audio and Video |
| 6 | Sampling rate | 8 ks/s for audio | ≈128 ks/s for audio | 8 ks/s for audio |
| 7 | Bit rate | 64 (or 128) kbps for audio | ≈128 kbps for audio | about 32 (or 64) kbps for audio |

**Fig. 8.30**  Comparison between various digital modulation schemes

time. In ADPCM, low frequencies are properly reproduced, but high frequencies get distorted. This distortion is easily audible in 11 kHz ADPCM files, but is almost impossible to recognise with 44 kHz ADPCM files. This implies that sampling rate is very important for sound reproduction. Figure 8.30 shows the comparison between various modulation schemes. In low quality audio applications, DM may be used with lower sampling rates.

## 8.11  Important Concepts and Formulae

- We can write the equation of the PAM signal as

$$\sum_{k=-\infty}^{k=\infty} m(kT) p_s(t - kT - \tau) \tag{8.66}$$

- The FT of a PAM signal is

$$\mathcal{F}\left\{ \sum_{k=-\infty}^{k=\infty} m(kT) p_s(t - kT - \tau) \right\} = \mathcal{F}\{p_s(t - \tau)\} \mathcal{F}\{m_\delta(t)\}$$

$$= 2\omega_s \tau e^{-j\omega\tau} \frac{\sin(\omega\tau)}{\omega\tau} \left[ \sum_{k=-\infty}^{k=\infty} M(\omega - k\omega_s) \right]$$

- The frequency response of the equaliser is

$$H(\omega) = \begin{cases} \left[ \frac{\sin(\omega\tau)}{\omega\tau} \right]^{-1} & -\omega_m \leq \omega \leq \omega_m \\ \text{anything} & \omega < -\omega_m \text{ and } \omega > \omega_m \end{cases}$$

- The quantised signal is

$$m_q(kT) = \Delta q \left\lfloor \frac{m(kT)}{\Delta q} \right\rfloor$$

- for a mid-rise quantiser

$$-\frac{\Delta}{2} \leq q \leq \frac{\Delta}{2}$$

$$q = 0 \qquad \text{elsewhere}$$

we also realise that $q$ should be uniformly distributed,

$$f_Q(q) = \begin{cases} = \frac{1}{\Delta} & -\frac{\Delta}{2} \leq q \leq \frac{\Delta}{2} \\ = 0 & \text{elsewhere} \end{cases}$$

the mean of this distribution is

$$m_Q = \int_{-\Delta/2}^{\Delta/2} q f_Q(q) dq$$

$$= \int_{-\Delta/2}^{\Delta/2} \frac{q}{\Delta} dq$$

$$= 0$$

while the variance,

$$\sigma_Q^2 = \int_{-\Delta/2}^{\Delta/2} \underbrace{q^2}_{\text{sq of error}} \underbrace{f_Q(q)dq}_{\text{probability}}$$

$$= \int_{-\Delta/2}^{\Delta/2} \frac{q^2}{\Delta} dq$$

$$= \frac{\Delta^2}{12}$$

- The SNR of a quantised signal is

$$\text{SNR} = \frac{P}{\sigma_Q^2} = \frac{P}{\left(\frac{\Delta^2}{12}\right)} = \frac{12P}{\left(\frac{2m_{\max}}{L}\right)^2} = \frac{3L^2 P}{m_{\max}^2}$$

- The rate of transmission in PCM is

$$R = SN \quad \text{bits/s} \tag{8.67}$$

where $S$ is the sampling rate in samples/s and $N$ is the number of bits per sample.
- The equation for the $\mu$ law is

$$v = \text{sgn}(u) \frac{\ln(1 + \mu|u|)}{\ln(1 + \mu)}$$

- The inverse transformation for the $\mu$ law is

$$u = \text{sgn}(v) \frac{e^{\ln(\mu+1)|v|} - 1}{\mu}$$

$$= \text{sgn}(v) \frac{(1 + \mu)^{|v|} - 1}{\mu}$$

- The compander function for the A-law is

$$v = \begin{cases} \text{sgn}(u) \frac{A|u|}{1 + \ln A} & 0 \le |u| \le \frac{1}{A} \\ \text{sgn}(u) \frac{1 + A|u|}{1 + \ln A} & \frac{1}{A} \le |u| \le 1 \end{cases}$$

where $A$ is the parameter.

## 8.12   Self-Assessment

### 8.12.1   Review Questions

1. Explain how proper reconstruction is obtained by passing a sampled signal through an LPF of the correct bandwidth (refer Fig. 7.6).
2. Why is it impractical to use sampling with impulses (delta functions)?
3. Why do we study sampling with delta functions?
4. Explain why does the term $e^{-j\omega\tau}$ in Eq. (8.5)

$$\mathcal{F}\left\{\sum_{k=-\infty}^{k=\infty} m(kT)p_s(t-kT-\tau)\right\} = \mathcal{F}\{p_s(t-\tau)\}\,\mathcal{F}\{m_\delta(t)\}$$

$$= 2\omega_s\tau e^{-j\omega\tau}\frac{\sin(\omega\tau)}{\omega\tau}\left[\sum_{k=-\infty}^{k=\infty} M(\omega-k\omega_s)\right]$$

plays no significant role in the reconstruction of the PAM signal.

5. Explain why the equaliser filter for PAM signal must have a frequency response given in Eq. (8.7):

$$H(\omega) = \begin{cases} \left[\frac{\sin(\omega\tau)}{\omega\tau}\right]^{-1} & -\omega_m \le \omega \le \omega_m \\ \text{anything} & \omega < -\omega_m \text{ and } \omega > \omega_m \end{cases}$$

6. In Fig. 8.3, what is the bandwidth of the reconstruction filter with respect to the bandwidth of the anti-aliasing filter and why?
7. Can we do TDM in pulse width modulation or pulse position modulation?
8. What is the advantages of using the $A$ and $\mu$ laws?
9. Where is Delta modulation used?
10. Explain with examples the distortions produced in delta modulation.

### 8.12.2   Numerical Problems

1. Plot the time-domain waveform

$$f(t) = \sum_{k=-\infty}^{\infty} \sin(2\pi \times 10^3 \times kT)p_s(t-kT-\tau)$$

where $T = 2\pi/\omega_s$, and $\omega_s = 2\pi \times 1.1 \times 10^3$ rad/s, $\tau = 10^{-4}$ s
2. Plot the magnitude of the FT of $f(t)$ of Problem 1.
3. Plot the time-domain waveform

$$f(t) = \sum_{k=-\infty}^{\infty} \sin(2\pi \times 10^3 \times kT) p_s(t - kT - \tau)$$

where $T = 2\pi/\omega_s$, and $\omega_s$ is equal to 1.1 times the Nyquist rate of sampling. $\tau = 10^{-4}$ s

4. Plot the magnitude of the FT of $f(t)$ of Problem 3.
5. 10 PAM each of $\omega_m = 2\pi \times 22 \times 10^3$ and $\tau = 10^{-6}$ s, waveforms are to be multiplexed using TDM. Find the time between two PAM pulses, $m_{aPAM}(t)$ and $m_{(a+1)PAM}(t)$.
6. If we require to quantise the signal

$$f(t) = 10 \sin(2\pi \times 10^4 t)$$

to have $\Delta q = 10^{-4}$ how many minimum number of bits are required?
7. In Problem 6, what is the output of the quantiser if the input is 2.5?
8. In Problem 6, what is the output quantisation SNR?
9. Convert the output of the quantiser from Problem 7 into a pulse code modulation string if $-10$ corresponds to $0000...$
10. For an analogue signal of bandwidth 4 MHz, converted into 1024 quantisation levels, what is the transmission rate in bits/s
11. Messages are typed with a keyboard containing 128 symbols. One bit is reserved to check for corrections. The bit rate is 1 MB/s. What is the fastest rate at which the messages may be sent out in symbols/s?
12. A $\mu$ law compander, $\mu = 255$, is used to compand signals of normalised values $0, 0.1, 0.2, ..., 1$. Find the output of the compander.
13. Prove that the inverse transformation of the $\mu$ law is

$$u = \text{sgn}(v) \frac{e^{\ln(\mu+1)|v|} - 1}{\mu}$$

$$= \text{sgn}(v) \frac{(1+\mu)^{|v|} - 1}{\mu} \tag{8.68}$$

14. Using Eq. (8.68), obtain the original values of the compander input taking values of the floor function operated on $v$ quantised to 0.001.
15. A $A$ law compander, $A = 256$, is used to compand signals of normalised values $0, 0.1, 0.2, ..., 1$. Find the output of the compander.
16. Prove that the inverse transformation of the $A$ law is

$$u = \text{sgn}(v) \begin{cases} \frac{|v|[1+\ln A]}{A} & |v| < \frac{1}{1+\ln A} \\ \frac{\exp\{|v|[1+\ln A]-1\}}{A} & 1 > |v| > \frac{1}{1+\ln A} \end{cases} \tag{8.69}$$

17. Using Eq. (8.69), obtain the original values of the compander input taking values of the floor function operated on $v$ quantised to 0.001. Compare the $\mu$ and $A$ law companders.

18. Convert 00011 into a Manchester line code.
19. If the probability of error is to be $10^{-20}$ and $N_0 = 10^{-10}$ J, Find the approximate value of the required energy per bit.
20. If $E_b/N_0 = 10$ calculate the probability of error, $P_e$.
21. In the case of delta modulation for the signal

$$f(t) = 10 \sin(2\pi \times 10^4 t)$$

and $\Delta = 0.01$ find the value of $T$ so that distortion does not take place.
22. Prove that the right-hand side of Eq. (7.11), is a periodic function with period $\omega_s$.

$$M_\delta(\omega) = M(\omega) * \left\{ \omega_s \sum_{k=-\infty}^{k=\infty} \delta\left(\omega - k\omega_s\right) \right\}$$

23. Show, with diagrams, why Eq. (7.13)

$$\frac{\omega_s}{2} \geq \omega_m$$

is the condition for proper reconstruction of the sampled signal.
24. Show why Eq. (7.14).

$$\omega_s - \omega_m \geq \text{BW}_{\text{LPF}} > \omega_m$$

is true
25. Find the frequency response (magnitude and phase) of the equaliser for a PAM signal sampled with pulses of the type $e^{-at} u(t)$ where $a \gg 1$.

# References

1. Deslauriers N, Avdiu BTO (2005) Naturally sampled triangle carrier PWM bandwidth limit and output spectrum. IEEE Trans Power Electron 20:100–106
2. Haykin S (1999) Communication systems. Wiley
3. Lathi BP (1998) Modern digital and analog communication systems, 3rd ed. Oxford University Press

# Chapter 9
# Baseband Pulse Transmission

In digital communication, we often send pulses over a channel with no carrier as it is in the case of analogue signals. One of the possibilities is that the information may be sent over dedicated lines as in the case of a wired LAN. This is called baseband transmission since no carrier is involved. When we transmit this digital data comprising a sequence of bits consisting of zeros and ones, we convert these into electrical signals, in the form of electrical pulses or symbols. Each of these pulses is called a *symbol*. For the case of non return to zero pulses (NRZ), a bit which is 1 is transmitted as $+A$ and a 0 is transmitted as $-A$, where $A > 0$. We transmit the pulses over a medium which may have two types of problems associated with it

1. The medium may be band-limited. This implies that some of the frequencies associated with the string of pulses may be cut off, and the fidelity of the signal may not be maintained. Square pulses may become rounded.
2. The medium may exhibit dispersion. This implies that different frequencies would travel at different velocities, and square pulses may show a spread when they reach the receiver.
3. Apart from these problems, there will further be an addition of noise to the signal which will tend to smear the signal.

## 9.1 Matched Filter

We now look at the detection of single symbol at the receiver end when noise is added to the signal. To this end, when we receive the signal, we pass it through a special filter which maximises the signal-to-noise ratio at a certain pont of time, $t = T_b$. If we observe Fig. 9.1a the input to a detection filter, the received signal, is $x(t)$. $x(t)$ is shown in Fig. 9.1b, and the output signal of the filter, $y(t)$, is given by

© The Author(s), under exclusive license to Springer Nature Singapore Pte Ltd. 2022    397
S. Bhooshan, *Fundamentals of Analogue and Digital Communication Systems*,
Lecture Notes in Electrical Engineering 785,
https://doi.org/10.1007/978-981-16-4277-7_9

**Fig. 9.1** Matched filter
details

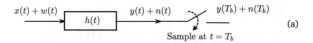

$$y(t) = h(t) * x(t) \tag{9.1}$$

The output signal is sampled at time $t = T_b$ as shown in Fig. 9.1a. The noise which accompanies the signal is white noise, $w(t)$. The output noise of the filter is $n(t)$ where the PSD of $n(t)$ is given by

$$P_{n(t)}(\omega) = \frac{N_0}{2} |H(\omega)|^2 \tag{9.2}$$

We want to design the filter such that we get a maximum signal-to-noise ratio among all possible filters for the output $y(t)$ at $t = T_b$. Maximising signal-to-noise ratio is important since the signal-to-noise ratio, $S/N$, is an indicator of how distorted the signal level is due to noise alone. The higher the signal power, and the lower the noise power, the more easily will the signal be extracted from the noise.

Now

$$
\begin{aligned}
y(T_b) &= \int_{-\infty}^{\infty} x(\tau) h(T_b - \tau) d\tau \\
&= \mathcal{F}^{-1} \{ X(\omega) H(\omega) \} \big|_{t=T_b} \\
&= \frac{1}{2\pi} \int_{-\infty}^{\infty} X(\omega) H(\omega) e^{j\omega T_b} d\omega
\end{aligned}
\tag{9.3}
$$

and

$$y^2(T_b) = \left| \frac{1}{2\pi} \int_{-\infty}^{\infty} X(\omega) H(\omega) e^{j\omega T_b} d\omega \right|^2 \tag{9.4}$$

while the signal-to-noise ratio at the output of the receiver is

$$\frac{y^2(T_b)}{E[n^2(t)]} = \frac{\left|\frac{1}{2\pi}\int_{-\infty}^{\infty} X(\omega)H(\omega)e^{j\omega T_b}d\omega\right|^2}{(N_0/2)\int_{-\infty}^{\infty}|H(\omega)|^2\,d\omega} \tag{9.5}$$

It is important to note that $E[n^2(T_b)] = E[n^2(t)] \; \forall \; t$, for the case of a WSS process. Now Schwarz's inequality states that for any two complex functions, Steele [3]

$$\left|\int a(\omega)b(\omega)d\omega\right|^2 \leq \int |a(\omega)|^2\,d\omega \times \int |b(\omega)|^2\,d\omega \tag{9.6}$$

The inequality has an equal to sign when $b(\omega) = ka^*(\omega)$. ($a(\omega)^*$ is the complex conjugate of $a(\omega)$)

So if $a(\omega) = X(\omega)e^{j\omega T_b}$ and $b(\omega) = H(\omega)$, then

$$\frac{y^2(T_b)}{E[n^2(t)]} \leq \frac{(1/2\pi)\int_{-\infty}^{\infty}|X(\omega)|^2\,d\omega \int_{-\infty}^{\infty}|H(\omega)|^2\,d\omega}{(N_0/2)\int_{-\infty}^{\infty}|H(\omega)|^2\,d\omega} \tag{9.7}$$

and the maximum value that this inequality attains is when

$$H_{opt}(\omega) = kX^*(\omega)e^{-j\omega T_b} \tag{9.8}$$

so

$$h_{opt}(t) = \frac{k}{2\pi}\int_{-\infty}^{\infty} X^*(\omega)e^{j\omega(t-T_b)}d\omega \tag{9.9}$$

but for a real signal, $X^*(\omega) = X(-\omega)$, so

$$h_{opt}(t) = \frac{k}{2\pi}\int_{-\infty}^{\infty} X(-\omega)e^{j\omega(t-T_b)}d\omega$$

$$= kx(T_b - t) \tag{9.10}$$

Now that we know the optimum impulse response of the receiving filter, the waveform of the optimum filter impulse response is shown in Fig. 9.1c for the pulse shown in Fig. 9.1b. $h_{opt}(t)$ is a reflected version of $x(t)$ and shifted to the right by $T_b$. For corroboration,

$$x(T_b - t)|_{t=0} = x(T_b)$$

and

$$x(T_b - t)|_{t=T_b} = x(0)$$

For further analysis, substituting Eq. (9.8) in (9.7),

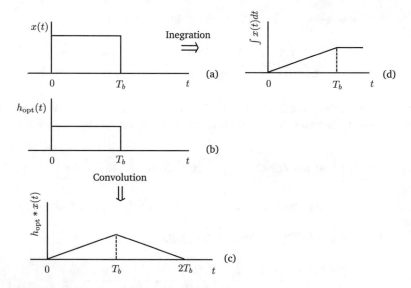

**Fig. 9.2** Matched filter approach for a rectangular pulse

$$\frac{y^2(T_b)}{E[n^2(t)]}\bigg|_{\max} = \frac{(1/2\pi)\int_{-\infty}^{\infty}|X(\omega)|^2\,d\omega\int_{-\infty}^{\infty}\left|H_{\text{opt}}(\omega)\right|^2 d\omega}{(N_0/2)\int_{-\infty}^{\infty}\left|H_{\text{opt}}(\omega)\right|^2 d\omega}$$

$$= \left(\frac{2}{N_0}\right)\left(\frac{1}{2\pi}\right)\underbrace{\int_{-\infty}^{\infty}|X(\omega)|^2\,d\omega}_{\text{Energy per symbol}} \tag{9.11}$$

where the equal to sign implies that we have used the optimum filter. But

$$E = \left(\frac{1}{2\pi}\right)\int_{-\infty}^{\infty}|X(\omega)|^2\,d\omega$$

is the energy per symbol by Eq. (2.132) (Parseval's or Raleigh's energy theorem). Therefore,

$$\frac{y^2(T_b)}{E[n^2(t)]}\bigg|_{\max} = \frac{2E}{N_0} \tag{9.12}$$

**Example 9.1** Discuss the matched filter for the rectangular pulse.

Consider the rectangular pulse shown in Fig. 9.2a

$$x(t) = \begin{cases} A & 0 \le t \le T_b \\ 0 & \text{elsewhere} \end{cases} \tag{9.13}$$

and the matched filter impulse response, $h(t)$, is given by

$$h(t) = kx(T_b - t)$$
$$= kx(t) \tag{9.14}$$

which applies only to the rectangular pulse, and which is shown in Fig. 9.2b. The convolution of the two functions is shown in Fig. 9.2c. The output is sampled at $t = T_b$ when the output reaches a maximum.

However, to obtain an impulse response as shown in Fig. 9.2b, the circuitry required would be quite complex. But only in the case of a rectangular pulse, we may employ another technique. Looking at the shape of the filter output, we realise that may simply use an integrator instead of using a matched filter and sample the waveform at $t = T_b$. An integrator, as we know, is much easier to implement. The manner in which we proceed is that the integrator is initialised to zero at $t = 0$, then integration is performed; at $t = T_b$, the output of the integrator is sampled after which the integrator is reinitialised.

## 9.2 Error Due to Noise

When we transmit a bit sequence using a NRZ waveform shown in Fig. 8.18e, a positive pulse is transmitted denoting a '1' and a negative pulse is transmitted denoting a '0'. These can be represented mathematically as

$$s(t) = \begin{cases} A & \text{for a 1} \\ -A & \text{for a 0} \end{cases}$$

Now the symbols are corrupted by additive white Gaussian noise.

$$x(t) = s(t) + w(t) = \begin{cases} +A + w(t) & \text{a ``1'' was sent} \\ -A + w(t) & \text{a ``0'' was sent} \end{cases} \tag{9.15}$$

where the symbol received is corrupted by $w(t)$ and which is added to the symbol in generation and transmission.

For example, if the sequence '1001' is transmitted as pulses, then a sequence of pulses consisting of a positive pulse followed by two negative pulses and then a positive pulse are generated. Figure 9.3 shows this sequence of pulses with the addition of white noise. White Gaussian noise is also shown in the same figure as $w(t)$. It is obvious that

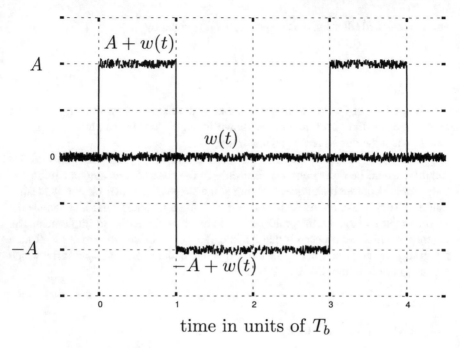

**Fig. 9.3** White noise and signal plus white noise

$$E[+A + w(t)] = A + E[w(t)] = A$$
$$\text{and } E[-A + w(t)] = -A + E[w(t)] = -A \qquad (9.16)$$

where the mean of the white noise is zero.

Now the incoming signal is integrated over the period ($T_b$) of the pulse and nor-malised, to it, for the symbol 1,

$$y(T_b) = \frac{1}{T_b}\left[\int_0^{T_b} \{A dt + w(t)\}\, dt\right]$$
$$= A + \frac{1}{T_b}\int_0^{T_b} w(t) dt$$

Now from Sect. 3.18, $y(T_b)$ is a Gaussian random variable. The Gaussian random variable $y$ has a mean of $A$ and a variance given by

$$E[(Y - A)^2] = E[Y^2(T_b) - 2Y(T_b)A - A^2]$$

$$= E\left[\left(A + \frac{1}{T_b}\int_0^{T_b} w(t)dt\right)^2 - 2\left(A + \frac{1}{T_b}\int_0^{T_b} w(t)dt\right)A - A^2\right]$$

$$= E\left[\frac{1}{T_b^2}\int_0^{T_b}\int_0^{T_b} w(t_1)w(t_2)dt_1dt_2\right]$$

$$= \frac{1}{T_b^2}\int_0^{T_b}\int_0^{T_b} E[w(t_1)w(t_2)]dt_1dt_2$$

$$= \frac{1}{T_b^2}\int_0^{T_b}\int_0^{T_b} R_W(t_1 - t_2)\, dt_1dt_2 \tag{9.17}$$

where $R_W(t_1 - t_2)$ is the autocorrelation function of white noise and

$$R_W(t_1 - t_2) = \frac{N_0}{2}\delta(t_1 - t_2) \tag{9.18}$$

so

$$\sigma_Y^2 = E[(Y - A)^2]$$

$$= \frac{N_0}{2T_b^2}\int_0^{T_b}\int_0^{T_b}\delta(t_1 - t_2)\, dt_1dt_2$$

$$= \frac{N_0}{2T_b^2} \tag{9.19}$$

The pdf of $Y$ given that the symbol is 1 is Gaussian. Therefore, from Eq. (3.150),

$$f_Y(y|\text{symbol }1) = \sqrt{\frac{T_b}{\pi N_0}}e^{-\frac{(y-A)^2}{N_0/T_b}} \tag{9.20}$$

Similarly, the pdf of $Y$ given that the symbol is 0 is

$$f_Y(y|\text{symbol }0) = \sqrt{\frac{T_b}{\pi N_0}}e^{-\frac{(y+A)^2}{N_0/T_b}} \tag{9.21}$$

If we plot these two probability distribution functions, we find that they cross over at $y = 0$ as shown in Fig. 9.4. This implies that sometimes a zero will be sent and it will be interpreted as a 1 and also the other way around. The greater the value of $A$, the chances of this happening will be reduced; also the smaller the value of $\sigma = \sqrt{2N_0/T_b}$, the same effect will be held. The block diagram of the system is shown in Fig. 9.5. Here, the decision-making device decides that the symbol is a 1 if the value of $y(T_b) > 0$ and 0 if $y(T_b) < 0$.

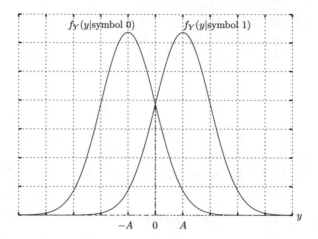

**Fig. 9.4** Probability of error calculation

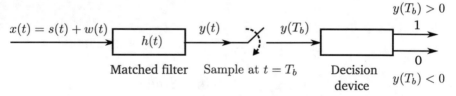

**Fig. 9.5** Block diagram of the decision-making whether the symbol is a 0 or a 1

To investigate the probability of error (from the graph),

$$P_{e0}(0 \text{ was sent assumed it to be 1}) = \sqrt{\frac{T_b}{\pi N_0}} \int_0^\infty e^{-\frac{(y+A)^2}{N_0/T_b}} \, dy$$

To do this integral, we make the following substitutions:

$$z = \frac{y+A}{\sqrt{N_0/T_b}}$$

$$\Rightarrow dz = \frac{dy}{\sqrt{N_0/T_b}}$$

and the integral becomes

$$P_{e0} = \frac{1}{\sqrt{\pi}} \int_{z_0}^\infty e^{-z^2} \, dz$$

where

$$z_0 = \sqrt{\frac{A^2 T_b}{N_0}}$$

We now set $E_b = A^2 T_b$ as the energy per bit. Then

$$z_0 = \sqrt{\frac{E_b}{N_0}}$$

$$P_{e0} = \frac{1}{\sqrt{\pi}} \int_{z_0}^{\infty} e^{-z^2} dz$$

$$= \frac{1}{2} \text{erfc} \left( \sqrt{\frac{E_b}{N_0}} \right)$$

where erfc is the complementary error function:

$$\text{erfc}(x) = \frac{2}{\sqrt{\pi}} \int_x^{\infty} e^{-z^2} dz$$

Similarly, from the graph 9.4, it is clear that

$$P_{e1}(1 \text{ was sent assumed it to be } 0) = P_{e1} = P_{e0}$$

To now calculate the probability that an error occurs

$$P_e = P_{e1} p_1 + P_{e0} p_0$$

where $p_1$ is the probability of occurrence of a 1 ($=1/2$), and $p_0$ is the probability of occurrence of a 0 ($=1/2$). Then

$$P_e = \frac{1}{2} \text{erfc} \left( \sqrt{\frac{E_b}{N_0}} \right)$$

The plot of $P_e$ versus $E_b/N_0$ is shown in Fig. 9.6. A good approximation to the erfc(.) is given by Chiani [1]

$$\text{erfc}(x) \cong \frac{1}{6} e^{-x^2} + \frac{1}{2} e^{-\frac{4}{3}x^2} \quad x > 0 \tag{9.22}$$

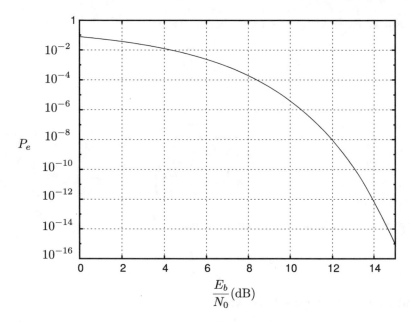

**Fig. 9.6**  Probability of error versus $E_b/N_0$ for NRZ coding

## 9.3  Intersymbol Interference

### 9.3.1  Introduction

In a digital communication system, we use symbols to communicate. In binary communication, these symbols are zero and one. Intersymbol interference (ISI) is a distortion of the signal whereby symbols spread in their width and interfere with the previous and subsequent symbols. ISI is the result of *multipath* propagation and channel *dispersion*. Figure 9.7a shows a typical spread of a single symbol as it passes through a channel, while Fig. 9.7b shows the effect of ISI on a series of pulses. The main reason why intersymbol interference arises is the *conflicting demand that the symbols be both time-limited and band-limited*. Note that if the channel is *not band-limited*, intersymbol interference *rarely* takes place; as when the channel is a two wire line with (theoretically) infinite bandwidth.

To illustrate this point, let us pass the signal

$$x(t) = \frac{2\sin t}{t}$$

whose Fourier Transform is

$$X(\omega) = \begin{cases} 2\pi & -1 \le \omega \le 1 \\ 0 & \text{elsewhere} \end{cases}$$

**Fig. 9.7**  Figure showing the effects of intersymbol interference

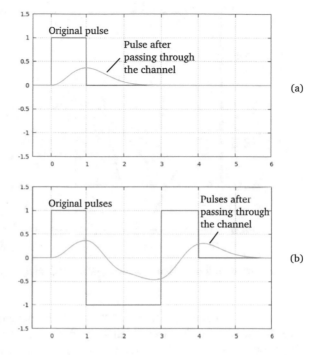

(a)

(b)

through a channel which has the spectrum

$$H_1(\omega) = \begin{cases} 1 & -1 \le \omega \le 1 \\ 0 & \text{elsewhere} \end{cases}$$

which implies that the signal will be undistorted:

$$y_1(t) = 2 \sin t / t$$

We next pass the same signal through the channel with only half the bandwidth

$$H_2(\omega) = \begin{cases} 1 & -0.5 \le \omega \le 0.5 \\ 0 & \text{elsewhere} \end{cases}$$

and we find that the output, $y_2(t)$, will be

$$y_2(t) = \frac{\sin(t/2)}{t/2}$$

This effect is shown in Fig. 9.8, where the pulse broadening is clearly visible.

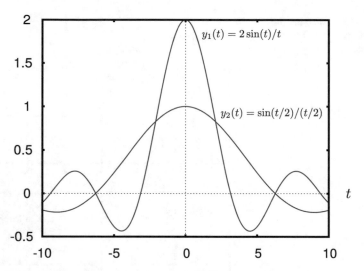

**Fig. 9.8** Two band-limited signals showing pulse broadening

One obvious solution to fight intersymbol interference is to slow down the symbol rate. In the slowed down case, each symbol when it spreads *does not reach its neighbour.* In this case, the impulse response is materially short enough so that when the spread takes place, this does not happen. However, we cannot slow down the symbol rate to be *lower than the Nyquist rate, since then we would not be able to reconstruct the original analogue signal.*

Another solution would be to separate symbols in time with guard time periods. But as the symbols get narrower, their spectra get wider, so due to the finite bandwidth of the channel, distortion again takes place. So we are stuck with the ISI problem. Let us now study this problem in some detail.

Let us look at the digital communication system shown in Fig. 9.9. In this block diagram, $h_t(t)$, $h_c(t)$ and $h_r(t)$ are the impulse responses of the transmit filter, the channel and the receive filter. $x(t)$ is a series of pulses (approximating impulses), of rate $R_b$ symbols/sec (which is generally bits/sec), so that the time between symbols is

$$T_b = \frac{1}{R_b} \ (\text{sec})$$

$$\omega_b = 2\pi R_b = \frac{2\pi}{R_b} \tag{9.23}$$

where $\omega_b$ is introduced as a radian measure. Therefore,

$$x(t) \cong \sum_k a_k \delta(t - kT_b)$$

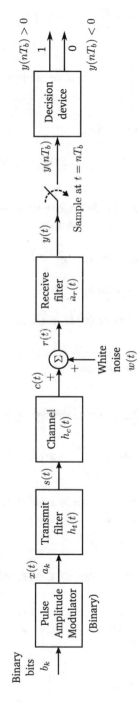

**Fig. 9.9** Block diagram of a typical digital communication system

and therefore,

$$s(t) = \sum_k a_k h_t(t - kT_b) \tag{9.24}$$

where

$$a_k = \begin{cases} +1 & \text{if the symbol is a 1} \\ -1 & \text{if the symbol is a 0} \end{cases}$$

after the symbols pass through the system,

$$y(t) = \sum_k a_k p(t - kT_b) + n(t) \tag{9.25}$$

where $n(t)$ is white noise after passing through the receive filter. Not considering the noise part at present, therefore, in the time domain,

$$p(t) = h_t(t) * h_c(t) * h_r(t)$$

which reflects in the transform domain as

$$P(\omega) = H_t(\omega) H_c(\omega) H_r(\omega) \tag{9.26}$$

The waveform $y(t)$ is sampled every $t = iT_b$ seconds, and the decision is has to made whether the symbol was a zero or a one.

   Now from Eq. (9.25),

$$y(t_i = iT_b) = \sum_k a_k p(iT_b - kT_b) + n(iT_b)$$

$$= a_i p(0) + \sum_{k,\ k \neq i} a_k p(iT_b - kT_b) + n(iT_b) \tag{9.27}$$

In this equation, the first term is what we want, the $i$th bit, and the other terms consist of intersymbol interference and noise.

### 9.3.2   *Nyquist Criterion and Nyquist Channel*

From examining Eq. (9.27), it is clear that $p[(i - k)T_b]$, $i \neq k$, is present as inter-symbol interference in $y(iT_b)$. Nyquist [2] suggested the following. Let $p(t)$ be designed in the following manner:

$$p(mT_b) = \begin{cases} 1 & m = 0 \\ 0 & m = \pm1, \pm2 \dots \end{cases}$$

then, in this case, going over to the frequency domain and sampling $p(t)$ at times $t = mT_b$,

$$P_\delta(\omega) = \int_{-\infty}^{\infty} p(t) \sum_{m=-\infty}^{\infty} \delta(t - mT_b)e^{-j\omega t} dt$$

$$= \omega_b \sum_{m=-\infty}^{\infty} \delta(\omega - m\omega_b)] * P(\omega)$$

$$= \omega_b \sum_{m=-\infty}^{\infty} P(\omega - m\omega_b) \tag{9.28}$$

where we have used Eq. (7.7). In this equation,

$$\omega_b = (2\pi / T_b)$$

Taking a different track,

$$P_\delta(\omega) = \int_{-\infty}^{\infty} p(t) \sum_{m=-\infty}^{\infty} \delta(t - mT_b)e^{-j\omega t} dt$$

$$= \int_{-\infty}^{\infty} [\sum_{m=-\infty}^{\infty} p(mT_b)\delta(t - mT_b)]e^{-j\omega t} dt$$

$$= \int_{-\infty}^{\infty} p(0)\delta(t)e^{-j\omega t} dt \quad [\because p(mT_b) = 0 \text{ for } m \neq 0] \tag{9.29}$$

$$= p(0) \tag{9.30}$$

So

$$\omega_b \sum_{m=-\infty}^{\infty} P(\omega - m\omega_b) = p(0) \tag{9.31}$$

or

$$\sum_{m=-\infty}^{\infty} P(\omega - m\omega_b) = \frac{T_b p(0)}{2\pi} = \text{a constant}$$

The next question which arises is how do we interpret these results? Let us take a look at Fig. 9.10, where we have drawn the function $P_\delta(\omega)$. As we know that $P_\delta(\omega)$ consists of the series of constant functions shifted to the right and left by $n\omega_b$ as shown in the figure.

$$P(\omega) = \begin{cases} k & -\omega_b/2 \leq \omega \leq \omega_b/2 \\ 0 & \text{elsewhere} \end{cases} \tag{9.32}$$

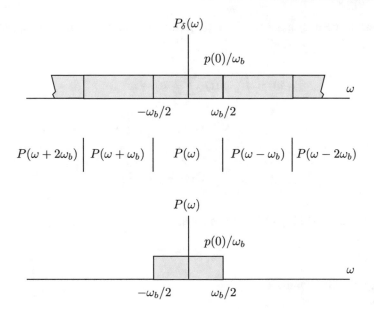

**Fig. 9.10** Spectrum of the sampled pulse

where $k$ is a constant. Taking the inverse FT of this function, we have

$$p(t) = \text{sinc}\left(\frac{\omega_b t}{2}\right)$$

This function is unity at $t = 0$ and has zeros at

$$\omega_b t_n = 2n\pi \quad n = \pm 1, \pm 2, \ldots$$
$$t_n = nT_b \quad n = \pm 1, \pm 2, \ldots$$

*but it is not time-limited.*

The shape of the pulse spectrum also defines the *ideal Nyquist channel*:

$$H_N(\omega) = \begin{cases} 1 & -W \leq \omega \leq W (= \omega_b/2) \\ 0 & \text{elsewhere} \end{cases}$$

and the pulse passes through the channel without distortion. Figure 9.11a shows the pulse $\text{sinc}(\omega_b t/2)$, where the time coordinate is shown in units of $T_b$. Notice that at times $t = nT_b$, $n \neq 0$, the function is always zero. In (b) part of the figure, the bit sequence 1001 is shown in terms of the Nyquist pulses, and below are shown the sampling times, where we may note that though the pulses interfere with each

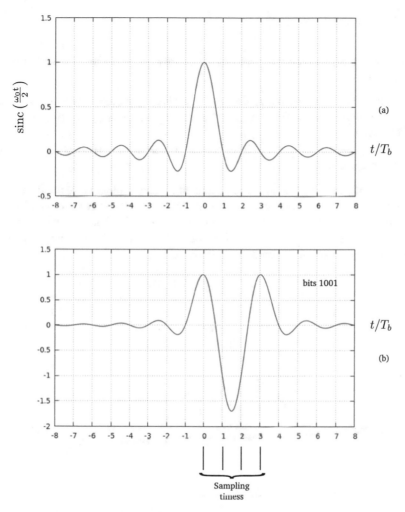

**Fig. 9.11** Nyquist's pulses, $p(t)$, **a** the function sinc($\omega_b t/2$), **b** Four bits, 1001 renderd as a sum of Nyquist pulses

other, we get the correct desired output of 1001. It is interesting to corroborate that at $t = 0$, $T_b$, $2T_b$ and $3T_b$, the function is always either $+1$ or $-1$, and at all other times of $t = nT_b$, it is zero.[1] The channel bandwidth is the minimum possible, namely

$$W = \omega_b/2 = \pi/T_b$$

---

[1] Explain why this statement is true (see Problem 1 in Review Questions).

### 9.3.3   A Simple Filter

Though the analysis which is given above is very good, the hurdles which we will face are

1. That *we will not be able to design these kinds of pulses, due to the sharpness of the frequency spectrum of* $P(\omega)$ at the edges, given by Eq. (9.26) shown in Fig. 9.10.
2. These pulses occupy an infinite extent in time from $t = -\infty$ to $t = \infty$ and, therefore, are non-causal.

However, we can use these results as a launching pad to obtain a realistic design.

Many functions satisfy the requirement that their value is zero (or small) at $t = nT_b$, $n \neq 0$. A particularly simple example is shown in Fig. 9.12, where $P(\omega)$ is shown in the upper part of the figure while $p(t)$ is shown in the lower part of the figure. The spectrum of the pulse is confined to

$$-1.5W \leq \omega \leq 1.5W$$

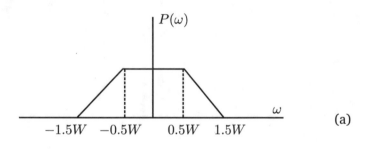

(a)

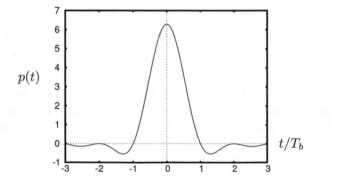

(b)

**Fig. 9.12**   Example function, **a** $P(\omega)$, with **b** $p(t)$

which implies that the channel bandwidth, $1.5W$, is $(3/2)(\omega_b/2) = 3\omega_b/4$. In the most general form, the filter must satisfy

$$P(\omega) + P(\omega - 2W) + P(\omega + 2W) = \begin{cases} K \text{ (constant)} & \text{for } -W \le \omega \le W \\ 0 & \text{otherwise} \end{cases}$$

(9.33)

and one of the solutions to this problem is that shown in the figure.
$p(t)$ is given by

$$p(t) = K \times \left[ -\frac{2\cos\left(\frac{3tW}{2}\right) - 2\cos\left(\frac{tW}{2}\right)}{t^2 W} \right] \quad W = \pi/T_b \qquad (9.34)$$

which has zeros at

$$t = nT_b, \; n \ne 0 \text{ and } n = \ldots -2, -1, 1, 2 \ldots$$

and sampling is done every $T_b$ seconds. The impulse response is also suitably very short, which is proportional to $1/t^2$.

## 9.3.4   Raised Cosine Filter

A frequency spectrum of a filter used in practice is the raised cosine function, which is given by

$$P(\omega) = \begin{cases} \frac{1}{2W}\left[1 + \cos\left(\frac{\pi\omega}{2W}\right)\right] & -2W \le \omega \le 2W \\ 0 & \text{elsewhere} \end{cases}$$

(9.35)

for which the time-domain pulse is

$$p(t) = K\frac{\pi^2 \sin(2tW)}{W\left(\pi^2 t - 4t^3 W^2\right)} \quad W = \pi/T_b \qquad (9.36)$$

$$= 4K\,\text{sinc}\,(tW)\left[\frac{\cos(tW)}{1 - \left(\frac{2tW}{\pi}\right)^2}\right] \qquad (9.37)$$

The spectrum and time-domain response are shown in Fig. 9.13. The interesting part about this pulse is that it decays as $1/t^3$ and the possibility of ISI is drastically minimised. In the last two examples, the channel bandwidth is more than the Nyquist bandwidth, $W = \omega_b/2$, the price which has to be paid for improved performance.

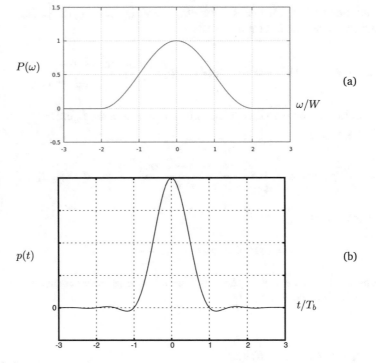

**Fig. 9.13** Raised cosine spectrum, **a** spectrum, **b** time-domain pulse

### 9.3.4.1   General Raised Cosine Filter

The general form of the raised cosine filter is

$$
P(\omega) =
\begin{cases}
\frac{1}{2W} & -\omega_1 \le \omega \le \omega_1 \\[2mm]
\frac{1}{4W}\left\{1 - \sin\left[\frac{\pi(|\omega|-W)}{2(W-\omega_1)}\right]\right\} & \omega_1 \le |\omega| \le 2W - \omega_1 \\[2mm]
0 & \text{elsewhere}
\end{cases}
\tag{9.38}
$$

where

$$
\alpha = 1 - \frac{\omega_1}{W}
$$

$$
\omega_1 = W(1 - \alpha)
$$

For $\alpha = 0$, $\omega_1 = W$, then the equation reduces to

$$
P(\omega) =
\begin{cases}
\frac{1}{2W} & -W \le \omega \le W \\[2mm]
0 & \text{elsewhere}
\end{cases}
$$

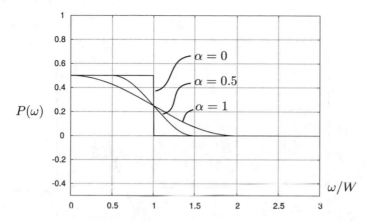

**Fig. 9.14** Generalised raised cosine spectrum

For $\alpha = 0.5$, $\omega_1 = 0.5W$, Eq. (9.38) is used. And for $\alpha = 1$, $\omega_1 = 0$, the equation is

$$P(\omega) = \begin{cases} \frac{1}{4W} \left\{ 1 - \sin\left[ \frac{\pi(|\omega|-W)}{2W} \right] \right\} & 0 \le |\omega| < 2W \\ 0 & \text{elsewhere} \end{cases}$$

The plots are shown in Fig. 9.14. The bandwidth of each filter is $W = \omega_b/2$ for $\alpha = 0$, $1.5W = 3\omega_b/4$ for $\alpha = 0.5$ and $2W = \omega_b$ for $\alpha = 1$, where

$$W = \frac{\pi}{T_b}$$

or

$$T_b = \frac{\pi}{W}$$

The impulse response is given by

$$p(t) = \left[ \frac{\sin(\pi t/T_b)}{\pi t/T_b} \right] \left[ \frac{\cos(\alpha \pi t/T_b)}{1 - (2\alpha t/T_b)^2} \right] \tag{9.39}$$

The figure showing the impulse response is shown in Fig. 9.15. We can conclude from the figure that as $\alpha$ increases, the response becomes more contained.

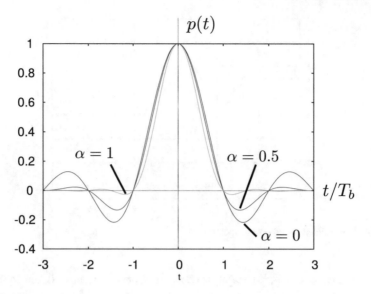

**Fig. 9.15** Raised cosine impulse response with $\alpha = 0, 0.5, 1$

## 9.4 Correlative Level Coding

### 9.4.1 Duobinary Signalling

Normally, the pulses which are transmitted on a channel are uncorrelated to each other. We now consider a new method of transmission where the pulses are correlated with the previous pulse to improve the performance of a system. For example, in *duobinary signalling*, we take a set of impulse like pulses, *which can be referred to a baseband binary PAM system* $\{a_k \equiv a(kT_b), \; k = 0, 1, \ldots\}$ *where*

$$a_k = \begin{cases} +1 & \text{for a 1} \\ -1 & \text{for a 0} \end{cases} \tag{9.40}$$

and obtain a new set of correlated pulses

$$c_k = a_k + a_{k-1} \tag{9.41}$$

This is shown in Fig. 9.16. The z transform of this equation is

$$C(z) = \left(1 + z^{-1}\right) A(z) \tag{9.42}$$

and the transfer function is

$$H(z) = 1 + z^{-1}$$

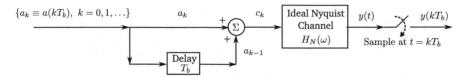

**Fig. 9.16** Duobinary signalling block diagram

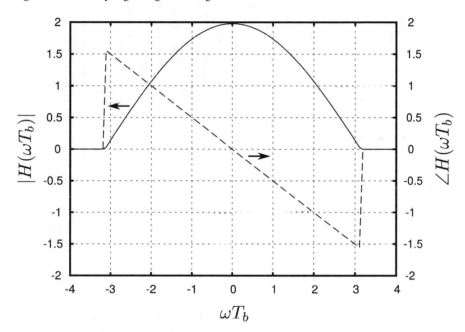

**Fig. 9.17** Duobinary frequency response

or

$$H(\omega T_b) = 1 + e^{-j\omega T_b}$$

$$= 2\cos\left(\frac{\omega T_b}{2}\right) e^{-j\omega \frac{T_b}{2}} \tag{9.43}$$

The frequency response is shown in Fig. 9.17.

If the pulses are like impulses, then

$$h(t) = \delta(t) + \delta(t - T_b) \tag{9.44}$$

is the impulse response of the correlator. These pulses then traverse an ideal Nyquist channel, the output of which in time domain is

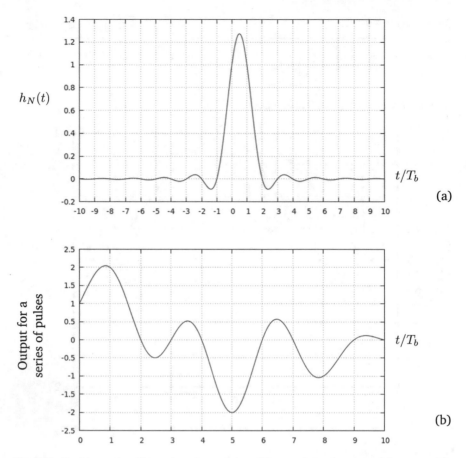

**Fig. 9.18** Duobinary signalling, **a** impulse response of the correlator plus channel, **b** output of the channel for the bit sequence 11010010

$$h_N(t) = \frac{\sin\left(\frac{\pi(t-T_b)}{T_b}\right)}{\frac{\pi(t-T_b)}{T_b}} + \frac{\sin\left(\frac{\pi t}{T_b}\right)}{\frac{\pi t}{T_b}}$$

$$= \frac{\sin\left(\frac{\pi t}{T_b}\right) T_b^2}{\pi t \, (T_b - t)} \tag{9.45}$$

$$= \mathrm{sinc}\left(\frac{\pi t}{T_b}\right) \frac{T_b}{T_b - t} \tag{9.46}$$

where $h_N(t)$ is the impulse response of the correlator plus the Nyquist channel.

The impulse response, $h_N(t)$, is shown in Fig. 9.18a where it is clear that the signal decays as $1/t^2$ and takes the value of $+1$ at the two points, $t = 0$ and $t = T_b$. At all other points, $t = kT_b$, $k = -1, \pm 2, \pm 3 \ldots$, the value is zero. The chief advantage

**Table 9.1** Table showing the detection of bits 11010010

| $k$ | 0 | 1 | 2 | 3 | 4 | 5 | 6 | 7 | – | – |
|---|---|---|---|---|---|---|---|---|---|---|
| Bits | 1 | 1 | 0 | 1 | 0 | 0 | 1 | 0 | – | – |
| $a_k$ | +1 | +1 | −1 | +1 | −1 | −1 | +1 | −1 | – | – |
| $y_k$ | 1 | 2 | 0 | 0 | 0 | −2 | 0 | 0 | −1 | 0 |
| $\langle a \rangle_k = y_k - \langle a \rangle_{k-1}$ | 1 | 1 | −1 | 1 | −1 | −1 | 1 | −1 | 0 | 0 |

of correlative level coding is that the channel bandwidth is maintained at the ideal Nyquist bandwidth, $W$.

**Example 9.2** As an example of the use of this coding, if we want to transmit the set of bits, 11010010, then we transmit

$$\{a_k\}, \quad k = 0 \ldots 7$$

We then receive

$$y(t) = \sum_{k=0}^{7} a_k h_N (t - k T_b) \tag{9.47}$$

shown in Fig. 9.18b.

Table 9.1 shows the transfer of the bit sequence shown in the second row. $y_k$ is read from the (b) part of the figure, which is $-2 \leq y_k \leq +2$, and $y_k$ is an integer. The third row shows $\langle a \rangle_k = y_k - \langle a \rangle_{k-1}$ where $\langle a \rangle_k$ is the estimate of $a_k$. The last two columns show that bits are no longer being transmitted, and $\langle a \rangle_k = 0$ for those columns.

The major drawback of this scheme is that if there is an error in any one of the estimates of $a_k$, then the error propagates for the later values of $\langle a \rangle_k$.

### 9.4.2 Modified Duobinary Signalling

The impulse response for which is

$$h_{\text{MDB}}(t) = \delta(t) - \delta(t - 2T_b) \tag{9.48}$$

The frequency response for modified duobinary signalling is

$$H(\omega T_b) = 1 - e^{-j2\omega T_b}$$
$$= 2j \sin(\omega T_b) e^{-j\omega T_b}$$

which is shown in Fig. 9.20.

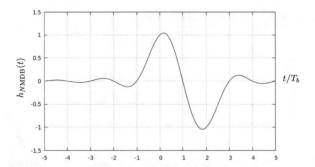

**Fig. 9.19**  Modified duobinary signalling impulse response, $h_{N\text{MDB}}(t)$

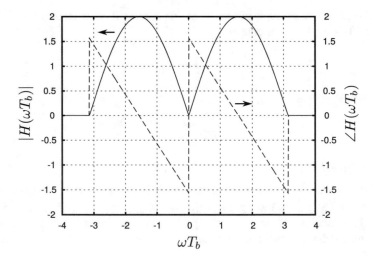

**Fig. 9.20**  Frequency response for modified duobinary signalling

The output of the ideal Nyquist channel is

$$
\begin{aligned}
h_{N\text{MDB}}(t) &= \frac{\sin\left(\frac{\pi t}{T_b}\right) T_b}{\pi t} - \frac{\sin\left(\frac{\pi (t-2T_b)}{T_b}\right) T_b}{\pi (t - 2T_b)} \\
&= -\frac{t T_b \sin\left(\frac{2\pi T_b - \pi t}{T_b}\right) - 2\sin\left(\frac{\pi t}{T_b}\right) T_b^2 + t \sin\left(\frac{\pi t}{T_b}\right) T_b}{2\pi t T_b - \pi t^2} \\
&= \frac{2\sin\left(\frac{\pi t}{T_b}\right) T_b^2}{\pi t \, (2T_b - t)} & (9.49) \\
&= 2\text{sinc}\left(\frac{\pi t}{T_b}\right)\left(\frac{T_b}{2T_b - t}\right) & (9.50)
\end{aligned}
$$

The impulse response, using the modified duobinary signalling, is given by Eq. (9.49), and a plot of the impulse response is shown in Fig. 9.19. The impulse response at the time instants $t = kT_b$ shows that it takes the values $+1$ and $-1$ at $k = 0$ and 2, while at all other integer values, it is zero. The advantage of the modified approach is that the dc value of the signal is zero since some communication channels cannot transmit a dc component (Fig. 9.20).

### 9.4.3 Precoding

The problem faced in duobinary and modified duobinary signalling is that if an error in the detection of the output is committed at any stage, it is carried forward and all the subsequent bits are in error. To take care of this problem, we do what is called as precoding of the data bits.

Referring to Fig. 9.21, the input to the precoder is the message bits, $\{b_k\}$, and the output is $\{d_k\}$. In duobinary signalling with precoding,

$$d_k = b_k \oplus d_{k-1} \tag{9.51}$$

where $\oplus$ is the *exclusive or* operation. One extra bit is required to be assumed, namely $d_{-1}$ which may added to the $\{d_k\}$ symbols which are generated from this Eq. (9.51). The output of the precoder is 0 or 1. These symbols are used as input to the PAM block, the output of which are the set of symbols $\{a_k\}$ which is given earlier by Eq. (9.40). The precoder circuit is shown in Fig. 9.22.

The truth table of the exclusive or operation is

| $a$ | $b$ | $a \oplus b$ |
|-----|-----|--------------|
| 0   | 0   | 0            |
| 1   | 0   | 1            |
| 0   | 1   | 1            |
| 1   | 1   | 0            |

**Example 9.3** Let us use the duobinary correlator with precoding for the bit sequence 11010010. Assuming $d_{-1} = 1$, then the output of the precoder is given in the following table:

| $k$ | $-1$ | 0 | 1 | 2 | 3 | 4 | 5 | 6 | 7 |
|-----|------|---|---|---|---|---|---|---|---|
| $b_k$ | | 1 | 1 | 0 | 1 | 0 | 0 | 1 | 0 |
| $d_k = b_k \oplus d_{k-1}$ | 1 | 0 | 1 | 1 | 0 | 0 | 0 | 1 | 1 |
| $a_k$ | $+1$ | -1 | $+1$ | $+1$ | $-1$ | $-1$ | $-1$ | $+1$ | $+1$ |
| $c_k = a_k + a_{k-1}$ | | 0 | 0 | $+2$ | 0 | $-2$ | $-2$ | 0 | $+2$ |
| $y_k$ | | 0 | 0 | $+2$ | 0 | $-2$ | $-2$ | 0 | $+2$ |
| $\langle b_k \rangle$ | | 1 | 1 | 0 | 1 | 0 | 0 | 1 | 0 |

where $\langle b_k \rangle$ is the estimate of $b_k$, the input binary bits.

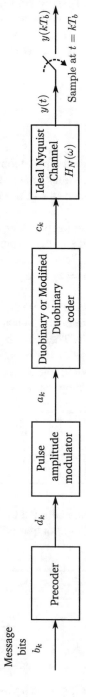

**Fig. 9.21** Duobinary signalling and modified duobinary signalling with precoding

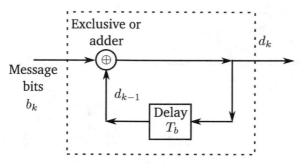

Precoder for duobinary correlator

**Fig. 9.22** Precoder

Using this scheme, the output $y_k$ has the possible outputs

$$y_k = \begin{cases} 0 & \text{if message bit } b_k = 1 \\ \pm 2 & \text{if message bit } b_k = 0 \end{cases} \tag{9.52}$$

For the modified duobinary scheme, instead of using Eq. (9.51), we use

$$d_k = b_k \oplus d_{k-2}$$

where two bits are required to be assumed, $d_{-2}$ and $d_{-1}$. The precoder circuitry is shown in Fig. 9.22. The output of the sampler after the Nyquist channel is then

$$y_k = \begin{cases} \pm 2 & \text{if message bit } b_k = 1 \\ 0 & \text{if message bit } b_k = 0 \end{cases} \tag{9.53}$$

Since we are assuming an exclusive or operation, these kinds of coding schemes are *nonlinear coding* schemes.

**Example 9.4** Let us use the modified duobinary correlator with precoding for the bit sequence 11010010 (Fig. 9.23). Assuming $d_{-1} = 1$, $d_{-2} = 1$, then the output of the precoder is given in the following table:

| $k$ | $-2$ | $-1$ | 0 | 1 | 2 | 3 | 4 | 5 | 6 | 7 |
|---|---|---|---|---|---|---|---|---|---|---|
| $b_k$ | | | 1 | 1 | 0 | 1 | 0 | 0 | 1 | 0 |
| $d_k = b_k \oplus d_{k-2}$ | 1 | 1 | 0 | 0 | 0 | 1 | 0 | 1 | 1 | 1 |
| $a_k$ | $+1$ | $+1$ | $-1$ | $-1$ | $-1$ | $+1$ | $-1$ | $+1$ | $+1$ | $+1$ |
| $c_k = a_k - a_{k-2}$ | | | $-2$ | $-2$ | 0 | $-2$ | 0 | 0 | $-2$ | 0 |
| $y_k$ | | | $-2$ | $-2$ | 0 | $-2$ | 0 | 0 | $-2$ | 0 |
| $\langle b_k \rangle$ | | | 1 | 1 | 0 | 1 | 0 | 0 | 1 | 0 |

where $\langle b_k \rangle$ is the estimate of $b_k$, the input binary bits.

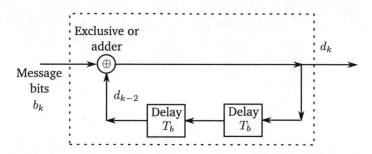

Precoder for modified duobinary correlator

**Fig. 9.23**   Precoder for modified duobinary correlator

### 9.4.4   Generalised Correlative Level Coding

Based on the schemes given above, we can setup a generalised correlative level scheme for bit transfer. The impulse response of the coder shown in Fig. 9.24 is given by

$$h(t) = \sum_{n=0}^{n=N} w_{k-n}\delta(t - nT_b) \quad n = 0, \ldots N; \quad w_{k-n} = 0, \pm 1$$

and the output of the Nyquist channel is

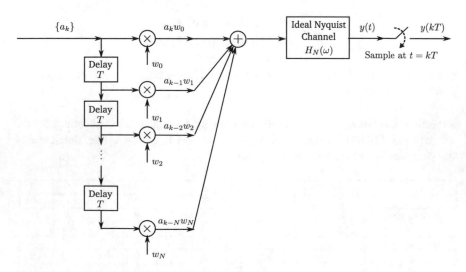

**Fig. 9.24**   Generalised  correlative level coding scheme

$$h_N(t) = \sum_{n=0}^{n=N} w_{k-n} \frac{\sin\left[\frac{\pi(t-nT_b)}{T_b}\right] T_b}{\pi(t-nT_b)} \quad n = 0, \ldots N; \quad w_{k-n} = 0, \pm 1$$

where we can adjust the weights to get any desired response. To get further refinements, we can add a precoder to the scheme.

## 9.5 Eye Diagram

The eye diagram is a tool to evaluate the overall performance of pulses sent over a channel which includes the addition of noise and intersymbol interference.

Figure 9.25 shows such a diagram. Essentially, the eye diagram consists of overlapping traces of the signal over a convenient time period. The various parameters are shown on the figure itself, the diagram being self explanatory.

- The slope of the eye, $m$, indicates the sensitivity to timing errors. The smaller the slope, the lower the possibility of timing errors.
- The width of the eye gives us an interval of time when the signal can be sampled.
- The signal is best sampled at the point shown on the diagram.
- The noise margin of the signal is the height of the eye opening.

The eye diagram using a raised cosine filter is shown in the lower part of the figure and shows us why it is considered superior to other filters (see Sect. 9.3.4).

## 9.6 Optimised Linear Receivers

So far we have considered intersymbol interference (Sect. 9.3) and interfering noise (Sect. 9.1) as two separate phenomena. In practice, these both are present.

### 9.6.1 Zero Forcing Equaliser

We first look at the zero forcing equaliser, which is a brute force method of equalisation. Referring to Fig. 9.9, we see that at the receiver end, the signal which emerges from the channel is $r(t)$ and the output of the receive filter is

$$y(t) = \int_{-\infty}^{\infty} r(\tau) h_r(t - \tau) d\tau$$

where $h_r(t)$ is the impulse response of the receive filter.

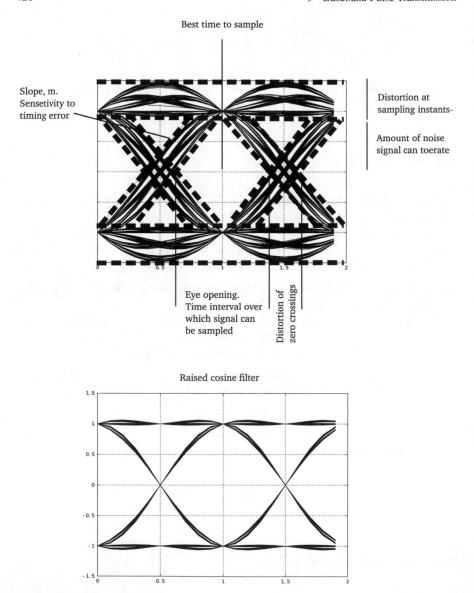

**Fig. 9.25** Eye diagram. Upper diagram, explanation. Lower diagram, developed from using a raised cosine filter

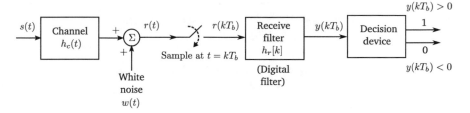

**Fig. 9.26** Zero forcing equaliser

Next, when we refer to Fig. 9.26, in digital format, this equation becomes

$$y(kT_b) = \sum_{m=-\infty}^{\infty} r(mT_b)h_r(kT_b - mT_b) \qquad (9.54)$$

and from Fig. 9.9,

$$c(t) = \sum_{k=-\infty}^{k=\infty} a_k h(t - kT_b) \qquad (9.55)$$

where $h(t)$ is the impulse response of the channel and all other earlier filters combined:

$$h(t) = h_t(t) * h_c(t) \qquad (9.56)$$

Normally, the convolution is over an infinite set of values. If the channel and earlier impulse responses convolved together is $h[k] \equiv h(kT_b)$, and the equaliser impulse response is $h_r[k] \equiv h_r(kT_b)$, then the impulse response of the combination is

$$f(kT_b) = \sum_{m=-\infty}^{\infty} h_r(mT_b)h(kT_b - mT_b)$$

$$\text{or } f[k] = \sum_{m=-\infty}^{\infty} h_r[m]h[k - m]$$

The zero forcing algorithm tries to eliminate intersymbol interference at decision time instants at the centre of the symbol interval. Using this in mind, we now force the impulse response to be such that its value at all times $f[k]$ is zero except at $t = 0$:

$$\text{or } f[k] = \sum_{m=-M}^{M} h_r[m]h[k - m]$$

which consists of $2M + 1$ equations in $2M + 1$ unknowns, $h_r[-M]$, $h_r[-M + 1]$, $h_r[0]$, $h_r[M - 1]$ and $h_r[M]$.

$$f[-M] = \sum_{m=-M}^{M} h_r[m]h[-M - m] = 0$$

$$f[-M + 1] = \sum_{m=-M}^{M} h_r[m]h[-M + 1 - m] = 0$$

$$\vdots \quad = \quad \vdots$$

$$f[0] = \sum_{m=-M}^{M} h_r[m]h[-m] = 1$$

$$\vdots \quad = \quad \vdots$$

$$f[M - 1] = \sum_{m=-M}^{M} h_r[m]h[M - 1 - m] = 0$$

$$f[M] = \sum_{m=-M}^{M} h_r[m]h[M - m] = 0$$

where $h_r[-M]$, $h_r[-M + 1]$, $\ldots h_r[0]$, $\ldots$, $h_r[M - 1]$ and $h_r[M]$ have to be determined. The channel coefficients being already characterised earlier. Written in full, this becomes

$$h[0]h_r[-M] + h[-1]h_r[-M + 1] + \cdots h[-2M + 1]h_r[M - 1] + h[-2M]h_r[M] = 0 \quad k = -M$$

$$h[1]h_r[-M] + h[0]h_r[-M + 1] + \cdots h[-2M + 2]h_r[M - 1] + h[-2M + 1]h_r[M] = 0 \quad k = -M + 1$$

$$\vdots \quad = 0$$

$$h[M]h_r[-M] + h[M - 1]h_r[-M + 1] + \cdots h[-M + 1]h_r[M - 1] + h[-M]h_r[M] = 1 \quad k = 0$$

$$\vdots \quad = 0$$

$$h[2M - 1]h_r[-M] + h[2M - 2]h_r[-M + 1] + \cdots h[0]h_r[M - 1] + h[-1]h_r[M] = 0 \quad k = M - 1$$

$$h[2M]h_r[-M] + h[2M - 1]h_r[-M + 1] + \cdots h[1]h_r[M - 1] + h[0]h_r[M] = 0 \quad k = M$$

which are $2M + 1$ equations in $2M + 1$ unknowns. These equations can be written in matrix form as

$$[\mathbf{f}] = [\mathbf{h}][\mathbf{h}_r]$$

where the matrix, $[\mathbf{h}]$,

$$[\mathbf{h}] = \begin{bmatrix} h[0] & h[-1] & \cdots & h[-M] & \cdots h[-2M+1] & h[-2M] \\ h[1] & h[0] & \cdots h[-M+1] & \cdots & h[-2M+2] & h[-2M+1] \\ \vdots & \vdots & \ddots & \vdots & \vdots & \vdots & \vdots \\ h[M] & h[M-1] & \cdots & h[0] & \cdots & h[-M+1] & h[-M] \\ \vdots & \vdots & \vdots & \vdots & \ddots & \vdots & \vdots \\ h[2M-1] & h[2M] & \cdots & h[M+1] & \cdots & h[0] & h[-1] \\ h[2M] & h[2M-1] & \cdots & h[M] & \cdots & h[1] & h[0] \end{bmatrix}$$

has to be inverted. The matrix has a Toeplitz form

$$\begin{bmatrix} a & b & c & \cdots & d \\ e & a & b & \ddots & \vdots \\ f & e & \ddots & \ddots & c \\ \vdots & \ddots & \ddots & a & b \\ g & \cdots & f & e & a \end{bmatrix}$$

whose inversion is easy through special algorithms. The inverted matrix has to pre-multiply

$$[\mathbf{f}] = \begin{bmatrix} 0 \\ 0 \\ \vdots \\ 1 \\ \vdots \\ 0 \\ 0 \end{bmatrix}$$

to give us the value of the receive filter coefficients

$$[\mathbf{h}_r] = \begin{bmatrix} h_r[-M] \\ h_r[-M+1] \\ \vdots \\ h_r[0] \\ \vdots \\ h_r[M-1] \\ h_r[M] \end{bmatrix}$$

The matrix [**h**] can be obtained from the response of the channel. The receive filter implementation is shown in Fig. 9.27.

In practice, zero forcing equalisation does not work for many applications, since though the channel impulse response may have infinite length, the impulse response

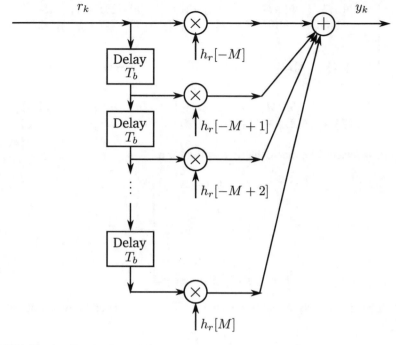

**Fig. 9.27** Receive filter implementation

of the equaliser also needs to be infinitely long, but is of finite length and noise added during transmission through the channel gets boosted. Furthermore, many channels have zeros in its frequency response which cannot be inverted.

## 9.6.2 Minimum Mean Squared Error (MMSE) Equalisation

To evaluate the presence of noise and intersymbol interference, we refer to Fig. 9.9. The binary bits are converted to impulses $a_k$ and

$$x(t) = \sum_k a_k \delta(t - kT_b)$$

where $T_b$ is the time between pulses. Also,

$$r(t) = \sum_k a_k h(t - kT_b) + w(t) \tag{9.57}$$

where $r(t)$ is the received signal and $w(t)$ is white noise. $h(t) = h_t(t) * h_c(t)$ is the convolution of the two impulse responses: the transmit filter and the channel impulse response. Also,

$$y(t) = \int_{-\infty}^{\infty} r(t - \tau) h_r(\tau) d\tau \qquad (9.58)$$

where $h_r(t)$ is the impulse response of the receive filter. We now define the error when we sample this signal at $t = iT_b$

$$e(iT_b) = y(iT_b) - a(iT_b) \qquad (9.59)$$

The square of the error is

$$e_i^2 = y_i^2 - 2y_i a_i + a_i^2 \qquad (9.60)$$

where $e_i \equiv e(iT_b)$ and so on. Forming the function

$$J = \frac{1}{2} E[e_i^2]$$

where $E[\ ]$ is the expected value, we now propose to minimise $J$. So

$$E[e_i^2] = E[y_i^2] - 2E[y_i a_i] + E[a_i^2] \qquad (9.61)$$

where

$$y(iT_b) = \int_{-\infty}^{\infty} r(iT_b - \tau) h_r(\tau) d\tau$$

$$= \int_{-\infty}^{\infty} \left[ \sum_k a_k h(iT_b - \tau - kT_b) + w(iT_b - \tau) \right] h_r(\tau) d\tau$$

$$= \sum_k a_k \int_{-\infty}^{\infty} h(iT_b - \tau - kT_b) h_r(\tau) d\tau + \int_{-\infty}^{\infty} w(iT_b - \tau) h_r(\tau) d\tau$$

$$= \alpha_i + n_i$$

where

$$\alpha_i = \sum_k a_k \int_{-\infty}^{\infty} h(iT_b - \tau - kT_b) h_r(\tau) d\tau$$

$$\text{and } n_i = \int_{-\infty}^{\infty} w(iT_b - \tau) h_r(\tau) d\tau$$

So

$$e_i = \alpha_i + n_i - a_i$$

$$\text{and } E\left[e_i^2\right] = E\left[\alpha_i^2\right] + E\left[n_i^2\right] + E\left[a_i^2\right] + 2E\left[\alpha_i n_i\right] - 2E\left[a_i n_i\right] - 2E\left[\alpha_i a_i\right]$$

Taking terms one at a time,

$$E\left[\alpha_i^2\right] = E\left[\sum_k a_k \int_{-\infty}^{\infty} h(kT_b - \tau)h_r(\tau)d\tau\right]^2$$

$$= \sum_k \sum_l E\left[a_k a_l\right] \int_{-\infty}^{\infty}\int_{-\infty}^{\infty} h(kT_b - \tau_1)h(lT_b - \tau_2)h_r(\tau_1)h_r(\tau_2)d\tau_1 d\tau_2$$

Now for a stationary signal, the expected value should be independent of the time, $t = iT_b$, and therefore, we may drop it. Also,

$$a_k = \begin{cases} 1 & \text{for bit 1} \\ -1 & \text{for bit 0} \end{cases}$$

so

$$E[a_j a_i] = \delta_{ij} = \begin{cases} 1 & i = j \\ 0 & i \neq j \end{cases}$$

since the separate pulses $a_k$ are such that

$$E\left[a_j a_j\right] = 1$$

and

$$E\left[a_i a_j\right] = 0 \quad j \neq i$$

since the pulses are uncorrelated. Proceeding with

$$E\left[\alpha_i^2\right] = \sum_k \sum_l E\left[a_k a_l\right] \int_{-\infty}^{\infty}\int_{-\infty}^{\infty} h(kT_b - \tau_1)h(lT_b - \tau_2)h_r(\tau_1)h_r(\tau_2)d\tau_1 d\tau_2$$

$$= \sum_k \int_{-\infty}^{\infty}\int_{-\infty}^{\infty} h(kT_b - \tau_1)h(kT_b - \tau_2)h_r(\tau_1)h_r(\tau_2)d\tau_1 d\tau_2$$

If we let

$$R_h(\tau_1, \tau_2) = \sum_k h(kT_b - \tau_1)h(kT_b - \tau_2)$$

which is called the *temporal autocorrelation function* of $h(kT_b)$, then

$$E\left[\alpha_i^2\right] = \int_{-\infty}^{\infty} \int_{-\infty}^{\infty} R_h(\tau_1, \tau_2) h_r(\tau_1) h_r(\tau_2) d\tau_1 d\tau_2$$

$$= \int_{-\infty}^{\infty} \int_{-\infty}^{\infty} R_h(\tau_1 - \tau_2) h_r(\tau_1) h_r(\tau_2) d\tau_1 d\tau_2$$

$$= \int_{-\infty}^{\infty} \int_{-\infty}^{\infty} R_h(\tau_2 - \tau_1) h_r(\tau_1) h_r(\tau_2) d\tau_1 d\tau_2$$

where, for a wide-sense stationary impulse response,

$$R_h(\tau_1, \tau_2) = R_h(\tau_1 - \tau_2) = R_h(\tau_2 - \tau_1)$$

Next we consider

$$E\left[n_i^2\right] = E\left[\left\{\int_{-\infty}^{\infty} w(iT_b - \tau) h_r(\tau) d\tau\right\}^2\right]$$

$$= \int_{-\infty}^{\infty} \int_{-\infty}^{\infty} E\left[w(iT_b - \tau_1) w(iT_b - \tau_2)\right] h_r(\tau_1) h_r(\tau_2) d\tau$$

$$- \frac{N_0}{2} \int_{-\infty}^{\infty} \int_{-\infty}^{\infty} \delta(\tau_1 - \tau_2) h_r(\tau_1) h_r(\tau_2) d\tau_1 d\tau_2$$

where the autocorrelation function for white noise is independent of $iT_b$

$$R_W(\tau_1, \tau_2) = \frac{N_0}{2} \delta(\tau_1 - \tau_2)$$

and the third term is

$$E\left[a_i^2\right] = 1$$

The next two terms are zero since essentially $n_i$ has zero mean and the signals are uncorrelated.

$$E\left[\alpha_i n_i\right] = E\left[a_i n_i\right] = 0$$

Finally, we take a look at the last term

$$E\left[\alpha_i a_i\right] = E\left[a_i \left\{\sum_k a_k \int_{-\infty}^{\infty} h(iT_b - \tau - kT_b) h_r(\tau) d\tau\right\}\right]$$

$$= \sum_k E\left[a_i a_k\right] \int_{-\infty}^{\infty} h(iT_b - \tau - kT_b) h_r(\tau) d\tau$$

$$= E\left[a_i a_i\right] \int_{-\infty}^{\infty} h(iT_b - \tau - iT_b) h_r(\tau) d\tau$$

$$= \int_{-\infty}^{\infty} h(-\tau) h_r(\tau) d\tau$$

So

$$J = \frac{1}{2} \int_{-\infty}^{\infty} \int_{-\infty}^{\infty} R_h(\tau_2 - \tau_1) h_r(\tau_1) h_r(\tau_2) d\tau_1 d\tau_2 - \frac{1}{2} \int_{-\infty}^{\infty} h(-\tau_1) h_r(\tau_1) d\tau_1$$

$$+ \frac{N_0}{4} \int_{-\infty}^{\infty} \int_{-\infty}^{\infty} \delta(\tau_1 - \tau_2) h_r(\tau_1) h_r(\tau_2) d\tau_1 d\tau_2$$

which is a functional. Using calculus of variations and setting $\tau_1 = t$,

$$\int_{-\infty}^{\infty} \left[ R_h(t - \tau_2) + \frac{N_0}{2} \delta(t - \tau_2) \right] h_r(\tau_2) d\tau_2 - h(-t) = 0$$

Taking the Fourier transform of this equation,

$$\left[ S_h(\omega) + \frac{N_0}{2} \right] H_r(\omega) = H^*(\omega)$$

$$H_r(\omega) = \frac{H^*(\omega)}{S_h(\omega) + \frac{N_0}{2}} \tag{9.62}$$

where $R_h(t) \Leftrightarrow S_h(\omega)$, and $H_r(\omega)$ is the transfer function of the receive filter, while

$$H(\omega) = H_t(\omega) H_c(\omega) \tag{9.63}$$

This equation suggests that two filters are required, the first filter is a matched filter whose impulse response is

$$H^*(\omega) \Leftrightarrow h(-t)$$

and the second filter

$$\frac{1}{S_h(\omega) + \frac{N_0}{2}} \Leftrightarrow T_r(t)$$

where $T_r(t)$ is a transversal filter. The full receive filter is shown in Fig. 9.28.

### 9.6.3  Adaptive Equalisation

The MMSE equaliser is good so long as the channel characteristics do not change with time. The fixed, matched filter and equaliser may not meet the needs of channels with changing parameters. To this end, we focus on Fig. 9.29. The figure shows a received signal $r(t)$ which is sampled to give $r_k$. Initially, this signal consists of symbols $a_k \equiv f_k$ a training signal which is sent through the communication channel and which is a known PN sequence; the signal is known to both the receiver and

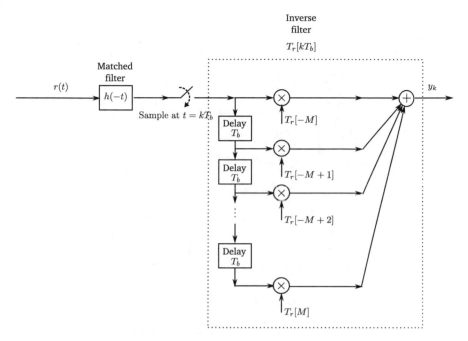

**Fig. 9.28** Receive filter MMSE equalisation

transmitter. Through $e_k$, the error signal, the weights $w_0$, $w_1$, ... $w_N$ are adjusted until the error signal is constant. That is,

$$e_k \cong \text{constant}$$

At the time when the channel is used for training, the symbols transmitted are $a_k \equiv f_k$ shown in Figs. 9.9 and 9.30. In Fig. 9.30, the switch is at position 1. Referring now to Eq. (8.65) and the following equations, where we applied the least mean squared algorithm (LMS), we find from those equations essentially can be written, in the present context, as

$$y_k = \sum_{n=0}^{N} w_n[i] r_{k-n}$$

$$e_k = a_k - y_k \tag{9.64}$$

where $a_k$ are the symbols which was transmitted (which, essentially are $f_k$), and $y_k$ is the output of the equaliser. $i$ is the $i^{\text{th}}$ iteration.

To obtain the $[i + 1]$ value of the coefficients, by the LMS algorithm, the pertinent equation is (see Eq. (8.65))

Sample at $t = kT_b$

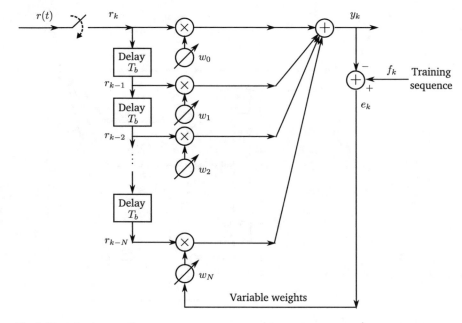

**Fig. 9.29**   Adaptive equaliser

$$w_n[i + 1] = w_n[i] + \mu r_{k-n} e_k \qquad 0 \le n \le N$$

which says that in the next iteration, the $n$th tap weight, $w_n[i + 1]$, is equal to the previous value of the tap weight, $w_n[i]$, plus the term

$$\mu r_{k-n} e_k$$

where $\mu$ is the step size, $r_{k-n}$ is the value of the input which multiplies $w_n[i]$ and $e_k$ is the current error. We now formalise the algorithm to evaluate the coefficients, $w_n$.

1. Set all weight coefficients equal to zero. For $i = 1$,

$$w_n[1] = 0$$

2. Calculate

$$y_k = \sum_{n=0}^{N} w_n[i] r_{k-n}$$

3. Next calculate

$$e_k = f_k - y_k$$

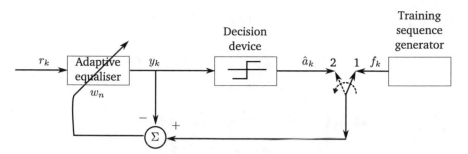

**Fig. 9.30**  Adaptive equaliser block diagram

4. Is $e_{k+1} - e_k \cong 0$? Stop. Else
5.

$$w_n[i + 1] = w_n[i] + \mu r_{k-n} e_k \qquad n = 0, 1, \ldots N$$

6. Go to Step 2.

After the training is over, the equalisation can be switched to the tracking mode, which is illustrated in Fig. 9.30 where the switch is connected to position 2. The incoming signal is fed to the adaptive equaliser, the output of which ($y_k$) goes into a decision device which gives an estimate of $a_k$ which is $\hat{a}_k$. This symbol is then used to give us the error

$$e_k = \hat{a}_k - y_k \tag{9.65}$$

As long as the error signal remains essentially a constant, we continue in the tracking mode; otherwise, we can go back to the training mode.

The LMS adaptive equaliser is one of the most well-known and popular algorithms. It executes quickly but converges very slowly, and its computational complexity grows linearly with the number of tap weights, $N$. Computationally, it is simple and particularly successful in the case when channel impulse response varies slowly as compared to the convergence time.

## 9.7  Important Concepts and Formulae

- The optimum matched filter FT of the impulse response is given by

$$H_{\text{opt}}(\omega) = kX^*(\omega)e^{-j\omega T_b}$$

where $X(\omega)$ is the input signal and $T_b$ is the pulse duration.

- The optimum impulse response of the matched filter is

$$h_{\text{opt}}(t) = \frac{k}{2\pi} \int_{-\infty}^{\infty} X(-\omega)e^{j\omega(t-T_b)}d\omega$$
$$= kx(T_b - t)$$

- The maximum SNR after passing through a matched filter is

$$\left.\frac{y^2(T_b)}{E[n^2(t)]}\right|_{\text{max}} = \frac{2E}{N_0}$$

- $P_e$ of NRZ coding is given by

$$P_e = \frac{1}{2}\text{erfc}\left(\sqrt{\frac{E_b}{N_0}}\right)$$

- The ideal pulse to avoid ISI is

$$P(\omega) = \begin{cases} k & -\omega_b/2 \leq \omega \leq \omega_b/2 \\ 0 & \text{elsewhere} \end{cases}$$

- The *ideal Nyquist channel* is defined by

$$H_N(\omega) = \begin{cases} 1 & -W \leq \omega \leq W(=\omega_b/2) \\ 0 & \text{elsewhere} \end{cases}$$

- A frequency spectrum of a filter of the raised cosine function is given by

$$P(\omega) = \begin{cases} \frac{1}{2W}\left[1 + \cos\left(\frac{\pi\omega}{2W}\right)\right] & -2W \leq \omega \leq 2W \\ 0 & \text{elsewhere} \end{cases}$$

For which the time-domain pulse is

$$p(t) = K\frac{\pi^2 \sin(2tW)}{W\left(\pi^2 t - 4t^3 W^2\right)} \quad W = \pi/T_b$$
$$= 4K\text{sinc}(tW)\left[\frac{\cos(tW)}{1 - \left(\frac{2tW}{\pi}\right)^2}\right]$$

- The general form of the raised cosine filter is

$$P(\omega) = \begin{cases} \frac{1}{2W} & -\omega_1 \leq \omega \leq \omega_1 \\ \frac{1}{4W}\left\{1 - \sin\left[\frac{\pi(|\omega|-W)}{2(W-\omega_1)}\right]\right\} & \omega_1 \leq |\omega| \leq 2W - \omega_1 \\ 0 & \text{elsewhere} \end{cases}$$

where

$$\alpha = 1 - \frac{\omega_1}{W}$$

$$\omega_1 = W(1 - \alpha)$$

- The impulse response is given by

$$p(t) = \left[\frac{\sin(\pi t/T_b)}{\pi t/T_b}\right]\left[\frac{\cos(\alpha\pi t/T_b)}{1 - (2\alpha t/T_b)^2}\right]$$

- In *duobinary signalling*, we take a set of impulse like pulses, *which can be referred to a baseband binary PAM system* $\{a_k \equiv a(kT_b),\ k = 0, 1, \ldots\}$ where

$$a_k = \begin{cases} +1 & \text{for a 1} \\ -1 & \text{for a 0} \end{cases}$$

and obtain a new set of correlated pulses

$$c_k = a_k + a_{k-1}$$

This is shown in Fig. 9.16. The $z$ transform of this equation is

$$C(z) = \left(1 + z^{-1}\right) A(z)$$

and the transfer function is

$$H(z) = 1 + z^{-1}$$

or

$$H(\omega T_b) = 1 + e^{-j\omega T_b}$$

$$= 2\cos\left(\frac{\omega T_b}{2}\right) e^{-j\omega\frac{T_b}{2}}$$

- In *modified duobinary* signalling, we proceed as earlier, but with

$$c_k = a_k - a_{k-2}$$

- The problem faced in duobinary and modified duobinary signalling is that if an error in the detection of the output is committed at any stage, it is carried forward and all the subsequent bits are in error.
- In duobinary signalling with precoding,

$$d_k = b_k \oplus d_{k-1}$$

where $\oplus$ is the *exclusive or* operation.
- The eye diagram is a tool to evaluate the overall performance of pulses sent over a channel which includes the addition of noise and intersymbol interference.

  - The slope of the eye, $m$, indicates the sensitivity to timing errors. The smaller the slope, the lower the possibility of timing errors.
  - The width of the eye gives us an interval of time when the signal can be sampled.
  - The signal is best sampled at the point shown on the diagram.
  - The noise margin of the signal is the height of the eye opening.

- The zero forcing equaliser is a brute force method of equalisation.

## 9.8   Self Assessment

### 9.8.1   Review Questions

1. Explain the statement in the Footnote 1.
2. What are the units of $N_0$?

### 9.8.2   Numerical Problems

1. For

$$x(t) = \begin{cases} A & 0 \le t \le T/4 \\ -A & T/4 < t \le T \\ 0 & \text{elsewhere} \end{cases}$$

show that the impulse response of the matched filter is

$$h(t) = \begin{cases} -A & 0 \le t \le 3T/4 \\ A & 3T/4 < t \le T \\ 0 & \text{elsewhere} \end{cases}$$

2. For Problem 1, find $x(t) * h(t)$ for $0 \le t \le T$. Show that it is

$$x(t) * h(t) = \begin{cases} -A^2 t & 0 \le t \le T/4 \\ -\frac{A^2 T - 2t A^2}{2} & T/4 \le t \le T/2 \\ 2t A^2 - A^2 T & T/2 \le t \le T \end{cases}$$

3. If

$$x(t) = A \sin\left(\frac{2\pi t}{T}\right) \qquad 0 \le t \le T$$

and zero elsewhere, what is the impulse response of the matched filter, $h(t)$, using the interval $0 \le t \le T$?

4. Show that for Problem 3

$$x(t) * h(t) = \begin{cases} -\frac{A^2 \sin\left(\frac{2\pi t}{T}\right) T - 2\pi t A^2 \cos\left(\frac{2\pi t}{T}\right)}{4\pi} & 0 \le t \le T \\ \frac{\pi\left(4A^2 \cos\left(\frac{2\pi t}{T}\right) T - 2t A^2 \cos\left(\frac{2\pi t}{T}\right)\right) + A^2 \sin\left(\frac{2\pi t}{T}\right) T}{4\pi} & T \le t \le 2T \end{cases}$$

5. Show that the effect of using a matched filter on a symbol $x(t)$ and sampling the output at $t = T_b$ is equivalent to having an output equal to

$$y(T_b) = k \int_0^{T_b} x^2(t)dt$$

where $k$ is a constant. Show that this is true for Problems 1 and 3.

6. For $N_0 = 10^{-3}$MKS units, estimate the amplitude in volts at the receiver for a bit rate of 1 Mbits/s to get a probability of error equal to $10^{-6}$?

7. At the receiver, the amplitude per bit is 1 mV for a bit rate of 1 Mbits/s. Practically, it is found that on an average, 1 bit in 10,000 bits is in error. Estimate $N_0$.

8. Obtain the impulse response expression for the frequency response shown in Fig. 9.31b:

$$p(t) = K \times \left[ -\frac{2\cos\left(\frac{3tW}{2}\right) - 2\cos\left(\frac{tW}{2}\right)}{t^2 W} \right] \qquad W = \pi/T_b$$

9. For

$$P(\omega) = \begin{cases} \frac{1}{2W}\left[1 + \cos\left(\frac{\pi\omega}{2W}\right)\right] & -2W \le \omega \le 2W \\ 0 & \text{elsewhere} \end{cases}$$

obtain analytically Eq. (9.36)

$$p(t) = K \frac{\pi^2 \sin(2tW)}{W\left(\pi^2 t - 4t^3 W^2\right)} \qquad W = \pi/T_b$$

$$= 4K \operatorname{sinc}(tW) \left[ \frac{\cos(tW)}{1 - \left(\frac{2tW}{\pi}\right)^2} \right]$$

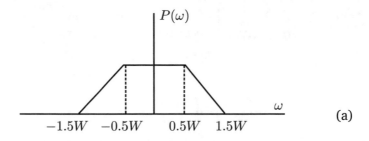

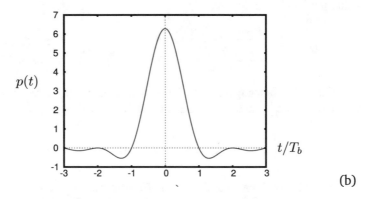

**Fig. 9.31**  Example function, **a** $P(\omega)$, with **b** $p(t)$

10. If we use a 100 kHz Nyquist channel and a general raised cosine filter with $\alpha = 0.75$, what is the maximum bit rate?
11. If we use a raised cosine filter with $T_b = 1\,\mu\text{s}$, what should be the bandwidth of the channel?
12. Show that the result of passing impulses through a Nyquist channel using a duobinary scheme is Eq. (9.45).

$$h_N(t) = \operatorname{sinc}\left(\frac{\pi t}{T_b}\right)\frac{T_b}{T_b - t}$$

13. For duobinary signalling, find the output for the sequence 00101101. Make a table as shown in Table 9.1
14. For modified duobinary signalling, find the output for the sequence 00101101. Make a table as shown in Table 9.1
15. Using precoding, for duobinary signalling, find the output for the sequence 00101101. Make a table as shown in Table 9.1
16. Using precoding, show that the duobinary scheme does indeed give as output Eq. (9.52):

$$y_k = \begin{cases} 0 & \text{if message bit } b_k = 1 \\ \pm 2 & \text{if message bit } b_k = 0 \end{cases}$$

17. Using precoding, for modified duobinary signalling, find the output for the sequence 00101101. Make a table as shown in Table 9.1

18. Using precoding, show that the modified duobinary scheme does indeed give as output Eq. (9.53):

$$y_k = \begin{cases} \pm 2 & \text{if message bit } b_k = 1 \\ 0 & \text{if message bit } b_k = 0 \end{cases}$$

19. A channel is slightly lossy with a transfer function

$$H(\omega) = \begin{cases} e^{-\alpha|\omega|} & -W \leq \omega \leq W \\ 0 & \text{elsewhere} \end{cases}$$

show that the impulse response of the channel is given by

$$h(t) = \frac{e^{-\alpha W} \left( 2t \sin(tW) - 2\alpha \cos(tW) + 2\alpha e^{\alpha W} \right)}{\pi \left( t^2 + \alpha^2 \right)}$$

20. A zero forcing algorithm is used with a tapped delay line filter with three coefficients: $h_r[-1]$, $h_r[0]$ and $h_r[1]$. The channel normalised impulse response is given by

$$h[n] = \frac{e^{-\frac{\pi}{10}} \left( 100n \sin(\pi n) - 10 \cos(\pi n) + 10 e^{\frac{\pi}{10}} \right)}{100n^2 + 1}$$

Apply the zero forcing algorithm to this system.

# References

1. Chiani M, Dardari DSMK (2003) New exponential bounds and approximations for the computation of error probability in fading channels. IEEE Trans Wirel Commun 4(2):840–845
2. Nyquist H (1928) Certain topics in telegraph transmission theory. Trans AIEE 47:617–644
3. Steele JM (2004) The Cauchy-Schwarz master class: an introduction to the art of mathematical inequalities. Cambridge University Press

# Chapter 10
# Passband Pulse Transmission

## 10.1 Introduction

Baseband pulse transmission is useful when we want to communicate over a medium which does not attenuate the baseband signal significantly, and we may employ some kind of time division multiplexing scheme as explained earlier in Sect. 8.2. This happens often in cases when communication is over short distances such as in cases of communication between a server and computer using a LAN. However, when there are a number of transmitters and receivers and, additionally, there is baseband pulse attenuation, we employ some kind of modulation using a carrier, as in the case of analogue communication. When we want to communicate with the help of a carrier, we should choose a medium *which does not attenuate the band of frequencies employed containing the carrier* and we may use *frequency division multiplexing* (FDM) as explained earlier in Sect. 5.12.

## 10.2 Carrier Modulations: Introduction

Figure 10.1 shows the very basic modes of carrier modulated pulses. In Fig. 10.1a, bits which we want to transmit are shown. In amplitude modulation of pulses known as *amplitude shift keying* (ASK), there are two amplitude levels: one level for bit zero and another for bit one. The signal is

$$s_{ASK}(t) = \begin{cases} A_1 \cos(\omega_{c0} t) & \text{for bit 1} \\ A_0 \cos(\omega_{c0} t) & \text{for bit 0} \end{cases} \tag{10.1}$$

In (a) part of the figure, the bits designated as one are of some convenient amplitude, $A_1$, while the bits designated as zero have zero amplitude ($A_0 = 0$). The width of

© The Author(s), under exclusive license to Springer Nature Singapore Pte Ltd. 2022    447
S. Bhooshan, *Fundamentals of Analogue and Digital Communication Systems*,
Lecture Notes in Electrical Engineering 785,
https://doi.org/10.1007/978-981-16-4277-7_10

**Fig. 10.1** **a** Amplitude shift keying (ASK), **b** phase shift keying (PSK) and **c** frequency shift keying (FSK)

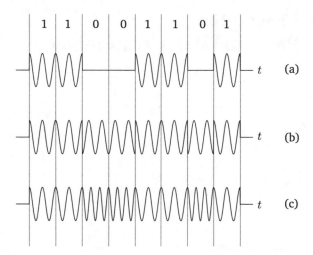

the bit, $T_b$, is related to the period of the carrier, $T_c$, by

$$T_b = nT_c \tag{10.2}$$

where $n$ is a positive integer. Which in simple terms means that the bit width is an integer number of carrier period widths, and the amplitude at the beginning and end of a bit are equal.

In phase modulation, called *phase shift keying* (PSK), the frequency and amplitude remains constant but the phase is changed. This is shown in Fig. 10.1b. When there is a transition between 0 and 1, there is a phase reversal of 180°. Also Eq. (10.2) holds. Then

$$s_{\text{PSK}}(t) = \begin{cases} A\cos(\omega_{c0}t) & \text{for bit 1} \\ A\cos(\omega_{c0}t + \pi) & \text{for bit 0} \end{cases} \tag{10.3}$$

And lastly in frequency modulation, called *frequency shift keying* (FSK), the amplitude is constant while the frequency is changed depending upon whether the bit is zero or one. The analytic expression of the signal is

$$s_{\text{FSK}}(t) = \begin{cases} A\cos(\omega_{c1}t) & \text{for bit 1} \\ A\cos(\omega_{c0}t) & \text{for bit 0} \end{cases} \tag{10.4}$$

As explained earlier, in all these cases, the bit width, $0 \le t \le T_b$, is related to the carrier frequency by

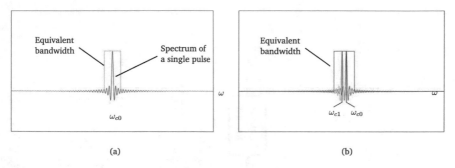

**Fig. 10.2** Frequency spectrum usage of **a** ASK and PSK and **b** FSK

$$\omega_{c0} T_b = \frac{2\pi T_b}{T_{c0}} = 2n\pi$$

$$\omega_{c1} T_b = \frac{2\pi T_b}{T_{c1}} = 2m\pi$$

where $m$ and $n$ are positive integers.

In passband transmission, *intersymbol interference plays a secondary role in the reception of data*, and the major concern is the probability of error due to the addition of noise when the signal transits the channel (see Sect. 3.16). An error occurs at the receiver when a 1 is transmitted but it is interpreted as a 0 and vice versa. This happens because the signal is corrupted by additive white noise. In passband transmission, the data occupies a band of frequencies as shown in Fig. 10.2. Since the spectrum is clustered around $\omega_{c0}$ or $\omega_{c1}$, as the case may be, and since both $\omega_{c0}$ and $\omega_{c1}$ are both very large, the ISI is negligible. Furthermore, in passband systems, the bandwidth is assumed to be *linear*: it is adequate with low distortion. However, the same is not the case with noise. Noise is present at the higher frequencies as well the lower frequencies and its role in the probability of error is very important.

## 10.3 Passband Systems: General Comments

Figure 10.3 shows a typical digital communication system. Consider a digital message source which outputs any one of $M$ *symbols*, as shown in the figure. The set of $M$ message symbols is

$$S = \{m_1, m_2, \dots, m_M\} \tag{10.5}$$

which is considered to be finite. For example, for $M = 2$, there may be only two symbols, 0 and 1. In another kind of communication, there may be four symbols, 00, 01, 10 and 11. This is the opposite of what is true in analogue communication where the source can transmit an infinite number of waveforms.

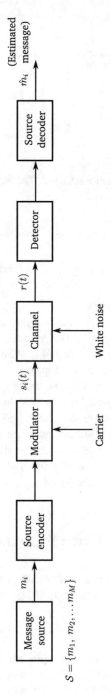

**Fig. 10.3**   A digital passband system

For each of these message symbols, $m_i$, there exists a probability that the message may be transmitted

$$\{p_1, \; p_2, \ldots, p_M\}$$

where $p_1$ is the probability that $m_1$ was transmitted, $p_2$ is the probability that $m_2$ was transmitted and so on. In many cases, each of these probabilities may be equal:

$$p_i = \frac{1}{M} \quad i = 1, 2, \ldots, M \tag{10.6}$$

Furthermore, for each symbol, a unique waveform or signal is used:

$$m_i \rightarrow s_i(t) \tag{10.7}$$

which occupies a time slot $nT \leq t \leq (n+1)T$. Also, $s_i(t) i = 1, \ldots, M$ is real and of finite energy

$$E_i = \int_0^T s_i^2(t) dt \tag{10.8}$$

The received signal has the property that

$$r_i(t) = s_i(t) + w(t) \tag{10.9}$$

where $w(t)$ is white Gaussian noise with zero mean and autocorrelation function

$$R_W(\tau_1, \tau_2) = \frac{N_0}{2} \delta(\tau_1 - \tau_2) \tag{10.10}$$

The channel is called an *additive white Gaussian noise* (AWGN) channel. Due to the presence of channel noise, the receiver has to decide after receiving the symbol as to which message symbol was sent. If an error was committed, the probability of such a mistake is given by

$$P_e = p_i P \left( \hat{m} \neq m_i | m_i \right)$$

where $\hat{m}$ is an estimate of the message symbol.

## 10.4  The Signal Vector Space-I

To get a theoretical insight into passband systems, we look at the generation of $M$ signals in such a way that they are unique in some sense. The transmitted signals are

$$s_i(t) i = 1, 2, \ldots, M; \quad 0 \leq t \leq T \tag{10.11}$$

In passband transmission, each signal which is transmitted may be *generated using N orthonormal functions which are generated from* $s_i(t)$

$$\phi_j(t), \quad j = 1, 2, \ldots, N; \qquad 0 \le t \le T \tag{10.12}$$

where $N \le M$.

Orthonormal functions satisfy the definition

$$\int_0^T \phi_i(t)\phi_j(t)dt \overset{\Delta}{=} \begin{cases} 1 & \text{for } i = j \\ 0 & \text{for } i \ne j \end{cases} \tag{10.13}$$

These orthonormal functions are chosen in such a way that we can write each of the signals $s_i(t)$ as a linear combination of the $\phi$s. Thus

$$s_1(t) = s_{11}\phi_1(t) + s_{12}\phi_2(t) + \cdots + s_{1N}\phi_N(t)$$
$$s_2(t) = s_{21}\phi_1(t) + s_{22}\phi_2(t) + \cdots + s_{2N}\phi_N(t)$$
$$\vdots$$
$$s_i(t) = s_{i1}\phi_1(t) + s_{i2}\phi_2(t) + \cdots + s_{iN}\phi_N(t)$$
$$\vdots$$
$$s_M(t) = s_{M1}\phi_1(t) + s_{M2}\phi_2(t) + \cdots + s_{MN}\phi_N(t) \tag{10.14}$$

To make this representation exact, the $\phi$s are generated in a *particular way* called the Gram–Schmidt orthonormalisation procedure.

### 10.4.1   Gram–Schmidt Orthonormalisation

We now take a look at the orhtonormalisation process. The first step is *orthogonalistion* of the functions. Let the *orthogonal* functions be the set

$$\mathcal{U} = \{u_1(t), u_2(t), \ldots, u_M(t)\}$$

which we have to find. Orthogonality means that

$$\int_0^T u_i(t)u_j(t) = \begin{cases} 0 & i \ne j \\ \|u_i(t)\|_f^2 & i = j \end{cases}$$

where $\|u_i(t)\|_f^2$ is called the *function norm* of the function $u_i(t)$. Some of these functions may be zero. We find these functions by means of an algorithm which we explain below. The first step in our algorithm is to first choose

$$u_1(t) = s_1(t)$$

We now write (the second step in our algorithm)

$$u_2(t) = s_2(t) - a_1 u_1(t)$$

where $a_1$ is an unknown constant. Then

$$\int_0^T u_1(t)u_2(t)dt = 0 = -a_1 \int_0^T u_1^2(t)dt + \int_0^T u_1(t)s_2(t)dt \qquad (10.15)$$

This equation ensures that $u_1$ is orthogonal to $u_2$. So

$$a_1 = \frac{\int_0^T u_1(t)s_2(t)dt}{\int_0^T u_1^2(t)dt}$$

So

$$u_2(t) = s_2(t) - \frac{\int_0^T u_1(t)s_2(t)dt}{\int_0^T u_1^2(t)dt} s_1(t)$$

We now proceed to the third step in our algorithm

$$u_3(t) = s_3(t) - b_1 u_1(t) - b_2 u_2(t)$$

If we take the inner product with either $u_1(t)$ or $u_2(t)$, we get

$$0 = \int_0^T u_1(t)s_3(t)dt - b_1 \int_0^T u_1^2(t)dt$$

$$\Rightarrow b_1 = \frac{\int_0^T u_1(t)s_3(t)dt}{\int_0^T u_1^2(t)dt}$$

$$\text{and } 0 = \int_0^T u_2(t)s_3(t)dt - b_2 \int_0^T u_2^2(t)dt$$

$$\Rightarrow b_2 = \frac{\int_0^T u_2(t)s_3(t)dt}{\int_0^T u_2^2(t)dt}$$

Generalising this result in the $i$th step in our algorithm,

$$u_i(t) = s_i(t) - \sum_{j=1}^{i-1} \frac{\int_0^T u_j(t)s_i(t)dt}{\int_0^T u_j^2(t)dt} u_j(t) \quad i = 1, 2, \ldots, M \qquad (10.16)$$

the process is stopped when we reach $N = M$. It should be noticed that some of the $u_i(t)$ will be zero. By the process of renumbering, $N \leq M$.

If $N < M$, then the $s_i(t)$ constitute a linearly dependent set of functions. If on the other hand $N = M$, then the $s_i(t)$ form a linearly independent set of functions. When we apply the algorithm, $u_{M+1}(t)$ is always zero (can you see why?), and the expansion in terms of the orthogonal functions is exact, and therefore, the expansion, Eq. (10.14), is also exact.

We now convert to the orthonormal functions, $\phi_i(t)$. Let

$$\|u_i(t)\|_f = \sqrt{\int_0^T u_i^2(t)dt}$$

Then

$$\phi_i(t) = \frac{u_i(t)}{\|u_i(t)\|_f}$$

$$\text{or } u_i(t) = \phi_i(t) \|u_i(t)\|_f$$

Then converting Eq. (10.16) using this equation,

$$u_i(t) = s_i(t) - \sum_{j=1}^{i-1} \frac{\int_0^T u_j(t)s_i(t)dt}{\int_0^T u_j^2(t)dt} u_j(t) \quad i = 1, 2, \ldots, N$$

$$\phi_i(t) \|u_i(t)\| = s_i(t) - \sum_{j=1}^{i-1} \frac{\int_0^T \phi_j(t) \|u_j(t)\|_f s_i(t)dt}{\int_0^T \phi_j^2(t) \|u_j(t)\|_f^2 dt} \phi_j(t) \|u_j(t)\|_f \quad i = 1, 2, \ldots, N$$

$$= s_i(t) - \sum_{j=1}^{i-1} \phi_j(t) \int_0^T \phi_j(t)s_i(t)dt$$

$$\text{or } \phi_i(t) = \frac{1}{\|u_i(t)\|_f} \left[ s_i(t) - \sum_{j=1}^{i-1} s_{ij}\phi_j(t) \right] \quad i = 1, 2, \ldots, N \tag{10.17}$$

**Example 10.1** Find the set of orthonormal functions for the signal set shown in Fig. 10.4.

Following the procedure, from Equation set (10.17),

$$u_1(t) = s_1(t)$$

and

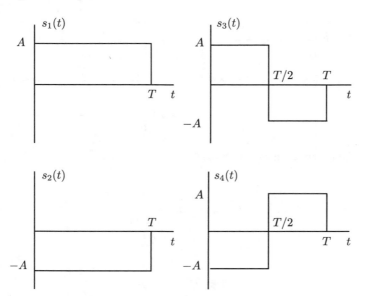

**Fig. 10.4** Signals $s_1(t), \ldots, s_4(t)$

$$u_2(t) = s_2(t) - \frac{\int_0^T u_1(t)s_2(t)dt}{\int_0^T u_1^2(t)dt}u_1(t)$$

$$= s_2(t) - \frac{\int_0^T s_1(t)s_2(t)dt}{\int_0^T s_1^2(t)dt}s_1(t)$$

$$= s_2(t) + \underbrace{\frac{\int_0^T s_1(t)s_1(t)dt}{\int_0^T s_1^2(t)dt}}_{=1}s_1(t)$$

$$= 0$$

Next we take a look at $u_3(t)$

$$u_3(t) = s_3(t) - \frac{\int_0^T u_1(t)s_1(t)dt}{\int_0^T u_1^2(t)dt}u_1(t) - \frac{\int_0^T u_2(t)s_2(t)dt}{\int_0^T u_2^2(t)dt}u_2(t)$$

$$= s_3(t)$$

and similarly

$$u_4(t) = 0$$

So after renumbering the functions,

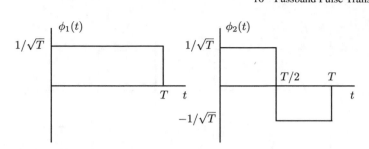

**Fig. 10.5** Orthonormal functions $\phi_1(t)$ and $\phi_2(t)$

$$\phi_1(t) = \frac{u_1(t)}{\|u_1(t)\|_f}$$

$$= \frac{s_1(t)}{A\sqrt{T}}$$

$$\text{and } \phi_2(t) = \frac{s_3(t)}{A\sqrt{T}}$$

The orthonormal functions, $\phi_1(t)$ and $\phi_2(t)$, are shown in Fig. 10.5. In this example, $M = 4$ and $N = 2$. Furthermore,

$$s_{11} = \int_0^T s_1(t)\phi_1(t)dt = A\sqrt{T}$$

$$s_{12} = \int_0^T s_1(t)\phi_2(t)dt = 0$$

$$\text{and } s_{21} = \int_0^T s_2(t)\phi_1(t)dt = -A\sqrt{T}$$

$$s_{22} = 0$$

and so on.

**Exercise 10.1** Find the set of orthonormal functions and other parameters for the signal set shown in Fig. 10.6.

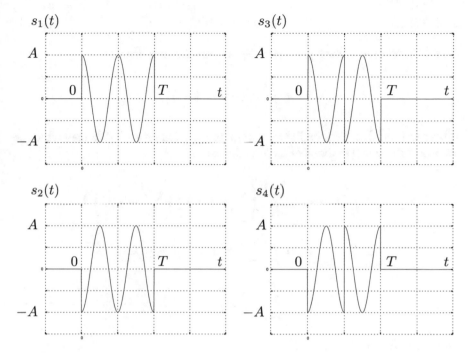

**Fig. 10.6** $s_1(t)$ to $s_4(t)$ for Exercise 10.1

## 10.5  The Signal Vector Space-II

Continuing from Eq. (10.14) and writing theses equations more concisely as

$$s_i(t) = \sum_{j=1}^{N} s_{ij}\phi_j(t) \tag{10.18}$$

where the vector $\mathbf{s}_i$ is defined by

$$\mathbf{s}_i = \begin{bmatrix} s_{i1} & s_{i2} & \cdots & s_{iN} \end{bmatrix} \quad i = 1, \cdots M \tag{10.19}$$

Similarly, we can form the vector of functions:

$$\boldsymbol{\phi}(t) = \begin{bmatrix} \phi_1(t) \\ \phi_2(t) \\ \vdots \\ \phi_N(t) \end{bmatrix} \tag{10.20}$$

so that the signal

$$s_i(t) = \mathbf{s}_i \bullet \boldsymbol{\phi}(t) \tag{10.21}$$

To find the energy of the signal $s_i(t)$, $E_i$, we perform the integration

$$E_i = \int_0^T s_i^2(t)dt = \|s_i(t)\|_f^2$$

where $\|s_i(t)\|_f^2$ is called the *function norm* of the function $s_i(t)$. We may write the energy of the signal in terms of the function norm.

$$E_i = \int_0^T s_i^2(t)dt = \int_0^T \left(\sum_{j=1}^N s_{ij}\phi_j(t)\right)\left(\sum_{k=1}^N s_{ik}\phi_k(t)\right)dt$$

$$= \int_0^T \sum_{k=1}^N \sum_{j=1}^N s_{ij}s_{ik}\phi_j(t)\phi_k(t)dt$$

$$= \sum_{k=1}^N \sum_{j=1}^N \int_0^T s_{ij}s_{ik}\phi_j(t)\phi_k(t)dt \tag{10.22}$$

Now since Eq. (10.13) is valid,

$$E_i = \sum_{k=1}^N \sum_{j=1}^N \int_0^T s_{ij}s_{ik}\phi_j(t)\phi_k(t)dt$$

$$= \sum_{k=1}^N \sum_{j=1}^N s_{ij}s_{ik}\delta_{jk}$$

$$= \sum_{j=1}^N s_{ij}s_{ij}$$

$$= \sum_{j=1}^N s_{ij}^2$$

$$= \mathbf{s}_i \mathbf{s}_i^T$$

$$= \|\mathbf{s}_i\|_v^2$$

where $\|\mathbf{s}_i\|_v$ is called the *vector norm* of the vector $\mathbf{s}_i$.

$$\|\mathbf{s}_i\|_v = \sqrt{s_{i1}^2 + s_{i2}^2 + \cdots s_{iN}^2} \tag{10.23}$$

This equation tells us that the energy in a signal is the magnitude squared of the vector characterising $s_i(t)$. In fact, this result is in accordance with the Raleigh energy theorem for the Fourier series given in Sect. 2.9. In other words, the vector $\mathbf{s}_i$ defines the signal $s_i(t)$. Furthermore,

$$\sqrt{E_i} = \sqrt{\int_0^T s_i^2(t)dt} = \|\mathbf{s}_i\|_v = \|s_i(t)\|_f \tag{10.24}$$

We may also define the Euclidean distance between two vectors, $\mathbf{s}_k$ and $\mathbf{s}_l$, through

$$\|\mathbf{s}_k - \mathbf{s}_l\|_v^2 = \sum_{j-1}^N \left(s_{kj} - s_{lj}\right)^2$$

$$= \int_0^T [s_k(t) - s_l(t)]^2 \, dt$$

Observing the previous equation, we observe that $\|\mathbf{s}_k - \mathbf{s}_l\|_v$ is the Euclidean distance $d_{kl} (= d_{lk})$ between vectors $\mathbf{s}_k$ and $\mathbf{s}_l$. To calculate the $s_{ij}$, we use the orthogonality property of the $\phi(t)$s. For example, from Equation set (10.14),

$$s_1(t) = s_{11}\phi_1(t) + s_{12}\phi_2(t) + \cdots \mid s_{1N}\phi_N(t)$$

$$\text{or } s_1(t)\phi_1(t) = s_{11}\phi_1^2(t) + s_{12}\phi_2(t)\phi_1(t) + \cdots + s_{1N}\phi_N(t)\phi_1(t)$$

Integrating this last equation,

$$\int_0^T s_1(t)\phi_1(t)dt = \underbrace{\int_0^T s_{11}\phi_1^2(t)dt}_{-s_{11}} + \underbrace{\int_0^T s_{12}\phi_2(t)\phi_1(t)dt + \cdots + \int_0^T s_{1N}\phi_N(t)\phi_1(t)dt}_{\text{All}=0}$$

$$s_{11} = \int_0^T s_1(t)\phi_1(t)dt$$

where all the terms shown in the underbracket $\underbrace{\cdots}_{\text{All}=0}$ are zero due to the orthonormality property (10.13), and

$$\int_0^T \phi_1^2(t)dt = 1$$

again due to the same property.

Generalising this result,

$$s_{ij} = \int_0^T s_i(t)\phi_j(t)dt \quad i = 1, \ldots, M; \quad j = 1, \ldots, N \tag{10.25}$$

The generation of $s_i(t)$ involves implementing Eq. (10.18), and the generation (on the output) of $s_{ij}$ involves Eq. (10.25). These two processes are shown in Fig. 10.7.

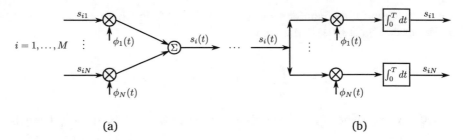

(a)                                                                          (b)

**Fig. 10.7** **a** Generation of $s_i(t)$ on transmission and **b** Generation of $s_{ij}$ upon reception. Refer Fig. 10.3

## 10.6  The Received Signal

Now the signal $s_i(t)$ is received as $r(t)$, the received signal. This received signal is the message signal corrupted with white noise,

$$r(t) = s_i(t) + w(t) \tag{10.26}$$

where $w(t)$ is white noise.

Referring to Fig. 10.7b, the received signal is subjected to a bank of integrators where integration is performed with $\phi_j(t)$. The transmitter and receiver are shown in Fig. 10.8.

Now,

$$\begin{aligned} r_j &= \int_0^T s_i(t)\phi_j(t)dt + \int_0^T w(t)\phi_j(t)dt \\ &= s_{ij} + w_j \quad j = 1, \ldots, N \end{aligned} \tag{10.27}$$

or

$$\mathbf{r} = \mathbf{s}_i + \mathbf{w} \tag{10.28}$$

Figure 10.9 shows the relation between the signal vector with the received vector. Each of these vectors is a $N$-dimensional vector.

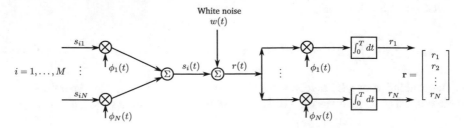

**Fig. 10.8** The received signal

**Fig. 10.9** Figure showing the received signal vector with the vector created from the signal which is sent. This diagram is in $N$-dimensional space corresponding the $N$-dimensional vector $s_i$ $i = 1, \ldots, M$,

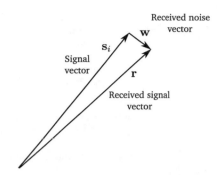

Let us compute the pdf of $R_j$. We start with Eq. (10.27). The average value of this equation is

$$E[s_{ij} + W_j] = s_{ij}$$

where we have used $W_j$ and not $w_j$ since it is a random variable. Note that

$$E[W_j] = E[\int_0^T W(t)\phi_j(t)dt]$$

$$= \int_0^T E[W(t)]\phi_j(t)dt$$

$$= 0$$

and the standard deviation of $R_j$ is given by $\sigma_{R_j}^2$ through

$$E\left[(R_j - s_{ij})^2\right] = E[W_j^2]$$

$$= E[\int_0^T \int_0^T W(u)\phi_j(u)W(t)\phi_j(t)dtdu]$$

$$= \int_0^T \int_0^T E[W(u)W(t)]\phi_j(u)\phi_j(t)dtdu$$

$$= \frac{N_0}{2} \int_0^T \int_0^T \delta(t - u)\phi_j(u)\phi_j(t)dtdu$$

$$= \frac{N_0}{2} \tag{10.29}$$

Since

$$E[W(u)W(t)] = \frac{N_0}{2}\delta(t - u) \tag{10.30}$$

Each $R_j$, therefore, has a pdf given by (see Eq. (3.150))

$$f_{R_j}(r_j|m_i) = \frac{1}{\sigma_{R_j}\sqrt{2\pi}} e^{-\left[\left(r_j - \mu_{R_j}\right)^2/\left(2\sigma_{R_j}^2\right)\right]} \qquad j = 1, \ldots, N; \ i = 1, \ldots, M$$

$$= \frac{1}{\sqrt{\frac{N_0}{2}}\sqrt{2\pi}} e^{-\left[\left(r_j - s_{ij}\right)^2/\left(2\frac{N_0}{2}\right)\right]}$$

$$= \frac{1}{\sqrt{N_0\pi}} \exp\left\{-\left[\left(r_j - s_{ij}\right)^2/N_0\right]\right\} \tag{10.31}$$

Now for a memoryless channel, $R_l$ and $R_m$, $l \neq m$, are independent of each other; therefore,

$$f_{\mathbf{R}}(\mathbf{r}|m_i) = \prod_{j=1}^{N} f_{R_j}(r_j|m_i) \qquad i = 1, \ldots, M$$

$$= (N_0\pi)^{-(N/2)} \exp\left\{-\sum_{j=1}^{N}\left[\left(r_j - s_{ij}\right)^2/N_0\right]\right\} \tag{10.32}$$

The conditional probabilities, $f_{\mathbf{R}}(\mathbf{r}|m_i)$ for each $m_i$ are called the *likelihood functions* and are also the *channel transitional probabilities*. It should be noted that for the noise,

$$\sum_{j=1}^{N} w_j \phi_j(t) \neq w(t) \tag{10.33}$$

since $\phi_j(t) \ \ j = 1, \ldots, N$ does not completely span the complete function space, but it does span the signal space $s_i(t)$. That is, rewriting (10.18),

$$s_i(t) = \sum_{j=1}^{N} s_{ij}\phi_j(t) \tag{10.34}$$

We now come to the decision-making process that when we receive the signal $r(t)$ and obtain $\mathbf{r}$ using a bank of integrators given in Fig. 10.8, we have to decide which signal it is among the signals

$$S = \{s_1(t), \ldots, s_M(t)\} \tag{10.35}$$

It is obvious that when the noise is small, the distance

$$d_i = \|\mathbf{s}_i - \mathbf{r}\|_v \tag{10.36}$$

will be small, and intuitively we may say that out of all $d_i$, we choose the smallest and decide that the smallest $d_i = d_k$ implies that the signal sent was $s_k(t)$ and consequently that the message sent was $m_k(t)$.

## 10.7  Probability of Error

To convert this intuitive result into a quantitative one, we investigate what is the probability that an error occurred. That is, if we decide that message $m_i$ was sent, when it was not

$$P_e = P(m_i|\mathbf{r}) \text{ where } m_i \text{ was not sent} \qquad (10.37)$$

which is the same as

$$P_e = 1 - P(m_i|\mathbf{r}) \text{ where } m_i \text{ was sent} \qquad (10.38)$$

so the optimum decision rule is

$$P(m_i \text{ sent}|\mathbf{r}) \geq P(m_j \text{ sent}|\mathbf{r}) \quad \forall \text{ values of } j \neq i \qquad (10.39)$$

This is called *the maximum a posteriori probability rule or MAP rule*. But we are interested more in the a priori probabilities. So we can say that using Baye's rule,

$$p_i \frac{f_\mathbf{R}(\mathbf{r}|m_i)}{f_\mathbf{R}(\mathbf{r})} \text{ is a maximum for } l = k \Rightarrow m_k \text{ was sent}$$

where $p_i$ is the probability of transmission of $s_i(t)$ which is generally to be assumed to be $1/M$. However, $f_\mathbf{R}(\mathbf{r})$ is independent of the message signal. Therefore,

$$f_\mathbf{R}(\mathbf{r}|m_i) \text{ is a maximum for } i = k \Rightarrow m_k \text{ was sent}$$

We now use the logarithm of this function since $f_\mathbf{R}(\mathbf{r}|m_i) > 0$. So

$$\ln[f_\mathbf{R}(\mathbf{r}|m_i)] \text{ is a maximum for } i = k \Rightarrow m_k \text{ was sent}$$

Therefore, using Eq. (10.32),

$$\sqrt{\sum_{j=1}^{N}(r_i - s_{ji})^2} \text{ is a minimum for } i = k \Rightarrow m_k \text{ was sent}$$

or this result is the same as

$$\|s_i - \mathbf{r}\|_v|_{\min} \forall \text{ values of } i \text{ is } k \Rightarrow \{\|s_k - \mathbf{r}\|_v \text{ is a minimum} \Rightarrow m_k \text{ was sent}\}$$

Now to calculate the probability that there was en error, we first assume that message $k$ was sent, the decision was that a different message was sent. The Euclidean distances

$$d_{ik} = \|s_k - s_i\|_v$$

determine the hyperplanes that separate $s_i$ from $s_k$. That is, if

$$d_i > d_k$$

where

$$d_i = \|s_i - r\|_v$$

then an error occurs. The error bounds occur at $d_{ik}/2$.

To define the probability of error, we partition the $N$-dimensional hyperspace into regions, $R_i$, such that if $r$ falls in $R_i$, then we decide that message $m_i$ was sent based on the Euclidean distance (10.36), and the average probability that an error occurred is

$$P_e = \frac{1}{M} \sum_{i=1}^{M} \int_{\cup R_j j \neq i} f_R(r|m_i) dr$$

However, to calculate this integral in even the most simplest cases for $N \geq 2$ is most difficult. We now proceed in the following manner deciding a lower bound.

We take the set of $s_i(t)$ given by (10.35) from which we compute $s_i$   $i = 1, \ldots, M$. We now compute the Euclidean distances

$$d_{ik} = \sqrt{\sum_{j=1}^{N} \left( s_{ij}^2 - s_{kj}^2 \right)} = \|s_i - s_k\|_v$$

for a given $i$ and for different values of $k$,   $k = 1, \ldots, M$. Then the probability of error is less than the sum of the error probabilities for each of the signal vectors

$$P_e(m_i) \leq \sum_{k=1}^{M, \, k \neq i} P_{e(i)}(k)$$

where $P_{e(i)}(k)$ is the tail probability that $s_k$ was selected as the transmitted signal, where instead it was $s_i$ which was transmitted. Figure 10.10 shows the region of integration for these tail probabilities (see Sect. 9.2 for similar relevant information).

The $P_{e(i)}(k)$s can be calculated by using

$$P_{e(i)}(k) = \frac{1}{\sqrt{\pi N_0}} \int_{d_{ik}/2}^{\infty} \exp\left\{ -\frac{x^2}{N_0} \right\} dx$$

$$= \frac{1}{2} \operatorname{erfc}\left( \frac{d_{ik}}{2\sqrt{N_0}} \right) \tag{10.40}$$

where erfc is the complementary error function:

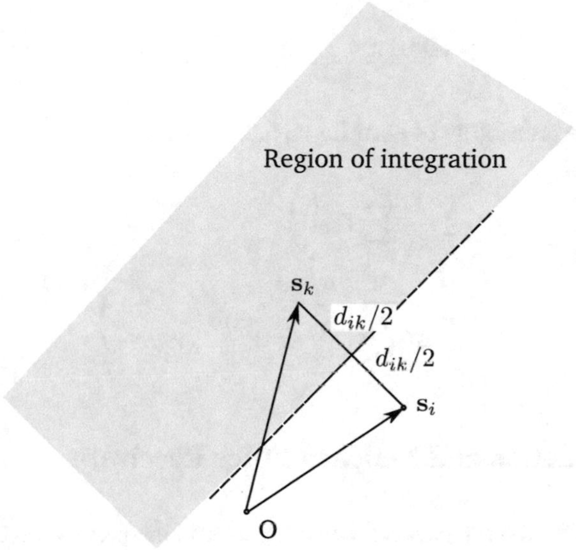

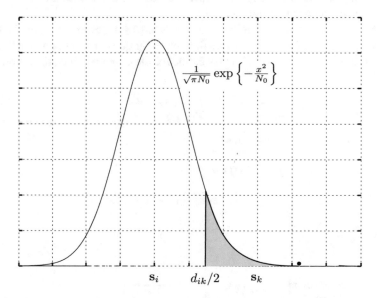

**Fig. 10.10**   Region of integration for $P_{e(i)}(k)$

**Fig. 10.11**   Region of integration to calculate $P_{e(i)}(k)$ from $\mathbf{s}_i$ to $\mathbf{s}_k$

$$\text{erfc}(x) = \frac{2}{\sqrt{\pi}} \int_x^\infty e^{-z^2} dz$$

And the region of integration is shown as the shaded part in Fig. 10.11. So

$$P_e(m_i) \leq \sum_{k=1}^{M,\,k\neq i} \frac{1}{2}\mathrm{erfc}\left(\frac{d_{ik}}{2\sqrt{N_0}}\right)$$

and the average probability of error for all values of $i$ is given by

$$P_e \leq \frac{1}{M}\sum_{i=1}^{M} P_{e(i)}(k)$$

$$= \frac{1}{M}\sum_{i=1}^{M}\left[\sum_{k=1}^{M,\,k\neq i}\frac{1}{2}\mathrm{erfc}\left(\frac{d_{ik}}{2\sqrt{N_0}}\right)\right] \qquad (10.41)$$

## 10.8   Correlation and Matched Filter Receivers

We have considered a typical correlation receiver in the previous sections. We now consider the matched filter receiver and show its equivalence with the correlation receiver considered earlier. The basic idea between a matched filter and correlation receiver has been considered in Sect. 9.1.

Let us show the equivalence between the use of integrators and matched filter receivers as shown in Fig. 10.12 where the two techniques of detection give the same value at sample instants $t = T$. To prove this, for the upper receiver, the output $y(T)$ at $t = T$ is

$$y(T) = \int_0^T s(t)\phi(t)dt \qquad (10.42)$$

Considering now the lower figure,

$$y(t) = s(t) * \phi(T - t)$$

$$= \int_{-\infty}^{\infty} s(\tau)\phi[T - (t - \tau)]d\tau$$

Now for $t = T$,

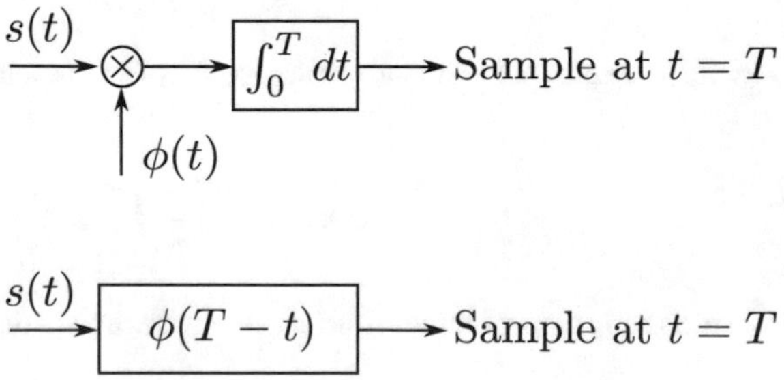

**Fig. 10.12** Correlative and matched filter receivers

$$y(T) = \int_{-\infty}^{\infty} s(\tau)\phi[T - (t - \tau)]d\tau \Big|_{t=T}$$

$$= \int_{-\infty}^{\infty} s(\tau)\phi[T - (T - \tau)]d\tau$$

$$= \int_{-\infty}^{\infty} s(\tau)\phi(\tau)d\tau$$

We also know that $s(t)$ is nonzero in the interval $[0, T]$ and zero outside this interval; therefore,

$$y(T) = \int_{0}^{T} s(\tau)\phi(\tau)d\tau$$

which is the same as Eq. (10.42).

## 10.9 Coherent Modulation Techniques

We now proceed to examine different modulation techniques applying the theory which we have developed to different digital systems. First, we will examine some of the coherent modulation techniques.

### 10.9.1 Coherent Binary PSK (BPSK)

We consider two symbols here,

$$s(t) = \begin{cases} s_1(t) = A\cos(\omega_c t) & \text{for symbol 1} \\ s_2(t) = A\cos(\omega_c t + \pi) = -A\cos(\omega_c t) & \text{for symbol 0} \end{cases}$$

where $0 \le t \le T_b$ and

$$T_b = n_c\left(\frac{2\pi}{\omega_c}\right) = n_c T$$

where $n_c$ is a positive integer.

For this set, we define the first (and only) orthogonal function

$$u_1(t) = A\cos(\omega_c t) \tag{10.43}$$

then

$$\int_{0}^{T_b} u_1^2(t)dt = \frac{T_b A^2}{2} \tag{10.44}$$

so from the orthogonal function, we derive the orthonormal function $\phi_1(t)$ as

$$\phi_1(t) = \left(\sqrt{\frac{T_b A^2}{2}}\right)^{-1} u_1(t) = \sqrt{\frac{2}{T_b}} \cos(\omega_c t) \tag{10.45}$$

so that

$$\int_0^{T_b} \phi_1^2(t)dt = 1$$

If we set the energy per bit as

$$E_b = \frac{T_b A^2}{2}$$

then

$$A = \sqrt{\frac{2E_b}{T_b}}$$

so

$$s_1(t) = \sqrt{\frac{2E_b}{T_b}} \cos(\omega_c t) = \sqrt{E_b}\phi_1(t)$$

$$\text{and } s_2(t) = -\sqrt{\frac{2E_b}{T_b}} \cos(\omega_c t) = -\sqrt{E_b}\phi_1(t) \tag{10.46}$$

In this case, $M = 2$ and $N = 1$.

So if the received signal vector is

$$\mathbf{r} = [r_1]$$

with only a single component,

$$s_{11} = \int_0^{T_b} s_1(t)\phi_1(t)dt = \sqrt{E_b}$$

$$s_{21} = \int_0^{T_b} s_2(t)\phi_1(t)dt = -\sqrt{E_b}$$

$$d_{12} = d_{21} = 2\sqrt{E_b} \tag{10.47}$$

Now the halfway point between $\sqrt{E_b}$ and $-\sqrt{E_b}$ is zero; therefore, if $r_1 > 0$, it implies that the transmitted signal was $s_1(t)$ (or that a 1 was transmitted). On the other hand, if $r_1 < 0$, then we would assume that $s_2(t)$ (or 0) was transmitted.

To calculate the probability of error, we use Eq. (10.41)

**Fig. 10.13** Details
concerning BPSK

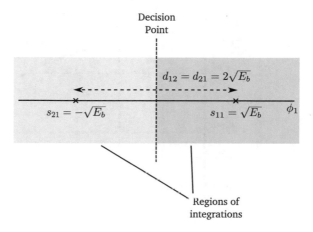

$$P_e \leq \frac{1}{2}\left[\frac{1}{2}\text{erfc}\left(\frac{d_{12}}{2\sqrt{N_0}}\right) + \frac{1}{2}\text{erfc}\left(\frac{d_{21}}{2\sqrt{N_0}}\right)\right]$$

$$= \frac{1}{2}\left[\frac{1}{2}\text{erfc}\left(\frac{\sqrt{E_b}}{\sqrt{N_0}}\right) + \frac{1}{2}\text{erfc}\left(\frac{\sqrt{E_b}}{\sqrt{N_0}}\right)\right] \qquad (10.48)$$

so

$$P_e = \frac{1}{2}\text{erfc}\left(\sqrt{\frac{E_b}{N_0}}\right) \qquad (10.49)$$

the equality holds since no approximations are involved (the analysis is exact since the union of integrations do not overlap). Figure 10.13 shows the details concerning BPSK. The average probability of error for different values of $E_b/N_0$ is shown in Fig. 10.14.

To modulate and demodulate BPSK, we use the block diagram shown in Fig. 10.15. The figure shows that the input bit stream is converted into positive and negative pulses which have a magnitude of $\sqrt{E_b}$, which is the energy per bit. We then multiply these pulses by $\phi_1(t) = \sqrt{2/T_b}\cos(\omega_c t)$ which gives us an output having the correct phase and magnitude information. On reception, the signal which is corrupted by noise is multiplied by $\phi_1(t)$ and integrated over the interval $[0, T_b]$ which gives $r_1$. Here, we are using coherent detection since if the LO of the transmitter and the LO of the receiver are not in phase, proper demodulation is not possible. This point is one of the most important points to be borne in mind.

### 10.9.1.1   Power Spectral Density of BPSK

We now take a look at the power spectral density of BPSK. We first have to understand that the power spectral density is centred at $\omega = \omega_c$ due to the presence of the carrier. However, we concentrate on the PSD as shifted to the baseband. We take the results

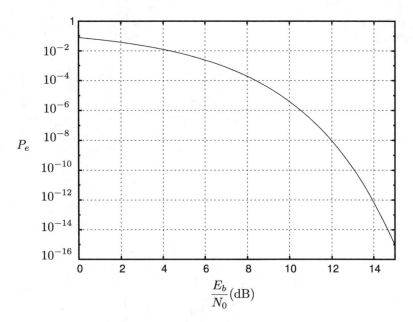

**Fig. 10.14**   Average probability of error for BPSK

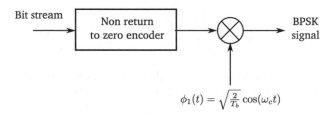

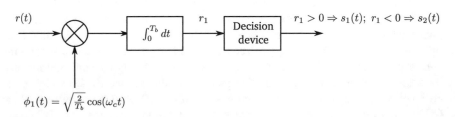

**Fig. 10.15**   Modulation and demodulation of BPSK

of Sect. 8.6.3 and write

$$S_y(\omega) = |P(\omega)|^2 S_x(\omega) \tag{10.50}$$

where

$$S_x(\omega) = \frac{1}{T_b}\left(R_0 + 2\sum_n R_n \cos n\omega T\right)$$

$$P(\omega) = T_b\sqrt{\frac{2E_b}{T_b}}\frac{\sin(\pi f T_b)}{(\pi f T_b)} \tag{10.51}$$

$$R_n = \lim_{N\to\infty}\frac{1}{N}\sum_n^N a_k a_{k+n} \qquad a_k = -1, +1 \tag{10.52}$$

In the case of BPSK,

$$R_0 = \lim_{N\to\infty}\frac{1}{N}\sum_n^N a_k^2$$

$$= \lim_{N\to\infty}\frac{1}{N}\times N$$

$$= 1$$

while

$$R_n = 0 \qquad \text{for } n \neq 0$$

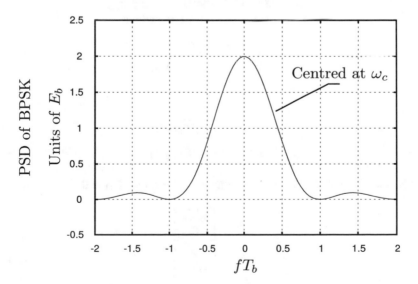

**Fig. 10.16** Power spectral density of BPSK

Therefore,

$$S_y(f) = 2E_b \frac{\sin^2{(\pi f T_b)}}{(\pi f T_b)^2}$$

The power spectral density is shown in Fig. 10.16.

## 10.9.2   Coherent Binary FSK (BFSK)

We now consider coherent binary FSK. The two signals are

$$s(t) = \begin{cases} s_1(t) = \sqrt{\frac{2E_b}{T_b}} \cos(\omega_{c1}t) & \text{for bit 1} \\ \text{and } s_2(t) = \sqrt{\frac{2E_b}{T_b}} \cos(\omega_{c2}t) & \text{for bit 0} \end{cases}$$

where

$$\omega_{c1} = \left(\frac{2\pi}{T_b}\right)(n_c + i)$$

$$\omega_{c2} = \left(\frac{2\pi}{T_b}\right)(n_c + j) \quad i, j = 1, \ldots; \ i \neq j \tag{10.53}$$

and $n_c$ is some positive integer. For these two signals,

$$\phi_1 = \begin{cases} \sqrt{\frac{2}{T_b}} \cos(\omega_{c1}t) & 0 \leq t \leq T_b \\ 0 & \text{elsewhere} \end{cases}$$

$$\phi_2 = \begin{cases} \sqrt{\frac{2}{T_b}} \cos(\omega_{c2}t) & 0 \leq t \leq T_b \\ 0 & \text{elsewhere} \end{cases} \tag{10.54}$$

so

$$s_1(t) = \sqrt{E_b}\phi_1(t)$$
$$\text{and } s_2(t) = \sqrt{E_b}\phi_2(t) \tag{10.55}$$

In this case, $M = 2$ and $N = 2$. So computing the value of $\mathbf{s}_1$ and $\mathbf{s}_2$,

$$\mathbf{s}_1 = \begin{bmatrix} \sqrt{E_b} \\ 0 \end{bmatrix}$$

$$\mathbf{s}_2 = \begin{bmatrix} 0 \\ \sqrt{E_b} \end{bmatrix} \tag{10.56}$$

The received signal vector is

$$\mathbf{r} = \begin{bmatrix} r_1 \\ r_2 \end{bmatrix}$$

so that if

$$\|\mathbf{r} - \mathbf{s}_1\|_v < \|\mathbf{r} - \mathbf{s}_2\|_v$$

we decide that $s_1(t)$ (or a 1) was sent, and if

$$\|\mathbf{r} - \mathbf{s}_1\|_v > \|\mathbf{r} - \mathbf{s}_2\|_v$$

then we decide that a zero was sent. And

$$\|\mathbf{s}_1 - \mathbf{s}_2\|_v = d_{12} = d_{21} = \sqrt{2E_b} \tag{10.57}$$

Using these values, we are in a position to calculate the probability of error, $P_e$,

$$P_e \leq \frac{1}{2}\left[\frac{1}{2}\mathrm{erfc}\left(\frac{d_{12}}{2\sqrt{N_0}}\right) + \frac{1}{2}\mathrm{erfc}\left(\frac{d_{21}}{2\sqrt{N_0}}\right)\right]$$
$$= \frac{1}{2}\left[\frac{1}{2}\mathrm{erfc}\left(\frac{\sqrt{E_b}}{\sqrt{2N_0}}\right) + \frac{1}{2}\mathrm{erfc}\left(\frac{\sqrt{E_b}}{\sqrt{2N_0}}\right)\right]$$

or

$$P_e = \frac{1}{2}\mathrm{erfc}\left(\sqrt{\frac{E_b}{2N_0}}\right) \tag{10.58}$$

The equality holds since no approximations are involved (the analysis is exact since the union of integrations does not overlap). Figure 10.17 shows the details concerning BFSK. The average probability of error for different values of $E_b/N_0$ is shown in Fig. 10.18. If we compare the average probability of error of BPSK with BFSK, we find that for the same $E_b/N_0$ BPSK has a much lower probability of error than BFSK, so it is much more superior than BFSK.

To modulate and demodulate BFSK, we use the block diagram shown in Fig. 10.19. The figure shows that the input bit stream is converted into positive pulses for the symbol 1 and no pulses for symbol zero. Each of the positive pulses has a magnitude of $\sqrt{E_b}$, the energy per bit. We then divide the input into two parts and one part is multiplied by $\phi_1(t) = \sqrt{2/T_b}\cos(\omega_{c1}t)$ for symbol 1 ($s_1(t)$), while the other part is passed through an inverter and then multiplied by $\phi_2(t) = \sqrt{2/T_b}\cos(\omega_{c2}t)$ for symbol 0 ($s_2(t)$). The original pulse stream has only positive symbols corresponding to symbol 1 and no pulses for symbol 0. After using an inverter, the 0 symbols have positive pulses and the 1 symbols have no pulses. After the multiplication, these two streams are added together to give BFSK.

On reception, the signal which is corrupted by noise is again divided into two parts, one part is multiplied by $\phi_1(t)$ and integrated over the interval $[0, T_b]$ while the

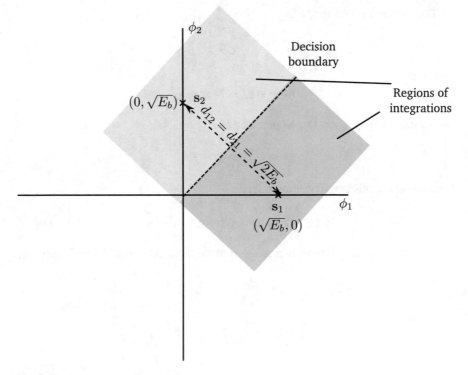

**Fig. 10.17**  Details concerning BFSK

second part is multiplied by $\phi_2(t)$ and similarly integrated over the interval $[0, T_b]$. The result of these two integrations is subtracted as shown in the figure. The output is then used as an input to decide whether $s_1(t)$ or $s_2(t)$ was transmitted. The PSD of BFSK is similar to BPSK with a few differences.

### 10.9.3   Coherent Quadrature PSK (QPSK)

In QPSK, the signals which we consider are as follows:

$$s_1(t) = \sqrt{\frac{2E_s}{T_s}} \cos(\omega_c t) \quad \text{for bits 00}$$

$$s_1(t) = \sqrt{\frac{2E_s}{T_s}} \sin(\omega_c t) \quad \text{for bits 01}$$

$$s_1(t) = -\sqrt{\frac{2E_s}{T_s}} \cos(\omega_c t) \quad \text{for bits 10}$$

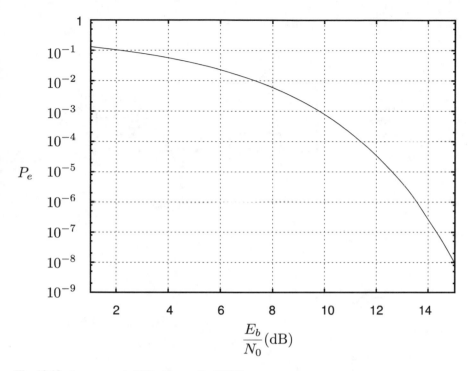

**Fig. 10.18** Average probability of error for BFSK

$$s_1(t) = -\sqrt{\frac{2E_s}{T_s}} \sin(\omega_c t) \qquad \text{for bits } 11$$

where $E_s$ is the energy per symbol, and $T_s$ is the symbol duration. For these signals,

$$\phi_1(t) = \sqrt{\frac{2}{T_s}} \cos(\omega_c t)$$

$$\phi_2(t) = \sqrt{\frac{2}{T_s}} \sin(\omega_c t) \qquad (10.59)$$

and therefore,

$$s_1(t) = \sqrt{E_s}\phi_1(t)$$
$$s_2(t) = \sqrt{E_s}\phi_2(t)$$
$$s_3(t) = -\sqrt{E_s}\phi_1(t)$$
$$s_4(t) = -\sqrt{E_s}\phi_2(t) \qquad (10.60)$$

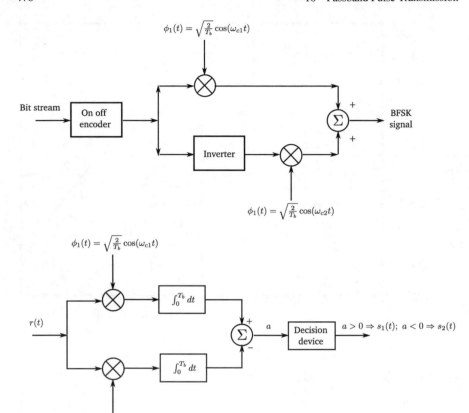

**Fig. 10.19** Modulation and demodulation of BFSK

In this case, $M = 4$ and $N = 2$. So

$$\mathbf{s}_1 = \begin{bmatrix} \sqrt{E_s} \\ 0 \end{bmatrix}$$

$$\mathbf{s}_2 = \begin{bmatrix} 0 \\ \sqrt{E_s} \end{bmatrix}$$

$$\mathbf{s}_3 = \begin{bmatrix} -\sqrt{E_s} \\ 0 \end{bmatrix}$$

$$\mathbf{s}_4 = \begin{bmatrix} 0 \\ -\sqrt{E_s} \end{bmatrix} \tag{10.61}$$

which are shown in Fig. 10.20.

To calculate the probability of error, we first need to use Eq. (10.41)

**Fig. 10.20** Diagram
showing $s_1, \ldots, s_4$ for
QPSK. Decision boundaries
are also shown for $s_1$. The
other decision boundaries for
$s_2, s_3$ and $s_4$ may be
calculated based on this
figure

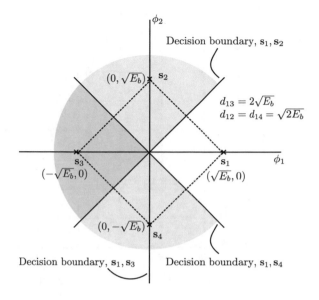

$$P_e \leq \frac{1}{M} \sum_{i=1}^{M} P_{e(i)}(k)$$

$$= \frac{1}{M} \sum_{i=1}^{M} \left[ \sum_{k=1}^{M,\, k \neq i} \frac{1}{2} \mathrm{erfc}\left( \frac{d_{ik}}{2\sqrt{N_0}} \right) \right]$$

From the symmetry of the diagram of QPSK constellation, and since the regions
of integrations overlap,

$$P_e \leq \sum_{k=1}^{M,\, k \neq i} \frac{1}{2} \mathrm{erfc}\left( \frac{d_{ik}}{2\sqrt{N_0}} \right) \tag{10.62}$$

where we may use any value of $i$. Using $i = 1$,

$$P_e \leq \frac{1}{2} \mathrm{erfc}\left( \frac{d_{12}}{2\sqrt{N_0}} \right) + \frac{1}{2} \mathrm{erfc}\left( \frac{d_{14}}{2\sqrt{N_0}} \right) + \frac{1}{2} \mathrm{erfc}\left( \frac{d_{13}}{2\sqrt{N_0}} \right)$$

$$= \mathrm{erfc}\left( \frac{d_{12}}{2\sqrt{N_0}} \right) + \frac{1}{2} \mathrm{erfc}\left( \frac{d_{13}}{2\sqrt{N_0}} \right) \tag{10.63}$$

since $d_{12} = d_{14}$. Or

$$P_e \leq \mathrm{erfc}\left( \sqrt{\frac{E_s}{2N_0}} \right) + \frac{1}{2} \mathrm{erfc}\left( \sqrt{\frac{E_s}{N_0}} \right) \tag{10.64}$$

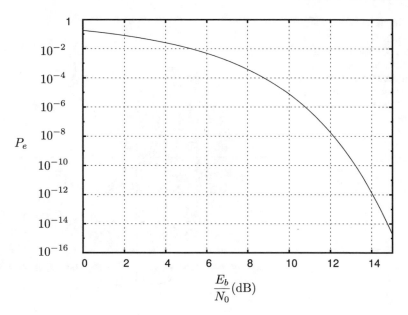

**Fig. 10.21**   Average probability of error, $P_e$, for QPSK

but each symbol represents two bits, so the energy per symbol is related to the energy per bit by

$$E_s = 2E_b \tag{10.65}$$

so

$$P_e \leq \text{erfc}\left(\sqrt{\frac{E_b}{N_0}}\right) + \frac{1}{2}\text{erfc}\left(\sqrt{\frac{2E_b}{N_0}}\right) \tag{10.66}$$

The average probability of error, $P_e$, plotted against $\sqrt{E_b/N_0}$ is shown in Fig. 10.21. In this equation,

$$\text{erfc}\left(\sqrt{\frac{E_b}{N_0}}\right) \gg \frac{1}{2}\text{erfc}\left(\sqrt{\frac{2E_b}{N_0}}\right)$$

and the second term may be neglected. Therefore,

$$P_e \approx \text{erfc}\left(\sqrt{\frac{E_b}{N_0}}\right) \tag{10.67}$$

A comparison with the average probability of error for BPSK (Fig. 10.14) shows that these two curves are almost Identical, and comparison with $P_e$ of BFSK shows far superior performance (Fig. 10.18).

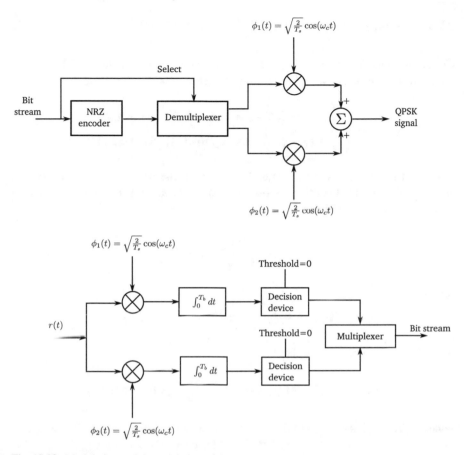

**Fig. 10.22**  Modulation and demodulation of QPSK

To modulate and demodulate QPSK, we use the block diagram shown in Fig. 10.22. The figure shows that the input bit stream is converted into positive or negative pulses for the symbols (00, 01, positive pulses) and (10, 11, negative pulses). The demultiplexer sends the pulses so produced to the correct arm. The carriers $\phi_1$ and $\phi_2$ then multiply these pulses as shown in the figure. The output is the sum of each arm of the diagram. The selection of the arm is done on the basis of two bits taken at a time.

On reception, the received signal is divided into two parts, one part is multiplied by $\phi_1(t)$ and integrated over the interval $[0, T_b]$ while the second part is multiplied by $\phi_2(t)$ and similarly integrated over the interval $[0, T_b]$. The result of these two integrations is subjected to two threshold decision devices. On the basis of these decisions, the multiplexer decides the two bits to be output.

## 10.9.4  Coherent Minimum Shift Keying (MSK)

Coherent minimum shift keying is a type of frequency modulation which includes phase information. Consider the FSK signal defined for $0 \le t \le T_b$ to be

$$
s(t) = \begin{cases} s_1(t) = \sqrt{\frac{2E_b}{T_b}} \cos\left[\omega_1 t + \theta(0)\right] & \text{for symbol 1} \\ s_2(t) = \sqrt{\frac{2E_b}{T_b}} \cos\left[\omega_2 t + \theta(0)\right] & \text{for symbol 0} \end{cases}
$$

where $\theta(0)$ is the phase at $t = 0$ and depends on the past history of the process such that the signal is continuous between bits. Considering the signal to be angle modulated, we may represent $s(t)$ as

$$
s(t) = \sqrt{\frac{2E_b}{T_b}} \cos[\theta(t)]
$$

where

$$
\theta(t) = \omega_c t \pm \frac{\pi h}{T_b} t + \theta(0) \qquad 0 \le t \le T_b
$$

At end of the $n$th bit, we will have

$$
\theta(nT_b - \Delta t) = \omega_c(nT_b - \Delta t) + \frac{b_a \pi h}{T_b}(nT_b - \Delta t) + \theta_a
$$

where $b_a$ is the plus or minus 1 during the $n$th bit and $\theta_a$ is the phase at the beginning of the $n$th bit. At the start of the next bit, we have

$$
\theta(nT_b + \Delta t) = \omega_c(nT_b + \Delta t) + \frac{b_a \pi h}{T_b} nT_b + \frac{b_b \pi h}{T_b}\Delta t + \theta_a
$$

where $b_b$ is equal to $\pm 1$ depending on the next bit. For the phase to be continuous,

$$
\theta(nT_b + \Delta t) - \theta(nT_b - \Delta t) = 2\omega_c \Delta t + \frac{(b_b + b_a)\pi h}{T_b}\Delta t
$$

must be zero in the limit as $\Delta t \to 0$ which is true.

Therefore, if consider this signal originating from the angle modulated signal,

$$
s(t) = \sqrt{\frac{2E_b}{T_b}} \cos\left[\omega_c t \pm \frac{\pi h}{T_b} t + \theta(0)\right] \qquad 0 \le t \le T_b
$$

where it will turn out that $h = (T_b/2\pi)(\omega_1 - \omega_2)$. The plus sign corresponds to the symbol 1, while the minus sign corresponds to the symbol 0. Then

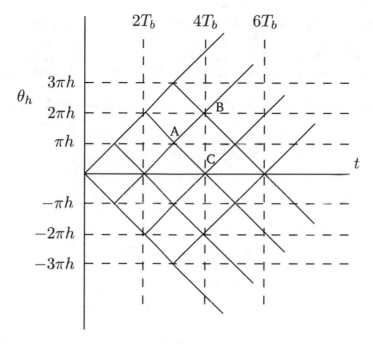

**Fig. 10.23**  Phase tree for continuous phase FSK

$$\theta(t) = \omega_c t \pm \frac{\pi h}{T_b} t + \theta(0)$$

$$\text{then } \omega(t) = \frac{d\theta}{dt} = \omega_c \pm \frac{\pi h}{T_b} \tag{10.68}$$

or

$$\omega_1 = \omega_c + \frac{\pi h}{T_b}$$

$$\omega_2 = \omega_c - \frac{\pi h}{T_b} \tag{10.69}$$

or

$$\omega_c = \frac{1}{2}(\omega_1 + \omega_2)$$

$$h = \frac{T_b}{2\pi}(\omega_1 - \omega_2) \tag{10.70}$$

$h$ is called the deviation ratio.

Let us consider only the phase information, dropping the term $\omega_c t$ and $\theta(0)$. All possible set of outcomes is shown in Fig. 10.23.

From the figure, we can see the possible outcomes of $\theta(t)$. For example, if we are at $A$ in the figure, at $t = 3T_b$, then $\theta(0) = \pi h$. Now if we transmit the symbol 1, then we go to $B$ where $\theta(t) = (\pi h/T_b)t + \pi h$; or if we transmit a symbol 0, then we go to $C$ where $\theta(t) = (-\pi h/T_b)t + \pi h$. The tree shows all possible pathways of $\theta(t)$.

In minimum shift keying (MSK), we choose $h = 1/2$ as a special case. In this case,

$$T_b(\omega_1 - \omega_2) = \pi$$

$$\omega_1 = \omega_2 + \frac{\pi}{T_b} \tag{10.71}$$

and

$$\omega_c = \frac{1}{2}(\omega_2 + \omega_2 + \frac{\pi}{T_b})$$

$$= \omega_2 + \frac{\pi}{2T_b} \tag{10.72}$$

or

$$T_b(f_1 - f_2) = \frac{1}{2} \tag{10.73}$$

The frequency difference is half the symbol rate:

$$f_1 - f_2 = \frac{1}{2} \times \frac{1}{T_b} \tag{10.74}$$

In MSK, Fig. 10.23 will be modified so that the phases along the y-axis will be in units of $\pi/2$ instead of $\pi h$ since $h = 1/2$.

We now look to transmit the sequence 1110011000, then we realise that this symbol sequence is graphed in Fig. 10.24. We can go through the symbol sequence in the following way:

$$
\begin{aligned}
\theta(t) &= \left[\omega_c + \frac{\pi}{2T_b}\right] t \quad 0 \leq t \leq T_b \quad \text{symbol 1} \\
&= \left[\omega_c + \frac{\pi}{2T_b}\right] t + \frac{\pi}{2} \quad T_b \leq t \leq 2T_b \quad \text{symbol 1} \\
&= \left[\omega_c + \frac{\pi}{2T_b}\right] t + \pi \quad 2T_b \leq t \leq 3T_b \quad \text{symbol 1} \\
&= \left[\omega_c - \frac{\pi}{2T_b}\right] t + \frac{3\pi}{2} \quad 3T_b \leq t \leq 4T_b \quad \text{symbol 0} \\
&\ \ \vdots
\end{aligned}
\tag{10.75}
$$

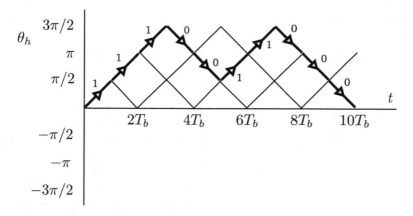

**Fig. 10.24** Phase trellis of the sequence 1110011000

and so on. In this way, we include phase information in MSK.

An MSK signal for the bits 01100 is shown in Fig. 10.25. In this figure, $\omega_1 = 2\pi/T_b$, $\omega_2 = 3\pi/T_b$ and $\omega_c = 5\pi/2T_b$. Notice that at the end of each bit, the phase becomes an even number of units of $\pi$ in $\cos[\theta(t)]$. In the final analysis, each bit is represented as

$$s_{1,2}(t) = \sqrt{\frac{2E_b}{T_b}} \cos\left(\frac{n\pi}{T_b}t \pm m\pi\right)$$

where $n$ = a positive integer or the next positive integer, and $m = 0$ or $1$.

Observing the cosine nature of the signals, we can choose the two functions, $\phi_1$ and $\phi_2$, as

$$\phi_1(t) = \sqrt{\frac{2}{T_b}} \cos(\omega_c t) \cos\left(\frac{\pi}{2T_b}t\right)$$

$$\phi_2(t) = \sqrt{\frac{2}{T_b}} \sin(\omega_c t) \sin\left(\frac{\pi}{2T_b}t\right) \qquad 0 \le t \le T_b$$

with

$$\int_0^{T_b} \phi_1(t)\phi_2(t)dt = 0$$

$$\int_0^{T_b} \phi_1^2(t)dt = 1$$

$$\int_0^{T_b} \phi_2^2(t)dt = 1$$

**Fig. 10.25** MSK signal

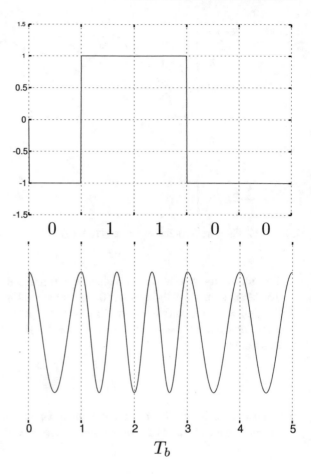

or we may write

$$s(t) = s_1\phi_1 + s_2\phi_2$$

where

$$s_1 = \int_{-T_b}^{T_b} s(t)\phi_1(t)dt$$

$$= \int_{-T_b}^{0} s(t)\phi_1(t)dt + \int_{0}^{T_b} s(t)\phi_1(t)dt$$

$$= \sqrt{E_b}\cos[\theta(0)]$$

$$= \begin{cases} \sqrt{E_b} & \theta(0) = 0 \\ -\sqrt{E_b} & \theta(0) = \pi \end{cases}$$

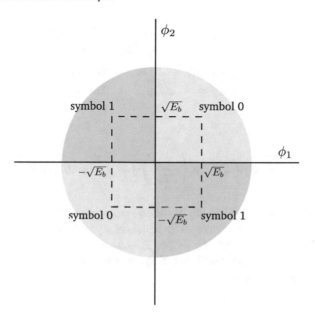

**Fig. 10.26** Constellation diagram for MSK

**Table 10.1** Table showing $(s_1, s_2)$ for MSK

| | 0 | 1 |
|---|---|---|
| $(s_1, s_2)(\sqrt{E_b})$ | (1, 1) | (−1, 1) |
| $(s_1, s_2)(\sqrt{E_b})$ | (−1, −1) | (1, −1) |

where integration is performed in $-T_b \le t \le T_b$. And

$$s_2 = \int_0^{2T_b} s(t)\phi_2(t)dt$$
$$= -\sqrt{E_b} \sin[\theta(T_b)]$$

where integration is performed in $0 \le t \le 2T_b$ and which gives

$$s_2 = \sqrt{E_b} \times \begin{cases} -1 & \theta(T_b) = \pi/2 \\ 1 & \theta(T_b) = -\pi/2 \end{cases}$$

so we have four values of $s_1$ and $s_2$ as shown in the constellation diagram shown in Fig. 10.26 and also shown in Table 10.1

We can then reevaluate the signals

$$s(t) = \begin{cases} \sqrt{E}\phi_1 + \sqrt{E}\phi_2 & \text{for symbol 0} \\ -\left(\sqrt{E}\phi_1 + \sqrt{E}\phi_2\right) & \text{for symbol 0} \\ -\sqrt{E}\phi_1 + \sqrt{E}\phi_2 & \text{for symbol 1} \\ \sqrt{E}\phi_1 - \sqrt{E}\phi_2 & \text{for symbol 1} \end{cases}$$

or in general

$$s(t) = s_1 \sqrt{\frac{2}{T_b}} \cos(\omega_c t) \cos\left(\frac{\pi}{2T_b}t\right) + s_2 \sqrt{\frac{2}{T_b}} \sin(\omega_c t) \sin\left(\frac{\pi}{2T_b}t\right)$$

where $s_1 \sqrt{2/T_b} \cos(\pi t/2T_b)$ is the in-phase component and $s_2 \sqrt{2/T_b} \sin(\pi t/2T_b)$ is the quadrature phase component. The details of $\phi_1$ and $\phi_2$ are given in Fig. 10.27.

In this figure, $\omega_c = 5\pi/2T_b$, $\omega_1 = 2\pi/T_b$ and $\omega_2 = 3\pi/T_b$. We notice that $\phi_1 + \phi_2$ is a sinusoidal function which has a frequency identical to $\omega_1$ while $\phi_1 - \phi_2$ has a frequency identical to $\omega_2$.

## 10.10   M-Ary PSK

For M-ary PSK, shown in Fig. 10.28, for $M = 8$

$$\phi_1(t) = \sqrt{\frac{2}{T_b}} \cos(\omega_c t)$$

$$\text{and } \phi_2(t) = \sqrt{\frac{2}{T_b}} \sin(\omega_c t)$$

and

$$s_i(t) = \sqrt{\frac{2E_b}{T_b}} \cos(\omega_c t + \frac{i\pi}{4}) \quad i = 0, 1, \ldots, 7$$

$$= \sqrt{\frac{2E_b}{T_b}} \left[ \cos(\omega_c t) \cos\left(\frac{i\pi}{4}\right) - \sin(\omega_c t) \sin\left(\frac{i\pi}{4}\right) \right]$$

And

$$s_{i1} = \sqrt{E_b} \cos\left(\frac{\pi i}{4}\right)$$

$$s_{i2} = -\sqrt{E_b} \sin\left(\frac{\pi i}{4}\right)$$

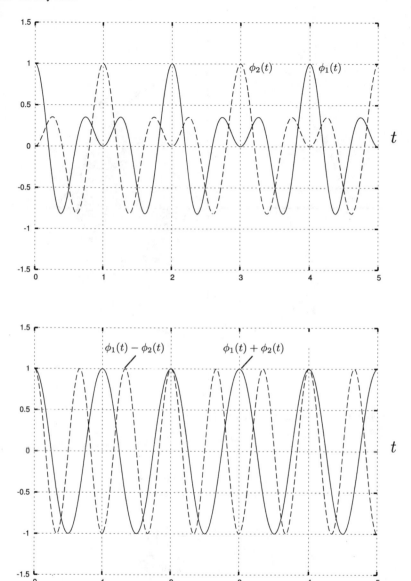

**Fig. 10.27** Plots of $\phi_1$, $\phi_2$ and $\phi_1 \pm \phi_2$

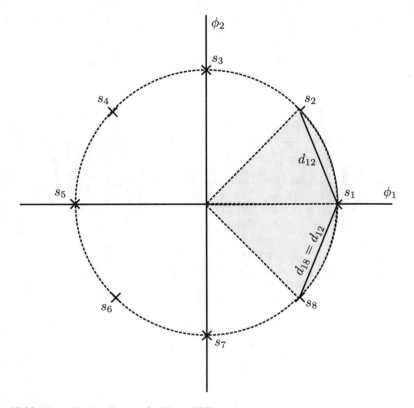

**Fig. 10.28** Constellation diagram for M-ary PSK

Therefore,

$$\mathbf{s}_i = \begin{bmatrix} s_{i1} \\ s_{i2} \end{bmatrix} = \begin{bmatrix} \sqrt{E_b} \cos\left(\frac{\pi i}{4}\right) \\ -\sqrt{E_b} \sin\left(\frac{\pi i}{4}\right) \end{bmatrix}$$

and

$$\mathbf{s}_{i+1} = \begin{bmatrix} s_{i1} \\ s_{i2} \end{bmatrix} = \begin{bmatrix} \sqrt{E_b} \cos\left[\frac{\pi(i+1)}{4}\right] \\ -\sqrt{E_b} \sin\left[\frac{\pi(i+1)}{4}\right] \end{bmatrix}$$

and

$$\mathbf{s}_i - \mathbf{s}_{i+1} = \begin{bmatrix} \sqrt{E_b} \cos\left(\frac{\pi i}{4}\right) - \sqrt{E_b} \cos\left(\frac{\pi(i+1)}{4}\right) \\ \left(\sqrt{Eb} \sin\left(\frac{\pi(i+1)}{4}\right) - \sqrt{Eb} \sin\left(\frac{\pi i}{4}\right)\right) \end{bmatrix}$$

$$\|\mathbf{s}_i - \mathbf{s}_{i+1}\|_v = d_{i,i+1} = 2\sqrt{E_b} \left|\sin\left(\frac{\pi}{8}\right)\right|$$

So the probability of error is given by using Eq. (10.62)

$$P_e = \text{erfc}\left[\sqrt{\frac{E_b}{N_0}} \times \sin\left(\frac{\pi}{8}\right)\right] \tag{10.76}$$

For the general case,

$$s_i(t) = \sqrt{\frac{2E_b}{T_b}} \cos\left(\omega_c t + \frac{i2\pi}{M}\right) \qquad i = 0, 1, \ldots, M-1$$

$$= \sqrt{\frac{2E_b}{T_b}}\left[\cos(\omega_c t)\cos\left(\frac{i2\pi}{M}\right) - \sin(\omega_c t)\sin\left(\frac{i2\pi}{M}\right)\right]$$

And

$$s_{i1} = \sqrt{E_b}\cos\left(\frac{2\pi i}{M}\right)$$

$$s_{i2} = -\sqrt{E_b}\sin\left(\frac{2\pi i}{M}\right)$$

Therefore,

$$\mathbf{s}_i = \begin{bmatrix} s_{i1} \\ s_{i2} \end{bmatrix} = \begin{bmatrix} \sqrt{E_b}\cos\left(\frac{\pi i}{4}\right) \\ -\sqrt{E_b}\sin\left(\frac{\pi i}{4}\right) \end{bmatrix}$$

and

$$\mathbf{s}_{i+1} = \begin{bmatrix} s_{i1} \\ s_{i2} \end{bmatrix} = \begin{bmatrix} \sqrt{E_b}\cos\left[\frac{2\pi(i+1)}{M}\right] \\ -\sqrt{E_b}\sin\left[\frac{2\pi(i+1)}{M}\right] \end{bmatrix}$$

and

$$\|\mathbf{s}_i - \mathbf{s}_{i+1}\|_v = d_{i,i+1} = 2\sqrt{E_b}\left|\sin\left(\frac{\pi}{M}\right)\right| \tag{10.77}$$

So the probability of error is given by using Eq. (10.62)

$$P_e = \text{erfc}\left[\sqrt{\frac{E_b}{N_0}} \times \sin\left(\frac{\pi}{M}\right)\right] \tag{10.78}$$

## 10.11  Noncoherent Schemes

### 10.11.1  Detection of Signals with Unknown Phase

Suppose a set of signals being transmitted is of the form

$$s_i(t) = \sqrt{\frac{2E}{T}} \cos \omega_i t \quad 0 \le t \le T$$

where $E$ is the energy of the signal, $T$ is the duration of the signal and $\omega_i$ are different values of frequency. $\omega_i = n\pi/T$ where $n$ is a positive integer. The $i$th received signal with the addition of white noise is

$$r_i(t) = x(t) + w(t) = \sqrt{\frac{2E}{T}} \cos(\omega_i t + \theta_i) + w(t) \quad 0 \le t \le T$$

$$= \sqrt{\frac{2E}{T}} \left[ \cos(\omega_i t) \cos \theta_i - \sqrt{\frac{2E}{T}} \sin(\omega_i t) \sin \theta_i \right] + w(t)$$

where $w(t)$ is white noise with PSD $N_0/2$. We now use two basis functions,

$$\phi_{Ii}(t) = \sqrt{\frac{2}{T}} \cos \omega_i t$$

$$\text{and } \phi_{Qi}(t) = \sqrt{\frac{2}{T}} \sin \omega_i t$$

then looking at the part without the noise,

$$\int_0^T x(t)\phi_{Ii}(t)dt = \sqrt{E} \cos \theta_i$$

$$\int_0^T x(t)\phi_{Qi}(t)dt = \sqrt{E} \sin \theta_i$$

where $x(t)$ is the received signal without the white noise. If we square these two functions and add them, we get the output as $E$. The block diagram of this processing is shown in Fig. 10.29.

### 10.11.2   General Noncoherent Orthogonal Modulation

In noncoherent orthogonal modulation, two orthogonal bandpass signals $s_1(t)$ (at $\omega_1$) or $s_2(t)$ (at $\omega_2$) are sent over a channel. Typically, the two signals have equal energy and are orthogonal to each other

$$\int_0^T s_1(t)s_2(t)dt = 0$$

At the receiver, they have an unknown phase, but are still orthogonal to each other. To separate the signals, we use a receiver where we employ two basis functions whose

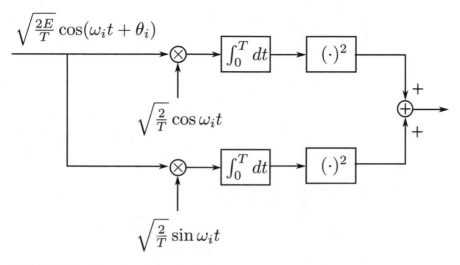

**Fig. 10.29** Processing for noncoherent signals

structure is in the interval $0 \leq t \leq T$ and with $i = 1, 2$.

$$\phi_{Ii}(t) = m(t) \cos \omega_i t$$
$$\text{and } \phi_{Qi}(t) = m(t) \sin \omega_i t$$

where the bandwidth of $m(t)$ is very much smaller than $\omega_i$. These two functions are in fact scaled versions of $s_1$ and $s_2$. With this in mind, the FTs of these signals are

$$\phi_{Ii}(\omega) = M(\omega) * \pi \left[ \delta \left( \omega - \omega_i \right) + \delta \left( \omega + \omega_i \right) \right]$$
$$= \pi \left[ M(\omega - \omega_i) + M(\omega + \omega_i) \right]$$
$$\text{and } \phi_{Qi}(\omega) = M(\omega) * j\pi \left[ \delta \left( \omega - \omega_i \right) - \delta \left( \omega + \omega_i \right) \right]$$
$$= j\pi \left[ M(\omega - \omega_i) - M(\omega + \omega_i) \right]$$

From observing these two FTs, we can see that they are the Hilbert transforms of each other. Also as per the properties of the Hilbert transforms, these two functions are also orthogonal to each other. The two basis functions are relabeled as $\phi_i(t)$ and $\phi_{ih}(t)$.

We now proceed to use a receiver as shown in Fig. 10.30. In this figure at the output, the larger of the two, $l_1$ or $l_2$, were chosen. We know that both the upper and lower halves of the figure have Gaussian distributed noise added to them. Therefore, the probability density function under consideration is the Raleigh distribution since the signal is on the output of the in-phase and quadrature phase signals (see Sect. 3.19.1).

Suppose $s_1(t)$ was transmitted, then in the upper half of the figure,

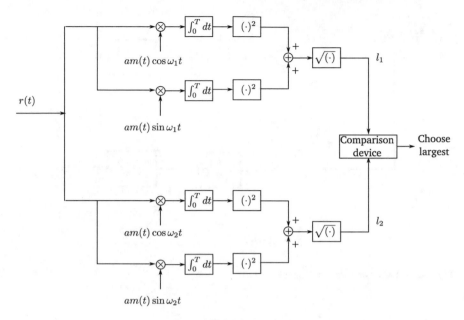

**Fig. 10.30**  General noncoherent orthogonal modulation receiver

$$r_{I1} = w_{I1} + \sqrt{E}$$
$$r_{Q1} = w_{Q1}$$
$$r_{I2} = w_{I2}$$
$$r_{Q2} = w_{Q2}$$

where $E$ is the energy of the signal, the subscripts $I$ and $Q$ refer to in-phase and quadrature phase components and the two subscripts 1 and 2 refer to the upper and lower arms of the figure. Then an error occurs when $s_1(t)$ was transmitted but $l_2 > l_1$. And we know that $l_2$ was the pure Raleigh distributed noise. With this in mind,

$$f_{L_2}(l_2) = \begin{cases} \frac{2l_2}{N_0}e^{-l_2^2/N_0} & l_1 \geq 0 \\ 0 & l_1 < 0 \end{cases}$$

Therefore, the probability that an error occurred is

$$P(l_2 > l_1|l_1) = \int_{l_1}^{\infty} \frac{2l_2}{N_0}e^{-l_2^2/N_0}dl_2$$
$$= \int_{l_1^2}^{\infty} \frac{1}{N_0}e^{-y/N_0}dy$$
$$= e^{-l_1^2/N_0}$$

We can from here go through some lengthy mathematics (which we omit) and come to the conclusion that the probability that an error was committed is

$$P_e = \frac{1}{2} \exp \left( -\frac{E}{2N_0} \right) \tag{10.79}$$

## 10.12   Noncoherent Binary Frequency Shift Keying

In noncoherent binary FSK the transmitted signal is (in the interval $0 \leq t \leq T_b$) for symbols 0 and 1, respectively,

$$s_1(t) = \sqrt{\frac{2E_b}{T_b}} \cos \omega_1 t$$

$$\text{and } s_2(t) = \sqrt{\frac{2E_b}{T_b}} \cos \omega_2 t \tag{10.80}$$

where $\omega_{1,2} = 2\pi n_{1,2}/T_b$ where $n_{1,2}$ are integers.

We have on reception used a noncoherent detection scheme as shown in Fig. 10.31. For this scheme, the probability of error based on our previously outlined theory is

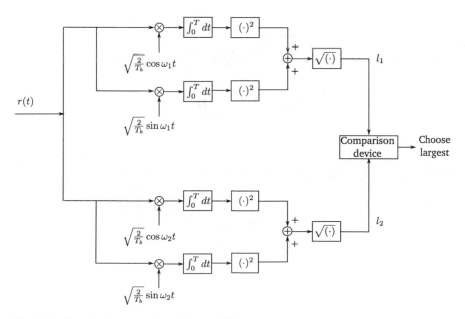

**Fig. 10.31**   Receiver for noncoherent binary FSK

$$P_e = \frac{1}{2} \exp\left(-\frac{E}{2N_0}\right)$$

## 10.13   Important Concepts and Formulae

1. Signal waveforms can be represented by

$$s_i(t) = \sum_{j=1}^{N} s_{ij}\phi_j(t)\ i = 1, \ldots, M$$

2. The vector $\mathbf{s}_i$ is defined by

$$\mathbf{s}_i = \begin{bmatrix} s_{i1} & s_{i2} & \cdots & s_{iN} \end{bmatrix}\quad i = 1, \cdots M$$

3. The $\phi$s are orthogonal to each other

$$\int_0^T \phi_i(t)\phi_j(t)dt = \delta_{ij}$$

4. The energy of a signal is defined by

$$E_i = \int_0^T s_i^2(t)dt = \|s_i(t)\|_f^2 = \|\mathbf{s}_i\|_v^2$$

5. The Euclidean distance between two signals is given by

$$\|\mathbf{s}_k - \mathbf{s}_l\|_v^2 = \sum_{j=1}^{N} \left(s_{kj} - s_{lj}\right)^2$$
$$= d_{kl}$$

6. The probability of error for the $i$th signal with respect to the $k$th signal is

$$P_{e(i)}(k) = \frac{1}{\sqrt{\pi N_0}} \int_{d_{ik}/2}^{\infty} \exp\left\{-\frac{x^2}{N_0}\right\} dx$$
$$= \frac{1}{2}\text{erfc}\left(\frac{d_{ik}}{2\sqrt{N_0}}\right)$$

## 10.14 Self Assessment

### 10.14.1 Review Questions

1. Why does ISI play a secondary role in passband pulse transmission?
2. Why do we require the frequency and bit width per bit in FSK to have the requirement,

$$\omega_b T = 2n\pi, \quad \text{where } n \text{ is a +ve integer?}$$

3. What is a linear channel?
4. Give an example when there are four message symbols.
5. Explain the equation

$$P_e = p_i P\left(\hat{m} \neq m_i | m_i\right)$$

where $p_i$ is the probability that symbol $m_i$ was transmitted and $\hat{m}$ was received.
6. If the transmitted signals are

$$s_i(t) \, i = 1, 2, \ldots, M; \quad 0 \leq t \leq T \qquad (10.81)$$

and the orthogonal functions are

$$\phi_j(t), \, j = 1, 2, \ldots, N; \quad 0 \leq t \leq T \qquad (10.82)$$

where $N \leq M$. Why is $N \leq M$?
7. Explain why

$$P_e = 1 - P(m_i | \mathbf{r}) \text{ where } m_i \text{ wassent}$$

8. Explain the equivalence of correlation and matched filter receivers.
9. What are coherent modulation schemes?

### 10.14.2 Numerical Problems

1. A binary source emits '0's and '1's with equal probability. In a sequence of 1 (8 bits) byte, what is the probability that

   (a) There are exactly four '0's and four '1's?
   (b) There are exactly three '0's and five '1's?
   (c) There are exactly two '0's and six '1's?
   (d) There are exactly one '0' and seven '1's?
   (e) At least one '0'.

2. We consider two symbols here,

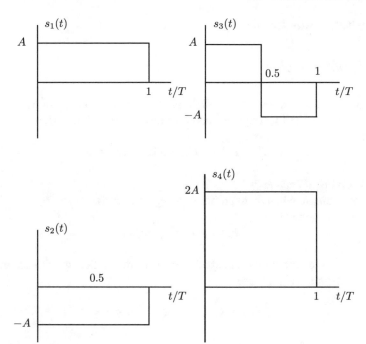

**Fig. 10.32** The low-pass waveforms for Problem 6

$$s(t) = \begin{cases} s_1(t) = A\cos(\omega_c t) & \text{for symbol 1} \\ s_2(t) = A\cos(\omega_c t + \pi) = -A\cos(\omega_c t) & \text{for symbol 0} \end{cases}$$

where $0 \le t \le T_b$ and

$$T_b = n_c \left(\frac{2\pi}{\omega_c}\right) = n_c T$$

where $n_c$ is a positive integer. Plot the waveform for the bit sequence 100110.

3. Plot the waveforms for BFSK for symbols 0 and 1.
4. Plot the waveforms for QPSK for symbols 0 and 1.
5. Plot the waveforms for MSK for symbol 001101.
6. The low-pass waveforms for communication are given in Fig. 10.32. Using the Gram–Schmidt orthonormalisation process, obtain the basis functions, $\phi_j(t)$.
7. For the signals of Problem 6, find the function norm of these signals.
8. For Problem 6, find $\mathbf{s}_i$ and compute the vector norm of the $\mathbf{s}_i$ you computed.
9. For Problem 6, draw the constellation diagram and then find the $d_{ik}$ the distance between the vectors from which you compute the probability of error.
10. For a set of waveforms,

$$s_i(t) = A\sin\left(\omega_c t + \frac{2\pi}{4}i\right) \quad i = 0,\ldots,3 \quad 0 < t < T_b$$

with

$$nT_c = T_b, n = 4$$

Using the Gram–Schmidt orthonormalisation process, obtain the basis functions, $\phi_j(t)$.

11. For the signals of Problem 10, find the function norm of these signals.
12. For Problem 10, find $\mathbf{s}_i$ and compute the vector norm of the $\mathbf{s}_i$ you computed.
13. For Problem 10, draw the constellation diagram and then find the $d_{ik}$ the distance between the vectors and compute the probability of error.
14. For the communication signal set (ASK)

$$s_i(t) = 0\ 1 \begin{cases} i = 1 & 0 \text{ for } 0 < t < T_b \\ i = 2 & A, \text{ for } 0 < t < T_b \end{cases}$$

Find the basis function.

15. For Problem 14, find $\mathbf{s}_1$ and $\mathbf{s}_2$.
16. For Problem 14, find the probability of error.
17. If in Problem 6, symbol 1 is emitted with probability 0.5, symbol 2 with probabilty 0.25, symbol 3 with probability 0.125 and symbol 4 with probability 0.125, then compute the probability of error.

# Chapter 11
# Spread Spectrum Modulation

## 11.1 Introduction

Spread spectrum techniques, which we will study now, are techniques by which a digital signal generated with a particular bandwidth is deliberately spread in the frequency domain, which gives the set of techniques this name. These techniques are used primarily to establish a means of secure communications, and in their early days, they were used by the military for their communication needs, since these techniques result in a signal which has enhanced resistance to natural interference, noise and jamming. These techniques are presently also used for civilian applications, especially mobile communications.

## 11.2 Pseudo Noise (PN) Sequences

A pseudo noise (PN) sequence is a periodic binary signal similar to noise (but which is not noise), which implies that a pseudo noise sequence satisfies the standard tests for statistical randomness. A PN sequence is generally generated by means of logic implemented hardware using a number of flip flops ($n$) which constitute a regular shift register, and a logic circuit as shown in Fig. 11.1. The output of the logic circuit is an input to the first flip flop.

The state of any flip flop is the state of the previous flip flop at the previous state

$$s_{j+1}(k + 1) = s_j(k) \tag{11.1}$$

where $j = 0, \ldots, m$ and $k \geq 0$. If the total number of shift registers are $n$, then the output consists of a maximum of $2^n$ states which appear on the output in random order. When these states are exhausted, the output repeats itself, which implies that the output is periodic. This happens when the *logic consists of modulo 2 adders*, whose characteristic is given by

© The Author(s), under exclusive license to Springer Nature Singapore Pte Ltd. 2022
S. Bhooshan, *Fundamentals of Analogue and Digital Communication Systems*,
Lecture Notes in Electrical Engineering 785,
https://doi.org/10.1007/978-981-16-4277-7_11

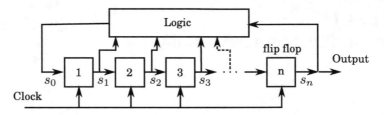

**Fig. 11.1**  A pseudo noise (PN) sequence generator

**Fig. 11.2**  PN sequence
generator for Example 11.1

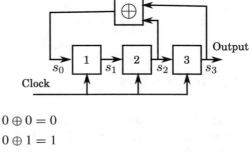

$$0 \oplus 0 = 0$$
$$0 \oplus 1 = 1$$
$$1 \oplus 0 = 1$$
$$1 \oplus 1 = 0 \qquad\qquad\qquad (11.2)$$

where $\oplus$ is modulo 2 addition operator.

**Example 11.1**  As an example, consider the PN sequence generator shown in
Fig. 11.2. In this case, let the initial states be $s_1 = 1$, $s_2 = 0$ and $s_3 = 1$. Based on
these states, $s_0 = 1$. After the clock pulse, $s_0 = 1$, $s_1 = 1$, $s_2 = 1$, $s_3 = 0$ and so on.
We now make a table of the states of the flip flops as given in Table 11.1.

**Table 11.1**  Table showing
the sequence of 0s and 1s on
the output of the PN sequence
generator shown in Fig. 11.2

| $s_0 s_1 s_2 s_3$ | Output |
|---|---|
| 1101 | 1 |
| 1110 | 0 |
| 0111 | 1 |
| 0011 | 1 |
| 1001 | 1 |
| 0100 | 0 |
| 1010 | 0 |
| 1101 | 1 |
| $\cdots$ | $\cdots$ |

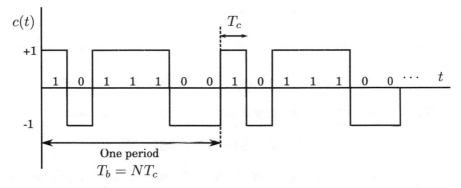

**Fig. 11.3** PN sequence of the example converted into a waveform

In this way, we find that the output (bit $s_3$) turns out to be

$$\underbrace{1011100}_{\text{Initial output}} \quad \underbrace{1011100}_{\text{repeated output}} \cdots$$

and then it is repeated. In this case, $m = 3$ and the PN sequence length is $N = 2^3 - 1 = 7$ states *where the state 000 is not allowed, since if this is the initial state, the circuit remains in the same state and does not go over to any other state.* That the sequence is repeated after 7 states is not a coincidence but it is a property of PN sequence generator in general. The PN sequence generated by such a circuit is called a *maximum length code*. If we convert this maximum length code into a series of pulses, we get the result shown in Fig. 11.3.

In this figure, the nomenclature adopted is that a single state is called a *chip*, and $N = 2^m - 1$, the period of the maximum length sequence is denoted as $T_b$. The time duration of the clock pulse in the PN sequence generator is $T_c$, the chip time duration.

### 11.2.1 Maximum Length Sequences

Maximum length (ML) [1] sequences are pseudo random in nature. As the length of the sequence grows larger, the probability of the appearance of a 0 or 1 is equally likely. Some properties of ML sequences are as follows:

1. *Balance property.* In each period of an ML sequence, the number of '1's is always greater than the number of '0's by one. For example in our sequence 1011100, there are four '1's and three '0's.
2. *Run property.* In each period of a ML sequence, *approximately* half the runs of both zeros and ones are of length one, one fourth are of length two, one eighth are of length three and so on. Since our sequence from Example 11.1 is too short, let us consider the 31 state ML sequence [2]

000010101110110001111001101001

four 1s are of run length one and four 0s are of run length 1 (=8) shown underlined;

0000<u>1010</u>1110<u>1</u>100011110011<u>0100</u>1

two 1s are of run length 2 and two 0s are of run length 2 (=4) shown underlined;

000010101110<u>11</u>0001111<u>001</u>101001

one set of 1s are of run length 3 and one set of 0s are of run length 3 (=6) shown underlined;

0000101011<u>101</u>1<u>000</u>11110011010 01

one set of 0s are of run length 4 (=4) (underlined)

<u>0000</u>10101110110001111001101001

and one set of 1s are run length 5 (=5) shown underlined

000010101110110001<u>11111</u>001101001

This property ensures that the ML sequences are truly (pseudo) random. The total number of run sequences is 16.
3. *Correlation property.* The autocorrelation function represented by $-1$s and $1$s (as shown in Fig. 11.3) is periodic and resembles a strain of the Kronecker delta. Thus, referring to the figure where $c(t)$ is shown, the autocorrelation function is[1]

$$R_c(\tau) = \frac{1}{T_b} \int_{-T_b/2}^{T_b/2} c(t)c(t+\tau)dt \qquad (11.3)$$

where $-T_b/2 \leq \tau \leq T_b/2$. Since the ML sequence is periodic, the autocorrelation function is also periodic. From this integral,

$$R_c(\tau) = \begin{cases} 1 - \frac{N+1}{NT_c}\tau & 0 \leq \tau \leq T_c \\ 1 + \frac{N+1}{NT_c}\tau & -T_c \leq \tau \leq 0 \\ -\frac{1}{N} & \text{elsewhere} \end{cases} \qquad (11.4)$$

which becomes close to the result of Example 3.28 of the random binary wave. As $N$ becomes large, the only difference being that the autocorrelation function

---

[1] The autocorrelation function of a function is $R(\tau) = \int_{-\infty}^{\infty} f(t)f(t+\tau)dt$ and is different from the autocorrelation function of a random process.

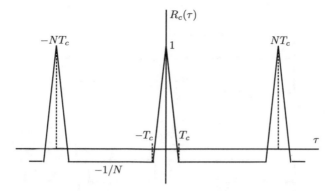

**Fig. 11.4** Autocorrelation function of a ML sequence

of an ML sequence is periodic while that of a random binary wave is not periodic. The autocorrelation function is shown in Fig. 11.4.

## 11.3   A Spread Spectrum Baseband Waveform

The creation of a spread spectrum waveform called *the direct sequence (DS) spread spectrum* or DSSS is done in the following manner. If we want to transmit a set of bits given by the waveform $b(t)$ and we have an PN sequence $c(t)$, then we multiply the two to give us

$$m(t) = c(t)b(t) \qquad (11.5)$$

We transmit this waveform over a baseband medium. The three waveforms are shown in Fig. 11.5. On reception, we demodulate the received waveform by multiplying $m(t)$ by $c(t)$ to give us $b(t)$. Thus is true since

$$m(t)c(t) = c(t)b(t)c(t) = c^2(t)b(t)$$
$$\text{and since } c^2(t) = 1$$
$$m(t)c(t) = b(t) \qquad (11.6)$$

From the theory of the Fourier transforms, the spectrum of $b(t)$ occupies the approximate bandwidth of $1/T_b$ while $c(t)$ occupies the approximate bandwidth of $N/T_b$, for one period. And if we consider the bandwidth of $c(t)b(t)$, it will be the convolution of $C(\omega)$ with $B(\omega)$ which will be[2] approximately $(N+1)/T_b$.

If we consider noise or an interference signal while the message signal passes through the channel, then the received signal is

---

[2] The increase of bandwidth by the addition of the bandwidths of two functions is a property of the convolution process as is in the case of Problem 9 of Sect. 2.19.1.

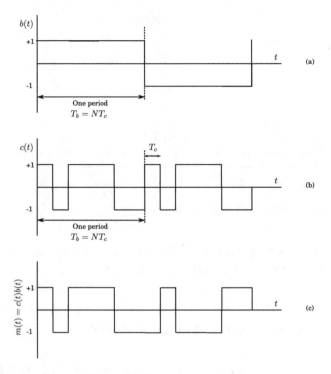

**Fig. 11.5**  Spread spectrum used in baseband

$$r(t) = c(t)b(t) + j(t) \tag{11.7}$$

and the demodulated signal is

$$r(t)c(t) = b(t) + c(t)j(t) \tag{11.8}$$

If the signal is observed more carefully, we see that $b(t)$ is a low-pass signal of bandwidth approximately $1/T_b$, while $c(t)j(t)$ is a signal of large bandwidth (at least $N/T_b$). The signal is, therefore, passed through an LPF to give us $b(t)$ plus narrowband noise of bandwidth $1/T_b$. The full system is shown in Fig. 11.6. The

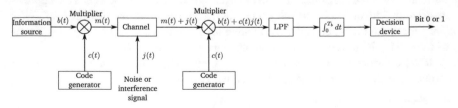

**Fig. 11.6**  Baseband spread spectrum communication

detection of the bit train $b(t)$ is accomplished by passing it through an integrator as shown in Fig. 11.6 and then a decision device. The integrator itself behaves like a low-pass filter.

Let us take a qualitative (and *not* a mathematically rigorous) look at the frequency spectrum of $b(t), c(t)$ and $m(t)$. The signal $b(t)$ has pulse width of $T_b$ so the spectrum will look like

$$b(t) \stackrel{\approx}{\Rightarrow} \left(\frac{T_b}{2}\right) \frac{\sin\left(\frac{\omega T_b}{2}\right)}{\left(\frac{\omega T_b}{2}\right)}$$

where the symbol $\stackrel{\approx}{\Rightarrow}$ 'has an FT approximately equal to', and which will have an effective bandwidth given by

$$\frac{\omega_{bw,\text{bit}} T_b}{2} \approx \pi$$

$$\omega_{bw,\text{bit}} \approx \frac{2\pi}{T_b}$$

Similarly, for one chip of $c_1(t)$,

$$c_1(t) \stackrel{\approx}{\Rightarrow} \left(\frac{T_c}{2}\right) \frac{\sin\left(\frac{\omega T_c}{2}\right)}{\left(\frac{\omega T_c}{2}\right)}$$

then

$$\left(\frac{T_b}{2}\right)^2 \int_{-\infty}^{\infty} \left[\frac{\sin\left(\frac{\omega T_b}{2}\right)}{\left(\frac{\omega T_b}{2}\right)}\right]^2 d\omega = T_b$$

which is the energy in one bit, and

$$\left(\frac{T_c}{2}\right)^2 \int_{-\infty}^{\infty} \left[\frac{\sin\left(\frac{\omega T_c}{2}\right)}{\left(\frac{\omega T_c}{2}\right)}\right]^2 d\omega = \left(\frac{T_b}{2}\right)^2 \left(\frac{1}{N^2}\right) \int_{-\infty}^{\infty} \left[\frac{\sin\left(\frac{\omega T_c}{2}\right)}{\left(\frac{\omega T_c}{2}\right)}\right]^2 d\omega$$

$$= \frac{T_b}{N}$$

is the energy in one chip. We now have to equate the two energies of $c(t) = \sum_{k=1}^{N} c_k(t - kT_c)$ where $c_k(t - kT_c)$ takes the values $\pm 1$, and $b(t)$ using the energy in one period (or bit) giving

$$c(t) \stackrel{\approx}{\Rightarrow} \left(\frac{T_b}{2\sqrt{N}}\right) \frac{\sin\left(\frac{\omega T_c}{2}\right)}{\left(\frac{\omega T_c}{2}\right)} = \left(\frac{T_b}{2\sqrt{N}}\right) \frac{\sin\left(\frac{\omega T_b}{2N}\right)}{\left(\frac{\omega T_b}{2N}\right)}$$

which will have an effective bandwidth

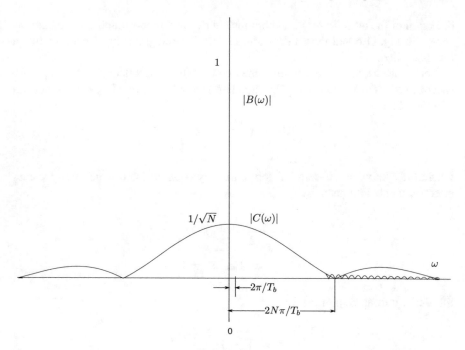

**Fig. 11.7**   Spreading effect of the PN sequence on a bit in spread spectrum communication

$$\omega_{bw,\text{code}} \cong \frac{2N\pi}{T_b}$$

From these results, we can see that multiplication by the PN sequence effectively has a spreading effect on the frequency domain since the bandwidth has increased by a factor of $N$.

Furthermore,[3]

$$M(\omega) \overset{\approx}{\Rightarrow} \left(\frac{T_b}{2}\right) \left[\frac{\sin\left(\frac{\omega T_b}{2}\right)}{\left(\frac{\omega T_b}{2}\right)}\right] * \left[\frac{T_c}{2\sqrt{N}} \frac{\sin\left(\frac{\omega T_b}{2N}\right)}{\left(\frac{\omega T_b}{2N}\right)}\right]$$

$$\overset{\approx}{\Rightarrow} \frac{T_c}{2\sqrt{N}} \frac{\sin\left(\frac{\omega T_b}{2N}\right)}{\left(\frac{\omega T_b}{2N}\right)}$$

The approximate spectra of $B(\omega)$ and $C(\omega)$ are shown in Fig. 11.7.[4]

---

[3] Section 11.8.2, Numerical Problem 8.

[4] See also Problem 7 which shows the spectral broadening for $N$ pulses.

## 11.4  DS Spread Spectrum with BPSK

We now take a look at the baseband spread spectrum and introduce a carrier. The waveforms are shown in Fig. 11.8. In (a), we see the PN code sequence, $c(t)$, (Fig. 11.5b) multiplied by the bit sequence, $b(t)$ (Fig. 11.5a) or

$$m(t) = c(t)b(t)$$

In Fig. 11.8b, the carrier waveform is shown, $A\cos(\omega_c t)$, while (c) shows $Am(t)\cos(\omega_c t)$.

$$s(t) = Am(t)\cos(\omega_c t)$$

In the process of demodulation, we first synchronise the carrier and multiply the incoming received signal by a synchronised local carrier. The received signal is

$$\begin{aligned} r(t) &= s(t) + j(t) \\ &= Am(t)\cos(\omega_c t) + j(t) \end{aligned} \tag{11.9}$$

where $j(t)$ is either a noise signal or a jamming signal. The output of carrier demodulation is

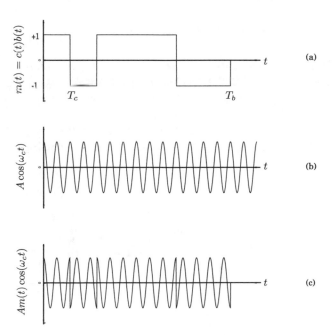

**Fig. 11.8** BPSK direct sequence spread spectrum waveforms

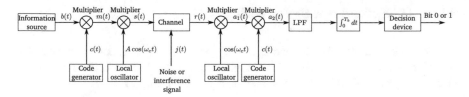

**Fig. 11.9** Block diagram of the direct sequence spread spectrum

$$r(t)\cos(\omega_c t) = a_1(t) = Am(t)\cos^2(\omega_c t) + j(t)\cos(\omega_c t)$$

$$= \frac{A}{2}m(t) + \frac{A}{2}m(t)\cos(2\omega_c t) + j(t)\cos(\omega_c t) \qquad (11.10)$$

After carrier synchronisation, the codes have to be synchronised in the next stage, and the signal $a_1(t)$ is multiplied by the PN sequence. Therefore,

$$a_2(t) = c(t)a_1(t)$$

$$= c(t)\left[\frac{A}{2}m(t) + \frac{A}{2}m(t)\cos(2\omega_c t) + j(t)\cos(\omega_c t)\right]$$

$$= c(t)\left[\frac{A}{2}c(t)b(t) + \frac{A}{2}c(t)b(t)\cos(2\omega_c t) + j(t)\cos(\omega_c t)\right]$$

$$= \frac{A}{2}b(t) + \frac{A}{2}b(t)\cos(2\omega_c t) + c(t)j(t)\cos(\omega_c t) \qquad (11.11)$$

The second term on the right of the equations, $(A/2)m(t)\cos(2\omega_c t)$, is in a higher frequency band ($\omega = 2\omega_c$) while $(A/2)m(t)$ is a baseband signal corrupted with wideband noise or a jammimg signal. The block diagram of direct sequence spread spectrum communication system is shown in Fig. 11.9.

Let us take a closer look at the jamming signal and its processing. The jamming signal is usually a narrowband signal at or near $\omega = \omega_c$ shown in Fig. 11.10a. When

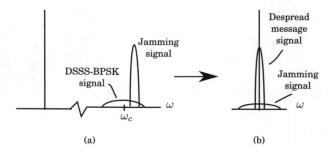

(a)                                                    (b)

**Fig. 11.10** Jamming signal processing in DSSS-BPSK communication. **a** Input to the receiver and **b** Output of the receiver

the jamming signal is multiplied by $c(t)$, it spreads the jamming signal as considered earlier. Another multiplication by $\cos \omega_c t$ then shifts it to the origin. The results of these operations is shown in Fig. 11.10b.

## 11.5 Signal Space Perspective

We now look at the DS BPSK communication system from the point of view of the signal space perspective.[5] To this end, let the mutually orthogonal signals be defined over the various chip intervals as[6]

$$
\phi_{Ik}(t) = \begin{cases} \sqrt{\frac{2}{T_c}} \cos(\omega_c t) & kT_c \leq t \leq (k+1)T_c \\ 0 & \text{elsewhere} \end{cases}
$$

$$
\phi_{Qk}(t) = \begin{cases} \sqrt{\frac{2}{T_c}} \sin(\omega_c t) & kT_c \leq t \leq (k+1)T_c \\ 0 & \text{elsewhere} \end{cases} \tag{11.12}
$$

That is, we define these orthogonal functions over a chip interval ($k$ taking on a particular value) and orthogonal to each other ($k$ taking different values):

$$
\int_0^{T_b} \phi_{Ak}(t)\phi_{Bl}(t)dt = \begin{cases} 0 & k \neq l; \ A = B \\ 0 & k = l; \ A \neq B \\ 0 & k \neq l; \ A \neq B \\ 1 & k = l; \ A = B \end{cases} \tag{11.13}
$$

where $A$ and $B$ take the values $I$ and $Q$. Furthermore, for a single bit interval, $0 \leq t \leq T_b$

$$
s(t) = \pm\sqrt{\frac{2E_b}{T_b}}c(t)\cos(\omega_c t) \tag{11.14}
$$

where the plus sign is used when the bit is 1 and minus sign is used when the bit is 0. Now

$$
c(t)\cos(\omega_c t) = \sum_{k=0}^{N-1} c_k \Delta_k(t) \cos[\omega_c(t - kT_c)] \tag{11.15}
$$

where

$$
\Delta_k(t) = \begin{cases} 1 & kT_c \leq t \leq (k+1)T_c \\ 0 & \text{elsewhere} \end{cases} \tag{11.16}
$$

---

[5] Given in Chap. 10.
[6] See Sect. 10.9.

and $c_k = \pm 1$ depending on the chip being 1 or 0, respectively, in the $k$th interval. Or

$$
\begin{aligned}
s(t) &= \pm \sqrt{\frac{2E_b}{T_b}} \sum_{k=0}^{N-1} c_k \Delta_k(t) \cos\left[\omega_c(t - kT_c)\right] \\
&= \pm \sqrt{\frac{2E_b}{T_b}} \sum_{k=0}^{N-1} c_k \sqrt{\frac{T_c}{2}} \phi_{Ik}(t) \\
&= \pm \sqrt{\frac{E_b}{N}} \sum_{k=0}^{N-1} c_k \phi_{Ik}(t)
\end{aligned}
\tag{11.17}
$$

Next we take a look at the jamming signal. Let the jamming power be written as (in general), without specifying its nature,

$$
\begin{aligned}
J &= \frac{1}{T_b} \int_0^{T_b} j^2(t)dt \\
&= \frac{1}{T_b} \int_0^{T_b} \left\{ \sum_{k=0}^{N-1} \left[ j_{Ik}\phi_{Ik}(t) + j_{Qk}\phi_{Qk}(t) \right] \right\}^2 dt \\
&= \frac{1}{T_b} \left\{ \sum_{k=0}^{N-1} \left[ j_{Ik}^2 + j_{Qk}^2 \right] \right\}
\end{aligned}
$$

making the assumption that

$$
j_{Ik}^2 = j_{Qk}^2
$$

since the jammer does not know the phase of the carrier, then

$$
J = \frac{2}{T_b} \sum_{k=0}^{N-1} j_{Ik}^2
$$

On the input, we can understand that signal power on the input would be

$$
S_i = \frac{E_b}{T_b}
$$

and the jamming signal power would be $\mathcal{J}$ which would be concentrated near $\omega_c$. So the input SNR would be

$$
(\text{SNR})_I = \frac{E_b}{\mathcal{J}T_b}
$$

On the output when we process the signal and pass it through a LPF, the jamming signal power spreads by a factor of $N = T_b/T_c$ and the *effective* jamming signal power is reduced to $\mathcal{J}/N$

$$\text{and (SNR)}_O = \frac{2E_b/T_b}{(\mathcal{J}/N)/T_b}$$

or

$$(\text{SNR})_O = 2N(\text{SNR})_I$$

where the processing gain (PG) is

$$\text{PG} = \frac{T_b}{T_c} = N$$

and the probability of error is

$$P_e = \frac{1}{2}\text{erfc}\left(\sqrt{\frac{E_b}{\mathcal{J}T_c}}\right)$$

comparing this result with Eq. (10.49), which is reproduced below,

$$P_e = \frac{1}{2}\text{erfc}\left(\sqrt{\frac{E_b}{N_0}}\right)$$

then

$$\frac{E_b}{N_0} = \frac{E_b}{\mathcal{J}T_c} = \frac{E_b/T_b}{\mathcal{J}T_c/T_b}$$

But

$$\frac{E_b}{T_b} = \mathcal{P}$$

where $\mathcal{P}$ is the signal power. So

$$\frac{E_b}{N_0} = N\left(\frac{\mathcal{P}}{\mathcal{J}}\right)$$

or

$$\frac{\mathcal{J}}{\mathcal{P}} = \text{jamming margin} = \frac{N}{E_b/N_0}$$

## 11.6  Frequency Hop Spread Spectrum

DSSS has many advantages over conventional digital communication[7] especially with regard to jamming interference signals. However, one of the fundamental limitations of this technology is the length of PN sequences which can be produced.

---

[7] Given in Chap. 10.

| **Table 11.2** PN sequence | $n$ | $2^n - 1$ |
|---|---|---|
| lengths and the number of flip | 3 | 7 |
| flops | 4 | 15 |
| | 5 | 31 |
| | 6 | 63 |
| | 7 | 127 |
| | 8 | 255 |
| | 9 | 1023 |

For example, if three flip flops are used, the PN sequence length is $2^3 - 1 = 7.$[8] Table 11.2 shows the number of flip flops required and the PN sequence lengths.

To remedy these problems, we use a slightly different approach to the use of a spread spectrum waveform. We use different frequencies to modulate the signal at different rates. If $R_s$ is the symbol rate (a symbol may be a 1 or a 0 or other combinations) and $R_h$ is the hop rate, then

1. In slow frequency hopping,

$$R_s = nR_h \qquad n = 1, 2, 3, \ldots$$

Therefore, many symbols may use the same frequency. For $n = 1$, one symbol would use a single frequency; for $n = 2$, two symbols would use a single frequency and so on

2. In fast frequency hopping, the hop rate is many times the symbol rate

$$R_h = nR_s \qquad n = 1, 2, 3, \ldots$$

and in this, during the transmission of a symbol, the frequency would change. Therefore, in this case, a single symbol may use two or more frequencies.

## 11.7   Important Concepts and Formulae

- *Balance property.* In each period of an ML sequence, the number of '1's is always greater than the number of '0's by one. For example, in our sequence 1011100, there are four '1's and three '0's.
- *Run property.* In each period of a ML sequence, *approximately* half the runs of both zeros and ones are of length one, one fourth are of length two, one eighth are of length three and so on.

---

[8] As in Example 11.1.

- *Correlation property*. The autocorrelation function represented by $-1$s and $1$s is periodic and resembles a strain of the Kronecker delta.
- The creation of a spread spectrum waveform called *the direct sequence (DS) spread spectrum* or DSSS is done in the following manner. If we want to transmit a set of bits given by the waveform $b(t)$ and we have an PN sequence $c(t)$, then we multiply the two to give us

$$m(t) = c(t)b(t)$$

- The spectrum of $b(t)$ occupies the approximate bandwidth of $1/T_b$ while $c(t)$ occupies the approximate bandwidth of $N/T_b$ for one period. And if we consider the bandwidth of $c(t)b(t)$, it will be the convolution of $C(\omega)$ with $B(\omega)$ which will be approximately $(N + 1)/T_b$.
- Let the mutually orthogonal signals be defined over the various chip intervals as

$$\phi_{Ik}(t) = \begin{cases} \sqrt{\frac{2}{T_c}} \cos(\omega_c t) & kT_c \le t \le (k+1)T_c \\ 0 & \text{elsewhere} \end{cases}$$

$$\phi_{Qk}(t) = \begin{cases} \sqrt{\frac{2}{T_c}} \sin(\omega_c t) & kT_c \le t \le (k+1)T_c \\ 0 & \text{elsewhere} \end{cases} \tag{11.18}$$

- That is, we define these orthogonal functions over a chip interval ($k$ taking on a particular value) and orthogonal to each other ($k$ taking different values):

$$\int_0^{T_b} \phi_{Ak}(t)\phi_{Bl}(t)dt = \begin{cases} 0 & k \ne l; \ A = B \\ 0 & k = l; \ A \ne B \\ 0 & k \ne l; \ A \ne B \\ 1 & k = l; \ A = B \end{cases} \tag{11.19}$$

where $A$ and $B$ take the values $I$ and $Q$. Furthermore, for a single bit interval, $0 \le t \le T_b$

- So the input SNR would be

$$(\text{SNR})_I = \frac{E_b}{\mathcal{J}T_b}$$

- On the output when we process the signal and pass it through a LPF, the jamming signal power spreads by a factor of $N = T_b/T_c$ and the *effective* jamming signal power is reduced to $\mathcal{J}/N$

$$\text{and } (\text{SNR})_O = \frac{2E_b/T_b}{(\mathcal{J}/N)/T_b}$$

or

$$(\text{SNR})_O = 2N(\text{SNR})_I$$

where the processing gain (PG) is

$$PG = \frac{T_b}{T_c} = N$$

and the probability of error is

$$P_e = \frac{1}{2}\text{erfc}\left(\sqrt{\frac{E_b}{\mathcal{J}T_c}}\right)$$

- then

$$\frac{E_b}{N_0} = \frac{E_b}{\mathcal{J}T_c} = \frac{E_b/T_b}{\mathcal{J}T_c/T_b}$$

But

$$\frac{E_b}{T_b} = \mathcal{P}$$

where $\mathcal{P}$ is the signal power. So

$$\frac{E_b}{N_0} = N\left(\frac{\mathcal{P}}{\mathcal{J}}\right)$$

## 11.8   Self Assessment

### 11.8.1   Review Questions

1. What is the importance of a PN sequence in spread spectrum techniques?
2. How do we generate PN sequences?
3. What is a maximum length sequence?
4. Explain the meaning of a symbol as opposed to a bit.

### 11.8.2   Numerical Problems

1. For the PN sequence generator of Fig. 11.11, find the PN sequence generated if the initial sequence is 010.
2. Plot the waveform of the PN sequence of Problem 1.
3. In a spread spectrum communication system, a PN sequence generator has five flip flops. If the chip rate is $10^6$ chips/s, find the chip duration, PN sequence length and PN sequence period.
4. Find the autocorrelation function of PN sequence of Problem 1.

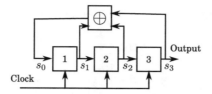

**Fig. 11.11** PN sequence generator for Problem 1

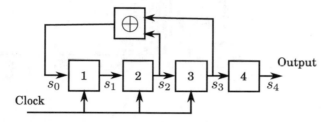

**Fig. 11.12** PN sequence generator for Problem 5

5. For the PN sequence generator of Fig. 11.12, find the PN sequence generated if the initial sequence is 1010.
6. For the PN sequence generated in Problem 5, show the balance and run properties.
7. To get an idea of how the spectrum is spread, consider a square wave of time duration $T$ as shown in Fig. 11.13. Find the energy spectral density of $N$ pulses each pulse of duration $T/N$. Show that the energy spectral density is given by

$$\Phi_e(\omega) = \frac{1}{2\pi} \left\{ \frac{\sin(\omega T/2N)}{\omega} \times \sin(\omega T/2) \right\}^2$$

What is the approximate bandwidth occupied by these pulses? **Hint**: Consider the pulse shape from $t = 0$ to $t = T/N$ which is repeatedly shifted to make the rest of the $N$ pulses. For $N = 10$, the energy spectral density is shown in Fig. 11.14

**Fig. 11.13** Square wave of limited duration

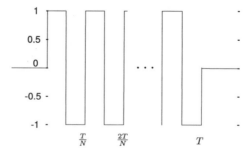

**Fig. 11.14** Energy spectral density for 10 pulses

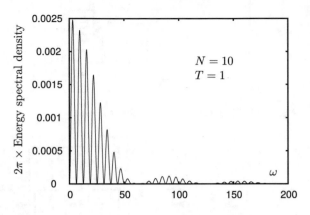

8. Show that

$$\left[ \frac{T_b \sin\left(\frac{\omega T_b}{2}\right)}{2\left(\frac{\omega T_b}{2}\right)} \right] * \left[ \frac{T_b \sin\left(\frac{\omega T_b}{2N}\right)}{2N\left(\frac{\omega T_b}{2N}\right)} \right] = \frac{T_b \sin\left(\frac{\omega T_b}{2N}\right)}{2N\left(\frac{\omega T_b}{2N}\right)}$$

where $N$ is an integer greater than or equal to 1. **Hint**: use time-domain functions.

# References

1. Golomb SW (1967) Coding theory. Holden-Day
2. Haykin S (1999) Communication systems. Wiley

# Chapter 12
# Introduction to Information Theory

## 12.1 Measure of Information

Information, from the engineering point of view, is different from the general definition of information as is generally understood. Suppose we have a page of a coded message in a substitution cipher. Now the first thing that a cryptologist will do is to find those letters which occur most frequently, that is, those letters whose probability of occurrence is high. That is, if the letter $h$ occurs most frequently, then he knows that $h$ is the letter $e$ since the letter $e$ occurs most frequently in English.

But having found the letter $e$, he also knows that he has moved forward in the decryption of the message only slightly. The relevant question we can ask is why this is so? And the answer is that since the probability of occurrence of $e$ is high, he has not made much headway in his search.

So, information has a connection with the probability of occurrence of a symbol. In fact, the higher the probability of occurrence of an event, the lower is its information content. Based on these ideas, the information content in the event

$$\{\text{occurrence of the letter } h\}$$

is

$$I\{h\} = \log_2 \frac{1}{P(h)} \quad \text{bits} \tag{12.1}$$

where $P(h)$ is the probability of the occurrence of the letter $h$.

So, if the probability of an occurrence of a symbol, $s$, is $p$, then the information contained in it is

$$I(s) = -\log_2 p \tag{12.2}$$

Thinking about this definition a bit, we realise that as $p \to 1, I \to 0$. That is, the event that is sure to happen contains no information at all: for example, the statement that 'the Sun will rise tomorrow' (which has a probability $\approx 1$) contains no information

© The Author(s), under exclusive license to Springer Nature Singapore Pte Ltd. 2022    517
S. Bhooshan, *Fundamentals of Analogue and Digital Communication Systems*,
Lecture Notes in Electrical Engineering 785,
https://doi.org/10.1007/978-981-16-4277-7_12

at all. On the other hand, the statement 'the Sun will *not* rise tomorrow' contains very high information content since the probability of this event is almost zero. Also, note that in this equation, $I$ will always be positive since $0 \leq p \leq 1$. Another important note is required here that the units of information is *bits* since the logarithm is taken as base 2 and bits is a short form of *bi*nary digi*ts*.

## 12.2  Entropy

Having defined a measure of information, let us consider a source of information with a finite alphabet,

$$S = \{s_1, \ldots, s_N\}$$

with probabilities of transmission

$$P(S = s_k) = p_k \quad k = 1, \ldots, N \tag{12.3}$$

and

$$\sum_{k=1}^{N} p_k = 1 \tag{12.4}$$

and by our earlier explanation

$$I(s_k) = -\log_2 p_k \tag{12.5}$$

**Example 12.1**  A source consists of four symbols with probabilities of transmission,

$$\left\{ \frac{1}{2}, \frac{1}{4}, \frac{1}{8}, \frac{1}{8} \right\}$$

Find the information contained of these symbols in bits.

$$I(s_1) = -\log_2 \frac{1}{2} = 1 \text{ bit}$$

$$I(s_2) = -\log_2 \frac{1}{4} = 2 \text{ bits}$$

$$I(s_3) = -\log_2 \frac{1}{8} = 3 \text{ bits}$$

$$I(s_4) = -\log_2 \frac{1}{8} = 3 \text{ bits}$$

Let us look at another parameter in information theory, namely, the *average information content of a source*. This parameter is called the *entropy of a source* and is defined as the average value of the information, $I(s_k)$. Based on this definition,

$$H(S) = E[I(s_k)]$$

$$= \sum_{k=1}^{N} p_k I(s_k)$$

$$= -\sum_{k=1}^{N} p_k \log_2 p_k$$

$$= -\frac{1}{\ln 2} \sum_{k=1}^{N} p_k \ln p_k \quad \text{bits} \tag{12.6}$$

As a point of interest, the entropy of the letters of the English language, as estimated by Shannon [4], is about 2.62 bits per letter. The implication is that, if we convert the English letters into bits in terms of an efficient code, then to transmit a six letter word, we would require about 16 bits.

Let us find out what is the distribution of the $p_k$ for maximising entropy of a source.[1] To do this, we use Lagrange multipliers. (Some explanation is given in Appendix J and [1]).

The object is to maximise

$$H(S) = -\sum_{k=1}^{N} p_k \log_2 p_k$$

with the condition that

$$\sum_{k=1}^{N} p_k = 1$$

Therefore, the object function, $O(p_1, \ldots p_N, \lambda)$, which we want to maximise is

$$O(p_1, \ldots, p_N, \lambda) = -\sum_{k=1}^{N} p_k \log_2 p_k + \lambda \left[ \sum_{k=1}^{N} p_k - 1 \right]$$

$$= -\frac{1}{\ln 2} \sum_{k=1}^{N} p_k \log_e p_k + \lambda \left[ \sum_{k=1}^{N} p_k - 1 \right]$$

We need to calculate $\partial O / \partial p_k \ k = 1, \ldots N$ and also $\partial O / \partial \lambda$ and equate these derivatives to 0. So

---

[1] See [3] for an alternative derivation.

$$\frac{\partial O}{\partial p_k} = -\frac{1}{\ln 2}\left[\log_e p_k + 1\right] + \lambda = 0 \quad N \text{ identical equations for } k = 1, \ldots, N$$

$$\frac{\partial O}{\partial \lambda} = \sum_{k=1}^{N} p_k - 1 = 0$$

which are $N + 1$ equations in $N + 1$ unknowns, which gives us a unique solution.

The solution to these equations must be such that the $\lambda$ parameter must be the same in each of the first set of $N$ equations since it is common to all these equations. Therefore, by inspection of these equations, *all the $p_k$s must be equal.*

$$p = p_k \quad k = 1, \ldots, N$$

which gives us $N$ equations each of which is

$$-\frac{1}{\ln 2}\left[\log_e p + 1\right] + \lambda = 0$$

Now looking at the $N + 1$th equation

$$\sum_{k=1}^{N} p_k = Np = 1$$

$$p_k = p = \frac{1}{N} \quad k = 1, \ldots, N$$

From this result, the entropy is maximised when all the $N$ symbols of the alphabet have identical frequency of occurrence, or a probability of $1/N$.

The maximum entropy of any source is then

$$H(S)_{max} = -\sum_{k=1}^{N} p_k \log_2 p_k \Bigg|_{p_k=1/N}$$

$$= -N\left[\frac{1}{N}\log_2 \frac{1}{N}\right]$$

$$= \log_2 N$$

From this result, we can conclude that the entropy of a source on $N$ symbols satisfies

$$0 \leq H(S) \leq \log_2 N \text{ bits} \tag{12.7}$$

In case of the English alphabet, $\log_2(26) = 4.7$ bits. This tells us that the maximum number of bits required to encode six letters would be about 29 bits if all the letters were equally likely to occur. If we use more than 29 bits, the coding algorithm would be inefficient.

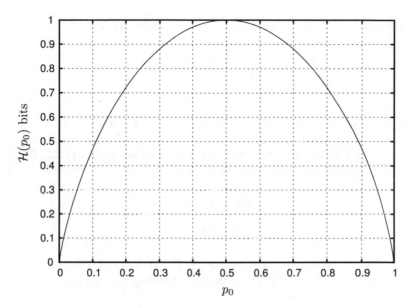

**Fig. 12.1** Entropy of a binary memoryless source

*Entropy as a measure of uncertainty.* From this result, Eq. (12.7), we may view entropy as a *measure of uncertainty* of a distribution. When one of the $p_k = 1$ and the others are zero, then there is no uncertainty:

$$H(S) = 0$$

On the other hand, when all the $p_k$s are equal, each equal to $1/N$ (the case for maximum uncertainty), the entropy is a maximum and equal to

$$H(S)|_{\max} = \log_2 N$$

**Example 12.2** Find the entropy of the source given, Example 12.1.

The entropy is given by

$$H(S) = -\sum p_k \log_2 p_k$$
$$H(S) = \frac{1}{2} \times 1 + \frac{1}{4} \times 2 + \frac{2}{8} \times 3 = 1.75$$

while

$$H(S)|_{\max} = 2 \text{ bits}$$

See Fig. 12.1.

**Example 12.3** Consider a source $S = \{s_1, s_2\}$ with probabilities

$$\{p_0, 1 - p_0\}$$

then this source has an entropy of

$$H(S) = \mathcal{H}(p_0) = -p_0 \log_2 p_0 - (1 - p_0) \log_2(1 - p_0) \qquad (12.8)$$

If we plot this function, we find that it has a single maximum at $p_0 = 0.5$. The graph at the ends shows that the entropy is zero since

$$\lim_{x \to 0} x \log_2 x = 0$$

$$\lim_{x \to 1} x \log_2 x = 0$$

and therefore, both terms on the right in Eq. (12.1) are zero. The entropy for this case is written as $\mathcal{H}(p_0)$ which is the entropy of a binary memoryless source and is called the *entropy function*.

**Example 12.4** Let $p_0 = 1/4; \ 1 - p_0 = 3/4$, then

$$
\begin{aligned}
H(S) = \mathcal{H}(p_0) &= -p_0 \log_2 p_0 - (1 - p_0) \log_2(1 - p_0) \\
&= -1/4 \times \log_2(1/4) - 3/4 \times \log_2(3/4) \\
&= 0.8113 \text{ bits}
\end{aligned}
$$

Again consider a source

$$S^2 = \{s_1 s_1, s_1 s_2, s_2 s_1, s_2 s_2\}$$

with probabilities
$$\{1/16, 3/16, 3/16, 9/16\}$$

which is derived from $S^1$. Then the entropy

$$
\begin{aligned}
H(S^2) &= -p_0^2 \log_2 p_0^2 - 2p_0(1 - p_0) \log_2 p_0(1 - p_0) - (1 - p_0)^2 \log_2(1 - p_0)^2 \\
&= -(1/16) \log_2(1/16) - 2 \times (3/16) \times \log_2(3/16) - 9/16 \times \log_2(9/16) \\
&= 0.25 + 0.9056 + 0.4669 \\
&= 1.6225 \\
&= 2 \times \mathcal{H}(p_0)
\end{aligned}
$$

We can prove that the source $S^n$ derived from $S^1$ has an entropy $H(S^n) = n\mathcal{H}(p_0)$.

In general, if a set of source symbols

$$S = \{s_1, \ldots, s_N\}$$

$$\text{with probabilities} = \{p_1, \ldots, p_N\} \quad \sum p_k = 1$$

has a source entropy $H(S)$, then the source derived from $S$,

$$S^n = \{\text{all possible } n \text{ symbols taken together } s_i s_j \ldots s_m\}$$

has an entropy $nH(S)$.

## 12.3 Source Coding

Data coding consists of two cases: source coding and channel coding. In this section, we will examine source coding which converts the symbols of a discrete source into binary in an efficient way so as to decrease the overall word length.

Given a discrete source of $N$ symbols, we have to encode these symbols into binary. An obvious approach is to encode *uniquely* those symbols which occur more frequently into fewer bits so that overall our messages may be shorter in length. The word "uniquely" has been used here to ensure that after encoding, each symbol should be decodable.

Thus, if our source is

$$S = \{s_1, \ldots, s_N\}$$

and $l_k$, $k = 1, \ldots, N$ are the lengths of the binary code words representing the symbols, $s_1, \ldots, s_N$. Then the average code word length, $\overline{L}$, is given by

$$\overline{L} = \sum_{k=1}^{N} l_k p_k \tag{12.9}$$

where $p_k$ is the probability of occurrence of symbol $s_k$.

Let $\overline{L}_{\min}$ be the minimum possible value of $\overline{L}$ then *coding efficiency* of our algorithm or process to construct the $l_k$ is given by

$$\eta = \frac{\overline{L}_{\min}}{\overline{L}} \tag{12.10}$$

where $0 < \eta \leq 1$, since $\overline{L} \geq \overline{L}_{\min}$.

The source coding theorem says that if $H(S)$ is the entropy of the source

$$\overline{L} \geq H(S) \tag{12.11}$$

That is the average length of binary code words cannot be made smaller than $H(S)$.

This statement implies that the minimum average is equal to the entropy of the source and it is not possible to make the average smaller than $H(S)$. Or

$$\overline{L}_{min} = H(S) \tag{12.12}$$

and therefore,

$$\eta = \frac{H(S)}{\overline{L}} \tag{12.13}$$

## 12.4   Data Coding and Compaction

In this section, we will look at practical ways of achieving overall decrease in the average length of code words representing the symbols of our source. There are two ways to encode the source: the first method which requires very little knowledge of the probability of occurrence of the source symbols is *fixed length* codes. For example, if we have to encode 10 symbols, then we require 4 bits to do this

$$s_1 \rightarrow 0000$$
$$s_2 \rightarrow 0001$$
$$\vdots$$
$$s_8 \rightarrow 0111$$
$$s_9 \rightarrow 1000$$
$$s_{10} \rightarrow 1001$$

We notice here that we use four bits per symbol and we have an average a word length of four. To decode the binary data, the decoder converts the binary symbols into blocks four bits long and then decodes each block in a unique way according to a lookup table.

Is there another way to encode the data? Take a look at the following scheme for four symbols

$$s_1 \rightarrow 0$$
$$s_2 \rightarrow 1$$
$$s_3 \rightarrow 00$$
$$s_4 \rightarrow 01$$

In this case, the coding is based on a variable length code, but the decoding is not unique. For example, the word $s_1 s_3$ is encoded as 000, but on decoding, we find that this may be misrepresented as $s_1 s_1 s_1$.

## 12.4.1   *Prefix Codes*

In a variable length code, we have to work out a unique method so that we can uniquely decode our encoded sample. To do this, we have to resort to what is known as prefix codes. In this method, *no code word is is a prefix of another code word.* Thus, let us look at the following scheme

$$s_1 \to 1$$
$$s_2 \to 01$$
$$s_3 \to 001$$
$$s_4 \to 0001$$

In this method, $s_1 s_3 s_4$ is encoded as 10010001. On decoding, the decoder decodes the message as discussed below. Starting from the left, we examine one bit at a time. The original code is 10010001. The first bit which is examined is 1, and 1 corresponds to $s_1$.

$$1 \to s_1$$

The next bit is 0, which is not a code word. Therefore, continuing in this way

$$0 \to \text{No code word}$$
$$00 \to \text{No code word}$$
$$001 \to s_3$$

Similarly

$$0001 \to s_4$$

So the code corresponds to $s_1 s_3 s_4$, which was the original code itself.

By looking at this method, we realise that *no code word is the prefix of another code word.* Also, every prefix code is uniquely decodable, but every code that is uniquely decodable does not have to be a prefix code, as shown in the fixed length code considered earlier.

We can analyse how to make prefix codes by observing Fig. 12.2. The tree in this case is a binary tree. As we consider the various branches, if a particular node is considered as a code word, that node is not the root of any more branches, and it ends there. For example, if we examine the figure, we realise that by examining each of the nodes, $s_1$, $s_2$, $s_3$ and $s_4$ are terminating nodes. We now consider the *Kraft–McMillan inequality*, which is an inequality related to the length of prefix codes.

For a set of symbols

$$\{s_1, s_2, \ldots, s_N\}$$

**Fig. 12.2** Tree for the prefix
code of our example

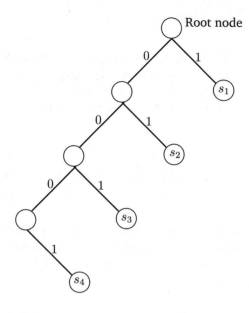

with probabilities

$$\{p_1, p_2, \ldots, p_N\}$$

and each symbol, $s_k$, $k = 1, \ldots, N$ is coded as the code word of length $l_k$, $k = 1, \ldots, N$, then

$$\sum_{n=1}^{N} \frac{1}{2^{l_k}} \leq 1$$

**Example 12.5** As an example for the code which we have just worked out

$$s_1 \rightarrow 1$$
$$s_2 \rightarrow 01$$
$$s_3 \rightarrow 001$$
$$s_4 \rightarrow 0001$$

we have,

$$\frac{1}{2^{l_1}} + \frac{1}{2^{l_2}} + \frac{1}{2^{l_3}} + \frac{1}{2^{l_4}} = \frac{1}{2} + \frac{1}{4} + \frac{1}{8} + \frac{1}{16}$$
$$= 0.9375$$
$$< 1$$

Next, we now look at the average code word length, $E[l_k]$,

$$\overline{L} = \sum_{k=1}^{N} p_k l_k$$

then given a discrete memoryless channel is bounded as follows:

$$H(\mathcal{L}) \leq \overline{L} \leq H(\mathcal{L}) + 1 \tag{12.14}$$

If we use an extended code for $\mathcal{S}^n$, then

$$nH(\mathcal{L}) \leq \overline{L} \leq nH(\mathcal{L}) + 1 \tag{12.15}$$

and therefore

$$H(\mathcal{L}) \leq \frac{\overline{L}}{n} \leq H(\mathcal{L}) + \frac{1}{n} \tag{12.16}$$

### 12.4.2 Huffman Code

The Huffman code algorithm describes a method to assign code words to different symbols in a systematic manner. Huffman codes are variable length prefix codes and the best way to understand the algorithm is through an example.

**Example 12.6** Let a set of six symbols with their probability of occurrence be shown in the symbol table below which are required to be encoded. The first step in the algorithm is to place the symbols in an order of decreasing probability:

$$s_1, \ p_1 = 0.4$$
$$s_2, \ p_2 = 0.2$$
$$s_3, \ p_3 = 0.1$$
$$s_4, \ p_4 = 0.1$$
$$s_5, \ p_5 = 0.1$$
$$s_6, \ p_6 = 0.1$$

We first allocate $s_5$ and $s_6$ a 0 and 1, respectively, and allocate them to node $A$ as shown in Fig. 12.3. We place node $A$ with a probability of 0.2 (=0.1+0.1) at the top of the symbol table in the position of probability 0.2.

The new table of symbols is

$$s_1, \ p_1 = 0.4$$
$$A, \ p_A = 0.2$$
$$s_2, \ p_2 = 0.2$$
$$s_3, \ p_3 = 0.1$$
$$s_4, \ p_4 = 0.1$$

**Fig. 12.3** Huffman coding algorithm

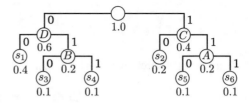

We next allocate $s_3$ and $s_4$ to node $B$ with $s_3$ allocated a 0 and $s_4$ allocated a 1. $B$ is of probability 0.2 (=0.1+0.1)

The new table is

$$s_1, \; p_1 = 0.4$$
$$B, \; p_B = 0.2$$
$$A, \; p_A = 0.2$$
$$s_2, \; p_2 = 0.2$$

We next combine the lowest two nodes, $s_2$ and $A$ to give the node $C$, and duly allocated a 0 and 1, respectively. Node $C$ is of probability 0.4 (=0.2+0.2)

The new table of values is

$$C, \; p_C = 0.4$$
$$s_1, \; p_1 = 0.4$$
$$B, \; p_B = 0.2$$

We next combine $s_1$ (allocated a 0) and $B$ (allocated a 1) to give $D$ with probability 0.6 (=0.2+0.4).

The new table is

$$D, \; p_D = 0.6$$
$$C, \; p_C = 0.4$$

And, finally, we combine $D$ and $C$ allocated as 0 and 1, respectively, with probability 1 (=0.4+0.6).

We now have the following codes corresponding to the various symbols (reading from the top of the tree downward to the symbol):

$$s_1, \; 00$$
$$s_2, \; 10$$
$$s_3, \; 010$$
$$s_4, \; 011$$
$$s_5, \; 110$$
$$s_6, \; 111$$

We can see that the symbols with the lowest probabilities have three-digit code and the symbols with higher probabilities have two-digit code as per our requirements. Furthermore, no code word is a prefix of another code word.

### 12.4.3  Lempel-Ziv (LZ) Coding

In Huffman coding, the probabilities of the source must be known before hand. Suppose these probabilities are not known? A completely different approach uses the fact that the symbols are already encoded into binary. For example, we encode the word "Newts" into a fixed length ASCII code reading as

$$01001110011001010111011101110100011100011...$$

In LZ data compaction we look have the previous binary string and divide it into blocks so that those do not occur in the earlier blocks. Blocks which do not appear earlier are separated by commas

$$01, 00, 11, 10, 011, 001, 010, 111, 0111, 01110, 100, 011100, 11...$$

Now we form a table as shown in Table 12.1.

The table needs an explanation. The row 'Number' is the numerical value of the sequence which is entered into a dictionary. Each sequence is directly written under that row. Each sequence, read from left to right is uniquely different from the previous sequences by only one digit (the last one), or each additional sequence is different from the numerical value of an earlier sequence by the addition of a single digit appearing after the comma. Thus, under number 8, we find 4,1. This means the sequence is 00 (under number 4) followed by 1. The code is generated as follows: the first three digits are the encoded value of the 'Number' followed by the last digit. For example, the code

$$\underbrace{111}_{7} 1$$

is actually 7,1.

In fact, this indeed is the procedure to decode the coded string. The full string is divided (in this case) into blocks of four bits. In each set of four bits, the first three bits refer to the numerical value of the dictionary entry and the last digit refers to the additional bit. Thus, the code 0100 is 2,0 (010 is 2), which is then 1 (entry under 2) and then 0, so that we have 10 as the decoded symbols and which is the entry under entry 6.

**Table 12.1**  Table for Lempel-Ziv coding example

| Number | 1 | 2 | 3 | 4 | 5 | 6 | 7 | 8 | 9 | 10 | 11 | ... |
|---|---|---|---|---|---|---|---|---|---|---|---|---|
| Sequences | 0 | 1 | 01 | 00 | 11 | 10 | 011 | 001 | 010 | 111 | 0111 | ... |
| Representations | - | - | 1,1 | 1,0 | 2,1 | 2,0 | 3,1 | 4,1 | 3,0 | 5,1 | 7,1 | ... |
| Code | | | 0011 | 0010 | 0101 | 0100 | 0111 | 1001 | 0110 | 1011 | 1111 | ... |

In actual practice, the codes are 12 bits each, which means we have $2^{12} = 4096$ dictionary entries. As each additional entry is made, the parsing of the original binary data becomes longer and longer. On an average, data compaction of about 53% is obtained.

## 12.5  Discrete Memoryless Channel

A discrete memoryless channel is a statistical description of a noisy channel where a discrete random variable $X$ is the input to the channel giving us another random variable $Y$ as the output. The output does not, in any way, depend on the previous input. Thus, the RV $X$ takes on various values of

$$X = \{x_1, x_2, \ldots x_M\} \tag{12.17}$$

and on the output we have the symbol set

$$\mathcal{Y} = \{y_1, y_2, \ldots y_N\} \tag{12.18}$$

where $M$ may be equal to $N$. The channel is shown in Fig. 12.4.

The output of the channel depends on the input with a certain probability given by

$$P(Y = y_m | X = x_n) = p(y_m | x_n) \tag{12.19}$$

which are called the *transition probabilities*.[2] We may conveniently represent the transition probabilities in terms of a $M \times N$ *transition probability matrix*

$$\mathbf{P} = \begin{bmatrix} p(y_1|x_1) & p(y_2|x_1) & \cdots & p(y_N|x_1) \\ p(y_1|x_2) & p(y_2|x_2) & \cdots & p(y_N|x_2) \\ \vdots & \vdots & \ddots & \vdots \\ p(y_1|x_M) & p(y_2|x_M) & \cdots & p(y_N|x_M) \end{bmatrix} \tag{12.20}$$

which is also called a *channel matrix* or *transition matrix*.

It is important to note that

$$\sum_{n=1}^{n=N} p(y_n|x_m) = 1 \quad \text{for } m = 1, 2, \ldots, M \tag{12.21}$$

since on the input any one $x_m$ will always result definitely in one of $y_n$s.

---

[2] See Example 3.8.

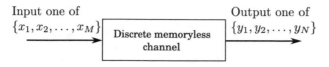

**Fig. 12.4** Discrete memoryless channel

If the probability of occurrence of $X$ is given by

$$P(X = x_m) = p(x_m) \qquad m = 1, \ldots M \tag{12.22}$$

then

$$p(x_m, y_n) = p(y_n|x_m)p(x_m)$$

$$\text{and therefore, } p(y_n) = \sum_{m=1}^{M} p(y_n|x_m)p(x_m) \tag{12.23}$$

which are the various probabilities that symbol $y_n$ was received. And similarly

$$p(x_m, y_n) = p(x_m|y_n)p(y_n) \tag{12.24}$$

where $p(x_m|y_n)$ are the probabilities that symbol $x_m$ was transmitted observing that $y_n$ was received.

### 12.5.1   Mutual Information

We now define a conditional entropy in the following manner

$$H(X|Y = y_n) = \sum_{m=1}^{M} p(x_m|y_n) \log_2 \left[ \frac{1}{p(x_m|y_n)} \right] \qquad n = 1, \ldots N$$

which is a random variable involving $y_n$. We now define an entropy which is defined by

$$H(X|\mathcal{Y}) = E[H(X|Y = y_n)]$$

Therefore

$$H(X|\mathcal{Y}) = \sum_{n=1}^{N} p(y_n) \left\{ \sum_{m=1}^{M} p(x_m|y_n) \log_2 \left[ \frac{1}{p(x_m|y_n)} \right] \right\}$$

$$= \sum_{n=1}^{N} \sum_{m=1}^{M} p(x_m|y_n) p(y_n) \log_2 \left[ \frac{1}{p(x_m|y_n)} \right]$$

$$= \sum_{n=1}^{N} \sum_{m=1}^{M} p(x_m, y_n) \log_2 \left[ \frac{1}{p(x_m|y_n)} \right] \qquad (12.25)$$

where we have used Eq. (12.24). We call $H(X|\mathcal{Y})$ the *conditional entropy* of $X$ given $\mathcal{Y}$. This parameter is a *measure of the uncertainty of $X$ observing $\mathcal{Y}$*. (See Sect. 12.2)

We now represent the quantity, $H(X) - H(X|\mathcal{Y})$ as the *mutual information*, which is the *uncertainty in $X$ having taken $\mathcal{Y}$ into consideration*:

$$I(X; \mathcal{Y}) = H(X) - H(X|\mathcal{Y}) \qquad (12.26)$$

similarly, we define

$$I(\mathcal{Y}; X) = H(\mathcal{Y}) - H(\mathcal{Y}|X) \qquad (12.27)$$

## 12.5.2   Properties of Mutual Information

1. *Mutual information is symmetric*. We start with the definition of entropy of $X$.

$$H(X) = \sum_{m=1}^{M} p(x_m) \log_2 \left[ \frac{1}{p(x_m)} \right]$$

$$= \sum_{m=1}^{M} \sum_{n=1}^{N} p(x_m, y_n) \log_2 \left[ \frac{1}{p(x_m)} \right] \qquad (12.28)$$

since

$$p(x_m) = \sum_{n=1}^{N} p(x_m, y_n) \qquad (12.29)$$

and therefore

$$H(X) - H(X|\mathcal{Y}) = \sum_{m=1}^{M} \sum_{n=1}^{N} \left\{ p(x_m, y_n) \log_2 \left[ \frac{1}{p(x_m)} \right] - p(x_m, y_n) \log_2 \left[ \frac{1}{p(x_m|y_n)} \right] \right\}$$

$$= \sum_{m=1}^{M} \sum_{n=1}^{N} p(x_m, y_n) \log_2 \left[ \frac{p(x_m|y_n)}{p(x_m)} \right] \qquad (12.30)$$

$$= \sum_{m=1}^{M} \sum_{n=1}^{N} p(x_m, y_n) \log_2 \left[ \frac{p(x_m, y_n)}{p(x_m) p(y_n)} \right] \qquad (12.31)$$

similarly

$$I(\mathcal{Y}; \mathcal{X}) = \sum_{m=1}^{M} \sum_{n=1}^{N} p(x_m, y_n) \log_2 \left[ \frac{p(y_n | x_m)}{p(y_n)} \right] \qquad (12.32)$$

$$= \sum_{m=1}^{M} \sum_{n=1}^{N} p(x_m, y_n) \log_2 \left[ \frac{p(x_m, y_n)}{p(y_n) p(x_m)} \right] \qquad (12.33)$$

therefore

$$I(\mathcal{Y}; \mathcal{X}) = I(\mathcal{X}; \mathcal{Y}) = \sum_{m=1}^{M} \sum_{n=1}^{N} p(x_m, y_n) \log_2 \left[ \frac{p(x_m, y_n)}{p(y_n) p(x_m)} \right]$$

$$= -\sum_{m=1}^{M} \sum_{n=1}^{N} p(x_m, y_n) \log_2 \left[ \frac{p(y_n) p(x_m)}{p(x_m, y_n)} \right] \qquad (12.34)$$

2. *Mutual information is always positive (or zero).* To prove this result, we look at the inequality

$$\log_e(x) \le x - 1 \qquad x \ge 0 \qquad (12.35)$$

with the equality holding for $x = 1$. To see this relationship graphically, see Fig. 12.5. Then

$$-\sum_{m=1}^{M} \sum_{n=1}^{N} p(x_m, y_n) \log_2 \left[ \frac{p(y_n) p(x_m)}{p(x_m, y_n)} \right] = -\frac{1}{\log_e 2} \sum_{m=1}^{M} \sum_{n=1}^{N} p(x_m, y_n) \log_e \left[ \frac{p(y_n) p(x_m)}{p(x_m, y_n)} \right]$$

$$\ge -\frac{1}{\log_e 2} \sum_{m=1}^{M} \sum_{n=1}^{N} p(x_m, y_n) \left[ \frac{p(y_n) p(x_m)}{p(x_m, y_n)} - 1 \right]$$

$$= -\frac{1}{\log_e 2} \sum_{m=1}^{M} \sum_{n=1}^{N} [p(y_n) p(x_m) - p(x_m, y_n)]$$

$$= 0 \qquad (12.36)$$

where we have used

$$\sum_{m=1}^{M} \sum_{n=1}^{N} p(y_n) p(x_m) = \sum_{m=1}^{M} p(x_m) \sum_{n=1}^{N} p(y_n) = 1$$

and $$\sum_{m=1}^{M} \sum_{n=1}^{N} p(x_m, y_n) = 1 \qquad (12.37)$$

Which gives the desired result

$$I(\mathcal{Y}; \mathcal{X}) = I(\mathcal{X}; \mathcal{Y}) \ge 0 \qquad (12.38)$$

The equality holds when

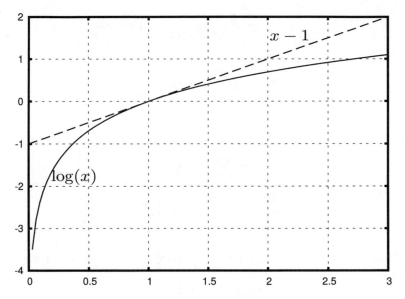

**Fig. 12.5** Graphs of $\log(x)$ and $x - 1$

$$p(x_m, y_n) = p(y_n)p(x_m) \tag{12.39}$$

that these two distributions involving $x_m$ and $y_n$ are independent random variables. The mutual information is a form of entropy, *so when the mutual information is zero, having observed $\mathcal{Y}$ gives us no clue to the uncertainty in $X$.*

3. *Mutual information is connected to the entropy of $X$ and $\mathcal{Y}$ through the relation*

$$I(X; \mathcal{Y}) = H(X) + H(\mathcal{Y}) - H(X, \mathcal{Y}) \tag{12.40}$$

To prove this result

$$
\begin{aligned}
I(\mathcal{Y}; X) = I(X; \mathcal{Y}) &= -\sum_{m=1}^{M}\sum_{n=1}^{N} p(x_m, y_n) \log_2 \left[ \frac{p(y_n)p(x_m)}{p(x_m, y_n)} \right] \\
&= -\sum_{m=1}^{M}\sum_{n=1}^{N} p(x_m, y_n) \left[ \log_2 \frac{1}{p(x_m, y_n)} - \log_2 \frac{1}{p(y_n)} - \log_2 \frac{1}{p(x_m)} \right] \\
&= -H(X, \mathcal{Y}) + \sum_{m=1}^{M}\sum_{n=1}^{N} p(x_m, y_n) \left[ \log_2 \frac{1}{p(y_n)} + \log_2 \frac{1}{p(x_m)} \right] \\
&= -H(X, \mathcal{Y}) + \sum_{n=1}^{N} p(y_n) \log_2 \frac{1}{p(y_n)} + \sum_{m=1}^{M} p(x_m) \log_2 \frac{1}{p(x_m)} \\
&= H(X) + H(\mathcal{Y}) - H(X, \mathcal{Y}) \tag{12.41}
\end{aligned}
$$

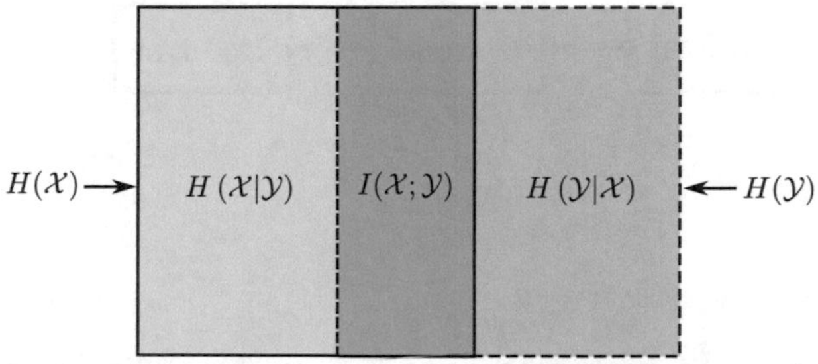

$$H(\mathcal{X}, \mathcal{Y}) = -I(\mathcal{X}; \mathcal{Y}) + H(\mathcal{X}) + H(\mathcal{Y})$$
$$H(\mathcal{X}) = I(\mathcal{X}; \mathcal{Y}) + H(\mathcal{X}|\mathcal{Y})$$
$$H(\mathcal{Y}) = I(\mathcal{X}; \mathcal{Y}) + H(\mathcal{Y}|\mathcal{X})$$

**Fig. 12.6** Set theoretic model of channel entropies

where we have used

$$H(X, \mathcal{Y}) = \sum_{m=1}^{M} \sum_{n=1}^{N} p(x_m, y_n) \log_2 \frac{1}{p(x_m, y_n)}$$

$$\sum_{m=1}^{M} \sum_{n=1}^{N} p(x_m, y_n) \log_2 \frac{1}{p(y_n)} = \sum_{n=1}^{N} p(y_n) \log_2 \frac{1}{p(y_n)}$$

and $$\sum_{m=1}^{M} \sum_{n=1}^{N} p(x_m, y_n) \log_2 \frac{1}{p(x_m)} = \sum_{m=1}^{M} p(x_m) \log_2 \frac{1}{p(x_m)}$$

therefore

$$H(X, \mathcal{Y}) = -I(X; \mathcal{Y}) + H(X) + H(\mathcal{Y}) \tag{12.42}$$

A set theoretic model of channel entropies is shown in Fig. 12.6, which encompasses the three relations, Eqs. (12.26), (12.27) and (12.42).

## 12.5.3 Channel Capacity

We now concentrate on the maximum number of bits (or symbols) which we can transfer every time we use a noisy channel, and which we shall call the channel capacity. Out of the two symbol sets, $X$ and $\mathcal{Y}$ the probabilities, $p(x_m)$ are independent of all the other probability distributions. Depending on which channel we use, the transition probabilities are channel dependent. Therefore, given a set of $p(x_m)$, we can, through the use of transition probabilities, obtain $p(y_n)$:

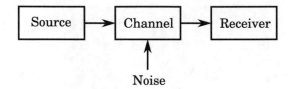

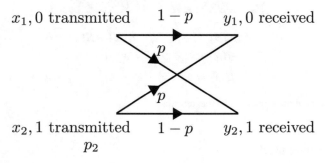

**Fig. 12.7**  Binary symmetric channel

$$p(y_n) = \sum_{m=1}^{M} p(y_n|x_m)p(x_m)$$

Now we consider the channel capacity which is defined as the greatest value of the mutual information subject to all possible $\{p(x_m)\}$ $m = 1, \ldots, M$, for a given channel. Or mathematically, the channel capacity is

$$C = \max_{\{p(x_m)\}} I(X; \mathcal{Y})$$

**Example 12.7**  Channel capacity for the binary symmetric channel. See Fig. 12.7.

In this case

$$p(y_1|x_1) = p(y_2|x_2) = 1 - p$$
$$\text{and } p(y_1|x_2) = p(y_2|x_1) = p \tag{12.43}$$

then

$$H(X) = p(x_1) \log_2 \frac{1}{p(x_1)} + p(x_2) \log_2 \frac{1}{p(x_2)} \tag{12.44}$$

we know that the entropy is maximised when

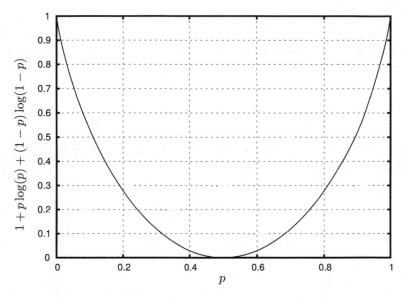

**Fig. 12.8** Plot of channel capacity versus $p$

$$p(x_1) = p(x_2) = \frac{1}{2}$$
$$H(X)|_{\max} = 1 \tag{12.45}$$

The channel capacity is then

$$C(p) = 1 + p \log_2 p + (1 - p) \log_2(1 - p) \tag{12.46}$$

plotting this function as a function of $p$ we obtain the results as shown in Fig. 12.8.

From the figure, it is clear that the capacity of the channel is 1 bit per channel use for $p = 0$, with no probability of error, and it is 0 bits per channel use when $p = 0.5$. *The channel capacity in essence is a measure of the effective number of bits which can be transferred without error (reliably) every time you use the channel.*

## 12.5.4 Channel Coding Theorem. Shannon's Theorem

In the channel coding theorem, we address the problem created by a noisy channel. Thus, for example, in the binary symmetric channel, for $p = 0.1$, on the average, every tenth bit is in error, and for $p = 0.05$, every twentieth bit is in error. How do we transfer information reliably with an arbitrarily low probability of error? Thus, for example, in a typical communication system, a probability of error of $10^{-8}$ may be a necessary requirement.

The solution to this problem is to include redundancy in our information at the expense of a much lower transfer rate than that which the channel capacity predicts. One solution is to divide the data into blocks of $k$ bits. Thus, for every $k$ bit of information, we add extra bits to make $n$ bits $(n > k)$.

$$\underbrace{b_1 b_2 \cdots b_k}_{k \text{ bits}} \underbrace{c_1 c_2 \cdots c_{n-k}}_{n-k \text{ bits}}$$

These $n - k$ bits are redundant bits which may be used to check the other bits to lower the probability of error. The ratio

$$r = \frac{k}{n} \tag{12.47}$$

is called the *code rate*. We now ask the important question: Can we lower the probability of error to an arbitrarily small value, $\epsilon$, and still maintain a code rate which not too small? The answer to this question is 'yes'.

The channel capacity so far has not taken time into consideration. Shannon's channel capacity theorem states that the value of $\epsilon$ can be made arbitrarily small as long as

$$r \le C \tag{12.48}$$

Let us take time into account. If $H(S)$ is the entropy of the source, then the source emits an average of $H(S)$ symbols (in bits) in time $T_s$. So the rate of the creation of symbols is $H(S)/T_s$ bits per second. These symbols are emitted through the channel in time $T_c$. Then the channel capacity theorem says

$$\frac{H(S)}{T_s} \le \frac{C}{T_c} \quad \text{bits/sec} \tag{12.49}$$

then we can reduce the probability of error to a value less than a small value $\epsilon$ as long as Eq. (12.49) is satisfied. From this equation, the code rate, $r$, is given by

$$r = \frac{H(S)T_c}{T_s} \tag{12.50}$$

**Example 12.8** Using a repetition code on a binary symmetric channel.

With an error probability $p = 0.1$, the probability of error with no repetition is

$$P_e = p = 0.1 \tag{12.51}$$

The channel capacity is, from Eq. 12.46,

$$C = 0.513$$

Suppose we want to transmit a 0, and instead we transmit a 000; on the other hand, when we want to transmit a 1, we transmit 111 over a binary symmetric channel. Therefore, when we want to transmit

$$01101\ldots$$

we transmit instead

$$000111111000111\ldots$$

On reception, we infer from blocks of three bits which we receive that if there are more numbers of zeros than there are ones then the bit was a zero otherwise it was a one; and if in the three bits we have more 1s than 0s then it is interpreted as a 1. In this case, $k = 1$ and $n = 3$.

Based on this scheme, the probability of error is

$$P_e = P(\text{three errors}) + P(\text{two errors})$$
$$= p^3 + 3p^2(1 - p)$$
$$= 0.028$$

and the code rate $r$ is

$$r = \frac{1}{3} = 0.3\dot{3} < 0.513$$

If, on the other hand, we transmit five 0s for every 0 and five 1s for every one, the probability of error is

$$P_e = p^5 + 5p^4(1 - p) + 10p^3(1 - p)$$
$$= 0.0086$$

and the code rate is

$$r = 0.2$$

As we keep increasing the number of repetitions, we find that the probability of error keeps decreasing and we can decrease it to a value less than a small value $\epsilon$.

## 12.6  Continuous Random Variables

We will take a look at continuous random variables from the point of view of information theory concepts. In the case of continuous random variables, the variables themselves and the channel supports continuous transfer. If $x$ is continuous, then the channel output will also be a continuous random $y$ and the transition probabilities will be $p(y|x)$. If we look at these concepts with more rigour, then the PDF of $x$ and $y$ would be $f_X(x)$ and $f_Y(y)$, respectively, and $p(y|x)$ would be $f_Y(y|x)$.

### 12.6.1  Differential Entropy

In  a similar way to describing discrete RVs, we define a differential entropy for continuous RVs given by

$$h(X) = -\int_{-\infty}^{\infty} f_X(x) \log_2 [f_X(x)] \, dx \qquad (12.52)$$

where $f_X(x)$ is the PDF of a continuous RV, $x$, and $h(X)$ is called the *differential or relative entropy*.

According to the definition of entropy, the absolute entropy is given by

$$H(X) = \lim_{\Delta p(x) \to 0} \left\{ - \sum_{k=-\infty}^{\infty} \Delta p_k(x) \log_2 [\Delta p_k(x)] \right\}$$

where

$$\Delta p_k(x) = f_X(x_k) \Delta x$$

is the probability that $x_k$ lies in the interval $[x_k - \Delta x/2,\ x_k + \Delta x/2]$. Therefore

$$
\begin{aligned}
H(X) &= \lim_{\Delta x \to 0} \left\{ - \sum_{k=-\infty}^{\infty} [f_X(x_k) \Delta x] \log_2 [f_X(x_k) \Delta x] \right\} \\
&= \lim_{\Delta x \to 0} \left\{ - \left[ \sum_{k=-\infty}^{\infty} [f_X(x_k) \Delta x] \log_2 [f_X(x)] + [f_X(x_k) \Delta x] \log_2 \Delta x \right] \right\} \\
&= -\int_{-\infty}^{\infty} f_X(x) \log f_X(x) dx - \lim_{\Delta x \to 0} \sum_{k=-\infty}^{\infty} [f_X(x_k) \Delta x] \log_2 \Delta x \\
&= h(X) - \lim_{\Delta x \to 0} \log_2 \Delta x \int_{-\infty}^{\infty} f_X(x) dx \qquad (12.53) \\
&= h(X) - \lim_{\Delta x \to 0} \log_2 \Delta x \qquad (12.54)
\end{aligned}
$$

where $\Delta x$ is a small interval, and $f_X(x_k) \Delta x$ is the probability occurrence of $X$ in that small interval. In the Eq. (12.53), we have equated the integral under the PDF to be 1. The term

$$\lim_{\Delta x \to 0} \left[ - \log_2 \Delta x \right]$$

is the *reference* to the absolute entropy, which turns out to be of a very large value. This is similar to potential in electromagnetic theory where only *differences* in potential are important. Therefore, the justification nomenclature employed: *differential* entropy, and we will always keep this in mind.

In general, the differential entropy may be positive or negative or even zero, since we are looking at only differences in the entropy (like *potential differences* in electromagnetic theory). Let us look at a particular example

**Example 12.9** Compare the differential entropies of the uniformly distributed RV to the Gaussian distributed RV having the same means and variances.

We can use the definition of the differential entropy to take a look at the uniformly distributed RV, in the interval $[-a, a]$ which has zero mean and variance.[3]

$$\sigma_X = \frac{a^2}{3}$$

then the differential entropy for the uniform distribution is

$$
\begin{aligned}
h(X) &= -\int_{-a}^{a} f_X(x) \log_2 [f_X(x)] \, dx \\
&= -\log_2 \left(\frac{1}{2a}\right) \int_{-a}^{a} f_X(x) \, dx \\
&= \log_2(2a)
\end{aligned}
$$

Then, we consider the Gaussian RV, with mean 0 and standard deviation $\sigma = \sqrt{a^2/3}$. The differential entropy is

$$
\begin{aligned}
h(Y) &= -\int_{-\infty}^{\infty} f_Y(y) \log_2 [f_Y(y)] \, dx \\
&= -\int_{-\infty}^{\infty} f_Y(y) \left[ \log_2 \left(\frac{1}{\sigma\sqrt{2\pi}}\right) + \log_2 \left\{ \exp\left(-\frac{y^2}{2\sigma^2}\right) \right\} \right] \\
&= -\int_{-\infty}^{\infty} f_Y(y) \left[ \log_2 \left(\frac{1}{\sigma\sqrt{2\pi}}\right) + \left(-\frac{y^2}{2\sigma^2}\right) \log_2 e \right] \\
&= \int_{-\infty}^{\infty} f_Y(y) \left[ \log_2 \left(\sigma\sqrt{2\pi}\right) + \left(\frac{y^2}{2\sigma^2}\right) \log_2 e \right] \\
&= \log_2 \left(\sigma\sqrt{2\pi}\right) + \frac{1}{2} \log_2 e \\
&= \frac{1}{2} \log_2 \left(2\pi\sigma^2 e\right) \\
&= \log_2 \left(\sqrt{\frac{2\pi a^2 e}{3}}\right)
\end{aligned}
\tag{12.55}
$$

---

[3] From Example 3.19 of Chap. 3.

where

$$f_Y(y) = \frac{1}{\sigma\sqrt{2\pi}} \exp\left\{-\frac{y^2}{2\sigma^2}\right\}$$

In this example

$$h(Y) > h(X) \tag{12.56}$$

Since

$$\sqrt{\frac{2\pi e}{3}} > 2 \quad \text{for } a > 0$$

Thus, when the entropy of two distributions *may be compared*, care must be taken that their *means and variances are the same*, and then only we may decide which has the larger entropy through their differential entropies.

We next compare differential entropies of any distribution, $f_X(x)$ with the Gaussian RV with distribution $f_Y(y)$, *both having the same mean and variance, m and* $\sigma^2$.

$$f_Y(y) = \frac{1}{\sigma\sqrt{2\pi}} \exp\left\{-\frac{(y-m)^2}{2\sigma^2}\right\}$$

Consider the relation by Eq. (12.35), reproduced here for the convenience of the student,

$$\log_e(x) \le x - 1 \quad x \ge 0$$

so, therefore

$$\int_{-\infty}^{\infty} f_X(x) \log_2\left[\frac{f_Y(x)}{f_X(x)}\right] dx \le \int_{-\infty}^{\infty} f_X(x)\left[\frac{f_Y(x)}{f_X(x)} - 1\right] dx$$

$$= \int_{-\infty}^{\infty} f_Y(x)dx - \int_{-\infty}^{\infty} f_X(x)dx$$

$$= 0$$

Notice here that since we are considering the logarithm of the ratio

$$\log_2\left[\frac{f_Y(x)}{f_X(x)}\right] = \log_2 f_Y(x) - \log_2 f_X(x)$$

the reference point or datum of the entropy is identical.

Or proceeding from where we left off

$$\int_{-\infty}^{\infty} f_X(y) \log_2 f_Y(y)dy - \int_{-\infty}^{\infty} f_X(x) \log_2 f_X(x)dx \le 0$$

$$\text{or } \int_{-\infty}^{\infty} f_X(x) \log_2 f_Y(x)dx \le \int_{-\infty}^{\infty} f_X(x) \log_2 f_X(x)dx$$

$$\Rightarrow -\int_{-\infty}^{\infty} f_X(x) \log_2 f_Y(x)dx \ge -\int_{-\infty}^{\infty} f_X(x) \log_2 f_X(x)dx$$

$$\Rightarrow \int_{-\infty}^{\infty} f_X(x) \left\{ \log_2 \left( \sigma \sqrt{2\pi} \right) + \left[ \frac{(x-m)^2}{2\sigma^2} \right] \log_2 e \right\} dx \geq h(X)$$

But

$$\int_{-\infty}^{\infty} f_X(x) dx = 1$$

$$\int_{-\infty}^{\infty} (x - m)^2 f_X(x) dx = \sigma^2$$

therefore

$$\log_2 \left( \sigma \sqrt{2\pi} \right) + \frac{1}{2} \log_2 e \geq h(X)$$

$$\text{or } \frac{1}{2} \log_2 \left( 2\pi \sigma^2 e \right) \geq h(X)$$

However

$$h(Y) = \frac{1}{2} \log_2 \left( 2\pi \sigma^2 e \right) \tag{12.57}$$

is the differential entropy of the Gaussian RV[4]. Therefore, the Gaussian RV has the largest differential entropy out of all other valid random variables having the same mean and variance. In other words, if we have any distribution with mean $m$ and variance $\sigma$, its entropy will always be less than that of the Gaussian RV with the same mean and variance.

## 12.6.2 Mutual Information

From the analogy with Eq. (12.31), reproduced here for convenience,

$$I(\mathcal{X}; \mathcal{Y}) = \sum_{m=1}^{M} \sum_{n=1}^{N} p(x_m, y_n) \log_2 \left[ \frac{p(x_m, y_n)}{p(x_m) p(y_n)} \right]$$

we define the mutual information for continuous random variables,

$$I(X; Y) = \int_{-\infty}^{\infty} \int_{-\infty}^{\infty} f_{XY}(x, y) \log_2 \left[ \frac{f_{XY}(x, y)}{f_X(x) f_Y(y)} \right] dx dy$$

From this definition, it is clear that

---

[4] See Example 12.9.

$$I(X; Y) = I(Y; X) \tag{12.58}$$

And it may be proved that

$$I(X; Y) \geq 0 \tag{12.59}$$

$$I(X; Y) = h(X) - h(X|Y) \tag{12.60}$$

$$= h(Y) - h(Y|X) \tag{12.61}$$

where $h(X)$ and $h(Y)$ are the differential entropies for $X$ and $Y$, while

$$
\begin{aligned}
h(X|Y) &= \int_{-\infty}^{\infty} \int_{-\infty}^{\infty} f_{XY}(x, y) \log_2 \left[ \frac{f_Y(y)}{f_{XY}(x, y)} \right] dxdy \\
&= \int_{-\infty}^{\infty} \int_{-\infty}^{\infty} f_{XY}(x, y) \log_2 \left[ \frac{1}{f_X(x|y)} \right] dxdy
\end{aligned}
\tag{12.62}
$$

similarly,

$$
\begin{aligned}
h(Y|X) &= \int_{-\infty}^{\infty} \int_{-\infty}^{\infty} f_{XY}(x, y) \log_2 \left[ \frac{f_Y(x)}{f_{XY}(x, y)} \right] dxdy \\
&= \int_{-\infty}^{\infty} \int_{-\infty}^{\infty} f_{XY}(x, y) \log_2 \left[ \frac{1}{f_Y(y|x)} \right] dxdy
\end{aligned}
\tag{12.63}
$$

### 12.6.3  Channel Capacity. Shanon–Hartley Theorem

We now take a look at the information capacity of a channel where the *only* distortion is caused by the addition of additive white Gaussian noise (AWGN)

$$y(t) = x(t) + w(t) \tag{12.64}$$

That is, the channel *ideal*, with the exception of the addition of white Gaussian noise of zero mean and power spectral density $N_0/2$. This also implies that in the absence of noise, $y(t) = x(t)$. Also, the signal transmitted is limited to $B$ hertz. Let us concentrate first on the channel capacity per sample, then later multiplying by $2B$ (applying the Nyquist criterion), we can get the channel capacity per second. It should also be noted that $x(t)$ is limited in power, that is

$$- M \leq x(t) \leq M \tag{12.65}$$

The mutual information, by the earlier definition is,

$$I(X; Y) = h(Y) - h(Y|X) \tag{12.66}$$

where

$$h(Y|X) = \int_{-\infty}^{\infty} \int_{-\infty}^{\infty} f_{XY}(y, x) \log_2\left(\frac{1}{f_Y(y|x)}\right) dx dy$$

$$= \int_{-\infty}^{\infty} f_X(x) dx \int_{-\infty}^{\infty} f_Y(y|x) \log_2\left(\frac{1}{f_Y(y|x)}\right) dy$$

$$= \int_{-\infty}^{\infty} f_Y(y|x) \log_2\left(\frac{1}{f_Y(y|x)}\right) dy \qquad (12.67)$$

Now since the channel is ideal and $x$ is given, the characteristics of $y$ are the same as that of the noise with a shift in the noise characteristics by a constant. This is true from Eq. 12.64.

Therefore

$$f_Y(y|x) = f_W(y - x)$$

$$\therefore, \int_{-\infty}^{\infty} f_Y(y|x) \log_2\left[\frac{1}{f_Y(y|x)}\right] dy = \int_{-\infty}^{\infty} f_W(y-x) \log_2\left[\frac{1}{f_Y(y|x)}\right] dy$$

$$= \int_{-\infty}^{\infty} f_W(z) \log_2\left[\frac{1}{f_W(z)}\right] dz$$

$$= h(W) \qquad (12.68)$$

where $f_W(\cdot)$ is the PDF of white Gaussian noise, and $z = y - x$ where $x$ is a constant.

Returning now to Eq. (12.66), we have to maximise this result. To do so, we must maximise $h(Y)$. The maximum value of the entropy of $Y$ is when $f_Y(y)$ is Gaussian. That is from Eq. (12.57)

$$h_{\max}(y) = \frac{1}{2} \log_2\left(2\pi e \sigma_Y^2\right) \qquad (12.69)$$

but, since $X$ and $W$ are uncorrelated

$$\sigma_Y^2 = \sigma_X^2 + \sigma_W^2$$
$$= S + N_0 B$$
$$= S + N$$

where $S$ is the finite signal power[5] and $N_0 B$ is the noise power. Therefore

$$h_{\max}(y) = \frac{1}{2} \log_2\left[2\pi e \left(S + N_0 B\right)\right]$$
$$= \frac{1}{2} \log_2\left[2\pi e \left(S + N\right)\right]$$

---

[5] Due to Expression (12.65).

and

$$h(W) = \frac{1}{2} \log_2 \left( 2\pi e \sigma_W^2 \right)$$

$$= \frac{1}{2} \log_2 \left( 2\pi e N_0 B \right)$$

or the maximum mutual information per sample is

$$I_{max}(X; Y) = \frac{1}{2} \log_2 \left[ 2\pi e \left( S + N_0 B \right) \right] - \frac{1}{2} \log_2 \left( 2\pi e N_0 B \right)$$

$$= \frac{1}{2} \log_2 \left[ \left( 1 + \frac{S}{N_0 B} \right) \right] \qquad \text{bits/sample}$$

$$= \frac{1}{2} \log_2 \left[ \left( 1 + \frac{S}{N} \right) \right] \qquad \text{bits/sample}$$

But there being $2B$ samples per second, the channel capacity, which is the maximum mutual information in bits per second, for a *Gaussian channel* is

$$C = B \log_2 \left[ \left( 1 + \frac{S}{N_0 B} \right) \right] \qquad \text{bits/s} \qquad (12.70)$$

$$= B \log_2 \left[ \left( 1 + \frac{S}{N} \right) \right] \qquad \text{bits/s} \qquad (12.71)$$

This expression tells us that channel capacity increases rapidly with bandwidth, but it is proportional to the logarithm of the signal power $S$. To examine this relation in more detail, we write it as

$$C = \frac{S}{N_0} \left\{ \frac{N_0 B}{S} \log_2 \left[ \left( 1 + \frac{S}{N_0 B} \right) \right] \right\}$$

$$= \frac{S}{N_0} \left\{ \chi \log_2 \left[ \left( 1 + \frac{1}{\chi} \right) \right] \right\} \qquad (12.72)$$

where

$$\chi = \frac{N_0 B}{S}$$

A plot of the function

$$f(\chi) = \chi \log_2 \left[ \left( 1 + \frac{1}{\chi} \right) \right]$$

is given in Fig. 12.9.

For a constant value of $S/N_0$ and allowing $B$ to go to infinity, we want to find the maximum channel capacity on the basis of bandwidth alone. This channel capacity, $C_B$, is given by

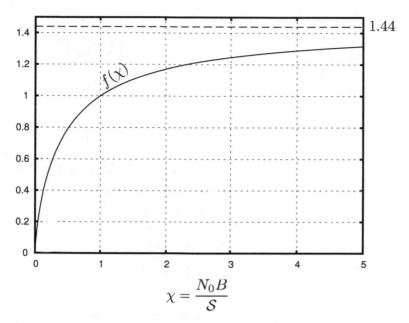

**Fig. 12.9** Plot of $f(\chi)$ versus $\chi$

$$C_B = \frac{S}{N_0} \lim_{\chi \to \infty} \chi \log\left[\left(1 + \frac{1}{\chi}\right)\right] = 1.44 \frac{S}{N_0} \tag{12.73}$$

which is also shown in the figure. From this example, it is clear that allowing the bandwidth to increase is only a single aspect of increasing the channel capacity. To increase channel capacity, keeping $B$ constant we have to look at the function

$$C = B \log_2\left[1 + \frac{\xi}{B}\right]$$

$$\text{where } \xi = \frac{S}{N_0}$$

Using $\xi = S/N_0$ as a parameter, the plot of the channel capacity is shown in Fig. 12.10. From the plots, it is clear that the channel capacity reaches an approximate maximum for

$$B_{max} \approx 10\left(\frac{S}{N_0}\right) \text{ (Hz)} \tag{12.74}$$

and its maximum value is approximately

$$C_{max} \approx 1.4\left(\frac{S}{N_0}\right) \text{ (bits/s)} \tag{12.75}$$

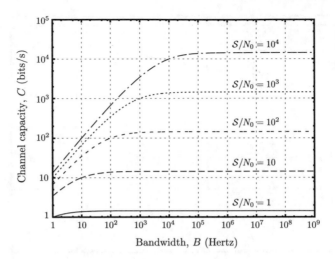

**Fig. 12.10** Plot of the channel capacity with $S/N_0$ as a parameter

Increasing the value of $B$ beyond $B_{max}$ gives diminishing returns in terms of channel capacity. It is instructive to compare these empirical rules with Eq. (12.73).

### 12.6.4 Sphere Packing

We can look at the channel capacity from another view point. Shannon posed a very interesting problem in his classic 1948 paper [2]: $f_i(t)$ is a communication signal of bandwidth $B$ hertz, of duration $T$ secs, then $f_i(t)$ is represented by a vector of $2BT$ samples, in $n = 2BT$ dimensional space, $\mathbb{R}^n$. Therefore (see Fig. 12.11),

$$f_i[1], \ f_i[2], \ \ldots f_i[n]$$

are the sample points corresponding to the vector $\mathbf{f}_i$. If the noise is represented by $w(t)$, then the signal plus noise is represented by the sample points,

$$f_i[1] + w_i[1], \ f_i[2] + w_i[2], \ldots, f_i[n] + w_i[n]$$

which corresponds to the vector $\mathbf{f}_i + \mathbf{w}_i$. The signal power for this particular signal is $S_i$ and the noise power is $N_i$, where

$$S_i = \frac{1}{T} \sum_{j=1}^{N} f_i^2[j]$$

**Fig. 12.11** Sphere packing details

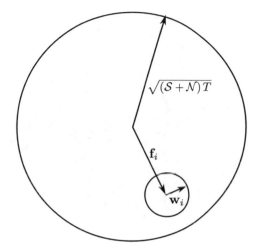

$$\mathcal{N}_i = \frac{1}{T} \sum_{j=1}^{N} w_i^2[j] \qquad (12.76)$$

or

$$\sum_{j=1}^{N} f_i^2[j] = \mathcal{S}_i T$$

$$\sum_{j=1}^{N} w_i^2[j] = \mathcal{N}_i T \qquad (12.77)$$

A signal will be correctly decoded is one which lies at the centre of a small sphere of radius $\mathcal{N}_i T = \mathcal{N}T$. The maximum number of such signals which can be decoded is by the use of a maximum likelihood receiver is equal to the number of small spheres which can be fitted into the larger sphere (Fig. 12.12a). If the maximum signal energy is $(\mathcal{S} + \mathcal{N})T$, then the corresponding radius of the big sphere is $\sqrt{(\mathcal{S} + \mathcal{N})T}$, and the radii of the small spheres are $\sqrt{\mathcal{N}T}$. If the noise level is high, the number of decodable spheres will reduce, and the channel capacity will fall. This is shown in Fig. 12.12b. Due to an increase in the noise power, the radius of the outer sphere increases and so do the radii of the smaller spheres.

Therefore, if $A_n$ is the proportionality constant to define the volume of a sphere in $n$ dimensional space, then

$$\text{Number of such vectors, } M = \frac{A_n R^n}{A_n r^n} = \left( \sqrt{\frac{\mathcal{S} + \mathcal{N}}{\mathcal{N}}} \right)^n$$

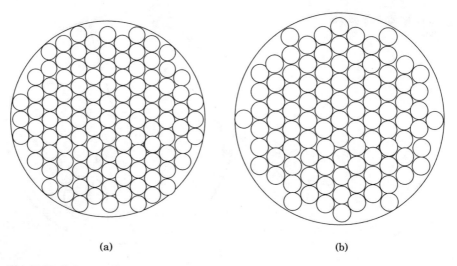

(a)                                                                  (b)

**Fig. 12.12**  Sphere packing

$$\text{or } M = \left(\frac{S+N}{N}\right)^{BT}$$

$$C = \frac{\log_2 M}{T} = B \log_2\left(1 + \frac{S}{N}\right) \text{ (bits/s)} \tag{12.78}$$

where $C$ is the channel capacity, where alternatively, we can look at $M$ as uniquely decodable code words using a maximum likelihood receiver.

### 12.6.5  Implications of Shanon's Theorem

Let us take a look at a demodulator whose input parameters are $S_i$, $N_i$ and $B_i$, where these parameters are the input signal power, noise power and input bandwidth, and similarly, the output parameters are $S_o$, $N_o$ and $B_o$. For all communication systems, at the receiver (Fig. 12.13)

$$C = B_i \log_2\left(1 + \frac{S_i}{N_i}\right) = B_o \log_2\left(1 + \frac{S_o}{N_o}\right) \tag{12.79a}$$

$$\text{or } \frac{S_o}{N_o} = \left(1 + \frac{S_i}{N_i}\right)^{B_i/B_o} - 1 \tag{12.79b}$$

$$= \left[1 + \frac{S_i}{N_0 B_o \, (B_i/B_o)}\right]^{B_i/B_o} - 1 \tag{12.79c}$$

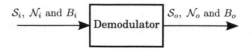

**Fig. 12.13**  Demodulator with $S_i$, $N_i$ and $B_i$; $S_o$, $N_o$ and $B_o$

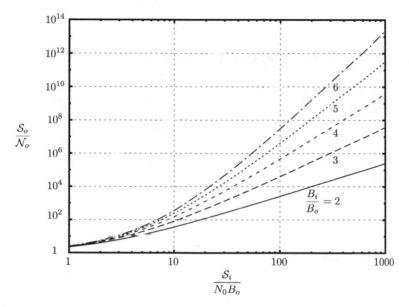

**Fig. 12.14**  Output SNR versus $S_i/(N_0 B_o)$

$B_i$ and $B_o$ are the input and output bandwidth, respectively. Eq. (12.79a) applies to digital systems, while the other two equations apply to analogue systems. This implies that the output SNR can be traded for the input SNR with respect to bandwidth. A plot of output SNR versus $S_i/(N_0 B_o)$ is given in Fig. 12.14.

Some explanation is required here. In the analogue case, $N_0$ is channel dependant, but $B_o$ is generally constant with the application. For example, for voice $B_o = 4\,\text{kHz}$, for music, it is 22 kHz and so on. If, in our modulation scheme, we can somehow manage $B_i > B_o$, we may be able to have an advantage in terms of the output SNR.

Applying this equation to Eq. (5.101),

$$\left(\frac{S_o}{N_o}\right)_{\text{DSB-SC}} = \frac{2S_i}{N_0 B}$$

we find that there is no improvement in signal-to-noise ratio since $B_i = B_o$. Similarly, for SSB/SC observing Eq. (5.108),

$$\left(\frac{S_o}{N_o}\right)_{\text{DSB-SC}} = \frac{2S_i}{N_0 B}$$

there is no improvement in the output SNR. However, in the case of FM, we find that there is an exchange of output SNR with bandwidth as seen in Eq. (6.114),

$$\frac{S_o}{N_o} == \frac{3k_f^2 \left\langle m^2(t) \right\rangle B}{\omega_m^3} \left( \frac{S_i}{N_i} \right)$$

## 12.7  Important Concepts and Formulae

- If the probability of an occurrence of a symbol, $s$, is $p$ then the information contained in it is
$$I(s) = -\log_2 p$$

- The *average information content of a source* is called the *entropy of a source* and is defined as the average value of the information, $I(s_k)$, where $s_k$ is the $k$th symbol whose probability of occurrence is $p_k$.

$$H(S) = E[I(s_k)]$$

- The average code word length is given by

$$\overline{L} = \sum_{k=1}^{N} l_k p_k \qquad (12.80)$$

where $p_k$ is the probability of occurrence of symbol $s_k$.
- *Coding efficiency* of our algorithm or process to construct the $l_k$ is given by

$$\eta = \frac{\overline{L}_{min}}{\overline{L}} \qquad (12.81)$$

where $0 < \eta \leq 1$.
- The source coding theorem says that if $H(S)$ is the entropy of the source

$$\overline{L} \geq H(S) \qquad (12.82)$$

That is, the average length of binary code words cannot be made smaller than $H(S)$.
- The minimum
$$\overline{L}_{min} = H(S) \qquad (12.83)$$

and therefore
$$\eta = \frac{H(S)}{\overline{L}} \qquad (12.84)$$

- Prefix codes: *no code word is is a prefix of another code word.*
- The *Kraft–McMillan inequality*, which is an inequality related to the length of prefix codes. For a set of code words

$$\{s_1, s_2, \ldots, s_N\}$$

with probabilities

$$\{p_1, p_2, \ldots, p_N\}$$

and each symbol, $s_k$, $k = 1, \ldots, N$ is coded as the code word of length $l_k$, $k = 1, \ldots, N$, then

$$\sum_{n=1}^{N} \frac{1}{2^{l_k}} \leq 1$$

- A discrete memoryless channel is a statistical description of a noisy channel where a discrete random variable $X$ is the input to the channel giving us another random variable $Y$ as the output. The output does not, in anyway, depend on the previous input.
- Given a conditional entropy,

$$H(X|Y = y_n) = \sum_{m=1}^{M} p(x_m|y_n) \log_2 \left[ \frac{1}{p(x_m|y_n)} \right] \qquad n = 1, \ldots N$$

which is a random variable involving $y_n$. We call $H(X|\mathcal{Y})$ the *conditional entropy* of $X$ given $\mathcal{Y}$.
- $H(X) - H(X|\mathcal{Y})$ is the *mutual information*, which is the *uncertainty in $X$ having taken $\mathcal{Y}$ into consideration*:

$$I(X; \mathcal{Y}) = H(X) - H(X|\mathcal{Y})$$

similarly, we define

$$I(\mathcal{Y}; X) = H(\mathcal{Y}) - H(\mathcal{Y}|X)$$

- *Mutual information is connected to the entropy of $X$ and $\mathcal{Y}$ through the relation*

$$I(X; \mathcal{Y}) = H(X) + H(\mathcal{Y}) - H(X, \mathcal{Y}) \tag{12.85}$$

- The channel capacity is

$$C = \max_{\{p(x_m)\}} I(X; \mathcal{Y})$$

- Then the channel capacity theorem (Shannon's theorem) says

$$\frac{H(S)}{T_s} \leq \frac{C}{T_c} \quad \text{bits/sec}$$

- The differential entropy for continuous RVs given by

$$h(X) = -\int_{-\infty}^{\infty} f_X(x) \log_2 [f_X(x)] \, dx \tag{12.86}$$

- The maximum mutual information in bits per second, for a *Gaussian channel* is

$$C = B \log_2 \left[ \left( 1 + \frac{S}{N_0 B} \right) \right] \quad \text{bits/s}$$

This expression tells us that channel capacity increases rapidly with bandwidth, but it is proportional to logarithm of the signal power $S$.

## 12.8  Self Assessment

### 12.8.1  Review Questions

1. If the probability of some event is high, why does it contain *less* information than an event whose probability is low?
2. Give a reason why is the measure of information is defined as

$$I = -\log_2 p$$

What are the units?
3. A source has symbols $\{s_k\} \, k = 1, \ldots, 6$ which are emitted with probabilities,

$$p_k = 0, \; k = 1, \ldots, 5$$

and

$$p_6 = 1$$

Find the entropy of the source. What are your comments on this source?
4. Show that for a source with symbols $\{s_k\} \, k = 1, \ldots, 2^N$, a fixed length code is as good a code as any.

## 12.8.2 Numerical Problems

1. Find the information for values of probability of various events with, $p = 0.1, 0.2, \ldots, 0.9$.
2. A source has symbols $\{s_k\}$ $k = 1, \ldots, 8$ which are emitted with probabilities,

$$p_k = \frac{1}{8}, \; k = 1, \ldots, 8$$

Find the entropy of the source.
3. A source has symbols $\{s_k\}$ $k = 1, \ldots, 6$ which are emitted with probabilities,

$$p_k = 0.1, \; k = 1, \ldots, 4$$

and

$$p_5 = p_6 = 0.3$$

Find the entropy of the source.
4. A source has symbols $\{s_k\}$ $k = 1, \ldots, 6$ which are emitted with probabilities,

$$p_k = \frac{1}{6}, \; k = 1, \ldots, 6$$

Find the entropy of the source, and compare it with the entropy obtained in Problem 3. What are your comments?
5. A source has symbols $\{s_k\}$ $k = 1, \ldots, 6$ which are emitted with probabilities,

$$p_k \neq \frac{1}{6}, \; k = 1, \ldots, 6$$

but

$$\sum_{k=1}^{6} p_k = 1$$

Compare the entropy of this source to that obtained in Problem 4. Give your comments.
6. A source has symbols $\{s_k\}$ $k = 1, \ldots, 16$ which are emitted with equal probabilities,

$$p_k = \frac{1}{16}, \; k = 1, \ldots, 16$$

Find a fixed length code to code the source. Show that this code satisfies the source coding theorem.
7. A source has symbols $\{s_k\}$ $k = 1, \ldots, 26$ which are emitted with equal probabilities,

$$p_k = \frac{1}{26}, \ k = 1, \ldots, 26$$

Find a fixed length code to code the source. Show that this code satisfies the source coding theorem.

8. For A source has symbols $\{s_k\} \ k = 1, \ldots, 4$ which are emitted with equal probabilities,

$$p_k = \frac{1}{4}, \ k = 1, \ldots, 4$$

Find a Huffman code to code the source, and also a fixed length code to code this source. Find the coding efficiency in both cases.

9. For A source has symbols $\{s_k\} \ k = 1, \ldots, 4$ which are emitted with equal probabilities,

$$p_k = \frac{1}{k^2}, \ k = 2, 3, 4$$

and

$$p_1 = 0.5764$$

Find a Huffman code to code the source, and also a fixed length code to code this source. Find the coding efficiency in both cases.

10. For the six symbols

$$
\begin{aligned}
s_1 \ & p_1 = 0.4 \\
s_2 \ & p_2 = 0.2 \\
s_3 \ & p_3 = 0.1 \\
s_4 \ & p_4 = 0.1 \\
s_5 \ & p_5 = 0.1 \\
s_6 \ & p_6 = 0.1
\end{aligned}
$$

placing the combined symbols at the lowest points of the symbol table as opposed to the highest points and find the new symbol codes. Compare with the symbol table in Sect. 12.4.2. Find the coding efficiency.

11. Code the following sequence by the LZ method

$$11100100110001110000\ldots$$

12. The joint probability distribution for $X = \{x_k\} \ k = 1, \ldots, 4$ and $Y = \{y_k\} \ k = 1, \ldots, 4$ is

$$
p(y_i, x_j) = 
\begin{array}{c|cccc}
 & x_1 & x_2 & x_3 & x_4 \\
\hline
y_1 & \frac{1}{4} & 0 & 0 & 0 \\
y_2 & 0 & \frac{1}{4} & 0 & 0 \\
y_3 & 0 & 0 & \frac{1}{4} & 0 \\
y_4 & 0 & 0 & 0 & \frac{1}{4}
\end{array}
$$

Calculate the entropies $\mathcal{H}(X)$, $\mathcal{H}(Y)$, $\mathcal{H}(X, Y)$, and $I(X; Y)$.

13. The joint probability distribution for $\mathcal{X} = \{x_k\}\, k = 1, \ldots, 4$ and $\mathcal{Y} = \{y_k\}\, k = 1, \ldots, 4$ is

$$p(y_i, x_j) = \frac{1}{16}, \; i = 1, \ldots, 4 \; j = 1, \ldots, 4$$

Calculate the entropies $\mathcal{H}(\mathcal{X}), \mathcal{H}(\mathcal{Y}), \mathcal{H}(\mathcal{X}, \mathcal{Y})$, and $I(\mathcal{X}; \mathcal{Y})$.

14. The joint probability distribution for $\mathcal{X} = \{x_k\}\, k = 1, \ldots, 4$ and $\mathcal{Y} = \{y_k\}\, k = 1, \ldots, 4$ is

$$p(y_i, x_j) = \begin{array}{c} \\ y_1 \\ y_2 \\ y_3 \\ y_4 \end{array} \begin{array}{cccc} x_1 & x_2 & x_3 & x_4 \\ \frac{1}{8} & \frac{1}{16} & \frac{1}{16} & 0 \\ \frac{1}{16} & \frac{1}{8} & 0 & \frac{1}{16} \\ \frac{1}{16} & 0 & \frac{1}{8} & \frac{1}{16} \\ 0 & \frac{1}{16} & \frac{1}{16} & \frac{1}{8} \end{array}$$

Calculate the entropies $\mathcal{H}(\mathcal{X}), \mathcal{H}(\mathcal{Y}), \mathcal{H}(\mathcal{X}, \mathcal{Y})$, and $I(\mathcal{X}; \mathcal{Y})$. From Problems 12, 13, and this one what do you infer?

15. If the random variable $X$ is constrained by the equation,

$$-a \leq X \leq a$$

then PDF,

$$f_X(x, y) = \begin{cases} \frac{x}{2a} & -a \leq x \leq a \\ 0 & \text{elsewhere} \end{cases}$$

maximises $h(X)$. Hint: use Lagrange multipliers to prove this[6] using $f_X(x)$ as a variable.

16. For the PDF

$$f_X(x, y) = \begin{cases} \frac{xy}{4a^2} & -a \leq x \leq a, \; -a \leq y \leq a \\ 0 & \text{elsewhere} \end{cases}$$

calculate $h(X), h(Y)$, and $I(X, Y)$.

17. For the binary erasure channel shown in Fig. 12.15, where a few of the bits are erased by the channel, find the channel capacity.

18. In the case where instead of a 1, $2n + 1$, 1s are transmitted, and instead of a 0, $2n + 1$, 0s are transmitted, show that the probability of error is,

$$P_e = \sum_{i=n+1}^{2n+1} \binom{2n+1}{i} p^i (1-p)^{2n+1-i}$$

---

[6] This is the continuous analogue of the optimisation considered in 12.2.

**Fig. 12.15** Binary erasure channel

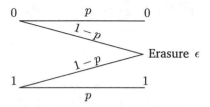

19. For a telephone channel of 4 KHz, and a signal-to-noise ratio of 20 dB, find the information capacity of the channel. What should be the $S/N$ for a transmission rate of 10 Kb/s

20. For a set of 100 symbols, and a bandwidth of 4 KHz, what should be the energy per bit for $N_0 = 10^{-6}$ MKS units, so that the channel capacity is 10 Kb/s?

# References

1. Bertsekas DP (1996) Constrained optimisation and Lagrange multiplier methods. Athena Scientific, Massachusettes
2. Shannon CE (1948) A mathematical theory of communication. The Bell Syst Tech J XXVII(3)
3. Haykin S (1999) Communication systems. Wiley
4. Shannon CE (1951) Prediction and entropy of printed English. Bell Syst Tech J 30(1):50–64

# Chapter 13
# Error Control Coding

## 13.1 Introduction

We have seen in earlier chapters how different modulation schemes can be used to improve error probabilities. Using the results of this chapter, we will reduce the error probability using error control coding. Error control coding is essentially adding $n - k$ bits to $k$ bits before transmission of an $n$ bit packet to reduce the probability of error as considered in Example 12.8. In that example, $k$ was equal to 1 and $n$ was 3 or 5.

Error control coding as applied to an alphabet consisting of two symbols, 0 and 1 are considered here. Two types of error control coding exist as applied to digital communication systems:

1. $n$ bits are received. Out of these $n$ bits, $k$ $(< n)$ bits are the message bits and $n - k$ bits are the error control bits. On processing these extra bits, a decision is made whether the message bits were in error or not. Further, the algorithm is also capable of making the *most probable* correction.
2. $n$ bits are received. Out of these $n$ bits, $k$ $(< n)$ bits are the message bits and $n - k$ bits are the error control bits. On processing these extra bits, a decision is made whether the message bits were in error or not. The algorithm is *not* capable of making a correction. Retransmission of the message bits is requested.

The idea behind error control codes is to develop a clever algorithm which takes the $k$ message bits and calculates the control bits. The channel encoder adds these control bits to the message bits and transmits the packet of $n$ bits. On reception, the decoder uses the $n - k$ control bits to check whether the message bits were in error or not. Then the decision is made to transmit the message bits again or apply the most probable correction. The main idea behind error control coding is that, it is more advantageous to do this rather than to increase the signal power to higher levels.

© The Author(s), under exclusive license to Springer Nature Singapore Pte Ltd. 2022     559
S. Bhooshan, *Fundamentals of Analogue and Digital Communication Systems*,
Lecture Notes in Electrical Engineering 785,
https://doi.org/10.1007/978-981-16-4277-7_13

## 13.2   Linear Block Codes

To introduce linear block codes, we consider the following simple example.

**Example 13.1**  The simplest method of checking a code word is to add one extra bit, called the parity check bit. The parity check bit is defined according to a *predefined rule*: for example, if the number of 1s in the first $k$ bits is even then the parity check bit is 0, and if the number of 1s in the first $k$ bits is odd, then the parity check bit is 1. Thus, for example, if we want to transmit

$$11101001 \quad k = 8$$

then we transmit

$$111010011 \quad k = 8, \; n = 9$$

the number of 1s in the first 8 bits is 5, so the parity check bit is 1. On reception, we count the number of 1s in the first 8 bits. If a single error has been committed, a 0 becomes a 1 or a 1 becomes a 0, then we can catch the error since the parity check bit is in error. For example, if the third bit is in error, we receive

$$110010011$$

then the parity check bit should be 0 but it is 1. So we know that an error has taken place. But, on the other hand, if two errors take place, then we are unable to detect the error, since the parity check bit is not in error. Here, for example, the first and third bits are in error:

$$010010011$$

the parity check bit is 1, which should be, therefore, we cannot detect the error.

To continue, we devise methods (a) to generate more parity check bits which we call *generalised parity check bits*, and (b) to *find and correct* single error and multiple errors using what is known in the literature as *linear block codes*.

A linear block code is one that on the modulo addition of two such valid codes gives us a third valid code. The rules for modulo 2 addition is the same as that for the exclusive or operation:

$$
\begin{aligned}
1 \oplus 1 &= 0 \\
1 \oplus 0 &= 1 \\
0 \oplus 1 &= 1 \\
0 \oplus 0 &= 0
\end{aligned}
\tag{13.1}
$$

where $\oplus$ represents modulo 2 addition. Thus, two code words

$$11101001 \text{ and } 10110101$$

are two valid code words, then

$$1\ 1\ 1\ 0\ 1\ 0\ 0\ 1$$
$$\oplus\ 1\ 0\ 1\ 1\ 0\ 1\ 0\ 1$$
$$=\ 0\ 1\ 0\ 1\ 1\ 1\ 0\ 0$$

should be a valid code word.

In what follows, $0 \times 1 = 1 \times 0 = 0 \times 0 = 0$ and $1 \times 1 = 1$.

**Example 13.2** Modulo 2 arithmetic.

Modulo 2 arithmetic involves modulo 2 addition and multiplication with no carry. Also, $1 - 1 = 1 \oplus 1 = 0, 0 - 0 = 0 \oplus 0 = 0, 1 - 0 = 1 \oplus 0 = 1$ and $0 - 1 = 0 \oplus 1 = 1$.

1. For example, in addition

$$1\ 1\ 1\ 1$$
$$+\ 1\ 1\ 0\ 1$$
$$=\ 0\ 0\ 1\ 0$$

2. In the case of subtraction

$$1\ 1\ 1\ 1$$
$$-\ 1\ 1\ 0\ 1$$
$$=\ 0\ 0\ 1\ 0$$

   becomes

$$1\ 1\ 1\ 1$$
$$+\ 1\ 1\ 0\ 1$$
$$=\ 0\ 0\ 1\ 0$$

3. Or multiplication

$$1\ 1\ 0\ 0\ 1$$
$$\times\ \ \ \ \ \ \ \ \ 1\ 0$$
$$0\ 0\ 0\ 0\ 0$$
$$1\ 1\ 0\ 0\ 1\ x$$
$$=\ 1\ 1\ 0\ 0\ 1\ 0$$

4. And division

$$1\ 1\ 0\ 0\ 1$$
$$\div\ \ \ \ \ \ 1\ 0$$

   then

$$
\begin{array}{r}
1\ 1\ 0\ 0 \\
\overline{\hphantom{1}} \ \overline{\hphantom{1}} \ \overline{\hphantom{1}} \ \overline{\hphantom{1}} \ \overline{\hphantom{1}} \\
10\ )\ 1\ 1\ 0\ 0\ 1 \\
-\ \underline{1\ 0}\ \downarrow\ \downarrow\ \downarrow \\
1\ 0\ \downarrow\ \downarrow \\
-\ \underline{1\ 0}\ \downarrow\ \downarrow \\
0\ 1 \\
-\ \underline{0\ 0} \\
\rightarrow\ 1
\end{array}
$$

Remainder

To check this calculation, we find that

$$11001 = 1100 \times 10 + 1$$

To proceed in a more systematic way, we define the code word **c** in the following way

$$
\begin{aligned}
\mathbf{c} &= \begin{bmatrix} m_1\ m_2\ \cdots\ m_k\ |\ b_1\ b_2\ \cdots\ b_{n-k} \end{bmatrix} \\
&= \begin{bmatrix} c_1\ c_2\ \cdots\ c_k\ c_{k+1}\ c_{k+2}\ \cdots\ c_n \end{bmatrix}
\end{aligned}
$$

where $\mathbf{m} = \begin{bmatrix} m_1\ m_2\ \cdots\ m_k \end{bmatrix}$    are the message bits

and $\mathbf{b} = \begin{bmatrix} b_1\ b_2\ \cdots\ b_{n-k} \end{bmatrix}$    are the parity bits      (13.2)

The parity bits are generated by using coefficients. Thus

$$\mathbf{b} = \mathbf{mP}$$

$$
\text{or } b_j = \sum_{i=1}^{k} m_i p_{ij}
$$

$$
= m_1 p_{1j} \oplus m_2 p_{2j} \oplus \cdots \oplus m_k p_{kj}
$$

where **b**, and **m** are row vectors of dimension $n - k$ and $k$, respectively; while **P** is a *coefficient matrix* of dimension $k \times (n - k)$ with entries

$$
p_{ij} = \begin{cases} 1 & \text{if } b_j \text{ depends on } m_i \\ 0 & \text{if } b_j \text{ does not depend on } m_i \end{cases}
$$

To obtain a code word, we define a generator matrix, **G**, such that

$$\mathbf{c} = \mathbf{mG} \qquad\qquad (13.3)$$

where

$$\mathbf{G} = \begin{bmatrix} 1 & 0 & \cdots & 0 & p_{11} & p_{12} & \cdots & p_{1,n-k} \\ 0 & 1 & \cdots & 0 & p_{21} & p_{22} & \cdots & p_{2,n-k} \\ \vdots & \vdots & \ddots & \vdots & \vdots & \vdots & \ddots & \vdots \\ 0 & 0 & \cdots & 1 & p_{k1} & p_{k2} & \cdots & p_{k,n-k} \end{bmatrix}$$

$$= \begin{bmatrix} \mathbf{I}_k & | & \mathbf{P} \end{bmatrix}_{k \times n}$$

where a $k \times n$ matrix has $k$ rows and $n$ columns. Furthermore

$$\mathbf{I}_k = \begin{bmatrix} 1 & 0 & \cdots & 0 \\ 0 & 1 & \cdots & 0 \\ \vdots & \vdots & \ddots & \vdots \\ 0 & 0 & \cdots & 1 \end{bmatrix}_{k \times k}$$

$$\text{and } \mathbf{P} = \begin{bmatrix} p_{11} & p_{12} & \cdots & p_{1,n-k} \\ p_{21} & p_{22} & \cdots & p_{2,n-k} \\ \vdots & \vdots & \ddots & \vdots \\ p_{k1} & p_{k2} & \cdots & p_{k,n-k} \end{bmatrix}_{k \times (n-k)}$$

so

$$\begin{aligned} \mathbf{c} &= \mathbf{m}_{1 \times k} \mathbf{G}_{k \times n} \\ &= \mathbf{m}_{1 \times k} \left[ (\mathbf{I}_k)_{k \times k} \mid \mathbf{P}_{k \times (n-k)} \right] \\ &= \left[ \mathbf{m}_{1 \times k} \mid \mathbf{m}_{1 \times k} \, \mathbf{P}_{k \times (n-k)} \right] \\ &= \left[ \mathbf{m}_{1 \times k} \mid \mathbf{b}_{1 \times (n-k)} \right]_{1 \times n} \end{aligned} \tag{13.4}$$

From this discussion, it is clear that $k$ bits are encoded at any time to give us a code word of $n$ bits. Such codes are called $(n, k)$ linear block codes with code rate $r = k/n$.

To prove that the addition of two valid code words results in another valid code word, we must realise that the total number of messages of length $k$ is $2^k$. Any one of these massages is a valid message. Or the modulo two sum of two messages, $\mathbf{m}_i$ and $\mathbf{m}_j$ is also a valid message, $\mathbf{m}_l$.

$$\mathbf{m}_i \oplus \mathbf{m}_j = \mathbf{m}_l \tag{13.5}$$

therefore

$$\begin{aligned} \left( \mathbf{m}_i \oplus \mathbf{m}_j \right) \mathbf{G} &= \mathbf{c}_i \oplus \mathbf{c}_j \\ \mathbf{m}_l \mathbf{G} &= \mathbf{c}_l \end{aligned} \tag{13.6}$$

is also a valid code word.

Let us now examine the matrix

$$\mathbf{H}^T = \begin{bmatrix} \mathbf{P}_{k \times (n-k)} \\ \mathbf{I}_{(n-k)} \end{bmatrix}_{n \times (n-k)} \tag{13.7}$$

which is a $n \times (n - k)$ matrix. Forming the product

$$\begin{aligned}
\mathbf{cH}^T &= \mathbf{c}_{1 \times n} \begin{bmatrix} \mathbf{P}_{k \times (n-k)} \\ \mathbf{I}_{(n-k)} \end{bmatrix}_{n \times (n-k)} \\
&= \begin{bmatrix} \mathbf{m}_{1 \times k} & \mathbf{b}_{1 \times (n-k)} \end{bmatrix} \begin{bmatrix} \mathbf{P}_{k \times (n-k)} \\ \mathbf{I}_{(n-k)} \end{bmatrix} \\
&= \mathbf{m}_{1 \times k} \mathbf{P}_{k \times (n-k)} \oplus \mathbf{b}_{1 \times (n-k)} \mathbf{I}_{(n-k)} \\
&= \mathbf{b} \oplus \mathbf{b} \\
&= \mathbf{0}
\end{aligned} \tag{13.8}$$

The last step in our equations can be corroborated through Equation set (13.1). The matrix $\mathbf{H}$ is called the *parity check matrix*.

We can take a step further and use Eq. (13.3) in the equation:

$$\mathbf{c} = \mathbf{mH}^T,$$

then

$$\mathbf{c}_{1 \times n} \mathbf{H}^T_{n \times (n-k)} = \mathbf{m}_{1 \times k} \mathbf{G}_{k \times n} \mathbf{H}^T_{n \times (n-k)}$$
$$\text{therefore } \mathbf{G}_{k \times n} \mathbf{H}^T_{n \times (n-k)} = \mathbf{0} \tag{13.9}$$

From these equations, it is clear that when we receive a code word $\mathbf{c}$ (*with no error*), then we form the matrix product $\mathbf{cH}^T$ which is then zero.

What happens when the received code word is in error? To analyse this situation, let $\mathbf{r}$ the received code word. Then

$$\mathbf{r} = \mathbf{c} \oplus \mathbf{e} \tag{13.10}$$

where $\mathbf{c}$ is the valid code word and $\mathbf{e}$ is the error. Then

$$\begin{aligned}
\mathbf{s} &= (\mathbf{c} \oplus \mathbf{e})\mathbf{H}^T \\
&= \mathbf{cH}^T \oplus \mathbf{eH}^T \\
&= \mathbf{0} \oplus \mathbf{eH}^T \\
&= \mathbf{eH}^T
\end{aligned} \tag{13.11}$$

where $\mathbf{s}$ is called the syndrome. The problem with obtaining the syndrome vector, $\mathbf{s}$, is that, many $\mathbf{e}$ vectors will give us the same syndrome, so how are we to proceed in a more sensible way? The way to proceed is to calculate the syndromes for 1 error in any position, which has the maximum probability, then two errors in any position, which has a lower probability and so on till all the syndrome vectors are covered. Notice that the syndrome vectors are of dimension $1 \times (n - k)$ giving us $2^{(n-k)}$ syndrome vectors and it may happen that all the syndrome vectors *will not be usable to correct the error*, but instead, *may be usable to detect the error*.

**Example 13.3** Repetition code.

Let us look at the coefficient matrix for a repetition code. When we transmit a 0, then the parity bits are 0000 and when we transmit a 1, then the parity bits are 1111. So the coefficient matrix **P** is

$$\mathbf{P} = \begin{bmatrix} 1 & 1 & 1 & 1 \end{bmatrix}$$

To be specific, when

$$\mathbf{m} = \begin{bmatrix} 0 \end{bmatrix}$$

then

$$\mathbf{b} = \mathbf{mP}$$
$$= \begin{bmatrix} 0 \end{bmatrix} \begin{bmatrix} 1 & 1 & 1 & 1 \end{bmatrix}$$
$$= \begin{bmatrix} 0 & 0 & 0 & 0 \end{bmatrix}$$

and similarly

$$\mathbf{m} = \begin{bmatrix} 1 \end{bmatrix}$$

then

$$\mathbf{b} = \mathbf{mP}$$
$$= \begin{bmatrix} 1 \end{bmatrix} \begin{bmatrix} 1 & 1 & 1 & 1 \end{bmatrix}$$
$$= \begin{bmatrix} 1 & 1 & 1 & 1 \end{bmatrix}$$

Using this coefficient matrix, the generator matrix, **G**, is

$$\mathbf{G} = \begin{bmatrix} 1 & | & 1 & 1 & 1 & 1 \end{bmatrix}$$

and

$$\mathbf{H}^T = \begin{bmatrix} 1 & 1 & 1 & 1 \\ \hline 1 & 0 & 0 & 0 \\ 0 & 1 & 0 & 0 \\ 0 & 0 & 1 & 0 \\ 0 & 0 & 0 & 1 \end{bmatrix}$$

with

$$\mathbf{GH}^T = 0$$

## *13.2.1   Syndrome Decoding*

Let us take a fresh look at the decoding problem for a $(n, k)$ linear code. Since $\mathbf{r}$ is $1 \times n$, theoretically there are $2^n$ possible received vectors. From Eq. (13.10), we see that the number of valid code words, $\mathbf{c}$, are $2^k$. The error vector, $\mathbf{e}$, also has $2^n$ possibilities. However, to a valid code word, $\mathbf{c}_j$, if we add any random vector, $\mathbf{e}$

$$\mathbf{r}_j = \mathbf{e} \oplus \mathbf{c}_j \quad j = 1, \ldots, 2^k \tag{13.12}$$

Then

$$\begin{aligned} \mathbf{s}_j &= \mathbf{r}_j \mathbf{H}^T \\ &= (\mathbf{e} \oplus \mathbf{c}_j) \mathbf{H}^T \\ &= \mathbf{e} \mathbf{H}^T \oplus \mathbf{0} \end{aligned} \tag{13.13}$$

which is independent of $j$. Let us rewrite Eq. (13.11),

$$\mathbf{s}_{1 \times (n-k)} = \mathbf{e}_{1 \times n} \mathbf{H}^T_{n \times (n-k)}$$

which consists of $n - k$ equations (number of values of $\mathbf{s}$) with $n$ variables (number of values of $\mathbf{e}$). Therefore, to these equations, there is no unique solution. Two $\mathbf{e}$ vectors will give us the same syndrome *provided they differ by a valid code word*. This is the most important point. The next question we ask is: by how much do the valid code words differ from each other? We have to define *a distance metric, $d$*, for valid code words.

The *Hamming distance, $d(\mathbf{c}_i, \mathbf{c}_j)$*, between two code words, $\mathbf{c}_i$ and $\mathbf{c}_j$ is the number of positions where the two code words differ. The *Hamming weight* of a code word $\mathbf{c}_i$, $w(\mathbf{c}_i)$ is its distance from $\mathbf{0}$. Thus, if

$$\mathbf{c}_l = \mathbf{c}_i \oplus \mathbf{c}_j$$

then the distance

$$d(\mathbf{c}_i, \mathbf{c}_j) = w(\mathbf{c}_l)$$

If a linear block code is to correct $t$ errors (with the highest probability), then around any code word, $\mathbf{c}_i$, we should be able to draw a (hyper) sphere of radius $t$. Also, the nearest code word, $\mathbf{c}_j$ also has a (hyper)sphere of radius $t$ around it. Then the minimum distance between $t$ error correcting code words must satisfy

$$d_{\min} \geq 2t + 1$$

When $\mathbf{r}$ falls in this radius, then it can be correctly decoded. Therefore, the errors which can be correctly decoded are those with weight

**Fig. 13.1** Figure concerning the minimum distance $d_{min}$

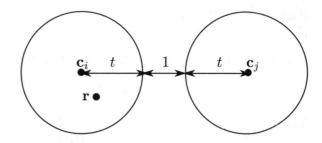

$$w(\mathbf{e}) \leq t$$
$$\text{or } d(\mathbf{r} \oplus \mathbf{c}_i) \leq t$$

This relation is diagrammatically shown in Fig. 13.1, and based on this relation, the following table can be drawn.

| $t$ | $d_{min}$ |
|---|---|
| 1 | 3 |
| 2 | 5 |
| 3 | 7 |

To calculate the value of $d_{min}$, we use Eq. (13.13). The equation says that the value of the syndrome is $\mathbf{0}$ if the error is the zero vector. Now it is should also be equal to zero when the error is such that we reach another code word. The distance is of minimum value when the weight of the error is of minimum value

$$d_{min} = w(\mathbf{e}_{min}) \tag{13.14}$$

To put it in another way, let

$$\mathbf{H}^T = \begin{bmatrix} \mathbf{h}_1^T \\ \mathbf{h}_2^T \\ \vdots \\ \mathbf{h}_n^T \end{bmatrix} \tag{13.15}$$

where $\mathbf{h}_i^T$ are the rows of $\mathbf{H}^T$ (and $\mathbf{h}_i$ are the columns of $\mathbf{H}$). So,

$$\mathbf{e} \bullet \mathbf{H}^T = \begin{bmatrix} e_1 & e_2 & \cdots & e_n \end{bmatrix} \begin{bmatrix} \mathbf{h}_1^T \\ \mathbf{h}_2^T \\ \vdots \\ \mathbf{h}_n^T \end{bmatrix} = \mathbf{0}$$

Or, some of these rows (where the $e_i = 1$) are added up in such a way that they give us the zero vector. The fewest rows when added up in such a manner give us $d_{min}$ corresponding to $\mathbf{e}_{min}$.

**Example 13.4** We now find the minimum distance for the repetition code, for Example 13.3.

From examining the code words, the repetition code has only code words,

$$\mathbf{c}_1 = \begin{bmatrix} 1 & 1 & 1 & 1 & 1 \end{bmatrix}$$

and

$$\mathbf{c}_2 = \begin{bmatrix} 0 & 0 & 0 & 0 & 0 \end{bmatrix}$$

and the minimum distance between these code words is 5.

$$d_{\min}(\mathbf{c}_1, \mathbf{c}_2) = 5$$

so

$$2t + 1 = 5$$

or

$$t = 2$$

so a maximum of two errors may be corrected.

Also, for this example,

$$\mathbf{H}^T = \begin{bmatrix} 1 & 1 & 1 & 1 \\ - & - & - & - \\ 1 & 0 & 0 & 0 \\ 0 & 1 & 0 & 0 \\ 0 & 0 & 1 & 0 \\ 0 & 0 & 0 & 1 \end{bmatrix}$$

so when all the rows are added up, they give us the zero vector (row 1, row 2, row 3, row 4 and row 5). This result is that these are the fewest rows, and therefore, $d_{\min} = 5$, or $t = 2$: We can correct up to two errors using this repetition code.

**Example 13.5**

The generator matrix for a (7,3) code is

$$\mathbf{G} = \begin{bmatrix} 1 & 0 & 0 & | & 0 & 1 & 1 & 1 \\ 0 & 1 & 0 & | & 1 & 0 & 1 & 0 \\ 0 & 0 & 1 & | & 1 & 1 & 0 & 1 \end{bmatrix}$$

and

$$\mathbf{H}^T = \begin{bmatrix} 0 & 1 & 1 & 1 \\ 1 & 0 & 1 & 1 \\ 1 & 1 & 0 & 1 \\ - & - & - & - \\ 1 & 0 & 0 & 0 \\ 0 & 1 & 0 & 0 \\ 0 & 0 & 1 & 0 \\ 0 & 0 & 0 & 1 \end{bmatrix}$$

Then when 4 rows are added up they give us the zero row (row 1, row 2, row 3, and row 7). Therefore,

$$d_{min} = 4$$

which implies that $t = 1$, or that only one error may be safely corrected. For example, for

$$\mathbf{e}_1 = \begin{bmatrix} 1 & 0 & 0 & 0 & 0 & 0 & 0 \end{bmatrix}$$

the syndrome is

$$\mathbf{s}_1 = \begin{bmatrix} 0 & 1 & 1 & 1 \end{bmatrix}$$

for

$$\mathbf{e}_2 = \begin{bmatrix} 0 & 1 & 0 & 0 & 0 & 0 & 0 \end{bmatrix}$$

the syndrome is

$$\mathbf{s}_1 = \begin{bmatrix} 1 & 0 & 1 & 1 \end{bmatrix}$$

and so on. A table of the errors and syndromes is depicted in Table 13.1. From the table, for rows 14 and 15, it is clear that for two errors, the syndrome is not unique. Therefore, only one error may be corrected reliably with the highest probability. A table of codes and weights is shown in Table 13.2.

Unfortunately, we do not know the error, but we know the syndrome instead. So we proceed as in Example 13.5, which gives the safest way to proceed (and occupies the minimum requirements for storage):

1. We find $d_{min}$ and then $t$.
2. Make a table of syndromes for up to and including $t$ errors starting with one error and going up to $t$ errors. The syndromes for these errors will be unique.
3. Calculate the syndrome vector from the received vector $\mathbf{r}$

$$\mathbf{s}_0 = \mathbf{r}\mathbf{H}^T$$

and obtain the error vector, $\mathbf{e}_0$ from the table corresponding to syndrome $\mathbf{s}_0$.
4. If the syndrome is not there, ask for retransmission of the data, since there are more than $t$ errors.
5. Since

**Table 13.1**  Table of errors and their syndromes

| No. | errors | syndromes |
|-----|--------|-----------|
| 1 | 0000000 | 0000 |
| 2 | 1000000 | 0111 |
| 3 | 0100000 | 1011 |
| 4 | 0010000 | 1101 |
| 5 | 0001000 | 1000 |
| 6 | 0000100 | 0100 |
| 7 | 0000010 | 0010 |
| 8 | 0000001 | 0001 |
| 9 | 1100000 | 1100 |
| 10 | 1010000 | 1010 |
| 11 | 1001000 | 1111 |
| 12 | 1000100 | 0011 |
| 13 | 1000010 | 0101 |
| 14 | 1000001 | 0110 |
| 15 | 0110000 | 0110 |
| $\vdots$ | $\vdots$ | $\vdots$ |

**Table 13.2**  Table for message vectors and their codes

| S. no. | Message | Code | Weight |
|--------|---------|------|--------|
| 1 | 000 | 0000000 | 0 |
| 2 | 001 | 0011101 | 4 |
| 3 | 010 | 0101010 | 3 |
| 4 | 011 | 0110111 | 5 |
| 5 | 100 | 1000111 | 4 |
| 6 | 101 | 1011010 | 4 |
| 7 | 111 | 1110000 | 3 |

$$\mathbf{r} = \mathbf{c}_0 \oplus \mathbf{e}_0$$
$$\mathbf{c}_0 = \mathbf{r} \oplus \mathbf{e}_0 \tag{13.16}$$

6. $\mathbf{c}_0$ is the most *probable code* we are looking for.

## 13.2.2   *Hamming Codes*

*Hamming codes* are codes which satisfy

$$n = 2^m - 1$$
$$k = 2^m - m - 1$$
$$n - k = m \qquad\qquad (13.17)$$

where $m \geq 3$.

**Example 13.6** For the generator matrix for a (7,4) Hamming code is

$$\mathbf{G} = \begin{bmatrix} 1\,0\,0\,0\,1\,0\,1 \\ 0\,1\,0\,0\,1\,1\,0 \\ 0\,0\,1\,0\,0\,1\,1 \\ 0\,0\,0\,1\,1\,1\,1 \end{bmatrix}$$

and

$$\mathbf{H}^T = \begin{bmatrix} 1\,0\,1 \\ 1\,1\,0 \\ 0\,1\,1 \\ 1\,1\,1 \\ 1\,0\,0 \\ 0\,1\,0 \\ 0\,0\,1 \end{bmatrix}$$

for this Hamming code

$$d_{\min} = 3$$

or $t = 1$, which is obtained by adding rows 1,4 and 6 of $\mathbf{H}^T$.

$$\begin{array}{r} 1\,0\,1 \\ 1\,1\,1 \\ \oplus\,0\,1\,0 \\ \hline = 0\,0\,0 \end{array}$$

The codes so obtained for the messages are given in Table 13.3, and the table of syndromes for single errors are given in Table 13.4.

Let us apply our knowledge to the received code word 0111001. The syndrome for this code word is 011. The error $\mathbf{e}_0 = [0010000]$, therefore, the code word sent was $\mathbf{c}_0 = [0101001]$ if only a single error took place in transmission.

## 13.2.3 Dual Codes

From the equation

$$\mathbf{G}_{k \times n} \mathbf{H}^T_{n \times (n-k)} = 0$$

**Table 13.3** Messages and code words for the (7,4) Hamming code

| Message | Code word | Message | code word |
|---------|-----------|---------|-----------|
| 0000 | 0000000 | 1000 | 1000101 |
| 0001 | 0001111 | 1001 | 1001010 |
| 0010 | 0010011 | 1010 | 1010110 |
| 0011 | 0011100 | 1011 | 1011001 |
| 0100 | 0100110 | 1100 | 1100011 |
| 0101 | 0101001 | 1101 | 1101110 |
| 0110 | 0110101 | 1110 | 1110000 |
| 0111 | 0111010 | 1111 | 1111111 |

**Table 13.4** Syndromes for single errors

| Syndrome No | Error | Syndrome |
|-------------|-------|----------|
| 5 | 1000000 | 101 |
| 6 | 0100000 | 110 |
| 3 | 0010000 | 011 |
| 7 | 0001000 | 111 |
| 4 | 0000100 | 100 |
| 2 | 0000010 | 010 |
| 1 | 0000001 | 001 |

we may find its transpose

$$\left\{ \mathbf{G}_{k \times n} \mathbf{H}^T_{n \times (n-k)} \right\}^T = \mathbf{H}_{(n-k) \times n} \mathbf{G}^T_{n \times k}$$
$$= \mathbf{0}$$

which implies that we may define a new linear $(n, n-k)$ block code with generator matrix $\mathbf{G}_d = \mathbf{H}_{n \times (n-k)}$ and parity check matrix $\mathbf{H}_d = \mathbf{G}_{k \times n}$. However in this case the parity bits are placed before the message bits, which can be remedied by a slight adjustment in the matrices.

**Example 13.7** From Example 13.5, we may form a (7,4) Hamming code with generator matrix

$$\mathbf{G}_d = \begin{bmatrix} 0 & 1 & 1 & 1 \\ 1 & 0 & 1 & 1 \\ 1 & 1 & 0 & 1 \\ \hline 1 & 0 & 0 & 0 \\ 0 & 1 & 0 & 0 \\ 0 & 0 & 1 & 0 \\ 0 & 0 & 0 & 1 \end{bmatrix}^T = \begin{bmatrix} 0 & 1 & 1 & 1 & 0 & 0 & 0 \\ 1 & 0 & 1 & 0 & 1 & 0 & 0 \\ 1 & 1 & 0 & 0 & 0 & 1 & 0 \\ 1 & 1 & 1 & 0 & 0 & 0 & 1 \end{bmatrix}_{4 \times 7}$$

therefore

$$\mathbf{G}' = \begin{bmatrix} 1\ 0\ 0\ 0\ 0\ 1\ 1 \\ 0\ 1\ 0\ 0\ 1\ 0\ 1 \\ 0\ 0\ 1\ 0\ 1\ 1\ 0 \\ 0\ 0\ 0\ 1\ 1\ 1\ 1 \end{bmatrix}_{4\times 7}$$

and parity check matrix

$$\mathbf{H}_d = \begin{bmatrix} 1\ 0\ 0\ |\ 0\ 1\ 1\ 1 \\ 0\ 1\ 0\ |\ 1\ 0\ 1\ 0 \\ 0\ 0\ 1\ |\ 1\ 1\ 0\ 1 \end{bmatrix}_{3\times 7}$$

and

$$\mathbf{H}' = \begin{bmatrix} 0\ 1\ 1\ 1\ |\ 1\ 0\ 0 \\ 1\ 0\ 1\ 0\ |\ 0\ 1\ 0 \\ 1\ 1\ 0\ 1\ |\ 0\ 0\ 1 \end{bmatrix}_{3\times 7}$$

where $\mathbf{G}'$ and $\mathbf{H}'$ are the generator and parity check matrix for new code.

## 13.3 Cyclic Codes

### 13.3.1 Introduction

Cyclic codes form an important subclass of linear block codes. The reason why the cyclic codes are so successful is that unlike systematic linear block codes, the code length can be made very large and can be implemented using simple shift registers.

Cyclic block codes have two important properties:

1. If $\mathbf{c}_1$ and $\mathbf{c}_2$ are two cyclic block codes, then $\mathbf{c}_1 \oplus \mathbf{c}_2$ is also a code word.
2. If $\mathbf{c}$ is a cyclic block code such that cyclic shifts of a code word results in another code word.

$$\mathbf{c}^{(0)} = [c_1, c_2, \cdots, c_n]$$
$$\mathbf{c}^{(1)} = [c_2, c_3, \cdots, c_n, c_1]$$
$$\mathbf{c}^{(2)} = [c_3, c_4, \cdots, c_n, c_1, c_2]$$
$$\vdots = \vdots$$
$$\mathbf{c}^{(n-1)} = \left[ c_n, c_1, c_2, \cdots, c_{n-2}, c_{n-1} \right] \tag{13.18}$$

are all the code words, and in general,

$$\mathbf{c}^{(i)} = \left[ c_{i+1}, c_{i+2}, \cdots c_{n-1}, c_n, c_1, c_2, \cdots, c_{i-1}, c_i \right] \tag{13.19}$$

is a code word. It is erroneous to think that there are $n$ distinct code words. One example is $[0000 \cdots 00]$ and another is $[1111 \cdots 11]$.

**Example 13.8** The three code words, $\mathbf{c}^{(1)} = [100]$, $\mathbf{c}^{(2)} = [010]$, and $\mathbf{c}^{(3)} = [001]$ form a cyclic code.

In what follows, $0 \times 1 = 1 \times 0 = 0 \times 0 = 0$; and $1 \times 1 = 1$. Also, $x^i \oplus x^i = 0$. The code vector $\mathbf{c}$ can be *represented by* the $n - 1$ degree polynomial

$$c(x) = c_1 x^{n-1} \oplus c_2 x^{n-2} \oplus \cdots \oplus c_{n-1} x \oplus c_n \qquad (13.20)$$

where $x$ is indeterminate.

**Example 13.9** For example, if

$$\mathbf{c}^{(0)} = [100110]$$

then the polynomial representation is

$$c^{(0)}(x) = x^5 \oplus x^2 \oplus x$$

Now, if multiply $c(x)$ by $x$ and divide by $x^n \oplus 1$, then the remainder is $\mathbf{c}^{(1)}$

$$
\begin{array}{r}
c_1 \\
\hline
x^n \oplus 1 \overline{)c_1 x^n \oplus c_2 x^{n-1} \oplus \cdots \oplus c_{n-1} x^2 \oplus c_n x} \\
c_1 x^{n-1} \qquad\qquad\qquad\qquad \oplus c_1 \\
\hline
\text{Rem} = c_2 x^{n-1} \oplus \cdots \oplus c_{n-1} x^2 \oplus c_n x \ \oplus c_1
\end{array}
$$

where

$$c_1 x^n \oplus c_1 x^n = 0$$
$$\text{and } c_i x^{n-i-1} \oplus 0 = c_i x^{n-i-1} \quad i = 2, \ldots, n$$

therefore

$$x^i c(x)/(x^n \oplus 1) \text{ has a remainder} = c^{(i)}(x)$$

or

$$x^i c(x) = q(x)(x^n \oplus 1) \oplus c^{(i)}(x) \qquad (13.21)$$

where $q(x)$ is some polynomial. In our example

$$xc(x) = c_1(x^n \oplus 1) \oplus c^{(1)}(x) \qquad (13.22)$$

### 13.3.2  Generator Polynomial

The polynomial $x^n \oplus 1$ plays an important role in cyclic codes. Let $g(x)$ be a factor of $x^n \oplus 1$, of degree $n - k$. That is,

$$x^n \oplus 1 = g(x)_{n-k} h(x)_k$$

where the subscripts are the maximum powers of the polynomials. We now express the polynomial $c(x)$ as

$$c(x)_{n-1} = m(x)_{k-1} g(x)_{n-k} \tag{13.23}$$

where $m(x)$ is our message polynomial corresponding to our message vector. There are $2^k$ message vectors $m(x)$ which give us $2^k$ polynomials $c(x)$. The structure of $g(x)$ is

$$g(x) = g_1 x^{n-k} \oplus g_2 x^{n-k-1} \oplus g_3 x^{n-k-2} \oplus \cdots \oplus g_{n-k} x \oplus g_{n-k+1}$$
$$g_1 = g_{n-k+1} = 1 \tag{13.24}$$

The second equation can be proved quite easily using the properties of the polynomial $x^n \oplus 1$ [1].

To prove that $c(x)$ is a cyclic polynomial, consider a particular polynomial

$$c(x) = c_1 x^{n-1} \oplus c_2 x^{n-2} \oplus \cdots \oplus c_{n-1} x \oplus c_n$$

with a particular message vector

$$m(x) = m_1 x^{k-1} \oplus m_2 x^{k-2} \oplus \cdots \oplus m_{k-1} x \oplus m_k$$

The generator matrix is fixed. Then

$$xc(x)_{n-1} = c_1 (x^n \oplus 1)_n \oplus c^{(1)}(x)_{n-1}$$
$$= c_1 h(x)_k g(x)_{n-k} \oplus c^{(1)}(x)_{n-1}$$
$$= xm(x)_{k-1} g(x)_{n-k}$$

which supports the idea that

$$c^{(1)}(x)_{n-1} = m^{(1)}(x)_{k-1} g(x)_{n-k}$$

where $m^{(1)}(x)$ is any one of the other $2^k$ message vectors.

We now systematise this knowledge to make the first $k$ bits to be message bits and the last $n - k$ bits are to be parity check bits. To do this, we write

$$c(x)_{n-1} = \left[ x^{(n-k)} m(x)_{k-1} \right]_{n-1} \oplus p(x)_{n-k-1}$$

where the first term has the highest power term of $x^{n-1}m_1$ and a lowest power term of $x^{(n-k)}m_k$. $p(x)$ is the remainder when we divide $x^{(n-k)}m(x)$ by $g(x)$. To prove this, we divide $x^{(n-k)}m(x)$ with $g(x)_{n-k}$ then the right-hand side is

$$\frac{\left[x^{(n-k)}m(x)_{k-1}\right]_{n-1}}{g(x)_{n-k}} = q(x)_{k-1} \oplus \frac{p(x)_{n-k-1}}{g(x)_{n-k}}$$

*this is always possible* since when we add

$$\frac{p(x)_{n-k-1}}{g(x)_{n-k}}$$

to both sides and using modulo 2 addition

$$\frac{p(x)_{n-k-1}}{g(x)_{n-k}} \oplus \frac{p(x)_{n-k-1}}{g(x)_{n-k}} = 0$$

then

$$\frac{\left[x^{(n-k)}m(x)_{k-1}\right]_{n-1} \oplus p(x)_{n-k-1}}{g(x)_{n-k}} = q(x)_{k-1}$$

or

$$\left[x^{(n-k)}m(x)_{k-1}\right]_{n-1} \oplus p(x)_{n-k-1} = g(x)_{n-k}q(x)_{k-1} \qquad (13.25)$$

where $p(x)_{n-k-1}$ gives us the parity check digits.

**Example 13.10**  Let us consider a (7,4) code, where $n = 7$ and $k = 4$ and $n - k = 3$. Then

$$x^7 \oplus 1 = (x \oplus 1)(x^3 \oplus x \oplus 1)(x^3 \oplus x^2 \oplus 1)$$

let the generator polynomial be

$$g(x) = 1x^3 \oplus 0x^2 \oplus 1x \oplus 1$$
$$= x^3 \oplus x \oplus 1$$

and let the message be [1010]. Corresponding to this message, we have the polynomial

$$m(x) = 1x^3 \oplus 0x^2 \oplus 1x \oplus 0$$
$$= x^3 \oplus x$$

therefore

$$x^3 m(x) = x^6 \oplus x^4$$

dividing $x^3 m(x)$ by $g(x)$ by modulo 2 division

$$x^3 \oplus 1 \quad [= q(x)]$$

$$x^3 \oplus x \oplus 1 \overline{)x^6 \oplus x^4}$$

$$\underline{x^6 \oplus x^4 \oplus x^3}$$

$$x^3$$

$$\underline{x^3 \oplus x \oplus 1}$$

$$x \oplus 1 \quad [= p(x) = 0x^2 \oplus 1x \oplus 1]$$

So the code word is

$$\mathbf{c} = [1010011]$$

If we have to represent our knowledge in terms of a generator matrix, we know from the previous sections that the generator matrix is $k \times n$. And the generator polynomial is $n - k$. So we consider using Eq. (13.23), the generation of a single code word

$$x^{k-1}g(x)_{n-k} = x^3(x^3 \oplus 0x^2 \oplus x \oplus 1)$$
$$= x^6 \oplus x^4 \oplus x^3$$

this is the first row of our generator matrix, and the subsequent rows can be obtained by cyclic shifts:

$$\mathbf{G'} = \begin{bmatrix} 1 & 0 & 1 & 1 & 0 & 0 & 0 \\ 0 & 1 & 0 & 1 & 1 & 0 & 0 \\ 0 & 0 & 1 & 0 & 1 & 1 & 0 \\ 0 & 0 & 0 & 1 & 0 & 1 & 1 \end{bmatrix}$$

which is not in the form we want. Adding the third and fourth rows to the first

$$\mathbf{G''} = \begin{bmatrix} 1 & 0 & 0 & 0 & 1 & 0 & 1 \\ 0 & 1 & 0 & 1 & 1 & 0 & 0 \\ 0 & 0 & 1 & 0 & 1 & 1 & 0 \\ 0 & 0 & 0 & 1 & 0 & 1 & 1 \end{bmatrix}$$

then fourth row to the second

$$\mathbf{G} = \begin{bmatrix} 1 & 0 & 0 & 0 & 1 & 0 & 1 \\ 0 & 1 & 0 & 0 & 1 & 1 & 1 \\ 0 & 0 & 1 & 0 & 1 & 1 & 0 \\ 0 & 0 & 0 & 1 & 0 & 1 & 1 \end{bmatrix} \tag{13.26}$$

which is in the systematic form. From the generator matrix (or by the method of division, but which is a more laborious process), we can systematically obtain all the code words as shown in Table 13.5. From the table, it is clear that there are four cyclic

**Table 13.5**  Table of code words of the generator matrix Eq. (13.26)

| Message | Code word | Message | Code word |
|---------|-----------|---------|-----------|
| 0000 | 0000000 (a) | 1000 | 1000101 (b) |
| 0001 | 0001011 (b) | 1001 | 1001110 (c) |
| 0010 | 0010110 (b) | 1010 | 1010011 (c) |
| 0011 | 0011101 (c) | 1011 | 1011001 (b) |
| 0100 | 0100111 (c) | 1100 | 1100010 (b) |
| 0101 | 0101100 (b) | 1101 | 1101001 (c) |
| 0110 | 0110001 (b) | 1110 | 1110100 (c) |
| 0111 | 0111010 (c) | 1111 | 1111111 (d) |

codes involved: (a) 0000000, (b) 0001011, (c) 0011101 and (d) 1111111 which are so labelled in the table.

### 13.3.3   Decoding

To decode the received vector, we know that any code word can be divided by $g(x)$. Therefore, if the received vector is **r**, then if **r** is a code word, then $r(x)$ can be exactly divided by $g(x)$ to give the message vector as per Eq. (13.23), and there is no remainder. Otherwise

$$\frac{r(x)}{g(x)} = m(x) \oplus \frac{s(x)}{g(x)} \tag{13.27}$$

where $s(x)$ is the syndrome polynomial and the remainder as well. Also

$$r(x) = c(x) \oplus e(x)$$

$$\text{or } \frac{r(x)}{g(x)} = \frac{c(x)}{g(x)} \oplus \frac{e(x)}{g(x)}$$

since the term on the left and the second term on the right both have remainders, we can calculate the syndromes by considering the error term only.

**Example 13.11**  Calculate the syndrome for the error term

$$\mathbf{e} = [1000000]$$

for Example 13.10. To proceed,

$$e(x) = x^6$$

q

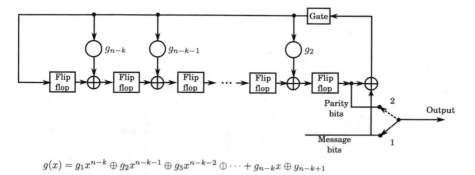

$$g(x) = g_1 x^{n-k} \oplus g_2 x^{n-k-1} \oplus g_3 x^{n-k-2} \oplus \cdots + g_{n-k} x \oplus y_{n-k+1}$$

**Fig. 13.2** Hardware to generate code words

$$
\begin{array}{r}
x^3 \oplus x \oplus 1 \\
\hline
x^3 \oplus x \oplus 1)\overline{x^6} \\
\underline{x^6 \oplus x^4 \oplus x^3} \\
x^4 \oplus x^3 \\
\underline{x^4 \oplus x^2 \oplus x} \\
x^3 \oplus x^2 \oplus x \\
\underline{x^3 \oplus x \oplus 1} \\
x^2 \oplus 1
\end{array}
$$

so the syndrome is

$$\mathbf{s} = [101]$$

In this way, we can calculate the other syndromes and form a table of syndromes.

### 13.3.4 Implementation

We now concentrate on how to produce the code words using hardware. The encoder given in Fig. 13.2 implements the code generator in terms of flip flops (used as delay elements), modulo 2 adders, multipliers, a gate and a switch. In the first phase, the gate is ON to allow the $k$ message bits to be shifted into the flip flops. Simultaneously, the switch is in position 1. The gate is then turned OFF and the switch is moved to position 2 to access the remaining $n - k$ bits which are the parity bits. The circuit is a form of a divider which divides $x^k m(x)_{k-1}$ by the generator polynomial $g(x)_{n-k}$ with coefficients $\left[g_1, g_2, \cdots, g_{n-k}, g_{n-k+1}\right]$ given in Equation set (13.24) (Table 13.6).

**Table 13.6** Table showing the sequential values at various points in the circuit. m:message bits, and Sw: switch position

| m | (a) | (b) | (c) | (d) | (e) | (f) | Gate | Sw | Code |
|---|-----|-----|-----|-----|-----|-----|------|----|------|
| 1 | 1 | 0 | 1 | 0 | 0 | 0 | ON | 1 | 1 |
| 0 | 0 | 1 | 1 | 1 | 1 | 0 | | | 0 |
| 1 | 0 | 0 | 0 | 1 | 1 | 1 | | | 1 |
| 0 | 1 | 0 | 1 | 0 | 0 | 1 | | | 0 |
| 0 | 0 | 1 | 1 | 1 | 1 | 0 | OFF | 2 | 0 |
| 0 | 1 | 0 | 1 | 1 | 1 | 1 | | | 1 |
| 0 | 1 | 1 | 0 | 1 | 1 | 1 | | | 1 |

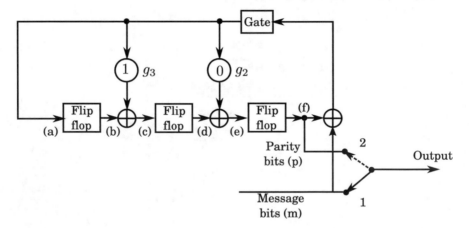

**Fig. 13.3** Implementation for Example 13.10

**Example 13.12** To see how this circuit works, consider Example 13.10 where the generator polynomial is

$$g(x) = 1x^3 \oplus 0x^2 \oplus 1x \oplus 1$$
$$= x^3 \oplus x \oplus 1$$

where

$$g_1 = 1; \ g_2 = 0; \ g_3 = 1; \ \text{and} \ g_4 = 1$$

and the message signal is [1010]. The implementation for this case is given in Fig. 13.3. Each flip flop acts like a unit delay. To properly understand the circuit, if the signal at the points shown is $s_{(i)}[n]$, then

**Table 13.7** Some BCH codes. For example when the coefficients are 11101001, then $g(x) = x^7 \oplus x^6 \oplus x^5 \oplus x^3 \oplus 1$

| $n$ | $k$ | $t$ | $g(x)$ coefficients |
|-----|-----|-----|---------------------|
| 7 | 4 | 1 | 1011 |
| 15 | 7 | 2 | 11101001 |
| 15 | 5 | 3 | 10100110111 |
| 31 | 16 | 3 | 1 000111110 101111 |

$$s_{(a)}[n] = m[n] \oplus s_{(f)}[n]$$
$$s_{(b)}[n] = s_{(a)}[n-1]$$
$$s_{(c)}[n] = s_{(b)}[n] \oplus s_{(a)}[n]$$
$$s_{(d)}[n] = s_{(c)}[n-1]$$
$$s_{(e)}[n] = s_{(d)}[n]$$
$$s_{(f)}[n] = s_{(e)}[n-1] \tag{13.28}$$

At the start, all signals are set to zero. The code generated is $\mathbf{c} = \lfloor 1010011 \rfloor$ which corroborates the results of Example 13.10. To calculate the syndrome, we use the same circuit as in Fig. 13.3.

### 13.3.5 Bose Chowdhari Hocquenghem (BCH) Codes

BCH codes are powerful cyclic codes with the following parameters:

$$n = 2^m - 1$$
$$n - k \leq mt$$
$$2t + 2 \geq d_{\min} \geq 2t + 1$$
$$t \leq 2^{m-1}$$

Some BCH codes are given in Table 13.7. BCH codes may be non primitive: those codes whose generator polynomials may be factored further into binary modulo 2 factors.

### 13.3.6 Reed Solomon Codes

Reed–Solomon codes belong to the class of non-binary cyclic codes. These codes are able to detect and correct multiple *symbol errors*. The Reed–Solomon $(n, k)$

**Table 13.8** Some commonly used CRC generator polynomials

| Code | $g(x)$ | $n-k$ |
|---|---|---|
| CRC-12 | $x^{12} \oplus x^{11} \oplus x^3 \oplus x^2 \oplus x \oplus 1$ | 12 |
| CRC-16 (USA) | $x^{16} \oplus x^{15} \oplus x^2 \oplus 1$ | 16 |
| CRC-16 (ITU) | $x^{16} \oplus x^{12} \oplus x^5 \oplus 1$ | 16 |

codes encode $m$ bit symbols into $n = 2^m - 1$ symbols or $m \times (2^m - 1)$ bits. A value of $m = 8$ is usually used. Reed–Solomon codes may also be used when there are multiple burst bit errors and error correction is a requirement.

For these codes

$$\text{Block length} : n = 2^m - 1 \text{ symbols}$$
$$\text{Message size} = k \text{ symbols}$$
$$\text{Parity check size} = n - k = 2t \text{ symbols}$$
$$d_{\min} = 2t + 1 \text{ symbols}$$

### 13.3.7   Cyclic Redundancy Check Codes

Cyclic redundancy codes are *error detection codes*. CRCs are very popular since they are easy to implement and analyse since the generator polynomials are short and have few terms. Some commonly used generator polynomials are given in Table 13.8.

CRCs are capable of burst error detection, and when an error is detected, an automatic repeat request is generated.

## 13.4   Convolutional Codes

### 13.4.1   Introduction

Convolutional codes are used when the bits come serially as opposed to when they come in blocks. Examining Fig. 13.4a, b, we can see that at the start of the convolution, the shift registers are both in a zero state, then for $j = 0$

$$m[j - 2] = m[-2] = 0$$
$$m[j - 1] = m[-1] = 0$$

and then if

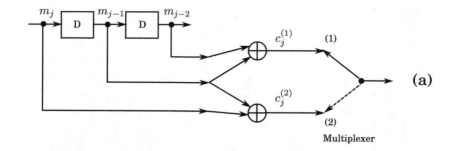

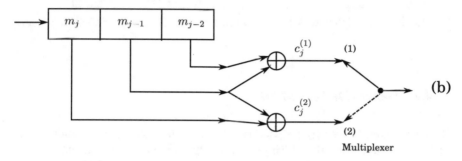

**Fig. 13.4** Convolutional encoder

$$m[0] = 0$$
$$c^{(1)}[0] = 0$$
$$c^{(2)}[0] = 0$$

or if

$$m[0] = 1$$
$$c^{(1)}[0] = 0$$
$$c^{(2)}[0] = 1$$

We proceed to enter the bits as they come and for each output, we get two bits. So for the input $\mathbf{m} = [1000101]$, the output is shown in the form of a table

| $j$ | $m[j]$ | $m[j-1]$ | $m[j-2]$ | $c^{(1)}[j]$ | $c^{(2)}[j]$ | Output Code, $\mathbf{c}$ |
|---|---|---|---|---|---|---|
| 0 | 1 | 0 | 0 | 0 | 1 | 01 |
| 1 | 0 | 1 | 0 | 1 | 1 | 01 11 |
| 2 | 0 | 0 | 1 | 1 | 0 | 01 11 10 |
| 3 | 0 | 0 | 0 | 0 | 0 | 01 11 10 00 |
| 4 | 1 | 0 | 0 | 0 | 1 | 01 11 10 00 01 |
| 5 | 0 | 1 | 0 | 1 | 1 | 01 11 10 00 11 |
| 6 | 1 | 0 | 1 | 1 | 1 | 01 11 10 00 11 11 |
| 7 | (0) | 1 | 0 | 1 | 1 | 01 11 10 00 11 11 11 |
| 8 | (0) | 0 | 1 | 1 | 0 | 01 11 10 00 11 11 11 10 |

Two trailing zeros (shown in brackets) are added to the message to clear the shift registers.

### 13.4.2   Generator Polynomial

The convolution coder may be visualised as an $M$ stage shift register with modulo 2 adders and a multiplexer which gives the output digits serially. For the example considered, for every input bit, $k = 1$, we get two output bits, $n = 2$, so the code rate is 1/2. In our example, the generator polynomial may be computed for the $i$th path as

$$g^{(i)}(D) = g_0^{(i)} + g_1^{(i)}D + \cdots + g_M^{(i)}D^M \tag{13.29}$$

This concept may be understood with the help of an example.

**Example 13.13**   Referring to Fig. 13.4, for this example, for path 1, and 2, $g(D)$ is given by

$$g^{(1)}(D) = D + D^2 \quad \text{for path (1)}$$
$$\text{and } g^{(2)}(D) = 1 + D \quad \text{for path (2)} \tag{13.30}$$

then for the message $\mathbf{m} = [1000101]$ the message polynomial is in terms of delays is,

$$m(D) = 1 + D^4 + D^6$$

so. In the D domain, the $D$-polynomials are multiplied since in the time domain they are convolved.

**Fig. 13.5** Code tree for
Fig. 13.4

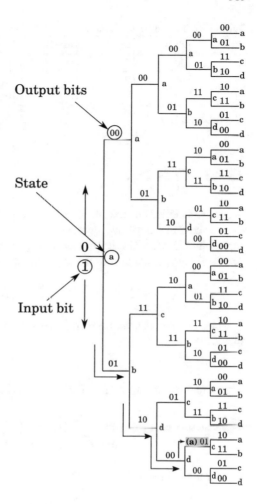

$$c^{(1)}(D) = g^{(1)}(D)m(D) = (D + D^2) + D^4(D + D^2) + D^6(D + D^2)$$
$$= D + D^2 + D^5 + D^6 + D^7 + D^8$$
$$= [011001111]$$

and $c^{(1)}(D) = g^{(2)}(D)m(D) = (1 + D) + D^4(1 + D) + D^6(1 + D)$
$$= 1 + D + +D^4 + D^5 + D^6 + D^7$$
$$= [110011110]$$

so the code generated is 01 11 10 00 01 11 11 11 10.

**Table 13.9**  Table of states, input, next state, and output

| State | Binary | Input | Next state | Output | Input | Next state | Output |
|-------|--------|-------|------------|--------|-------|------------|--------|
| $a$ | 00 | 0 | $a$ | 00 | 1 | $b$ | 01 |
| $b$ | 01 | 0 | $c$ | 11 | 1 | $d$ | 10 |
| $c$ | 10 | 0 | $a$ | 10 | 1 | $b$ | 11 |
| $d$ | 11 | 0 | $c$ | 01 | 1 | $d$ | 00 |

### 13.4.3  Code Tree

We may visualise our output in terms of a code tree, as shown in Fig. 13.5. The tree may be read in the following manner: At any node, the upward option represents an input bit of 0 while the lower option represents an input bit of 1. For example, the point **(a)** (shown shaded) in the figure represents an input of

$$m[0] = 1; \ m[1] = 1; \ m[2] = 1; \ \text{and } m[3] = 0$$

since to reach **(a)**, we take the lower option thrice and the upper option only once. Or

$$\mathbf{m} = [1110\cdots] \tag{13.31}$$

The output bits are shown on the branch as indicated.

$$c^{(1)}[j] = m[j-1] + m[j-2]$$
$$\text{and } c^{(2)}[j] = m[j] + m[j-1] \tag{13.32}$$

and for the values of Eq. (13.31)

$$c^{(1)}[3] = 0$$
$$\text{and } c^{(2)}[3] = 1 \tag{13.33}$$

and the total output code at **(a)** is $\mathbf{c} = [01 \ 10 \ 00 \ 01 \ldots]$ reading the output bits as we traverse the tree to reach **(a)**.

### 13.4.4  State Diagram

We next consider the encoder through a state diagram. If we represent the states of the shift registers as given the Table 13.9, then Fig. 13.6 is the state diagram of the encoder. Table 13.9 summarises the binary representation of the state, the input, the

**Fig. 13.6** State diagram description of an convolutional encoder

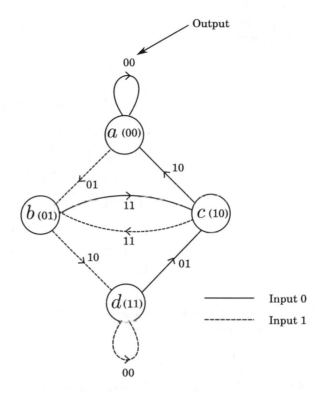

next state and the output. Each state has two outputs, corresponding to the two inputs 0 and 1. In the code tree also, the states are shown.

## 13.4.5 Trellis

We may alternatively show the states in terms of a trellis. Figure 13.7 shows the trellis diagram. To use the trellis diagram, if the input is

$$\mathbf{m} = [1110 \cdots]$$

the we start from state $a$, (present state) the initial state, we then go to state $b$ (next state, input 1) with an output 01. From $b$ (present state), we go to $d$ (next state, input 1) with output 10. From $d$ (present state), we go to $d$ (next state, input 1) with output 00, and then $d$ (present state, input 0) to $c$ (next state) with output 01. So far the output is

$$\mathbf{c} = [01 \ 10 \ 00 \ 01 \ \ldots]$$

**Fig. 13.7** Trellis diagram
for the encoder

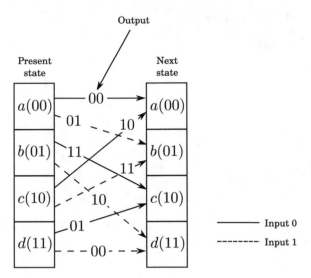

If the message code ends here, then we have additional outputs 10 00 to take care of the trailing zeros in the message.

## 13.4.6   Maximum Likelihood (ML) Decoding

We now come to the decoding of convolutional codes. In what we have studied so far, we have, **m**, the message words, **c**, the code words and, **r**, the received words. If the number of bits in the message word is $k$, then there are $2^k$ message words and $2^k$ code words. Ideally, when we compare the received word, **r**, the highest probability of the correct code word is when the Hamming distance

$$\text{Min} \left[ d_H \{\mathbf{r}, \mathbf{c}_k\} \right] \ \forall \ k$$

is minimum for the correct code word, and $d_H$ is the Hamming distance. This implies that we have to store $2^k$ code words in the receiver, which increases exponentially with $k$. Thus, to take an oversimplified example, based on Fig. 13.4, if $k = 1$, then the two message vectors are $m_1 = 0$ and $m_2 = 1$, the two code words are $c_1 = 00$ and $c_2 = 01$. If the received code word is 00, then interpret that $m_1$ was sent. If the received code word was 11, then we conclude that $m_2$ was sent.

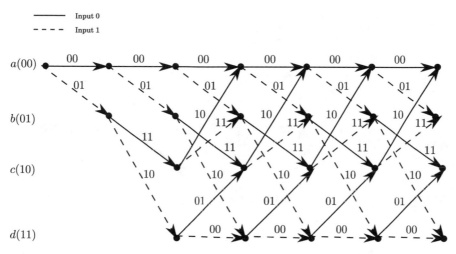

**Fig. 13.8** Full Trellis

### 13.4.7 Viterbi Algorithm

The Viterbi algorithm uses ML decoding but keeping the paths which are the most probable. The trellis is shown in Fig. 13.8.

Thus, let the message vector be

$$\mathbf{m} = [1110(0)(0)]$$

then the code word is

$$\mathbf{c} = [01\ 10\ 00\ 01\ 10\ 00]$$

and the received word is

$$\mathbf{r} = [11\ 10\ 00\ 10\ 00\ 00]$$

where the error has been committed in the first digit. Looking at the trellis, the two surviving paths are shown with the number of errors in brackets. The next two digits are 10, and the paths are shown as in the next stage. Looking at the next two bits (00), the surviving paths are shown and so on. From the trellis diagram, it is clear that the message word was $\mathbf{m} = [111000]$. After deleting the final two zeros, the message word was $\mathbf{m} = [1110]$. The process is shown in the upper diagram of Fig. 13.9.

Let us take another example: Let the message signal be the same, but the received word be

$$\mathbf{r} = [01\ 11\ 01\ 01\ 00\ 00]$$

with two errors. The trellis diagram is shown in the lower diagram of Fig. 13.9.

The algorithm may be stated as follows:

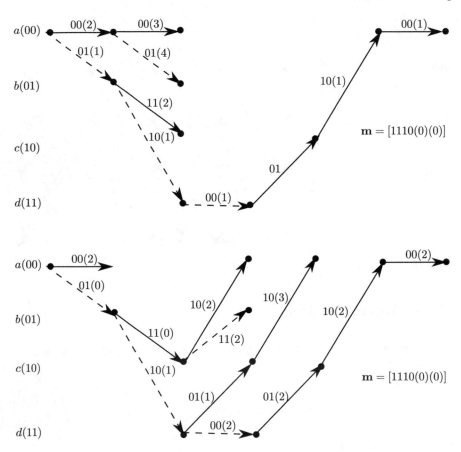

**Fig. 13.9**  Application of the Viterbi algorithm

1. Start with node $a(00)$.
2. Proceed with the next output, going to the next node(s) keeping the lowest two or three errors (or metrics).
3. Keep doing this till one path will clearly emerge as the 'winner,' with the lowest number of errors.

It is important to remember that the Viterbi algorithm *breaks down when two or more paths have the same number of errors when we reach the final output point*.

## 13.5  Important Concepts and Formulae

- Error control coding as applied to an alphabet consisting of two symbols, 0 and 1 are considered here. Two types of error control coding exist as applied to digital communication systems:

  1. $n$ bits are received. Out of these $n$ bits, $k$ $(< n)$ bits are the message bits and $n - k$ bits are the error control bits. On processing these extra bits, a decision is made whether the message bits were in error or not. Further, the algorithm is also capable of making the *most probable* correction.
  2. $n$ bits are received. Out of these $n$ bits, $k$ $(< n)$ bits are the message bits and $n - k$ bits are the error control bits. On processing these extra bits, a decision is made whether the message bits were in error or not. The algorithm is *not* capable of making a correction. Retransmission of the message bits is requested.

- The code word $\mathbf{c}$ in the following way

$$
\begin{aligned}
\mathbf{c} &= \begin{bmatrix} m_1 \; m_2 \; \cdots \; m_k \mid b_1 \; b_2 \; \cdots \; b_{n-k} \end{bmatrix} \\
&= \begin{bmatrix} c_1 \; c_2 \; \cdots \; c_k \; c_{k+1} \; c_{k+2} \; \cdots \; c_n \end{bmatrix}
\end{aligned}
$$

where $\mathbf{m} = \begin{bmatrix} m_1 \; m_2 \; \cdots \; m_k \end{bmatrix}$  are the message bits

and $\mathbf{b} = \begin{bmatrix} b_1 \; b_2 \; \cdots \; b_{n-k} \end{bmatrix}$  are the parity bits       (13.34)

- The parity bits are generated by using coefficients. Thus

$$
\mathbf{b} = \mathbf{mP}
$$

$$
\text{or } b_j = \sum_{i=1}^{k} m_i \, p_{ij}
$$

$$
= m_1 p_{1j} \oplus m_2 p_{2j} \oplus \cdots \oplus m_k p_{kj}
$$

where $\mathbf{b}$, and $\mathbf{m}$ are row vectors of dimension $n - k$ and $k$, respectively; while $\mathbf{P}$ is a *coefficient matrix* of dimension $k \times (n - k)$ with entries

$$
p_{ij} = \begin{cases} 1 & \text{if } b_j \text{ depends on } m_i \\ 0 & \text{if } b_j \text{ does not depend on } m_i \end{cases}
$$

- To obtain a code word, we define a generator matrix, $\mathbf{G}$, such that

$$
\mathbf{c} = \mathbf{mG}
$$

where

$$G = \begin{bmatrix} 1 & 0 & \cdots & 0 & p_{11} & p_{12} & \cdots & p_{1,n-k} \\ 0 & 1 & \cdots & 0 & p_{21} & p_{22} & \cdots & p_{2,n-k} \\ \vdots & \vdots & \ddots & \vdots & \vdots & \vdots & \ddots & \vdots \\ 0 & 0 & \cdots & 1 & p_{k1} & p_{k2} & \cdots & p_{k,n-k} \end{bmatrix}$$

$$= \begin{bmatrix} \mathbf{I}_k & | & \mathbf{P} \end{bmatrix}_{k \times n}$$

where a $k \times n$ matrix has $k$ rows and $n$ columns. Furthermore

$$\mathbf{I}_k = \begin{bmatrix} 1 & 0 & \cdots & 0 \\ 0 & 1 & \cdots & 0 \\ \vdots & \vdots & \ddots & \vdots \\ 0 & 0 & \cdots & 1 \end{bmatrix}_{k \times k}$$

$$\text{and } \mathbf{P} = \begin{bmatrix} p_{11} & p_{12} & \cdots & p_{1,n-k} \\ p_{21} & p_{22} & \cdots & p_{2,n-k} \\ \vdots & \vdots & \ddots & \vdots \\ p_{k1} & p_{k2} & \cdots & p_{k,n-k} \end{bmatrix}_{k \times (n-k)}$$

- Any one of these massages is a valid message. Or the modulo two sum of two messages, $\mathbf{m}_i$ and $\mathbf{m}_j$ is also a valid message, $\mathbf{m}_l$.

$$\mathbf{m}_i \oplus \mathbf{m}_j = \mathbf{m}_l \tag{13.35}$$

- therefore

$$\left( \mathbf{m}_i \oplus \mathbf{m}_j \right) \mathbf{G} = \mathbf{c}_i \oplus \mathbf{c}_j$$
$$\mathbf{m}_l \mathbf{G} = \mathbf{c}_l \tag{13.36}$$

- Forming the product

$$\mathbf{cH}^T = \mathbf{c}_{1 \times n} \begin{bmatrix} \mathbf{P}_{k \times (n-k)} \\ \mathbf{I}_{(n-k)} \end{bmatrix}_{n \times (n-k)}$$

$$= \begin{bmatrix} \mathbf{m}_{1 \times k} & \mathbf{b}_{1 \times (n-k)} \end{bmatrix} \begin{bmatrix} \mathbf{P}_{k \times (n-k)} \\ \mathbf{I}_{(n-k)} \end{bmatrix}$$

$$= \mathbf{m}_{1 \times k} \mathbf{P}_{k \times (n-k)} \oplus \mathbf{b}_{1 \times (n-k)} \mathbf{I}_{(n-k)}$$

$$= \mathbf{b} \oplus \mathbf{b}$$

$$= \mathbf{0} \tag{13.37}$$

- We proceed to find

$$\mathbf{c}_{1 \times n} \mathbf{H}^T_{n \times (n-k)} = \mathbf{m}_{1 \times k} \mathbf{G}_{k \times n} \mathbf{H}^T_{n \times (n-k)}$$

$$\text{therefore } \mathbf{G}_{k \times n} \mathbf{H}^T_{n \times (n-k)} = \mathbf{0} \qquad (13.38)$$

$$\begin{aligned}
\mathbf{s}_j &= \mathbf{r}_j \mathbf{H}^T \\
&= (\mathbf{e} \oplus \mathbf{c}_j) \mathbf{H}^T \\
&= \mathbf{e} \mathbf{H}^T \oplus \mathbf{0}
\end{aligned}$$

- *Hamming codes* are codes which satisfy

$$\begin{aligned}
n &= 2^m - 1 \\
k &= 2^m - m - 1 \\
n - k &= m
\end{aligned} \qquad (13.39)$$

where $m \geq 3$.
- We may define a new linear $(n, n-k)$ block code with generator matrix $\mathbf{G}_d = \mathbf{H}_{n \times (n-k)}$ and parity check matrix $\mathbf{H}_d = \mathbf{G}_{k \times n}$, which gives us the dual code.
- Cyclic codes form an important subclass of linear block codes. The reason why the cyclic codes are so successful is that unlike systematic linear block codes, the code length can be made very large and can be implemented using simple shift registers.

Cyclic block codes have two important properties:

1. If $\mathbf{c}_1$ and $\mathbf{c}_2$ are two cyclic block codes, then $\mathbf{c}_1 \oplus \mathbf{c}_2$ is also a code word.
2. If $\mathbf{c}$ is a cyclic block code such that cyclic shifts of a code word results in another code word.

$$\begin{aligned}
\mathbf{c}^{(0)} &= [c_1, c_2, \cdots, c_n] \\
\mathbf{c}^{(1)} &= [c_2, c_3, \cdots, c_n, c_1] \\
\mathbf{c}^{(2)} &= [c_3, c_4, \cdots, c_n, c_1, c_2] \\
\vdots &= \vdots \\
\mathbf{c}^{(n-1)} &= [c_n, c_1, c_2, \cdots, c_{n-2}, c_{n-1}]
\end{aligned} \qquad (13.40)$$

are all the code words, and in general,

$$\mathbf{c}^{(i)} = [c_{i+1}, c_{i+2}, \cdots c_{n-1}, c_n, c_1, c_2, \cdots, c_{i-1}, c_i] \qquad (13.41)$$

is a code word. It is erroneous to think that there are $n$ distinct code words. One example is $[0000 \cdots 00]$ and another is $[1111 \cdots 11]$.

- In the following, $+$ represents $\oplus$.
  The polynomial $x^n + 1$ plays an important role in cyclic codes. Let $g(x)$ be a factor of $x^n + 1$, of degree $n - k$. That is,

$$x^n + 1 = g(x)_{n-k} h(x)_k$$

- To decode the received vector in cyclic codes, we know that any code word can be divided by $g(x)$. Therefore, if the received vector is $\mathbf{r}$, then if $\mathbf{r}$ is a code word, then $r(x)$ can be exactly divided by $g(x)$ to give the message vector as per Eq. (13.23), and there is no remainder.
- BCH codes are powerful cyclic codes with the following parameters:

$$n = 2^m - 1$$
$$n - k \leq mt$$
$$2t + 2 \geq d_{\min} \geq 2t + 1$$
$$t \leq 2^{m-1}$$

- For RS codes

$$\text{Block length} : n = 2^m - 1 \text{ symbols}$$
$$\text{Message size} = k \text{ symbols}$$
$$\text{Parity check size} = n - k = 2t \text{ symbols}$$
$$d_{\min} = 2t + 1 \text{ symbols}$$

- Cyclic redundancy codes are *error detection codes.*
- Convolutional codes are used when the bits come serially as opposed to when they come in blocks.

## 13.6   Self Assessment

### 13.6.1   Review Questions

1. Show how a parity check code *detects* single errors, but does not detect *two or more* errors.
2. In your opinion, is it better to correct errors with an error correcting code or ask for retransmission?

## 13.6.2 Numerical Problems

1. In a code with 4 message bits, the number of ones are counted and appended after the message bits. Write all the code words for this scheme. Is this an error correcting code or error detecting code?

2. If $k$ message bits each of length $k$ bits are generated and parity bits are added as shown below

$$
\begin{array}{ccccc}
m_{11} & m_{12} & \cdots & m_{1k} & p_{r1} \\
m_{21} & m_{22} & \cdots & m_{2k} & p_{r2} \\
\vdots & \vdots & \ddots & \vdots & \vdots \\
m_{k1} & m_{k2} & \cdots & m_{kk} & p_{rk} \\
p_{c1} & p_{c2} & \cdots & p_{ck} &
\end{array}
$$

where $m_{rc}$ are the message bits in row $r$ and column $c$. $p_{rj}$ are the parity bits of row $j$, while $p_{cj}$ are the parity bits of column $j$. The full set of data (message bits and parity bits) is then sent over the channel. Show that by this method minimum of 1 error can be detected and corrected, *if there is no error in the parity bits*.

3. Obtain the generator matrix, **G**, when we use a repetition code of 5 bits. Also, obtain the parity check matrix, **H**, and show that

$$ \mathbf{GH}^T = \mathbf{0} $$

4. For the generator matrix

$$
\mathbf{G} = \begin{bmatrix}
1 & 0 & 0 & 0 & 1 & 0 & 0 \\
0 & 1 & 0 & 0 & 1 & 1 & 0 \\
0 & 0 & 1 & 0 & 0 & 1 & 1 \\
0 & 0 & 0 & 1 & 1 & 0 & 1
\end{bmatrix}
$$

show that it is a Hamming code. From the **H** matrix, find $d_{\min}$.

5. For Example 4, show that

$$ \mathbf{GH}^T = \mathbf{0} $$

6. For Example 4, find the dual code.

7. For a repetition code with $n = 6$ and $k = 1$, find all the possible messages. List all possible received messages which are in error which can be corrected.

8. For the factorisation, (where $+$ represents $\oplus$)

$$ 1 + x^9 = (1 + x)(1 + x + x^2)(1 + x^3 + x^6) $$

find the possible generator polynomials.

9. Using the generator polynomial

$$ g(x) = 1 + x^3 + x^6 $$

for $n = 9$, (and where $+$ represents $\oplus$), find the code for

$$\mathbf{m} = [101]$$

10. Using the generator polynomial

$$g(x) = 1 + x + x^2$$

for $n = 9$, (and where $+$ represents $\oplus$), find the code for

$$\mathbf{m} = [1010010]$$

11. The generator polynomial for a (15,13) cyclic code is

$$g(x) = 1 + x + x^2$$

(where $+$ represents $\oplus$), find the code for

$$\mathbf{m} = [1010010001001]$$

12. A convolutional encoder has the following characteristics:

$$g^{(1)}(D) = 1 + D^2$$
$$\text{and } g^{(2)}(D) = D + D^2$$

draw the block diagram of the encoder.
13. The message signal for the encoder of Problem 12 is

$$\mathbf{m} = [1100\ldots]$$

draw the code tree for this system and find the output for the message signal given.
14. The message signal for the encoder of Problem 12 is

$$\mathbf{m} = [1100\ldots]$$

draw the state diagram for this system and find the output for the message signal given.
15. The message signal for the encoder of Problem 12 is

$$\mathbf{m} = [1100\ldots]$$

draw the trellis diagram for this system and find the output for the message signal given.

16. Use the Viterbi algorithm to decode the coded signal for the code generated by the system of Problem 12.

# Reference

1. Haykin S (1999) Communication systems. Wiley

# Appendix A
# Mathematical Reference

## General

### Important Constants

$e = 2.7183$
$\pi = 3.1461$

### Taylor's Series Expansion

1. About $x = 0$, $f(x) = f(0) + f'(0)x + f''(0)\frac{x^2}{2} \ldots$
2. About $x = a$, $f(x) = f(a) + f'(a)(x - a) + f''(a)\frac{(x-a)^2}{2} \ldots$

## Complex Variables

### General

If $\quad z_1 = a + jb = r_1 e^{j\theta_1} = r_1(\cos\theta_1 + j\sin\theta_1) \quad$ and $\quad z_2 = c + jd = r_2 e^{j\theta_2} = r_2(\cos\theta_2 + j\sin\theta_2)$
where $j = \sqrt{-1}$, $r_1 = \sqrt{a^2 + b^2}$, $r_2 = \sqrt{c^2 + d^2}$, $\theta_1 = \tan^{-1}(b/a)$, and $\theta_2 = \tan^{-1}(d/c)$ then

1. $z_1 + z_2 = (a + jb) + (c + jd) = (a + c) + j(b + d)$
2. $z_1 - z_2 = (a + jb) - (c + jd) = (a - c) + j(b - d)$

© The Editor(s) (if applicable) and The Author(s), under exclusive license to Springer Nature Singapore Pte Ltd. 2022
S. Bhooshan, *Fundamentals of Analogue and Digital Communication Systems*,
Lecture Notes in Electrical Engineering 785,
https://doi.org/10.1007/978-981-16-4277-7

3. $z_1 z_2 = (a + jb) \times (c + jd) = ac + jbc + jad + bdj^2 = (ac - bd) + j(bc + ad) = r_1 r_2 e^{j(\theta_1 + \theta_2)}$

4. $z_1/z_2 = \frac{(a+jb)}{(c+jd)} = \left(\frac{ac+bd}{c^2+d^2}\right) + j\left(\frac{bc-ad}{c^2+d^2}\right) = (r_1/r_2)e^{j(\theta_1 - \theta_2)}$.

## *Inequalities*

$r_1$, $r_2$ are the *absolute values or magnitudes* of $z_1$, $z_2$, respectively, ($|z_1|$, $|z_2|$). The absolute value has three important properties:

1. $|z| \geq 0$, where $|z| = 0$ if and only if $z = 0$
2. $|z + w| \leq |z| + |w|$ (triangle inequality)
3. $|zw| = |z||w|$.

## *Complex Conjugates*

The complex conjugate of the complex number $z^* = (x + jy)^* = x - jy$. $z^*$ has the following properties:

1. $(z + w)^* = z^* + w^*$
2. $(zw)^* = z^* w^*$
3. $(z/w)^* = z^*/w^*$
4. $(z^*)^* = z$
5. $\Re(z) = \frac{1}{2}(z + z^*)$
6. $\Im(z) = \frac{1}{2}(z - z^*)$
7. $|z| = |z^*|$
8. $|z|^2 = zz^*$.

## *Euler's Identity*

$\cos\theta + j\sin\theta = e^{i\theta}$.

## Trigonometry

### *Basic Formulae*

1. $\sin^2\theta + \cos^2\theta = 1$
2. $\tan^2\theta + 1 = \sec^2\theta$

3. $1 + \cot^2 \theta = \csc^2 \theta$
4. $\csc \theta = 1/\sin \theta$; $\sec \theta = 1/\cos \theta$; $\tan \theta = 1/\cot \theta$
5. $\tan \theta = \sin \theta / \cos \theta$
6. $\sin(\pi/2 - \theta) = \cos \theta$; $\cos(\pi/2 - \theta) = \sin \theta$; $\tan(\pi/2 - \theta) = \cot \theta$
7. $\sin(-\theta) = -\sin \theta$; $\cos(-\theta) = \cos \theta$; $\tan(-\theta) = -\tan \theta$.

## Sum and Difference Formulae

1. $\cos(a \pm b) = \cos a \cos b \mp \sin a \sin b$
2. $\sin(a \pm b) = \sin a \cos b \pm \cos a \sin b$
3. $\tan(a \pm b) = (\tan a \pm \tan b)/(1 \mp \tan a \tan b)$.

## Double Angle Formulae

1. $\sin(2a) = 2 \sin a \cos a = 2 \tan a/(1 + \tan^2 a)$
2. $\cos(2a) = \cos^2 a - \sin^2 a = 1 - 2 \sin^2 a = 2 \cos^2 a - 1 = (1 - \tan^2 a)/(1 + \tan^2 a)$
3. $\tan(2a) = (2 \tan a)/(1 - \tan^2 a)$.

## Half Angle Formulae

1. $\sin(a) = \sqrt{\frac{1 - \cos(2a)}{2}}$; $\sin^2(a) = \left[\frac{1 - \cos(2a)}{2}\right]$
2. $\cos(a) = \sqrt{\frac{1 + \cos(2a)}{2}}$; $\cos^2(a) = \left[\frac{1 + \cos(2a)}{2}\right]$
3. $\tan(a) = \sqrt{\frac{1 - \cos(2a)}{1 + \cos(2a)}}$; $\tan^2(a) = \left[\frac{1 - \cos(2a)}{1 + \cos(2a)}\right]$.

## Product to Sum Formulae

1. $\sin(a) \sin(b) = \frac{1}{2}[\cos(a - b) - \cos(a + b)]$
2. $\cos(a) \cos(b) = \frac{1}{2}[\cos(a - b) + \cos(a + b)]$
3. $\sin(a) \cos(b) = \frac{1}{2}[\sin(a - b) - + \sin(a + b)]$.

## Sum and Difference to Product

1. $\sin(a) + \sin(b) = 2\sin\left(\frac{a+b}{2}\right)\cos\left(\frac{a-b}{2}\right)$
2. $\sin(a) - \sin(b) = 2\cos\left(\frac{a+b}{2}\right)\sin\left(\frac{a-b}{2}\right)$
3. $\cos(a) + \cos(b) = 2\cos\left(\frac{a+b}{2}\right)\cos\left(\frac{a-b}{2}\right)$
4. $\cos(a) - \cos(b) = 2\sin\left(\frac{a+b}{2}\right)\sin\left(\frac{a-b}{2}\right)$
5. $A\sin(a) + B\cos(a) = C\sin(a+\phi)$ where $C = \sqrt{A^2 + B^2}$ and $\phi = \tan^{-1}\left(\frac{B}{A}\right)$.

## Triangle Formulae

If in a triangle of sides $a, b$ and $c$, the angles opposite these sides are $A, B$ and $C$; then

1. $a/\sin A = b/\sin B = c/\sin C$
2. $a^2 = b^2 + c^2 - 2bc\cos A$
3. $b^2 = a^2 + c^2 - 2ac\cos B$
4. $c^2 = b^2 + a^2 - 2ba\cos C$.

## Powers of the Trigonometric Functions

1. $\sin^3\theta = \frac{1}{4}(3\sin\theta - \sin 3\theta)$
2. $\cos^3\theta = \frac{1}{4}(3\cos\theta + \cos 3\theta)$
3. $\sin^4\theta = \frac{1}{8}(\cos(4\theta) - 4\cos(2\theta) + 3)$
4. $\cos^4\theta = \frac{1}{8}(\cos(4\theta) + 4\cos(2\theta) + 3)$.

## Differentiation

$c$ is a constant, $f$ and $g$ are functions.

## Rules

1. $(cf)' = cf'$
2. $(f + g)' = f' + g'$
3. $(fg)' = f'g + fg'$
4. $\left(\frac{1}{f}\right)' = \frac{-f'}{f^2}$

5. $\left(\frac{f}{g}\right)' = \frac{f'g - fg'}{g^2}$

6. $[f(g)]' = f'g'$

7. *Derivative of inverse function:* $(f^{-1})' = \frac{1}{f'(f^{-1})}$ [example $(\ln x)' = \frac{1}{e^{\ln x}} = \frac{1}{x}$]

8. *Generalised power rule* $(f^g)' = f^g \left(g' \ln f + \frac{g}{f} f'\right)$.

## *Differentiation of Functions*

1. $c' = 0$

2. $x' = 1$

3. $(cx)' = c$

4. $(x^c)' = cx^{c-1}$

5. $\left(\frac{1}{x}\right)' = \left(x^{-1}\right)' = -x^{-2} = -\frac{1}{x^2}$

6. $\left(\frac{1}{x^c}\right)' = \left(x^{-c}\right)' = -cx^{-c-1} = -\frac{c}{x^{c+1}}$

7. $(c^x)' = c^x \ln c, \quad c > 0$

8. $(e^x)' = e^x$

9. $\left(\log_c x\right)' = \frac{1}{x \ln c}, \quad c > 0, c \neq 1$

10. $(\ln x)' = \frac{1}{x}, \quad x > 0$

11. $(x^x)' = x^x(1 + \ln x)$

12. $(\sin x)' = \cos x$

13. $(\arcsin x)' = \frac{1}{\sqrt{1-x^2}}$

14. $(\cos x)' = -\sin x$

15. $(\arccos x)' = \frac{-1}{\sqrt{1-x^2}}$

16. $(\tan x)' = \sec^2 x = \frac{1}{\cos^2 x}$

17. $(\arctan x)' = \frac{1}{1+x^2}$

18. $(\sec x)' = \sec x \tan x$

19. $(\text{arcsec } x)' = \frac{1}{|x|\sqrt{x^2-1}}$

20. $(\csc x)' = -\csc x \cot x$

21. $(\text{arccsc } x)' = \frac{-1}{|x|\sqrt{x^2-1}}$

22. $(\cot x)' = -\csc^2 x = \frac{-1}{\sin^2 x}$

23. $(\text{arccot } x)' = \frac{-1}{1+x^2}$

24. $(\sinh x)' = \cosh x = \frac{e^x + e^{-x}}{2}$

25. $(\text{arcsinh } x)' = \frac{1}{\sqrt{x^2+1}}$

26. $(\cosh x)' = \sinh x = \frac{e^x - e^{-x}}{2}$

27. $(\text{arccosh } x)' = \frac{1}{\sqrt{x^2-1}}$

28. $(\tanh x)' = \text{sech}^2 x$

29. $(\text{arctanh } x)' = \frac{1}{1-x^2}$

30. $(\text{sech } x)' = -\tanh x$

31. $(\text{sech } x)' = -\tanh x \text{ sech } x$

32. $(\text{arcsech } x)' = \frac{-1}{x\sqrt{1-x^2}}$

33. $(\operatorname{csch} x)' = -\coth x \operatorname{csch} x$

34. $(\operatorname{arccsch} x)' = \frac{-1}{x\sqrt{1+x^2}}$

35. $(\coth x)' = -\operatorname{csch}^2 x$

36. $(\operatorname{arccoth} x)' = \frac{1}{1-x^2}$.

# Integration

## *Common Substitutions*

1. $\int f(ax+b)dx = \frac{1}{a} \int f(u)du$ where $u = ax + b$

2. $\int f\left(\sqrt[n]{ax+b}\right) dx = \frac{n}{a} \int u^{n-1} f(u)du$ where $u = \sqrt[n]{ax+b}$

3. $\int f\left(\sqrt{a^2-x^2}\right) dx = a \int f(a\cos u)\cos u\, du$ where $x = a\sin u$

4. $\int f\left(\sqrt{x^2+a^2}\right) dx = a \int f(a\sec u)\sec^2 u\, du$ where $x = a\tan u$

5. $\int f\left(\sqrt{x^2-a^2}\right) dx = a \int f(a\tan u)\sec u \tan u\, du$ where $x = a\sec u$.

## *Indefinite Integrals*

$a, b, c$ are constants, $u, v, w$ are functions of $t$

1. $\int a\, dt = at$

2. $\int af(t)dx = a \int f(t)dt$

3. $\int (u \pm v \pm w \pm \cdots) dx = \int u\, dt \pm \int v\, dt \pm \int w\, dt \pm \cdots$

4. $\int u\, dv = uv - \int v\, du$

5. $\int f(at)dt = \frac{1}{a} \int f(u)du$

6. $\int F\{f(t)\}dt = \int F(u)\frac{dxt}{du} du = \int \frac{F(u)}{f'(t)} du$

7. $\int t^n dx = \begin{cases} \frac{t^{n+1}}{n+1} & n \neq -1 \\ \ln t & n = -1 \end{cases}$

8. $\int e^{-at} dx = -\frac{e^{-at}}{a}$

9. $\int a^{-bt} dx = -\frac{a^{-bt}}{b\log a}$

10. $\int \sin(at+b)\, dt = -\frac{1}{a}\cos(at+b)$

11. $\int \cos(at+b)\, dt = \frac{1}{a}\sin(at+b)$

12. $\int \frac{f'(t)}{f(t)} dt = \ln[f(t)]$

13. $\int \tan(at+b) = \frac{\ln[\sec(at+b)]}{a} = -\frac{\ln[\cos(at+b)]}{a}$

14. $\int \cot(at+b) = \frac{\ln[\sin(at+b)]}{a}$

15. $\int \sec(at+b) = \frac{\ln[\tan(at+b)+\sec(at+b)]}{a}$

16. $\int t^n e^{-at} dt \quad n = $ positive integer; use $\int u\, dv = uv - \int v\, du$

17. $\int \frac{1}{t^2+a^2} dt = \frac{1}{a}\arctan\left(\frac{t}{a}\right)$

18. $\int \frac{t}{t^2\pm a^2} dt = \frac{\log(t^2\pm a^2)}{2}$

19. $\int \frac{t^2}{t^2+a^2} dt = t - a\arctan\left(\frac{t}{a}\right)$

20. $\int \frac{1}{t^2-a^2} dt = \frac{\log(t-a)}{2a} - \frac{\log(t+a)}{2a}$

21. $\int \frac{t^2}{t^2-a^2} dt - \frac{a\log(t+a)}{2} + \frac{a\log(t-a)}{2} + t$

22. $\int \frac{1}{\sqrt{t^2+a^2}} dt = \sinh^{-1}\left(\frac{x}{|a|}\right)$

23. $\int \frac{1}{\sqrt{t^2-a^2}} dt = \log\left(2\sqrt{t^2-a^2}+2t\right)$

24. $\int \frac{t}{\sqrt{t^2\pm a^2}} dt = \sqrt{t^2\pm a^2}$

25. $\int \frac{1}{\sqrt{a^2-t^2}} dt = \arcsin\left(\frac{t}{a}\right)$

26. $\int \frac{1}{\sqrt{t^2+a^2}} dt = \sinh^{-1}\left(\frac{t}{|a|}\right)$

27. $\int \frac{1}{(a^2-t^2)^{\frac{3}{2}}} dt = \frac{t}{a^2\sqrt{a^2-t^2}}$

28. $\int \frac{1}{(t^2-a^2)^{\frac{3}{2}}} dt = -\frac{t}{a^2\sqrt{t^2-a^2}}$

29. $\int \frac{1}{(t^2+a^2)^{\frac{3}{2}}} dt = \frac{t}{a^2\sqrt{t^2+a^2}}$

30. $\int \frac{t}{(t^2+a^2)^{\frac{3}{2}}} dt = -\frac{1}{\sqrt{t^2+a^2}}$

31. $\int \frac{t^2}{(t^2+a^2)^{\frac{3}{2}}} dt = \sinh^{-1}\left(\frac{t}{|a|}\right) - \frac{t}{\sqrt{t^2+a^2}}$

32. $\int \frac{t^3}{(t^2+a^2)^{\frac{3}{2}}} dt = \frac{2a^2}{\sqrt{a^2+x^2}} + \frac{t^2}{\sqrt{a^2+t^2}}$

33. $\int \frac{1}{\sqrt{t^2+bt+c}} dt = \begin{cases} \log\left(t+\frac{b}{2}\right) & \text{for } 4c-b^2=0 \\ \sinh^{-1}\left(\frac{2t+b}{\sqrt{4c-b^2}}\right) & \text{for } 4c-b^2>0 \\ \log\left(2\sqrt{t^2+bt+c}+2t+b\right) & \text{for } 4c-b^2<0 \end{cases}$

34. $\int \frac{1}{\sqrt{-t^2+bt+c}} dt = \begin{cases} -\sin^{-1}\left(\frac{b-2t}{\sqrt{4c+b^2}}\right) & \text{for } b^2+4c>0 \\ -j\sinh^{-1}\left(\frac{2t-b}{\sqrt{-4c-b^2}}\right) & \text{for } b^2+4c<0 \\ j\ln\left(x-\frac{b}{2}\right) & \text{for } b^2+4c=0 \end{cases}$

35. $\int \frac{1}{(t^2+bt+c)^{\frac{3}{2}}} dt = \begin{cases} \frac{4t}{(4c-b^2)\sqrt{t^2+bt+c}} + \frac{2b}{(4c-b^2)\sqrt{t^2+bt+c}} & \text{for } 4c-b^2\neq 0 \\ -\frac{1}{2\left(t+\frac{b}{2}\right)^2} & \text{for } 4c-b^2=0 \end{cases}$

36. $\int \frac{1}{(-t^2+bt+c)^{\frac{3}{2}}} dt = \begin{cases} \frac{2b}{(-4c-b^2)\sqrt{-t^2+bt+c}} - \frac{4t}{(-4c-b^2)\sqrt{-t^2+bt+c}} & \text{for } 4c+b^2\neq 0 \\ -\frac{j}{2\left(t-\frac{b}{2}\right)^2} & \text{for } 4c+b^2=0 \end{cases}$

37. $\int \frac{1}{t^2+bt+c} dt = \begin{cases} \left(\sqrt{4c-b^2}\right)^{-1} 2\arctan\left(\frac{2t+b}{\sqrt{4c-b^2}}\right) & \text{for } 4c-b^2>0 \\ \left(\sqrt{b^2-4c}\right)^{-1} \log\left(\frac{2t-\sqrt{b^2-4c}+b}{2t+\sqrt{b^2-4c}+b}\right) & \text{for } 4c-b^2<0 \end{cases}$

38. $\int e^{\beta x} \sin(\alpha t+\phi) dt = \frac{e^{\beta x}[\beta\sin(\alpha t+\phi)-\alpha\cos(\alpha t+\phi)]}{\beta^2+\alpha^2}$

39. $\int e^{\beta x} \cos(\alpha t+\phi) dt = \frac{e^{\beta x}[\alpha\sin(\alpha t+\varphi)+\beta\cos(\alpha t+\varphi)]}{\beta^2+\alpha^2}$.

# Appendix B
# Set Theory

## Basic Definitions

1. $S$ is the sample space
2. $A$, $B$, $C$ are events $A$, $B$, $C \subset S$
3. $x$ are elementary outcomes
4. $\phi$ is the empty set
5. $A \subset B$ (subset) means that every element $x \in A \Rightarrow x \in B$, but if $x \in B \nRightarrow x \in A$
6. $A = B$ (equality) means that every element $x \in A \Rightarrow x \in B$, and every $x \in B \Rightarrow x \in A$
7. $A = B$ (equality) also means that $A \subset B$ and $B \subset A$
8. $A \cap B = \{x : x \in A \text{ and } x \in B\}$ (intersection)
9. $A \cap B = B \cap A$ ($\cap$ is commutative)
10. $(A \cap B) \cap C = A \cap (B \cap C)$ ($\cap$ is associative)
11. $A \cup B = \{x : x \in A \text{ or } x \in B\}$ (union)
12. $A \cup B = B \cup A$ ($\cup$ is commutative)
13. $(A \cup B) \cup C = A \cup (B \cup C)$ ($\cup$ is associative)
14. $(A \cap B) \cup C = (A \cup C) \cap (A \cup B)$ (distributive property)
15. $(A \cup B) \cap C = (A \cup C) \cap (A \cup B)$ (distributive property)
16. $A^c = \{x : x \notin A\}$ (complement)
17. $A - B = \{x : x \in A \text{ and } x \notin B\}$ (difference)
18. $A \cap B = \phi$ where $\phi = \{\}$ (mutually exclusive)
19. $A \cup A^c = S$
20. $A \cap A^c = \phi$
21. $(A \cap B)^c = A^c \cup B^c$ (DeMorgan's law 1)
22. $\left\{ \bigcap_{i=1}^n A_i \right\}^c = \bigcup_{i=1}^n A_i^c$
23. $(A \cup B)^c = A^c \cap B^c$ (DeMorgan's law 2)
24. $\left\{ \bigcup_{i=1}^n A_i \right\}^c = \bigcap_{i=1}^n A_i^c$.

© The Editor(s) (if applicable) and The Author(s), under exclusive license to Springer    607
Nature Singapore Pte Ltd. 2022
S. Bhooshan, *Fundamentals of Analogue and Digital Communication Systems*,
Lecture Notes in Electrical Engineering 785,
https://doi.org/10.1007/978-981-16-4277-7

## Venn Diagrams

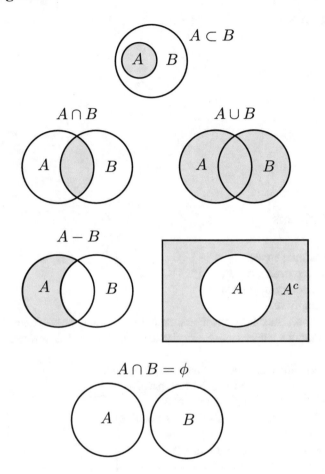

# Appendix C
# Important Formulae

## Fourier Series

### *Trigonometric Fourier Series*

For a periodic function, $x_T(t)$, of time period $T$, given by

$$x_T(t) = a_0 + \sum_{k=1}^{\infty} a_k \cos\left\{k\left(\frac{2\pi}{T}\right)t\right\} + \sum_{k=1}^{\infty} b_k \sin\left\{k\left(\frac{2\pi}{T}\right)t\right\}$$

where

$$a_0 = \frac{1}{T}\int_{-T/2}^{T/2} x_T(t)dt$$

and

$$a_m = \frac{2}{T}\int_{-T/2}^{T/2} x_T(t)\cos(m\omega_0 t)dt \qquad m = 1, 2, \ldots$$

$$b_m = \frac{2}{T}\int_{-T/2}^{T/2} x_T(t)\sin(m\omega_0 t)dt \qquad m = 1, 2, \ldots$$

### *Exponential Fourier Series*

For a periodic function, $x_T(t)$, of time period $T$, given by

$$x_T(t) = \sum_{k=-\infty}^{\infty} X_k \exp\left\{k\left(\frac{2\pi}{T}\right)t\right\}$$

© The Editor(s) (if applicable) and The Author(s), under exclusive license to Springer Nature Singapore Pte Ltd. 2022
S. Bhooshan, *Fundamentals of Analogue and Digital Communication Systems*,
Lecture Notes in Electrical Engineering 785,
https://doi.org/10.1007/978-981-16-4277-7

**Table C.1** Properties of the Fourier transform

| Property | $x(t)$ | $X(\omega)$ |
|---|---|---|
| Definition | $x(t) = \frac{1}{2\pi} \int_{-\infty}^{\infty} X(\omega) d\omega$ | $X(\omega) = \int_{-\infty}^{\infty} x(t) dt$ |
| Time shifting | $x(t - t_0)$ | $X(\omega) e^{-j\omega t_0}$ |
| Frequency shifting | $x(t) e^{j\omega_0 t}$ | $X(\omega - \omega_0)$ |
| Time multiplication | $x(at)$ | $\frac{1}{|a|} X\left(\frac{\omega}{a}\right)$ |
| Duality | $X(t)$ | $2\pi x(-\omega)$ |
| Duality | $\frac{1}{2\pi} X(-t)$ | $x(\omega)$ |
| Differentiation | $dx(t)/dt$ | $j\omega X(\omega)$ |
| $d^n/dt^n$ | $d^n x(t)/dt^n$ | $(j\omega)^n X(\omega)$ |
| Frequency differentiation | $(-jt)x(t)$ | $dX(\omega)/d\omega$ |
| $d^n/dt^n$ | $(-jt)^n x(t)$ | $d^n X(\omega)/d\omega^n$ |
| Time integration | $\int_{-\infty}^{t} x(\tau) d\tau$ | $\frac{1}{j\omega} X(\omega) + \pi X(0)\delta(\omega)$ |
| Parceval's theorem | $\int_{-\infty}^{\infty} x(t) x^*(t) dt = \frac{1}{2\pi} \int_{-\infty}^{\infty} X(\omega) X^*(\omega) d\omega$ | |
| Convolution–time | $x(t) * y(t)$ | $X(\omega) Y(\omega)$ |
| Convolution–frequency | $2\pi y(t) x(t)$ | $X(\omega) * Y(\omega)$ |
| Time integration | $\int_{-\infty}^{t} x(\tau) d\tau$ | $\frac{X(\omega)}{j\omega} + \pi X(0)\delta(\omega)$ |
| Frequency integration | | |

where

$$X_k = \frac{1}{T} \int_{-T/2}^{T/2} x_T(t) \exp\left\{-k\left(\frac{2\pi}{T}\right)t\right\} dt$$

# Fourier Transform

See Tables C.1 and C.2.

**Table C.2** Some Fourier transform pairs

| Time function | Fourier transform |
|---|---|
| $\delta(t)$ | $1$ |
| $1$ | $2\pi\delta(\omega)$ |
| $\text{sgn}(t)$ | $\frac{2}{j\omega}$ |
| $\frac{j}{\pi t}$ | $\text{sgn}(\omega)$ |
| $u(t)$ | $\pi\delta(\omega) + \frac{1}{j\omega}$ |
| $\frac{1}{2}\delta(t) + \frac{j}{\pi t}$ | $u(\omega)$ |
| $e^{-at}u(t)\ (a>0)$ | $\frac{1}{a+j\omega}$ |
| $e^{at}u(-t)\ (a>0)$ | $\frac{1}{a-j\omega}$ |
| $e^{-a|t|}\ (a>0)$ | $\frac{2a}{a^2+\omega^2}$ |
| $te^{-at}u(t)\ a>0$ | $\frac{1}{(a+j\omega)^2}$ |
| $t^n e^{-at}u(t)\ (a>0)$ | $\frac{n!}{(a+j\omega)^{n+1}}$ |
| $e^{j\omega_0 t}$ | $2\pi\delta(\omega-\omega_0)$ |
| $\cos\omega_0 t$ | $\pi[\delta(\omega-\omega_0)+\delta(\omega+\omega_0)]$ |
| $\sin\omega_0 t$ | $j\pi[\delta(\omega+\omega_0)-\delta(\omega-\omega_0)]$ |
| $\text{sgn}(t) = 2u(t) - 1$ | $\frac{2}{j\omega}$ |
| $\cos\omega_0 t\ u(t)$ | $\frac{\pi}{2}[\delta(\omega-\omega_0)+\delta(\omega+\omega_0)] + \frac{j\omega}{\omega_0^2-\omega^2}$ |
| $\sin\omega_0 t\ u(t)$ | $\frac{\pi}{2j}[\delta(\omega-\omega_0)-\delta(\omega+\omega_0)] + \frac{\omega_0}{\omega_0^2-\omega^2}$ |
| $e^{-at}\sin\omega_0 t\ u(t)\ (a>0)$ | $\frac{\omega_0}{(a+j\omega)^2+\omega_0^2}$ |
| $e^{-at}\cos\omega_0 t\ u(t)\ (a>0)$ | $\frac{a+j\omega}{(a+j\omega)^2+\omega_0^2}$ |
| $\text{rect}\left(\frac{t}{\tau}\right) = p_s(t)\ (\text{for } \tau/2)$ | $\tau\,\text{sinc}\left(\frac{\omega\tau}{2}\right)$ |
| $\frac{W}{\pi}\,\text{sinc}(Wt)$ | $\text{rect}\left(\frac{\omega}{2W}\right)$ |
| $\text{Tri}\left(\frac{t}{\tau}\right)$ | $\frac{\tau}{2}\,\text{sinc}^2\left(\frac{\omega\tau}{4}\right)$ |
| $\frac{W}{2\pi}\,\text{sinc}^2\left(\frac{Wt}{2}\right)$ | $\text{Tri}\left(\frac{w}{2W}\right)$ |
| $\sum_{n=-\infty}^{\infty}\delta(t-nT)$ | $\omega_0\sum_{n=-\infty}^{\infty}\delta(\omega-n\omega_0)$ |
| $e^{-t^2/2\sigma^2}$ | $\sigma\sqrt{2\pi}\,e^{-\sigma^2\omega^2/2}$ |

# Appendix D
# Butterworth Filters

## Lowpass Butterworth Filters

This is not a book on filter design, but we will try to present a simple method to design the well known 'Butterworth LPF'. The 'normalised' design is one where the cutoff frequency is at $\omega = 1$ and $R_L = R_i = 1\,\Omega$. For this design, starting from the load resistor, backward towards the load, the elements are numbered $C_1$, $L_2$, etc., (as per Fig. 4.5) and the element values are given by

$$C_k = 2\sin\left[\frac{2k-1}{2n}\pi\right] \quad k = 1, 3, 5, \ldots$$

$$L_k = 2\sin\left[\frac{2k-1}{2n}\pi\right] \quad k = 2, 4, 6, \ldots \tag{D.1}$$

If, however, the first element is a series inductor (instead of a capacitor) then the elements are given by

$$L_k = 2\sin\left[\frac{2k-1}{2n}\pi\right] \quad k = 1, 3, 5, \ldots$$

$$C_k = 2\sin\left[\frac{2k-1}{2n}\pi\right] \quad k = 2, 4, 6, \ldots \tag{D.2}$$

if we construct such a filter, then the frequency responses of the first five filters are given in Fig. D.1.

If we were to compare the results of this design with the previous design, we can first of all notice that at $\omega = 1$ r/s the output for $n = 1$ to $n = 5$ the output ratio $|V_{out}/V_{in}|$ is uniformly equal to 0.35, the 3 db value. When we observe the response for $\omega = 2$ r/s, the output ratios for $n = 1$ to $n = 4$ are 0.225, 0.125, 0.075 and 0.0225, respectively. The response is better than the previous design. At a third

© The Editor(s) (if applicable) and The Author(s), under exclusive license to Springer Nature Singapore Pte Ltd. 2022
S. Bhooshan, *Fundamentals of Analogue and Digital Communication Systems*,
Lecture Notes in Electrical Engineering 785,
https://doi.org/10.1007/978-981-16-4277-7

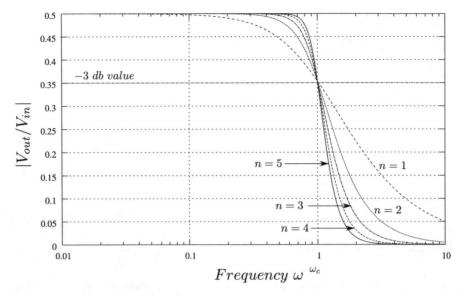

**Fig. D.1** Frequency responses of the first five lowest order Butterworth LPFs. $n = 1$ to $n = 5$

frequency, $\omega = 3$, the output ratios for $n = 1$ to 4 are about 0.15, 0.075, 0.0225 and 0, respectively. Overall the design is better.

The Butterworth filter frequency response is given by

$$|V_{out}/V_{in}| = \frac{0.5}{\sqrt{1 + \omega^{2n}}} \tag{D.3}$$

where $n$ is the order of the filter.

We have designed a low pass filter, but the design is limited for three reasons

1. The cutoff frequency is around $\omega = 1$
2. The load resistor is 1 $\Omega$
3. The values of the elements are unrealistic: $L = 1$ H; $C = 1$ F etc.

How do we design for practical values? Let us design for $R_L = 20\ \Omega$ and $f_c = 10$ kHz. We use the following *(first)* transformation to go from the 'normalised' design to the practical one.

$$\Omega = \frac{\Omega_c \omega}{\omega_c} \tag{D.4}$$

where $\Omega$ and $\Omega_c$ are the frequency and cutoff frequency of the normalised design; and $\omega$ and $\omega_c$ are the frequency and cutoff frequency of the practical design. Therefore

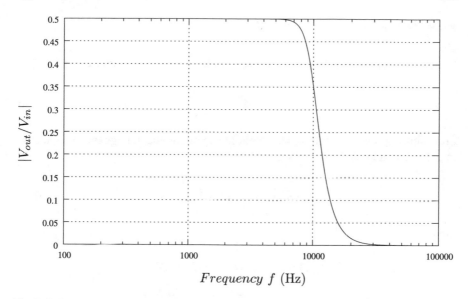

**Fig. D.2** Frequency response for a fifth order Butterworth filter with $f_c = 10^4$ and $R_L = 20 \, \Omega$

$$\Omega L = \frac{\Omega_c}{\omega_c}(\omega L)$$

$$\Omega C = \frac{\Omega_c}{\omega_c}(\omega C) \qquad\qquad (\text{D}.5)$$

using this transformation, a 1 H inductor in the normalised Butterworth LPF ($\Omega_c = 1$) is replaced by $(1/2\pi f_c) \times 1$ H, which is 15.592 $\mu$H, and capacitor of 1 F is replaced by 15.592 $\mu$F. The second transformation is an impedance transformation: each inductor and resistor is multiplied by 20 while each capacitor is *divided* by 20. Therefore, overall, using the elements of the normalised design

1. Each inductor is multiplied by $3.183 \times 10^{-4}$ (a 1 H inductor becomes 0.3183 mH)
2. Each capacitor is multiplied by $7.957 \times 10^{-7}$ (a 1 F capacitor becomes 0.7957 $\mu$F)
3. Each resistor is multiplied by 20 (a 1 $\Omega$ resistor becomes 20 $\Omega$).

Using these transformations, we apply them to the fifth order Butterworth filter for the case being considered and the results are shown in Fig. D.2.

## Pole-Zero Plot of Butterworth Filters

If we take a look at the frequency response of the normalised Butterworth filter

$$|H(j\omega)|^2 = \frac{1}{1 + \omega^{2n}} \tag{D.6}$$

From this frequency response, we would like to obtain the voltage transfer function in the $s$ (Laplace) domain, we set

$$s = j\omega \Rightarrow \omega = \left(\frac{s}{j}\right) \tag{D.7}$$

then

$$|H(j\omega)H(-j\omega)| = |H(s)H(-s)|$$
$$= \frac{1}{1 + (j\omega)^n \, (-j\omega)^n}$$
$$= \frac{1}{1 + (s)^n \, (-s)^n}$$
$$= \frac{1}{1 + (-1)^n \, (s)^{2n}}$$

if we look at the denominator of these transfer functions

$$s^2 - 1 \text{ for } n = 1$$
$$s^4 + 1 \text{ for } n = 2$$
$$s^6 - 1 \text{ for } n = 3$$

$$\vdots$$

$$s^{2n} + (-1)^n$$

and so on. For example for $n = 3$ the roots are

$$s_{\pm 1} = \pm 1$$
$$s_{\pm 2} = \pm 0.5 + 0.8660 j$$
$$s_{\pm 3} = \pm 0.5 - 0.8660 j$$

since the poles have to be in the left half plane for reasons of stability

$$H(s) = \frac{1}{(s+1)(s+0.5+j0.8660)(s+0.5-j0.8660)}$$

$$= \frac{1}{(s+1)\left[(s+0.5)^2+0.75\right]}$$

$$= \frac{1}{(s+1)(s^2+s+1)}$$

For more detail, the student is referred to a standard text on filter design.

## Highpass Butterworth Filters

We now proceed to high pass filters. We can design a high pass filter from a low pass one by another frequency transformation. If $\Omega$ and $\Omega_c$ are the frequency and cutoff frequency of the normalised LPF design; and $\omega$ and $\omega_c$ are the frequency and cutoff frequency of the practical HPF design. Then the transformation is

$$\Omega = \Omega_c \frac{\omega_c}{\omega} \tag{D.8}$$

and the elements change due to the transformation as

$$\Omega L = \Omega_c \omega_c \left(\frac{L}{\omega}\right) = \frac{1}{\frac{\omega}{\Omega_c \omega_c L}}$$

$$\Omega C = \Omega_c \omega_c \left(\frac{C}{\omega}\right) = \frac{1}{\frac{\omega}{\Omega_c \omega_c C}} \tag{D.9}$$

which means that each inductor $L$ of the LP normalised design is replaced by a capacitor

$$C_{HP} = 1/(\Omega_c \omega_c L) \tag{D.10}$$

and each capacitor $C$ of the LP normalised design is replaced by an inductor

$$L_{HP} = 1/\Omega_c \omega_c C \tag{D.11}$$

The resistances are left as they were. To change the impedance level of the circuit, we multiply all inductors and resistors by $R_{LPr}/R_{Lno}$ and *divide* all capacitors by this ratio. Here $R_{LPr}$ and $R_{Lno}$ are the load resistors of the practical HPF and normalised LPF.

We can of course go in two steps, first from the normalised LPF design to the normalised HPF design using Eq. (D.8), and then use the frequency scaling Eq. (D.5) to go to the final frequency. We adopt the second approach for illustration using (D.8)

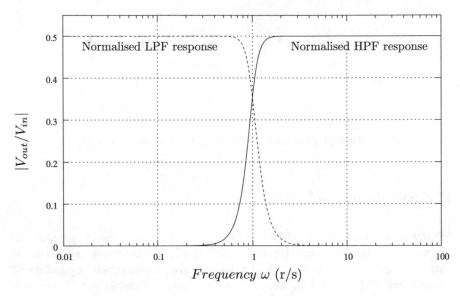

**Fig. D.3** LP to HP transformation for a Butterworth normalised filter

$$\Omega = \frac{1}{\Omega'} \tag{D.12}$$

where $\Omega'$ is the normalised HPF frequency and $\omega_c = \Omega_c = 1$. By Eqs. (D.10) and (D.11) each inductor is replaced by a capacitor of reciprocal value and each capacitor is replaced by an inductor of reciprocal value, as given in Eqs (D.10) and (D.11). The result of this transformation is shown in Fig. D.3.

**Example D.1** Now if we are to design for an actual filter, for $R_L = 20\ \Omega$ and $f_c = 10$ kHz (as in the earlier example), we apply the frequency scaling and resistance denormalisation as earlier:

1. Each inductor of the normalised HP design is multiplied by $3.183 \times 10^{-4}$ (a 1 H inductor becomes 0.3183 mH)
2. Each capacitor of the normalised HP design is multiplied by $7.957 \times 10^{-7}$ (a 1 F capacitor becomes 0.7957 μF)
3. Each resistor of the normalised HP design is multiplied by 20 (a 1 Ω resistor becomes 20 Ω).

## Bandpass Butterworth Filters

However, from the 'normalised' *Butterworth LPF* design, we may move to any center frequency and any load resistance in the following manner.

1. Start with the BPF design with center frequency $\omega_0$ and the desired $Q$.
2. Move to the 'normalised' (with cut off near 1 r/s) LPF design implementation with the Ls, Rs and Cs clearly specified.
3. If we call $\Omega$ the frequency of the LPF then $\Omega$ is substituted by

$$\Omega = \frac{\Omega_c}{FBW} \left\{ \frac{\omega}{\omega_0} - \frac{\omega_0}{\omega} \right\} \tag{D.13}$$

$\Omega$, $\Omega_c$ are the frequency and cutoff frequency of the LPF. $FBW = Q^{-1} = \{\omega_0/\Delta\omega\}^{-1}$ (FBF is the fractional band width) and $\omega$ is the frequency of the BPF. If we use this formula, then for an inductor $L$, in the LPF design

$$j\Omega L \rightarrow j\frac{\Omega_c Q}{\omega_0}\omega L + \frac{\Omega_c Q\omega_0 L}{j\omega} \tag{D.14}$$

which means that the inductor is replaced by a series combination of an inductor $(L_{BP})$ and capacitor $(C_{BP})$ with

$$L_{BP} = \frac{\Omega_c Q L}{\omega_0} = \frac{\Omega_c L}{\Delta\omega}$$

$$C_{BP} = (\Omega_c Q\omega_0 L)^{-1} = \left( \frac{\Omega_c \omega_0^2 L}{\Delta\omega} \right)^{-1} \tag{D.15}$$

4. Now we go to the capacitors of the normalised LPF design. If $C$ is such a capacitor, then

$$j\Omega C \rightarrow j\Omega_c Q \left\{ \frac{\omega}{\omega_0} - \frac{\omega_0}{\omega} \right\} C = \frac{j\Omega_c \omega}{\Delta\omega}C + \frac{\Omega_c \omega_0^2}{j\omega\Delta\omega}C$$

which means that the capacitor is replaced by a parallel combination of an inductor $(L_{BP})$ and capacitor $(C_{BP})$ with

$$C_{BP} = \frac{\Omega_c Q C}{\omega_0} = \frac{\Omega_c C}{\Delta\omega}$$

$$L_{BP} = (\Omega_c Q\omega_0 C)^{-1} = \left( \frac{\Omega_c \omega_0^2 C}{\Delta\omega} \right)^{-1} \tag{D.16}$$

5. After these replacements, if the load resistor is say 20 $\Omega$ then multiply each L by 20 and divide each C by 20.

After doing these operations, the desired filter is ready.

**Example D.2** Taking an example as before: we want to design for a BPF with $f_0 = 10$ KHz ($\omega_0 = 2\pi f_0$ r/s) and $\Delta f = 1000$ Hz ($\Delta\omega = 2\pi \Delta f$ r/s). After converting, we get the frequency response of the filter of Fig. D.4.

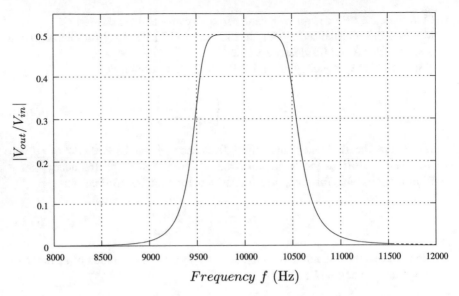

**Fig. D.4**  Frequency response of a BPF with $f_0 = 10^4$ Hz and $\Delta f = 10^3$ Hz

# Appendix E
# A Note About Inductors

Generally, capacitors and resistors are off the shelf components, but inductors have to be constructed. If $L$ is in $\mu$H, $N$ is the number of turns employed, the length "$l$" of the inductor in inches, the coil radius, $r$, is in inches and $p$, the pitch, in inches. (Pitch is the distance between the center of one turn and the next then $Np$ is the coil length.) Then a very simple formula obtained by Wheeler for the value of the inductor is

$$L = \frac{r^2 N^2}{10l + 9r} \qquad (E.1)$$

If we look at the equivalent circuit of an inductor shown in Fig. E.1b, the input impedance is

$$\begin{aligned}
Z_{in} &= R_c + \frac{j\omega L_0/(G_e + j\omega C_0)}{j\omega L_0 + 1/(G_e + j\omega C_0)} \\
&= R_c + \frac{j\omega L_0}{1 + j\omega L_0(G_e + j\omega C_0)} \\
&= R_c + \frac{j\omega L_0}{1 + j\omega L_0 G_e - \omega^2 L_0 C_0} \qquad (E.2)
\end{aligned}$$

where $R_c$ is the series resistor whose frequency dependence is given by

$$R_c(\omega) = R_{c0}[1 + (\omega\mu\sigma r^2)^{1/2}] \qquad (E.3)$$

As this equation tells us, the series resistance increases with frequency. In this equation, $\mu = \mu_r\mu_0$ is the permeability of the wire, $\sigma$ is conductivity of the wire and $r$ is the conductor radius. The factor

$$\sqrt{\omega\mu\sigma} = \sqrt{2}/\text{skin depth}$$

© The Editor(s) (if applicable) and The Author(s), under exclusive license to Springer
Nature Singapore Pte Ltd. 2022
S. Bhooshan, *Fundamentals of Analogue and Digital Communication Systems*,
Lecture Notes in Electrical Engineering 785,
https://doi.org/10.1007/978-981-16-4277-7

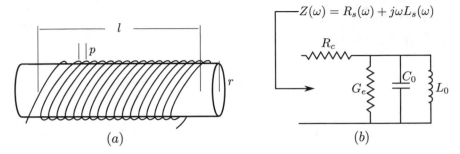

**Fig. E.1** Equivalent circuit of an inductor and inductor parameters

To reduce skin-effect resistance, the conductor is generally made of copper or it is silver or gold plated. $L_0$ is the low frequency inductance. $L_0$ is frequency dependent to some extent owing to skin effect, but the dependency is slight, especially if the conductor is non magnetic. $G_e$ is a conductance associated with magnetic and dielectric material losses (that is, for example, if a ferrite core is used). $C_0$ is the effect of the distributed capacitance of the inductor. The input series resistance and series inductance

$$Z(\omega) = R_s(\omega) + j\omega L_s(\omega) \tag{E.4}$$

where

$$R_s(\omega) = R_c + \frac{\omega^2 L_0^2 G_e}{\left(1 - \omega^2 L_0 C_0\right)^2 + (\omega L_0 G_e)^2}$$

$$L_s(\omega) = \frac{L_0 \left(1 - \omega^2 L_0^2 C_0^2\right)}{\left(1 - \omega^2 L_0 C_0\right)^2 + (\omega L_0 G_e)^2} \tag{E.5}$$

From these equations, it is clear that the inductor resonates at a frequency $\omega_r = 1/\sqrt{L_0 C_0}$. The $Q$ or quality factor of the coil is given by

$$Q = \frac{\omega L_0}{R_c} \tag{E.6}$$

To measure the self capacitance of the coil, $C_0$, the coil is placed in parallel with a capacitance $C_1$ and the resonance frequency is measured (at frequency $f = f_1$). The coil is then placed in parallel with capacitor $C_2$, so that the resonance frequency is at $f_2 = 2f_1$, and the capacitance used, $C_2$, is measured. Then

$$C_0 = (C_1 - 4C_2)/3$$

The self capacitance of the coil is then calculated using this formula.

**Example E.1** A 90 µH coil is resonated at 909 kHz with a 340 pf capacitor. Then coil is again resonated at twice that frequency, 1818 kHz with an 80 pf capacitor. Self capacitance $C_0$ is, therefore, $= (340 - 4 \times 80)/3 = 6.6$ pf. For a derivation of the formula, see the September 2007 issue of the XSS Newsletter.[1]

The reasoning behind this formula is as follows:

$$\omega_1 = \frac{1}{\sqrt{L_0(C_0 + C_1)}}$$

$$\omega_2 = 2\omega_1 = \frac{1}{\sqrt{L_0(C_0 + C_2)}}$$

so

$$\frac{2}{\sqrt{L_0(C_0 + C_1)}} = \frac{1}{\sqrt{L_0(C_0 + C_2)}}$$

or

$$4L_0(C_0 + C_2) = L_0(C_0 + C_1)$$

$$4C_0 + 4C_2 = C_0 + C_1$$

$$3C_0 = C_1 - 4C_2$$

or

$$\boxed{C_0 = (C_1 - 4C_2)/3} \tag{E.7}$$

---

[1] http://www.midnightscience.com/formulas-calculators.html#formulas5.

# Appendix F
# Logarithmic Units

The decibel (dB) is a logarithmic function to the base 10, that is the ratio of two values of a physical parameter, usually the power. Thus, if $P_1$ is the power of one quantity and $P_0$ the power of the other quantity, then

$$C \text{ (dB)} = 10 \log_{10} \left( \frac{P_1}{P_0} \right) \quad P_1, \ P_0 \text{ are powers} \qquad (F.1)$$

is the ratio $P_1/P_0$ expressed in decibels or dB.

If $A$ and $B$ are voltages or currents, then the formula would be

$$C = 20 \log_{10} \left( \frac{A}{B} \right) \quad A, \ B \text{ are voltages or currents} \qquad (F.2)$$

This is so since the voltage or current is proportional the square root of the power

$$v \propto \sqrt{P}$$
$$i \propto \sqrt{P}$$

Sometimes a power is expressed in dBm or dBW. In this case, the power, $P$, is expressed in mW or W and then

$$C \text{ (dBm)} = 10 \log_{10}[P \text{ (mW)}]$$
$$C \text{ (dBW)} = 10 \log_{10}[P \text{ (W)}] \qquad (F.3)$$

Also, the voltage gain of an amplifier is expressed as

$$G \text{ (dB)} = 20 \log_{10} \left( \frac{V_o}{V_i} \right)$$

where $V_o$ is the output voltage and $V_i$ is the input voltage.

© The Editor(s) (if applicable) and The Author(s), under exclusive license to Springer 625
Nature Singapore Pte Ltd. 2022
S. Bhooshan, *Fundamentals of Analogue and Digital Communication Systems*,
Lecture Notes in Electrical Engineering 785,
https://doi.org/10.1007/978-981-16-4277-7

In communication engineering, the signal to noise ratio in dB is

$$\text{SNR (dB)} = 10 \log_{10} \left( \frac{S}{N} \right)$$

where $S$ is the signal power and $N$ is the noise power.

# Appendix G
# Wiener-Khinchin Theorem

Starting from Eq. (3.113), we make a change of variables

$$u = x$$
$$v = x - y \tag{G.1}$$

or

$$x = u$$
$$y = v - u \tag{G.2}$$

where the Jacobian is

$$\begin{vmatrix} \frac{\partial x}{\partial u} & \frac{\partial x}{\partial v} \\ \frac{\partial y}{\partial u} & \frac{\partial y}{\partial v} \end{vmatrix} = \begin{vmatrix} 1 & 0 \\ -1 & 1 \end{vmatrix} = 1$$

and the integration limits are, $-T/2 \le x \le T/2$, $T/2 \le y \le T/2$. Which gives us the region bounded by

$$u = -\frac{T}{2}$$
$$u = \frac{T}{2}$$
$$v = -\frac{T}{2} + u$$
$$v = \frac{T}{2} + u \tag{G.3}$$

The region of integration is shown in Fig. G.1. Substituting the Jacobian and the new variables

$$E\left[|X_T(\omega)|^2\right] = \int \int_{\mathcal{R}} R_X(v) e^{-j\omega v} du dv \tag{G.4}$$

© The Editor(s) (if applicable) and The Author(s), under exclusive license to Springer Nature Singapore Pte Ltd. 2022
S. Bhooshan, *Fundamentals of Analogue and Digital Communication Systems*,
Lecture Notes in Electrical Engineering 785,
https://doi.org/10.1007/978-981-16-4277-7

where $\mathcal{R}$ is the region of integration shown in the figure. The integration can split into two parts: $v < 0$ and $v > 0$. For $v < 0$

$$
\begin{aligned}
I_- &= \int_{v=-T/2}^{0} \left[ \int_{-T/2}^{T/2+v} du \right] R_X(v) e^{-j\omega v} dv \\
&= \int_{v=-T/2}^{0} [T + v] R_X(v) e^{-j\omega v} dv \\
&= T \int_{v=-T/2}^{0} R_X(v) e^{-j\omega v} dv + \underbrace{\int_{v=-T/2}^{0} v R_X(v) e^{-j\omega v} dv}_{I_{-B}}
\end{aligned}
$$

and for $v > 0$

$$
\begin{aligned}
I_+ &= \int_{v=0}^{T/2} \left[ \int_{-T/2+v}^{T/2} du \right] R_X(v) e^{-j\omega v} dv \\
&= \int_{v=0}^{T/2} [T - v] R_X(v) e^{-j\omega v} dv \\
&= T \int_{v=0}^{T/2} R_X(v) e^{-j\omega v} dv - \underbrace{\int_{v=0}^{T/2} v R_X(v) e^{-j\omega v} dv}_{I_{+B}}
\end{aligned}
$$

where the two functions $I_{-B}$ and $I_{+B}$ will be considered next. If we look at these two integrals, we realise that we may write the sum as

$$
I_{-B} + I_{+B} = - \int_{v=-T/2}^{T/2} |v| R_X(v) e^{-j\omega v} dv
$$

Going back, we find that the total integral, $I$ is

$$
I = I_- + I_+
$$

We now define, the *power spectral density*, (PSD) as

$$
\begin{aligned}
S_X(\omega) &= \lim_{T \to \infty} \frac{E\left[|X_T(\omega)|^2\right]}{T} \\
&= \lim_{T \to \infty} \left\{ \int_{v=0}^{T/2} R_X(v) e^{-j\omega v} dv + \int_{v=-T/2}^{0} R_X(v) e^{-j\omega v} dv + \frac{I_{-B} + I_{+B}}{T} \right\}
\end{aligned}
$$

the last term on the right can be shown to be zero (as $T \to \infty$), (*only if the numerator integrals of this term are bounded as* $T \to \infty$), while the first two terms add up to the FT of the auto-correlation function. So

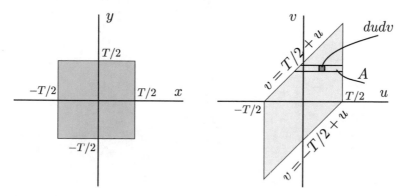

**Fig. G.1** Regions of integrations

$$S_X(\omega) \Leftrightarrow R_X(\tau) \tag{G.5}$$

which is also called the Weiner–Khinchin theorem.

# Appendix H
# Mathematical Details for Sect. 3.14

When the input to a linear system is a random process, we would like to compute the autocorrelation function of the output. To this end, we consider (with respect to Fig. H.1)

$$E[Y(t_1)Y(t_2)] = E\left[\int_{-\infty}^{\infty} h(\tau_1)x(t_1 - \iota_1)d\tau_1 \times \int_{-\infty}^{\infty} h(\tau_2)x(t_2 - \tau_2)d\tau_2\right]$$

$$\text{or } R_Y(t_1, t_2) = E\left[\int_{-\infty}^{\infty}\int_{-\infty}^{\infty} h(\tau_1)x(t_1 - \tau_1)h(\tau_2)x(t_2 - \tau_2)d\tau_1 d\tau_2\right]$$

$$= \int_{-\infty}^{\infty}\int_{-\infty}^{\infty} h(\tau_1)h(\tau_2)E\left[x\underbrace{(t_1 - \tau_1)}_{a}x\underbrace{(t_2 - \tau_2)}_{b}\right]d\tau_1 d\tau_2 \quad \text{(H.1)}$$

since the random process, $X(t)$ is WSS

$$E\left[x\underbrace{(t_1 - \tau_1)}_{a}x\underbrace{(t_2 - \tau_2)}_{b}\right] = R_X(a - b)$$

$$= R_X(t_1 - t_2 - \tau_1 + \tau_2) \quad \text{(H.2)}$$

therefore

$$R_Y(t_1, t_2) = \int_{-\infty}^{\infty}\int_{-\infty}^{\infty} h(\tau_1)h(\tau_2)R_X(t_1 - t_2 - \tau_1 + \tau_2)d\tau_1 d\tau_2$$

$$\text{or } R_Y(\tau) = \int_{-\infty}^{\infty}\int_{-\infty}^{\infty} h(\tau_1)h(\tau_2)R_X(\tau - \tau_1 + \tau_2)d\tau_1 d\tau_2 \quad \text{(H.3)}$$

where $\tau = t_1 - t_2$. which tells us that $Y(t)$ is also WSS. If we set $\tau = 0$ ($t_1 = t_2 = t$) in this equation, we get the value of $E[Y(t)^2]$.

© The Editor(s) (if applicable) and The Author(s), under exclusive license to Springer Nature Singapore Pte Ltd. 2022
S. Bhooshan, *Fundamentals of Analogue and Digital Communication Systems*,
Lecture Notes in Electrical Engineering 785,
https://doi.org/10.1007/978-981-16-4277-7

**Fig. H.1** Input and output of a linear system

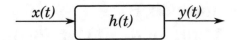

$$E[Y(t)^2] = \int_{-\infty}^{\infty} \int_{-\infty}^{\infty} h(\tau_1)h(\tau_2)R_X(-\tau_1 + \tau_2)d\tau_1 d\tau_2$$

$$= \frac{1}{2\pi} \left[ \int_{-\infty}^{\infty} \int_{-\infty}^{\infty} \left\{ \int_{-\infty}^{\infty} H(\omega)e^{j\omega\tau_1} d\omega \right\} h(\tau_2)R_X(-\tau_1 + \tau_2)d\tau_1 d\tau_2 \right]$$

$$= \frac{1}{2\pi} \left[ \int_{-\infty}^{\infty} H(\omega)d\omega \int_{-\infty}^{\infty} h(\tau_2)d\tau_2 \int_{-\infty}^{\infty} e^{j\omega\tau_1} R_X(-\tau_1 + \tau_2)d\tau_1 \right]$$

in this equation setting $\tau = \tau_2 - \tau_1$, then

$$e^{j\omega\tau_1} = e^{-j\omega(\tau_2-\tau_1)} \times e^{j\omega\tau_2}$$

and therefore

$$E[Y(t)^2] = \frac{1}{2\pi} \left[ \int_{-\infty}^{\infty} H(\omega)d\omega \underbrace{\int_{-\infty}^{\infty} h(\tau_2)e^{j\omega\tau_2}d\tau_2}_{H^*(\omega)} \underbrace{\int_{-\infty}^{\infty} e^{-j\omega\tau} R_X(\tau)d\tau}_{S_X(\omega)} \right]$$

$$= \frac{1}{2\pi} \left[ \int_{-\infty}^{\infty} H(\omega)H^*(\omega)S_X(\omega)d\omega \right]$$

$$= \frac{1}{2\pi} \left[ \int_{-\infty}^{\infty} |H(\omega)|^2 S_X(\omega)d\omega \right] \tag{H.4}$$

which is a very significant result. From this equation, it is clear that $|H(\omega)|^2 S_X(\omega)$ is the power spectral density of the output since $E[Y(t)^2]$ is the average power of the output signal.

# Appendix I
# Narrowband Noise Proofs

1. The in phase and quadrature phase components have the same average power. To prove this result, we use Equation set (3.164)

$$\begin{aligned}
E\left[n_I(t)^2\right] &= E\left\{[n_{\text{NB}}(t)\cos(\omega_c t) + n_{\text{NB}h}(t)\sin(\omega_c t)]^2\right\} \\
&= E[n_{\text{NB}}(t)^2 \cos^2(\omega_c t)] + E[n_{\text{NB}h}(t)^2 \sin^2(\omega_c t)] \\
&\quad + 2E[n_{\text{NB}h}(t)\sin(\omega_c t)n_{\text{NB}}(t)\cos(\omega_c t)] \\
&= \mathbb{P}_{\text{NB}}\cos^2(\omega_c t) + \mathbb{P}_{\text{NB}}\sin^2(\omega_c t) \\
&\quad + 2E[n_{\text{NB}h}(t)n_{\text{NB}}(t)]\sin(\omega_c t)\cos(\omega_c t) \\
&= \mathbb{P}_{\text{NB}}\cos^2(\omega_c t) + \mathbb{P}_{\text{NB}}\sin^2(\omega_c t) \\
&= \mathbb{P}_{\text{NB}}
\end{aligned}$$

where we have used Eq. (3.171).

2. The two random processes, $n_I(t)$ and $n_Q(t)$ are also uncorrelated. To prove this, we find the average value of the equation obtained by multiplying Eqs. (3.165) and (3.166)

$$\begin{aligned}
E\{n_{\text{NB}}(t)n_{\text{NB}h}(t)\} &= E\left\{n_I^2(t)\cos(\omega_c t)\sin(\omega_c t) - n_Q^2(t)\cos(\omega_c t)\sin(\omega_c t)\right. \\
&\quad \left. + n_I(t)n_Q(t)\left[\cos^2(\omega_c t) - \sin^2(\omega_c t)\right]\right\}
\end{aligned}$$

Now in this expression, $E[n_I^2(t)] = E[n_Q^2(t)]$, therefore, the first two terms cancel. And since $E\{n_{\text{NB}}(t)n_{\text{NB}h}(t)\} = 0$, therefore

$$E\left[n_I(t)n_Q(t)\right] = 0 \tag{I.1}$$

© The Editor(s) (if applicable) and The Author(s), under exclusive license to Springer Nature Singapore Pte Ltd. 2022
S. Bhooshan, *Fundamentals of Analogue and Digital Communication Systems*,
Lecture Notes in Electrical Engineering 785,
https://doi.org/10.1007/978-981-16-4277-7

# Appendix J
# The Method of Lagrange Multipliers

When we want to optimise a mathematical function, given a set of equality constraints, the method of Lagrange multipliers may be used to find the local maxima and minima of the given function subject to equality constraints. In this Appendix, only a special case will be considered and the general case left for the student to study from other texts.

Consider the problem of optimising

$$f(x_1, x_2, \ldots, x_n)$$

given that

$$g(x_1, x_2, \ldots, x_n) = k$$

where $k$ is a constant.

To apply the method of Lagrange multipliers, we form the object function

$$O(x_1, x_2, \ldots, x_n, \lambda) = f(x_1, x_2, \ldots, x_n) + \lambda g(x_1, x_2, \ldots, x_n)$$

and we form the $n + 1$ equations

© The Editor(s) (if applicable) and The Author(s), under exclusive license to Springer
Nature Singapore Pte Ltd. 2022
S. Bhooshan, *Fundamentals of Analogue and Digital Communication Systems*,
Lecture Notes in Electrical Engineering 785,
https://doi.org/10.1007/978-981-16-4277-7

$$\frac{\partial O}{\partial x_1} = \frac{\partial f}{\partial x_1} + \lambda \frac{\partial g}{\partial x_1} = 0 \dots (1)$$

$$\frac{\partial O}{\partial x_2} = \frac{\partial f}{\partial x_2} + \lambda \frac{\partial g}{\partial x_2} = 0 \dots (2)$$

$$\vdots \quad = \quad \vdots$$

$$\frac{\partial O}{\partial x_i} = \frac{\partial f}{\partial x_i} + \lambda \frac{\partial g}{\partial x_i} = 0 \dots (i)$$

$$\vdots \quad = \quad \vdots$$

$$\frac{\partial O}{\partial x_n} = \frac{\partial f}{\partial x_n} + \lambda \frac{\partial g}{\partial x_n} = 0 \dots (n)$$

$$\frac{\partial O}{\partial \lambda} = g - k = 0 \dots (n+1)$$

solving these $n + 1$ equations for the $n + 1$ unknowns, $x_1, x_2, \dots, x_n$, and $\lambda$, we obtain the optimised point. Taking an extremely simple example, suppose we want to find that cuboid of maximum volume with the area of the faces equal to 600 m$^2$. Then let $x_1, x_2, x_3$ be the three lengths, and we have to optimise

$$x_1 x_2 x_3$$

given

$$2(x_1 x_2 + x_2 x_3 + x_3 x_1) = 600 \tag{J.1}$$

Then

$$O = x_1 x_2 x_3 + \lambda \left[ 2(x_1 x_2 + x_2 x_3 + x_3 x_1) - 600 \right]$$

Taking the derivatives

$$\frac{\partial O}{\partial x_1} = x_2 x_3 + 2\lambda x_2 + 2\lambda x_3 = 0$$

$$\frac{\partial O}{\partial x_2} = x_1 x_3 + 2\lambda x_1 + 2\lambda x_3 = 0$$

$$\frac{\partial O}{\partial x_3} = x_1 x_2 + 2\lambda x_2 + 2\lambda x_1 = 0$$

$$\frac{\partial O}{\partial \lambda} = 2(x_1 x_2 + x_2 x_3 + x_3 x_1) - 600 = 0$$

From the first three equations, since $\lambda$ is common to all of them, and since they exhibit circular symmetry,

$$x_1 = x_2 = x_3 = x$$

and from the fourth equation

$$x = 10$$

Hence, the volume is optimised for a cube of side 10, and is equal to 1000 m$^3$.

To show that this is indeed a maximum, let $x_1 = 10.1$, $x_2 = 9.05$ and $x_3 = 10.89$. For these dimensions (to satisfy Eq. J.1), the volume is 995.4 m$^3$ which is less than 1000. Therefore, the solution is a maximum.

# Index

© The Editor(s) (if applicable) and The Author(s), under exclusive license to Springer   639
Nature Singapore Pte Ltd. 2022
S. Bhooshan, *Fundamentals of Analogue and Digital Communication Systems*,
Lecture Notes in Electrical Engineering 785,
https://doi.org/10.1007/978-981-16-4277-7